ANNUAL REPORTS IN MEDICINAL CHEMISTRY

Volume 19

Academic Press Rapid Manuscript Reproduction

ANNUAL REPORTS IN MEDICINAL CHEMISTRY

Volume 19

Sponsored by the Division of Medicinal Chemistry of the American Chemical Society

Editor-in-Chief: ***DENIS M. BAILEY***
STERLING-WINTHROP RESEARCH INSTITUTE
RENSSELAER, NEW YORK

SECTION EDITORS

BARRIE HESP • WILLIAM T. COMER • FRANK C. SCIAVOLINO
BEVERLY A. PAWSON • ROBERT W. EGAN • RICHARD C. ALLEN

ACADEMIC PRESS, INC. 1984
(Harcourt Brace Jovanovich, Publishers)
ORLANDO SAN DIEGO NEW YORK LONDON
TORONTO MONTREAL SYDNEY TOKYO

ACADEMIC PRESS, INC.
Orlando, Florida 32887

United Kingdom Edition published by
ACADEMIC PRESS, INC. (LONDON) LTD.
24/28 Oval Road, London NW1 7DX

LIBRARY OF CONGRESS CATALOG CARD NUMBER: 66-26843

ISBN 0-12-040519-9

PRINTED IN THE UNITED STATES OF AMERICA

84 85 86 87 9 8 7 6 5 4 3 2 1

CONTENTS

II. PHARMACODYNAMIC AGENTS

Section Editor: William T. Comer, Bristol-Myers Research and Development, 345 Park Avenue, New York, New York

III. CHEMOTHERAPEUTIC AGENTS

Section Editor: Frank C. Sciavolino, Pfizer Central Research, Groton, Connecticut

IV. METABOLIC DISEASES AND ENDOCRINE FUNCTION

Section Editor: Beverly A. Pawson, Roche Research Center, Hoffmann–La Roche Inc., Nutley, New Jersey

V. TOPICS IN BIOLOGY

Editor: Robert W. Egan, Merck Institute for Therapeutic Research, Rahway, New Jersey

CONTRIBUTORS

PREFACE

Volume 19 of *Annual Reports in Medicinal Chemistry* contains 30 chapters organized in the traditional format of previous volumes—in sections titled CNS Agents, Pharmacodynamic Agents, Chemotherapeutic Agents, Metabolic Diseases and Endocrine Function, Topics in Biology, and Topics in Chemistry and Drug Design. In addition, this year the editors experiment with a new section that summarizes worldwide first market introductions of therapeutic agents during the previous calendar year.

In the first six sections are included a mix of literature updates on a broad sweep of drug research areas, as well as reviews of highly specialized and newly emerging technologies. Topics covered for the first time include CNS autoreceptors, interleukin 2, endogenous natriuretic factors, and enzymic methods in organic synthesis. Phospholipases and collagenases, touched on in earlier Annual Reports, receive in-depth coverage in separate chapters in this volume. Reviews on recombinant DNA technology and the biology of leukotrienes bring the reader up to date in these highly important and active areas.

In the new section, Worldwide Market Introductions, the chapter "To Market, to Market" presents an aspect lacking in the previous 18 volumes of Annual Reports—an organized follow-up on market introduction, the ultimate endpoint of the many agents reviewed. The accumulation of data for this chapter proved formidable and while a number of sources were used, it is inevitable that we missed some drugs. We hope we have not made a major omission!

Finally, I want to extend my gratitude to the section editors and contributors whose time and effort made Volume 19 possible, and in particular to Martha Johnson, whose assistance was invaluable in preparing the final copy.

Denis M. Bailey
Rensselaer, New York
May 1984

Section I - CNS Agents

Editor: Barrie Hesp, Stuart Pharmaceuticals, Division of ICI Americas, Wilmington, DE 19897

Chapter 1. Analgesics

O. William Lever, Jr., Kwen-Jen Chang and John D. McDermed
Wellcome Research Laboratories, Burroughs Wellcome Co.
Research Triangle Park, NC 27709

Efforts toward alleviation of pain continue at an intensive pace at the clinical, pre-clinical and basic research levels. In this review, recent research developments in opioid and non-opioid analgesia are presented, along with selected clinical highlights.

Opioid Receptors and Ligands

Endogenous opioids and their multiple receptors continue as the focus of vigorous research. The proceedings of the 1983 International Narcotic Research Conference were published,[1] and reviews were recently provided on opioid peptides,[2,3] opioid receptors and their functional roles,[4-6] pharmacology and therapeutic implications of enkephalins and endorphins,[7-9] cardiovascular effects of endogenous opioids,[10] opioids and the adrenal-pituitary axis,[11] opioids and the hippocampus,[12] substance P in nociception,[13] potential therapeutic roles for opioid antagonists[14], roles of opioid peptides in appetite control,[15] and autoradiography of opioid receptors.[16]

Biosynthesis and Metabolism of Opioid Peptides - Opioid peptide biosynthesis was summarized in our previous review,[17] and additional overviews of the biosynthesis[18-22] and distribution[20,21,23-25] of endogeneous opioids have appeared. Several inhibitors of enkephalin convertase, a carboxypeptidase which may liberate enkephalins from immediate precursors, were described.[26,27] Degradation of enkephalins by peptidases continues to be studied,[28,29] as are the analgesic properties of inhibitors of enkephalin metabolism.[28-34] Thiorphan, *N*-[(*R,S*)-3-mercapto-2-benzylpropanoyl]glycine, is an enkephalinase inhibitor (K_i = 3.5 nM) which also inhibits (K_i = 140 nM) angiotensin-converting enzyme (ACE).[32] By contrast, the *retro-inverso* isomer, *retro*-thiorphan [(*R,S*)-HS-CH_2-CH(CH_2Ph)-*NH-CO*-CH_2CO_2H], does not inhibit ACE but retains potent enkephalinase inhibition (K_i = 6 nM) and has an *in vivo* analgesic potency (i.c.v.; mouse hot plate, mouse writhing) similar to that of thiorphan.[32] Thiorphan was reported to reduce postmyelographic side effects (headache, nausea, sciatica) when administered (i.v.) to patients prior to myelography.[33] Oral administration of D-phenylalanine, a carboxypeptidase inhibitor, was found to be of benefit in patients with certain chronic pain complaints; the drug was most effective after 3-5 weeks of therapy.[34]

Opioid Receptors and Pharmacology - The previously defined opioid receptor sub-types (μ,δ,κ,ε,σ) continue to be studied, and an additional binding site, designated the λ site, was recently described in rat brain.[35] The λ

ISBN 0-12-040519-9

sites have a regional distribution in brain different from that of μ receptors and have a high affinity for 4,5-epoxymorphinans (e.g., morphine, naloxone) and low affinity for non-epoxymorphinan μ ligands, κ ligands, δ ligands, and "universal" (μ,δ,κ) ligands.[35] Evidence that σ receptors may differ from phencyclidine receptors was provided by the report of a naloxone-inaccessible σ receptor in rat brain and spinal cord which stereoselectively binds (+)-ethylketocyclazocine and (+)-*N*-allylnormetazocine [(+)-SKF 10,047] and has drug selectivity and regional distribution different from those of μ, δ and phencyclidine receptors.[36] Additional studies on σ agonists and on their phencyclidine-like behavioral effects were reported.[37-41] The benzomorphan Mr 2034 [(-)-*N*-(2-tetrahydrofurfuryl)normetazocine], originally proposed as a κ ligand, is more accurately described as a universal (μ,δ,κ,σ) opioid ligand.[42]

The nature of opioid binding continues to be explored.[43-48] In rat brain and spinal cord, μ and δ sites predominate, and κ sites represent only 10% of total binding.[48] Both δ and μ receptors mediate antinociception.[49-52] The antinociceptive profiles of μ and κ agonists were reported to be distinguishable *in vivo*,[53,54] and the affinities of κ agonists for μ and δ receptors *in vivo* in the rat were assessed.[55] Respiratory depression is caused by opioid agonists but may be mediated by a receptor population different from that producing antinociception.[56-59] Evidence suggesting that the μ sites in mouse vas deferens and guinea pig ileum are different,[60,61] and support for the possibility that opioid agonists and antagonists interact with different sites on μ receptors,[62] were also presented. The antitussive effects of opioids appear to arise from interaction with pharmacological receptors which are less stereoselective and less sensitive to naloxone than the analgesic receptors.[63] Attempts to purify opioid receptors have progressed slowly.[64,65] Additional data were cited for the coexistence of allosterically coupled morphine and enkephalin receptors within an opioid receptor complex in rat brain.[66,67]

Newer Endogenous Opioid Peptides - The number of endogenous opioid peptides continues to grow rapidly.[3] Several fragments of proenkephalin A have received recent attention. A μ-selective octapeptide fragment, Tyr-Gly-Gly-Phe-Met-Arg-Arg-Val-NH_2, has been isolated from bovine brain[68] and from human pheochromocytoma[69] and has been designated both as metorphamide[68] and as adrenorphin.[69] Peptide E, a 25-amino acid fragment, binds preferentially to μ and κ receptors, but has minimal analgesic potency (i.c.v., mouse tail flick) *in vivo*.[70,71] The C-terminal heptapeptide fragment, Met^5-enkephalin-Arg^6-Phe^7, has been found in human and rabbit plasma,[72] rat brain[73-75] and spinal cord,[76,77] and lung of several species.[78] A unique high affinity binding site was described in rat lung membranes.[78] The role of this peptide as a Met-enkephalin precursor or a transmitter is speculative. The fragment Met^5-enkephalin-Arg^6-Gly^7-Leu^8 is a μ- and δ-selective ligand which binds poorly to κ receptors; synthetic extension to incorporate a Lys^9 residue improved κ affinity 10-fold but did not alter μ or δ binding or the μ,δ-selectivity.[79]

Synthetic Opioid Peptide Analogues - Structure-activity studies on analogues of the exceptionally μ-selective ligand morphiceptin (Tyr-Pro-Phe-Pro-NH_2) were reported.[60] One of the most potent analgesics was PL017 (Tyr-Pro-NMePhe-D-Pro-NH_2).[60] A peptide containing the PL017 sequence at the N-terminus and the dyn-A (6-17) sequence at the C-terminus had high affinity for both μ and κ sites.[80] Analogues of β-endorphin containing the dynorphin (5-13) or (6-13) sequences showed excitatory behavioral effects in the mouse but had no analgesic properties in the mouse tail flick assay (i.c.v.).[81] By contrast, a β-endorphin analogue with a dermorphin sequence (Tyr-D-Ala-Phe-Gly-Tyr-Pro-Ser) in residues

(1-7) was several-fold more potent than β-endorphin (mouse tail flick, i.c.v.).[82] A number of additional studies on the dermorphins, potent opioid heptapeptides (e.g., Tyr-D-Ala-Phe-Gly-Tyr-Pro-Ser-NH_2) originally isolated from the skin of South American frogs, and their synthetic analogues were reported.[83-88] A heptapeptide is not required for high activity. The dermorphin (1-4) analogue H_2N-C(NH)-Tyr-D-Ala-Phe-Gly-NH-adamantyl, which contains a guanidino N-terminus and an N-adamantyl amide moiety, is a powerful μ-agonist (mouse tail flick, i.c.v.) with potency 30,000 and 1000 times that of Met-enkephalin and morphine, respectively.[83]

New δ-selective ligands have emerged, including deltakephalin (Tyr-D-Thr-Gly-Phe-Leu-Thr),[89] certain enkephalin dimers,[17,90] and cyclic enkephalin analogues containing disulfide-linked penicillamine and cystine residues.[91] Possibly the most highly δ-selective ligands reported to date are the bis-penicillamine 14-membered cyclic disulfide-containing enkephalin analogues DPDPE (1a) and DPLPE (1b), which gave δ/μ selectivity ratios in isolated tissues of 3164 and 1088, respectively.[92] By contrast, the 14-membered cyclic enkephalin analogue 2, as well as a number of partial *retro-inverso* modifications of this cyclic structure, are μ-selective ligands.[93] Other approaches to conformational restriction were reported,[94,95] as were enkephalin analogues incorporating dehydroamino acids,[96,97] cyclopropyl-containing amino acids,[98,99] p-nitro substitution in Phe^4,[100] ketomethylene replacement of the Tyr-Gly and Gly-Gly amide linkages,[101] modifications at the C-terminus,[102] S-oxidized methionine,[103] and phosphonate analogues of the C-terminal residues.[104]

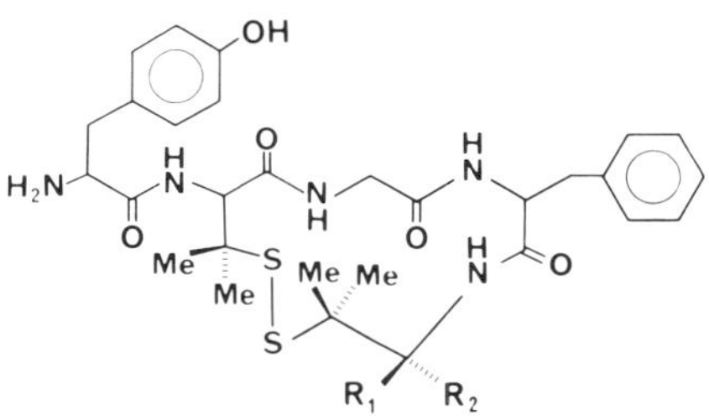

1a $R_1 = CO_2H$, $R_2 = H$ (DPDPE)
1b $R_1 = H$, $R_2 = CO_2H$ (DPLPE)

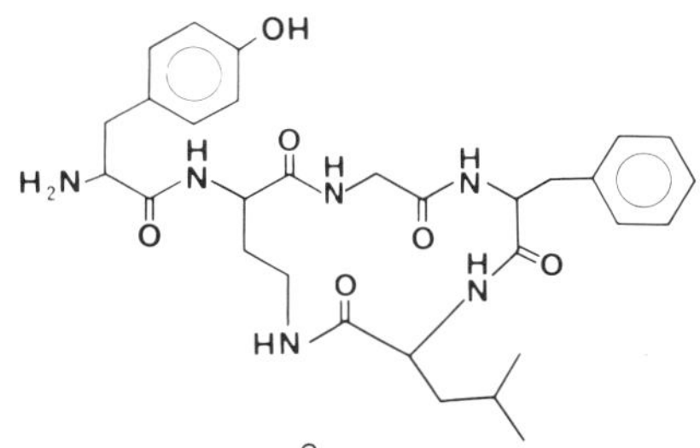

2

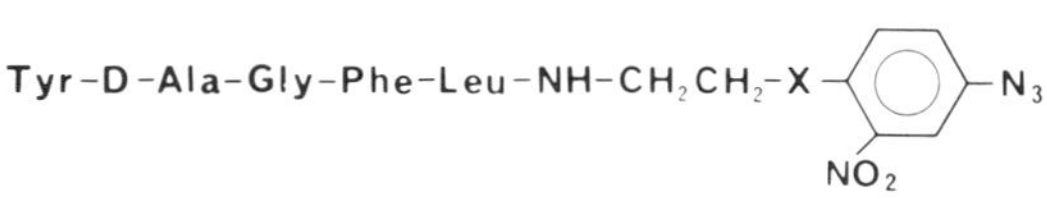

3 X = NH
4 X = $NHCO(CH_2)_2NH$

New Ligands, SAR, Structural Studies - Irreversible opioid ligands are of interest as receptor probes. The photolabile enkephalin derivatives 3 and 4,[105] the chloromethyl ketone Tyr-D-Ala-Gly-Phe-Leu-CH_2Cl,[106] and the fentanyl isothiocyanate 5[107] all appear to label δ receptors selectively, whereas the etonitazene isothiocyanate 6[107] appears to inactivate only μ receptors. A variety of epoxymorphinans with alkylating groups attached through amino groups at C-6 are opioid affinity labels.[108] Both β-CNA (7) and α-CNA (8) produce irreversible antagonism at μ, κ and δ receptors, and α-CNA appears to have an additional irreversible agonist property in the guinea pig ileum assay.[109] Further studies in the ileum with β-FNA (9) and β-FOA (10) led to the suggestion that μ agonists and antagonists interact with distinct receptor sites.[62] Comparisons of isolated tissue data for various epimeric antagonist affinity labels indicated that the chirality at C-6 is an important determinant of receptor alkylating ability.[110] The configuration at C-6 was shown to determine the ring C solution conformations of the oxymorphamines: ring C is a chair in the β-epimer 11 and a twist boat in the α-epimer 12.[111]

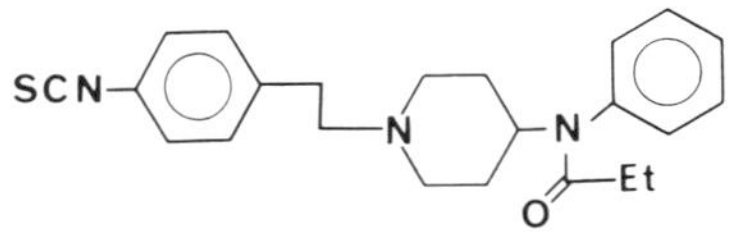

5

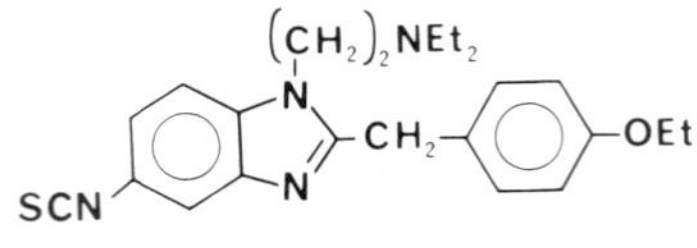

6

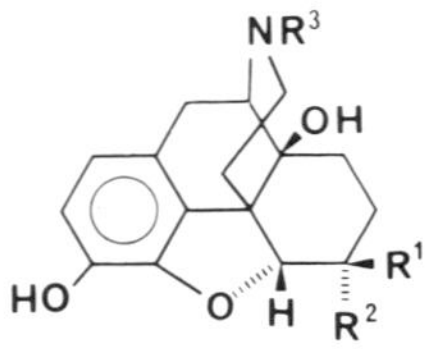

7 $R^1 = N(CH_2CH_2Cl)_2$, $R^2 = H$, $R^3 = CH_2$-cyclopropyl (β-CNA)

8 $R^1 = H$, $R^2 = N(CH_2CH_2Cl)_2$, $R^3 = CH_2$-cyclopropyl (α-CNA)

9 $R^1 = NHCOCH = CHCO_2Me$, $R^2 = H$, $R^3 = CH_2$-cyclopropyl (β-FNA)

10 $R^1 = NHCOCH = CHCO_2Me$, $R^2 = H$, $R^3 = Me$ (β-FOA)

11 $R^1 = NH_2$, $R^2 = H$, $R^3 = Me$

12 $R^1 = H$, $R^2 = NH_2$, $R^3 = Me$

Of four pentazocine analogues in which the *N*-alkyl group was folded back into 6- or 7-membered rings, compound 13 was the most potent (.5 X pentazocine, mouse writhing).[112,113] Conjugate addition of alkyl groups to the 8-position of codeinone gave a series of analgesics in which the 8-α-epimers were generally more potent (mouse tail flick) than the 8-β-epimers.[114] Syntheses of two isomeric groups of thiazolomorphans were reported, of which 14 was the most potent (10 X codeine, mouse writhing).[115,116] Compound 14 is a 9-β-epimer and is somewhat more potent than the 9-α-epimer, in contrast with precedents from the benzomorphans.

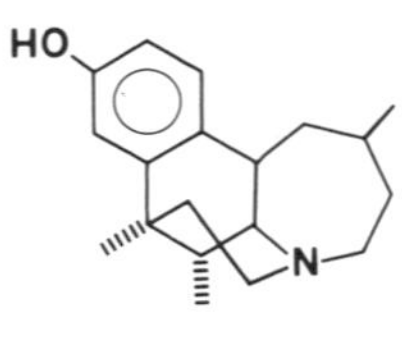

13

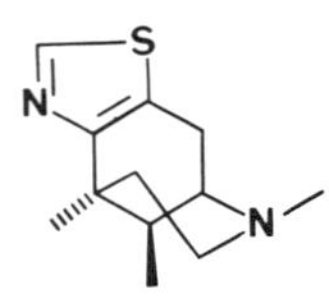

14

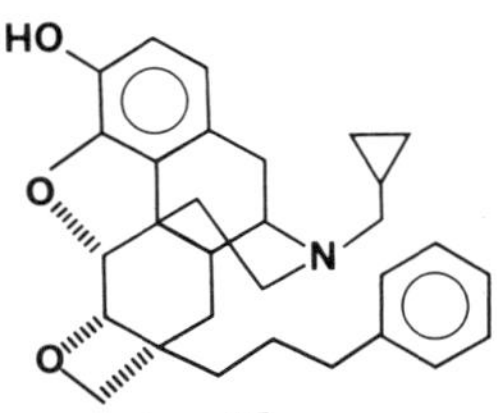

15

There were reports of further modifications of ring C of morphine and codeine.[117-119] From a large series of active compounds oxetane 15 was about 700 X more potent (mouse writhing, s.c.) than morphine.[119] Translocation of the hydroxyl group from the 4- to the 1-position destroyed activity in *N*-methylmorphinan-6-ones.[120] Syntheses of 6,14-exo-ethenomorphinans,[121] 2'-thiobenzomorphans,[122] ring-C-contracted morphinanones,[123] and more exotic heterocyclic analogs[124] were described without indications of biological activity. A series of fentanyl-enkephalin hybrids were essentially inactive in isolated tissue assays.[125]

The roles of hydrophobic and hydrophilic (amphiphilic) domains of the β-endorphin surface in relation to physicochemical, biochemical, analgesic and binding activities were explored.[126,127] Studies on four large synthetic peptides designed to simulate or modify the surface properties of β-endorphin led to the conclusion that exact residue homology may be less important to bioactivity than the gross physicochemical properties of the binding surface. In a complementary approach, a 40-residue peptide designed to mimic the binding site of opioid receptors was synthesized.[128] It displayed modest but significant affinity for several opioid peptides (Leu-enkephalin, $K_D = 5.8 \times 10^{-5}$) and discriminated them from non-opioids.

Non-Opioid Analgesia

Direct modulation of opioid pathways has been the major focus of modern analgesic research, but other modalities for antinociception appear to involve such pathways only indirectly or not at all. Stress, of which traditional Chinese acupuncture and modern electroacupuncture may be considered special cases, is one such mode. Early reports of naloxone blockade of acupuncture analgesia suggested mediation by endogenous opioids.[129] However, more recent studies contradict this hypothesis.[130] A superficially related clinical procedure for control of chronic pain is electrical transcutaneous or spinal cord nerve stimulation. A double-blind placebo-controlled study indicated that such analgesia is not naloxone-reversible.[131] The neurophysiological mechanisms mediating stress-induced autoanalgesia are complex and depend subtly on the parameters of the applied stress.[132] Electrical stress-induced analgesia in animals apparently involves both opioid and non-opioid substrates.[132,133] Front foot shocks in rats produced naloxone-blocked morphine-tolerant analgesia (tail flick) unaffected by hypophysectomy, and therefore describable as opioid, non-hormonal.[133] However, similar shocks to the hind feet gave analgesia unaffected by naloxone, morphine tolerance, or hypophysectomy and therefore non-opioid, non-hormonal.[133] Moreover, the footshock regimen influenced the nature of antinociception: prolonged intermittent shocks produced opioid-dependent analgesia, whereas brief continuous shocks produced non-opioid analgesia (rat tail flick).[134] Complete lesions of the raphe nucleus attenuated the latter but not the former,[134] though other studies have implicated the raphe nucleus in mediation of both types of analgesia.[135]

GABA-ergic mechanisms have been implicated in antinociception. In normal volunteers the GABA agonist THIP (16) produced analgesia against dental pain without tolerance, respiratory depression, or other characteristics of opiate therapy.[136] In an open study of cancer patients with chronic pain refractory to mild analgesics, THIP (20 mg i.m.) gave subjective pain control, but side effects (blurred vision, sedation) were dose-limiting.[137] The GABA-ergic benzodiazepine midazolam (17) given intrathecally to anaesthetized dogs strongly attenuated nociceptive sympathetic reflexes to electrical nerve stimulation.[138] Baclofen and other GABA agonists have also been found to be central antinociceptives in experimental animals.[139-141] These effects were generally naloxone-insensitive, though indirect effects on opioid substrates may be present.[142] GABA analgesia may[143] or may not[144] be cross-tolerant to morphine.

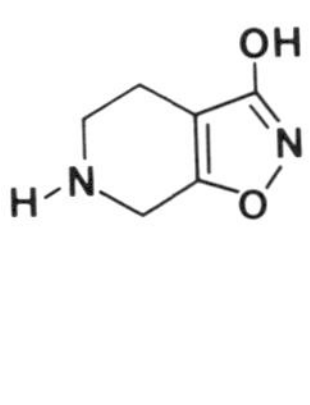

16

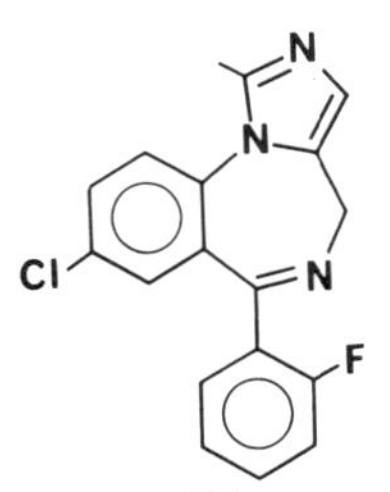

17

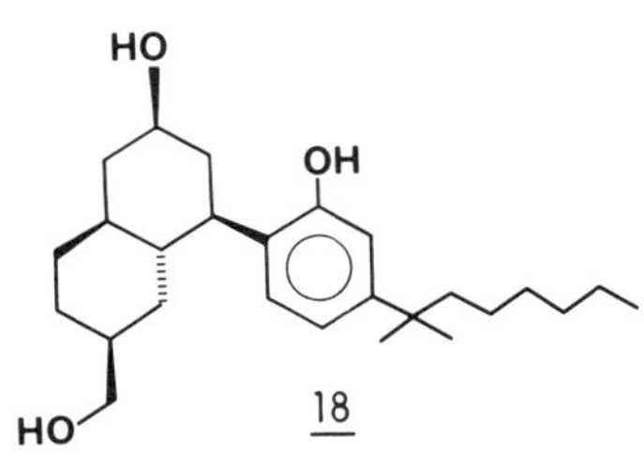

18

Intrathecal administration of the α-agonist clonidine attenuated nociceptive responses in rats and cats.[145] Attempts to characterize further the receptors involved as α_1 or α_2 gave inconclusive results.[146] Studies with monoamine depletors indicated that intact spinal adrenergic and serotonergic fibers are involved in a complex way in expression of morphine analgesia in rats.[147] Adrenergic systems may mediate foot shock analgesia (rats, tail flick).[148] A study in terminal cancer patients indicated effective control of chronic pain by epidural labetalol, a mixed α- and β-antagonist, though the result might have been due to a local anaesthetic effect.[149]

Several studies indicate that opioids activate serotonergic neurons and increase 5-HT turnover in the spinal cord.[150-152] Intrathecal 5-HT produced antinociception in mice (tail flick).[153] The role of serotonergic neurons in opioid analgesia may be model-dependent.[152]. Central cholinergic fibers may also mediate analgesia,[154] and both muscarinic[155] and nicotinic[156] agonists produce analgesia in rodents by central mechanisms. There is evidence that cholinergic connections mediate the expression of both endogenous and exogenous opioid analgesia.[154,157] Meptazinol is one drug in clinical use (*vide infra*) which appears to take fortuitous advantage of the opioid-cholinergic connection. In rodent antinociceptive models it displays pharmacology appropriate for an opiate partial agonist and a cholinomimetic.[158]

The analgesic properties of cannabinoid derivatives are well known, though no such drugs have yet been demonstrated to be substantially free from the CNS side effects to which *cannabis sativa* owes its popularity as a substance of abuse.[159,160] However, cannabinoid analgesia has appeal due to its opioid-independent mechanism and its consequent failure to induce dependence.[161] A series of preliminary reports[162-165] have described the syntheses of analgesics modelled on 9-nor-9-β-hydroxyhexahydrocannabinol; compound 18 was the most potent member (>60 X morphine).[162]

Clinical Highlights

Opioid Analgesics - Clinical evaluation of the newer opioid agonist-antagonist drugs continues, and their analgesic efficacy,[166] pharmacokinetics,[167] receptor pharmacology,[168] and dependence profiles[169] were recently reviewed. A review on the pharmacology and therapeutic efficacy of nalbuphine[170] and a compilation of clinical experiences with injectable meptazinol[171] were provided. Human pharmacokinetic data for meptazinol were reported.[172,173] Efficacy studies with epidural pentazocine,[174] epidural buprenorphine,[175] and parenteral dezocine[176] for postoperative pain relief were reported. Oral ciramadol was superior to codeine for relief of advanced cancer pain.[177] The clinical pharmacokinetics[178] and applications in anesthesia[179] of fentanyl and its derivatives were reviewed. Several additional reports on enkephalin analogues appeared.[180-184] For relief of severe episiotomy pain, metkephamid acetate (140 mg, i.m.) was superior to mepiridine (100 mg, i.m.) but minor side effects were more frequent; a lower dose (70 mg, i.m.) of metkephamid was ineffective.[180] The preclinical pharmacology of metkephamid was described.[181] Intrathecal administration of D-Ala^2-D-Leu^5-enkephalin (DADL) to a single patient showed that DADL could provide a powerful, long-lasting analgesic effect, and also is capable of producing profound, naloxone-reversible respiratory depression.[182]

Non-Steroidal Antiinflammatory Drugs (NSAIDS) - Therapeutic applications,[185-187] clinical pharmacokinetics[188], and mechanism of action[189] of the peripherally-acting NSAIDS were reviewed, and the pharmacology of diflunisal[190] and suprofen[191] were discussed in detail. Ibuprofen (400 mg, p.o.) was more effective than zomepirac (100 mg, p.o.) or aspirin (600 mg, p.o.) against episiotomy pain,[192] and rectal indomethacin (100 mg, eight hourly) provided effective analgesia following thoracotomy.[193] A dose-response study with i.v. indoprofen for postoperative pain was reported,[194] and in women with moderate to severe cancer pain, indoprofen (400 mg, i.v.) appeared equivalent to morphine (10 mg, i.m.).[195] Additional studies with diflunisal,[196,197] flurbiprofen,[198] fendosal,[199] and diclofenac[200] were reported. Therapeutic applications of NSAIDs in primary dysmenorrhea were reviewed.[201,202] In preclinical studies, propionic acids 19[203] and 20[204] were described as effective NSAIDs with reduced ulcerogenic properties.

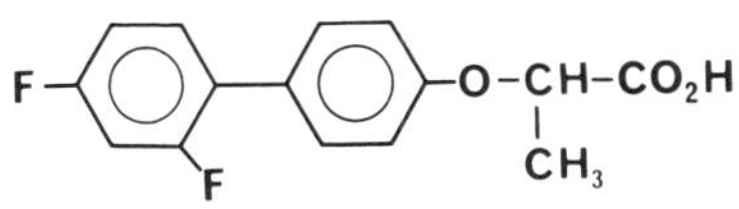

19

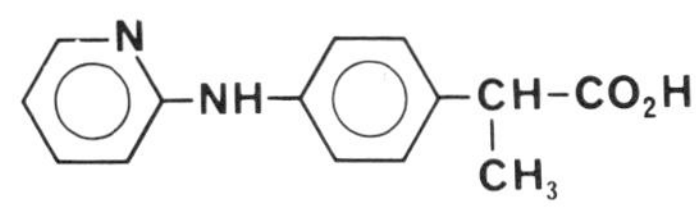

20

Migraine - Headache is the most common pain syndrome for which analgesic medication is applied and is associated with a large number of disorders with different etiologies. Comments here will be restricted to vascular headache, particularly migraine, which is probably a symptom of a reactive compensatory mechanism occurring late in a poorly understood pathogenetic sequence. A provocative review of this topic is available.[205] A classical migraine attack begins with a prodromal phase which involves regional cerebral vasoconstriction leading to hypoxia, ischemia,[206-208] and possibly reduced neuronal 5-HT synthesis.[208] Onset of the painful headache phase is marked by vasodilation,[206] which may reflect a defective vascular homeostatic mechanism.[209] Partly because of a dearth of animal models, treatment has evolved empirically from clinical experience and has traditionally consisted of symptom management.[210,211]

Most drug treatment in the past focused acutely on the painful vasodilation phase. The efficacy of dihydroergotamine[212] apparently results from its vasoconstrictor effects. NSAIDs inhibit prostacyclin biosynthesis and reduce the intensity, but not the frequency, of attacks.[213,214] Newer therapies have addressed prevention of attacks. The prophylactic effects of β-blockers have been established, though their mechanism of action remains speculative. In a long-term placebo-controlled study of propranolol in 245 migraineurs, more than 70% of the subjects had fewer and less severe attacks, and 46% reported that their improvement was maintained after discontinuation of medication.[215] Nadolol was shown to be similarly prophylactic, whereas pindolol and alprenolol were not.[216] Calcium channel blockers are the most recent additions to migraine prophylaxis. Clinical studies with flunarizine, a Ca^{++} blocker with anti-vasoconstrictive and anti-hypoxic effects in animals, indicate that it is clearly prophylactic, though its effect on severity of attacks is marginal.[217,218] Small clinical studies with nimodipine, nifedipine, and verapamil indicate efficacy for these also.[219] Nimodipine appears to produce the fewest side effects, probably because of its high selectivity for cerebral vasculature. These agents appear to prevent migraine attacks by attenuating Ca^{++}-dependent cerebral vasoconstriction, irrespective of the nature of the triggering constrictive stimulus.[220]

References

1. F. Porreca and T.F. Burks, Eds., Life Sci., 33, Suppl. I, pp 1-782 (1983).
2. "Opioid Peptides", J. Hughes, Ed., Brit. Med. Bull., 39, 1-106 (1983).
3. J.W. Thompson, Brit. Med. J., 288, 259 (1984).
4. A. Grossman and V. Clement-Jones, Clin. Endocrinol. Metab., 12, 31 (1983).
5. J.A. Gilbert and E. Richelson, Mol. Cell. Biochem., 55, 83 (1983).
6. A. Herz, J. Neural. Transm. (Suppl.), 18, 227 (1983).
7. D.L. Copolov and R.D. Helme, Drugs, 26, 503 (1983).
8. F.E. Bloom, Ann. Rev. Pharmacol. Toxicol., 23, 151 (1983).
9. V. Clement-Jones and G.M. Besser, Brit. Med. Bull., 39, 95 (1983).
10. J.W. Holaday, Biochem. Pharmacol., 32, 573 (1983); Ann. Rev. Pharmacol. Toxicol., 23, 541 (1983).
11. S. Mousa and D. Couri, Subst. Alcohol Actions Misuse, 4, 1 (1983).
12. W.A. Corrigall, Pharmacol. Biochem. Behav., 18, 255 (1983).
13. B. Pernow, Pharmacolog. Rev., 35, 85 (1983).

14. L.F. McNicholas and W.R. Martin, Drugs, 27, 81 (1984).
15. (a) J.E. Morley and A.S. Levine, Lancet, I, 398 (1983); (b) J.E. Morley, A.S. Levine, G.K. Yim and M.T. Lowy, Neurosci. Biobehav. Rev., 7, 281 (1983).
16. J.K. Wamsley, Pharmacolog. Rev., 35, 69 (1983).
17. O.W. Lever, Jr., K.-J. Chang and J.D. McDermed, Ann. Rep. Med. Chem., 18, 51 (1983).
18. J.L. Marx, Science, 220, 395 (1983).
19. R.V. Lewis and A.S. Stern, Ann. Rev. Pharmacol. Toxicol., 23, 352 (1983).
20. S. Udenfriend and D.L. Kilpatrick, Arch. Biochem. Biophys., 221, 309 (1983).
21. H. Imura, Y. Nakai, K. Nakao, S. Oki, I. Tanaka, H. Jingami, T. Yoshimasa, T. Tsukada, Y. Ikeda, M. Suda and M. Sakamoto, J. Endocrinol. Invest., 6, 139 (1983).
22. J. Hughes, Brit. Med. Bull., 39, 17 (1983).
23. S.F. Atweh and M.J. Kuhar, Brit. Med. Bull., 39, 47 (1983).
24. D.G. Smyth, Brit. Med. Bull., 39, 25 (1983).
25. A.C. Cuello, Brit. Med. Bull., 39, 11 (1983).
26. L.D. Fricker and S.H. Snyder, J. Biol. Chem., 258, 10950 (1983).
27. L.D. Fricker, T.H. Plummer, Jr. and S.H. Snyder, Biochem. Biophys. Res. Commun., 111, 994 (1983).
28. J.-C. Schwartz, S. de la Baume, C.C. Yi, P. Chaillet, H. Marcais-Collado and J. Costentin, J. Neural Transm. (Suppl.), 18, 235 (1983).
29. "Degradation of Endogenous Opioids: Its Relevance in Human Pathology and Therapy," S. Ehrenpreis and F. Sicuteri, Eds., Raven Press, New York, 1983.
30. P. Chaillet, H. Marcais-Collado, J. Constentin, C.C. Yi, S. de la Baume and J.-C. Schwartz, Eur. J. Pharmacol., 86, 329 (1983).
31. G. Borin, P. Giusti and L. Cima, Life Sci., 33, 1575 (1983).
32. B.P. Roques, E. Lucas-Soroca, P. Chaillet, J. Costentin and M.-C. Fournie-Zaluski, Proc. Natl. Acad. Sci. USA, 80, 3178 (1983).
33. P. Floras, A.M. Bidabe, J.M. Caille, G. Simonnet, J.M. Lecomte and M. Sabathie, Am. J. Neuroradiol., 4, 653 (1983).
34. R.C. Balagot, S. Ehrenpreis, J. Greenberg and M. Hyodo, in Ref. 29, p. 207.
35. J. Grevel and W. Sadee, Science, 221, 1198 (1983).
36. S.W. Tam, Proc. Natl. Acad. Sci., 80, 6703 (1983).
37. D.B. Vaupel, Eur. J. Pharmacol., 92, 269 (1983).
38. B.L. Slifer and R.L. Balster, J. Pharmacol. Exp. Ther., 225, 522 (1983).
39. H.E. Shannon, J. Pharmacol. Exp. Ther., 225, 144 (1983).
40. J.M. White and S.G. Holtzman, Psychopharmacol., 80, 1 (1983), J. Pharmacol. Exp. Ther., 224, 95 (1983).
41. J.M. Moerschbaecher and D.M. Thompson, J. Pharmacol. Exp. Ther., 226, 738 (1983).
42. N. Johnson and G.W. Pasternak, Life Sci., 33, 985 (1983).
43. J.E. Leysen, W. Gommeren and C.J. Niemegeers, Eur. J. Pharmacol., 87, 209 (1983).
44. J.W. Villiger, L.J. Ray and K.M. Taylor, Neuropharmacol., 22, 447 (1983).
45. J.-C. Meunier, Y. Kouakou, A. Puget and C. Moisand, Mol. Pharmacol., 24, 23 (1983).
46. H.H. Osborne and A. Herz, Eur. J. Pharmacol., 86, 373 (1983).
47. E. Castanas, P. Girand, K. Audigier, R. Drissi, F. Bondouresque, B. Conte-Devolx and C. Oliver, Mol. Pharmacol., 25, 38 (1984).
48. K.J. Mack, A. Killian and J.A. Weyhenmeyer, Life Sci., 34, 281 (1983).
49. S.L. Hart, H. Slusarczyk and T.W. Smith, Eur. J. Pharmacol., 95, 283 (1983).
50. G.S. Ling and G.W. Pasternak, Brain Res., 271, 152 (1983).
51. T.L. Yaksh, J. Pharmacol. Exp. Ther., 226, 303 (1983).
52. L.F. Tseng, Life Sci., 32, 2545 (1983).
53. N. Upton, R.D. Sewell and P.S. Spencer, Arch. Int. Pharmacodyn. Ther., 262, 199 (1983).
54. S.J. Ward and A.E. Takemori, J. Pharmacol. Exp. Ther., 224, 525 (1983).
55. P.L. Wood, D. Sanschagrin, J.W. Richard and M. Thakur, J. Pharmacol. Exp. Ther., 226, 545 (1983).
56. A. Pazos and J. Florez, Eur. J. Pharmacol., 87, 309 (1983).
57. S.J. Ward and A.E. Takemori, Eur. J. Pharmacol., 87, 1 (1983).
58. G.S. Ling, K. Spiegel, S.L. Nishimura and G.W. Pasternak, Eur. J. Pharmacol., 86, 487 (1983).
59. J.W. Holaday, G.W. Pasternak and A.I. Faden, Neurosci. Lett., 37, 199 (1983).
60. K.-J. Chang, E.T. Wei, A. Killian and J.K. Chang, J. Pharmacol. Exp. Ther., 227, 403 (1983).
61. L.M. Sayre, P.S. Portoghese and A.E. Takemori, Eur. J. Pharmacol., 90, 159 (1983).
62. P.S. Portoghese and A.E. Takemori, J. Med. Chem., 26, 1341 (1983).
63. T.T. Chau, F.E. Carter and L.S. Harris, J. Pharmacol. Exp. Ther., 226, 108 (1983).
64. T. Chow and R.S. Zukin, Mol. Pharmacol., 24, 203 (1983).
65. T.M. Cho, B.L. Ge, C. Yamato, A.P. Smith and H.H. Loh, Proc. Natl. Acad. Sci. USA, 80, 5176 (1983).
66. R. Rothman and T.C. Westfall, J. Neurobiol., 14, 341 (1983).
67. R. Rothman and T.C. Westfall, Neurochem. Res., 8, 913 (1983).
68. E. Weber, F.S. Esch, P. Bohlen, S. Paterson, A.D. Corbett, A.T. McKnight, H.W. Kosterlitz, J.D. Barchas and C.J. Evans, Proc. Natl. Acad. Sci. USA, 80, 7362 (1983).
69. H. Matsuo, A. Miyata and K. Mizuno, Nature, 305, 721 (1983).
70. M. Westphal, C.H. Li, E.P. Heimer and J. Meienhofer, Biochem. Biophys. Res. Commun., 114, 1084 (1983).
71. E.P. Heimer, T.J. Lambros, A.M. Felix, G. Fleminger, C.H. Li, M. Westphal and J. Meienhofer, Arch. Biochem. Biophys., 225, 518 (1983).
72. J. Chou, J. Tang and E. Costa, Life Sci. 32, 2589 (1983).
73. H.Y. Yang, P. Panula, J. Tang and E. Costa, J. Neurochem., 40, 969 (1983).
74. R.G. Williams and G.J. Dockray, Neurosci., 9, 563 (1983).
75. P. Giraud, E. Castanas, G. Patey, C. Oliver and J. Rossier, J. Neurochem., 41, 154 (1983).
76. E.A. Majane, M.J. Iadarola and H.-Y.T. Yang, Brain Res., 264, 336 (1983).
77. J. Tang, J. Chou, H.-Y.T. Yang and E. Costa, Neuropharmacol., 22, 1147 (1983).
78. J. Tang, J. Chou, A.Z. Zhang, H.-T. Yang and E. Costa, Life Sci., 32, 2371 (1983).
79. J.M. Zajac, N. Ling, J. Rossier and B.P. Roques, Eur. J. Pharmacol., 90, 147 (1983).
80. K.-J. Chang and J.-K. Chang, Life Sci., 33, Suppl. I, 267, (1983).
81. C.H. Li, D. Yamashiro, P. Ferrara, L.-F. Tseng and E.L. Way, Int. J. Peptide Protein Res., 21, 331 (1983).
82. D. Yamashiro, P. Nicolas and C.H. Li, Int. J. Peptide Protein Res., 21, 219 (1983).
83. S. Salvadori, G.P. Sarto and R. Tomatis, Eur. J. Med. Chem., 18, 489 (1983).
84. S. Salvadori, M. Marastoni, R. Tomatis and G. Sarto, Farmaco., 28, 153 (1983).
85. K. Darlak, Z. Grzonka, P. Janicki, A. Czlonkowski and S.W. Gumulka, J. Med. Chem., 26, 1445 (1983).
86. R. de Castiglione and G. Perseo, Int. J. Peptide Protein Res., 21, 471 (1983).
87. G. Feuerstein and A.I. Faden, J. Pharmacol. Exp. Ther., 226, 151 (1983).

88. E.C. Degli-Uberti, G. Trasforini, S. Salvadori, R. Tomatis, A. Margutti, M. Bianconi, C. Rotola and R. Pansini, J. Clin. Endocrinol. Metab., 56, 1032 (1983).
89. J.-M. Zajac, G. Gacel, F. Petit, P. Dodey, P. Rossignol and B.P. Roques, Biochem. Biophys. Res. Commun., 111, 390 (1983).
90. D. Rodbard, T. Costa, Y. Shimohigashi and S. Krumins, J. Recept. Res., 3, 21 (1983).
91. H.I. Mosberg, R. Hurst, V.J. Hruby, J.J. Galligan, T.F. Burks, K. Gee and H.I. Yamamura, Life Sci., 32, 2565 (1983).
92. H.I. Mosberg, R. Hurst, V.J. Hruby, K. Gee, H.I. Yamamura, J.J. Galligan and T.F. Burks, Proc. Natl. Acad. Sci. USA, 80, 5871 (1983).
93. J.M. Berman, M. Goodman, T.M.-D. Nguyen and P.W. Schiller, Biochem. Biophys. Res. Commun., 115, 864 (1983).
94. T.S. Sudha and P. Balaram, Int. J. Peptide Protein Res., 21, 381 (1983).
95. T. Deeks, P.A. Crooks and R.D. Waigh, J. Med. Chem., 26, 762 (1983).
96. Y. Shimohiagashi and C.H. Stammer, J. Chem. Soc., Perk. I, 803 (1983).
97. Y. Shimohigashi, C.H. Stammer, T. Costa and P.F. Vonvoigtlander, Int. J. Peptide Protein Res., 22, 489 (1983).
98. H. Kimura, C.H. Stammer, Y. Shimohigashi, C. Ren-Lin and J. Stewart, Biochem. Biophys. Res. Commun., 115, 112 (1983).
99. N. Muthukumaraswamy, A.R. Day, D. Pinon, C.S. Liao and R.J. Freer, Int. J. Peptide Protein Res., 22, 305 (1983).
100. J.-L. Fauchere, S. Pfenninger, K.Q. Do, C. Lemieux and P.W. Schiller, Helv. Chim. Acta, 66, 1053 (1983).
101. R.G. Almquist, C.M. Olsen, E.T. Uyeno and L. Toll, J. Med. Chem., 27, 115 (1984).
102. L. Nadasdi, D. Yamashiro, C.H. Li and P. Huidobro-Toro, Int. J. Peptide Protein Res., 21, 344 (1983).
103. J.A. Kiritsy-Roy, S.K. Chan and E.T. Iwamoto, Life Sci., 32, 889 (1983).
104. L. Kupczyk-Subotkowska and P. Mastalerz, Int. J. Peptide Protein Res., 21, 485 (1983); P. Mastalerz, L. Kupczyk-Subotkowska, Z.S. Herman and G. Laskawiec, Naturwiss., 69, 46 (1982).
105. C. Zioudrou, D. Varoucha, S. Loukas, N. Nicolaou, R.A. Streaty and W.A. Klee, J. Biol. Chem., 258, 10934 (1983).
106. M. Szucs, S. Benyhe, A. Borsodi, M. Wolleman, G. Jancso, J. Szecsi and K. Medzihradszky, Life Sci., 32, 2777 (1983).
107. K.C. Rice, A.E. Jacobson, T.R. Burke, Jr., B.S. Bajwa, R.A. Streaty and W.A. Klee, Science, 220, 314 (1983).
108. P.S. Portoghese and A.E. Takemori, in "The Chemical Regulation of Biological Mechanisms," A.M. Creighton and S. Turner, Eds., Royal Society of Chemistry Special Publication No. 42, p. 181 (1983).
109. L.M. Sayre, A.E. Takemori and P.S. Portoghese, J. Med. Chem., 26, 503 (1983).
110. L.M. Sayre, D.L. Larson, D.S. Fries, A.E. Takemori and P.S. Portoghese, J. Med. Chem., 26, 1229 (1983).
111. R.C. Crouch, A.V. Bhatia and O.W. Lever, Jr., Tetrahedron Lett., 24, 4801 (1983).
112. M. Hori, T. Kataoka, H. Shimizu, E. Imai, Y. Suzuki and N. Kawamura, Heterocycles, 20, 1979 (1983).
113. M. Hori, T. Kataoka, H. Shimizu, E. Imai, Y. Suzuki, N. Kawamura, H. Fujimura, M. Nozaki and M. Niwa, Chem. Pharm. Bull. Jap., 31, 2520 (1983).
114. F. Ghozland, P. Maroni, I. Viloria and J. Cros, Eur. J. Med. Chem., 18, 22 (1983).
115. K. Katsuura, M. Ohta and K. Mitsuhashi, Chem. Pharm. Bull. Jap., 30, 4378 (1982).
116. K. Katsuura, K. Yamaguchi, S. Sakai and K. Mitsuhashi, Chem. Pharm. Bull. Jap., 31, 1518 (1983).
117. D.L. Leland and M.P. Kotick, J. Org. Chem., 48, 1813 (1983).
118. M.P. Kotick, J. Org. Chem., 48, 1819 (1983).
119. M.P. Kotick, D.L. Leland, J.O. Polazzi, J.F. Howes and A. Bousquet, J. Med. Chem., 26, 1050 (1983).
120. H. Schmidhammer and A. Brossi, J. Org. Chem., 48, 1469 (1983).
121. A. Ghosh, D. Portlock, H. Dalzell, C. Malmberg, P. Herlihy and R. Razdan, J. Org. Chem, 48, 4137 (1983).
122. M. Hori, M. Ban, E. Imai, N. Iwata, Y. Baba, H. Fujimura, M. Nozaki and M. Niwa, Heterocycles, 20, 2359 (1983).
123. M. Bös, W. Fleischhacker and P. Tschany, Heterocycles, 20, 1077 (1983).
124. J. Bosch, A. Domingo and R. Granados, J. Heterocyclic Chem., 20, 887 (1983).
125. M.Y.H. Essawi and P.S. Portoghese, J. Med. Chem., 26, 348 (1983).
126. J.W. Taylor, R.J. Miller and E.T. Kaiser, J. Biol. Chem., 258, 4464 (1983).
127. J.P. Blanc, J.W. Taylor, R.J. Miller and E.T. Kaiser, J. Biol. Chem., 258, 8277 (1983).
128. W. Kullmann, J. Med. Chem., 27, 106 (1984).
129. D.J. Mayer, D.D. Price and A. Rafii, Brain Res., 121, 368 (1977).
130. C.R. Chapman, C. Benedetti, Y. Colpitts and R. Gerlach, Pain, 16, 13 (1983).
131. T.B. Freeman, J.N. Campbell and D.M. Long, Pain, 17, 189 (1983).
132. G.W. Terman, J.W. Lewis and J.C. Liebeskind, Brain Res., 260, 147 (1983).
133. L.R. Watkins and D.J. Mayer, Science, 216, 1185 (1982).
134. J.T. Cannon, J.W. Lewis, V.E. Weinberg and J.C. Liebeskind, Brain Res., 269, 231 (1983).
135. L.R. Watkins, E. Young, I. Kinsheck and D.J. Mayer, Soc. Neurosci. Abstr., 7, 340 (1981).
136. T. Lindeburg, S. Folsgard, H. Sillesen, E. Jacobsen and H. Kehlet, Acta Anaesthesiol. Scand., 27, 10 (1983).
137. M. Kjaer and H. Nielsen, Br. J. Clin. Pharmacol., 16, 477 (1983).
138. D. Niv, J.G. Whitwam and L. Loh, Br. J. Anaesthesiol., 55, 541 (1983).
139. J. Sawynok, Naunyn-Schmiedeberg's Arch. Pharmacol., 323, 54 (1983).
140. T.F. Murray, W. McGill and D.L. Cheney, Eur. J. Pharmacol., 90, 179 (1983).
141. P. Krogsgaard-Larsen, P. Jacobsen, E. Falch and H. Hjeds, in "Chemical Regulation of Biological Mechanisms," A.M. Chreighton and S. Turner, Eds., Royal Soc. Chem. Special Publ. 42, London, 1982, p. 92.
142. F.V. DeFeudis, Pharmacol. Res. Commun., 14, 383 (1982).
143. T. Andree, D.A. Kendall and S.J. Enna, Life Sci., 32, 2265 (1983).
144. S.P. Sivam and I.K. Ho, Neuropharmacol., 22, 767 (1983).
145. S. Yasuoka and T.L. Yaksh, Neuropharmacol., 22, 309 (1983).
146. J.R. Howe, J.Y. Wang and T.L. Yaksh, J. Pharmacol. Exp. Ther., 224, 552 (1983).
147. Y. Kuraishi, Y. Harada, S. Aratani, M. Satoh and H. Takagi, Brain Res., 273, 245 (1983).
148. W.T. Chance, Life Sci., 33, 2241 (1983).
149. E. Margaria, M. Gagliardi, L. Palieri, S. Treves and E. Fanzago, Int. J. Clin. Pharmacol. Ther. Toxicol., 21, 47 (1983).
150. J.W. Commisiong, J. Neurochem., 41, 1313 (1983).
151. C.A. Johnston and K.E. Moore, J. Neural Transm., 57, 65 (1983).
152. O.-G. Berge, K. Hole and S.-O. Ogren, Brain Res., 271, 51 (1983).
153. J.L.K. Hylden and G.L. Wilcox, Life Sci., 33, 789 (1983).
154. R. Dirksen and G.M. Nijhuis, Eur. J. Pharmacol., 91, 215 (1983).
155. J.A. Romano and T.M. Shih, Neuropharmacol., 22, 827 (1983).
156. M.D. Aceto, H. Awaya, B.R. Martin and E.L. May, Br. J. Pharmacol., 79, 869 (1983).

157. J.W. Lewis, J.T. Cannon and J.C. Liebeskind, Brain Res., 270, 289 (1983).
158. D.J. Bill, J.E. Hartley, R.J. Stephens and A.M. Thompson, Br. J. Pharmacol., 79, 191 (1983).
159. R.K. Razdan and J.F. Howes, Med. Res. Rev., 3, 119 (1983).
160. L. Lemberger, Ann. Rev. Pharmacol. Toxicol., 20, 151 (1980).
161. L.S. Harris, in "The Therapeutic Potential of Marihuana," S. Cohen and R.C. Stillman, eds., Plenum, New York, 1976, p. 299.
162. L. Melvin, M. Johnson and G. Milne, Am. Chem. Soc. 186th Nat. Mtg. Abstr., Medi. 22, 1983.
163. M. Johnson, J. Eggler, L. Melvin and G. Milne, Am. Chem. Soc. 186th Nat. Mtg. Abstr., Medi. 1, 1983.
164. L. Melvin, M. Johnson and G. Milne, Am. Chem. Soc. 186th Nat. Mtg. Abstr., Medi. 2, 1983.
165. L. Melvin, M. Johnson, C. Harbert, G. Milne and A. Weissman, J. Med. Chem., 27, 67 (1984).
166. E.M. Zola and D.C. McLeod, Drug Intell. Clin. Pharm. 17, 411 (1983).
167. R.E.S. Bullingham, H.J. McQuay and R.A. Moore, Clin. Pharmacokinet., 8, 332 (1983).
168. S.R. Hameroff, Contemp. Anesth. Pract., 7, 27 (1983).
169. W.L. Woolverton and C.R. Schuster, Pharmacol. Rev., 35, 33 (1983).
170. J.E. Errick and R.C. Heel, Drugs, 26, 191 (1983).
171. "Clinical Experience with Injectable Meptazinol--a New Strong Analgesic. Proceedings of a Symposium," Postgrad. Med. J., 59 (Suppl. I), pp. 1-94 (1983).
172. G. Davies, A.J. Sinclair, S.J. Warrington, R. Franklin, T. Frost, A. Latham and P.J. Robson, Eur. J. Clin. Pharmacol., 23, 535 (1982).
173. H.M. Norbury, R.A. Franklin and D.F. Graham, Eur. J. Clin. Pharmacol., 25, 77 (1983).
174. P.K. Kalia, R. Madan, R. Saksena, R.K. Batra and G.R. Gode, Anesth. Analg., 62, 949 (1983).
175. J. Cahill, D. Murphy, D. O'Brien, J. Mulhall and G. Fitzpatrick, Anaesthesia, 38, 760 (1983).
176. B.T. Finucane, J.B. Floyd, D.J. Petro and R. Braswell, Can. Anaesth. Soc. J., 29, 500 (1982).
177. J.E. Stambaugh, Jr., J. McAdams and G. Freeland, Clin. Pharmacol. Ther., 33, 198 (1983).
178. L.E. Mather, Clin. Pharmacokinet., 8, 422 (1983).
179. J.D. Borel, Contemp. Anesth. Pract., 7, 1 (1983).
180. S.S. Bloomfield, T.P. Barden and J. Mitchell, Clin. Pharmacol. Ther., 34, 240 (1983).
181. R.C. Frederickson, C.J. Parli, G.W. DeVane and M.D. Hynes, Natl. Inst. Drug Abuse Monogr. Ser., 43, 150 (1983).
182. B.M. Onofrio and T.L. Yaksh, Lancet, I, 1386 (1983).
183. N. Gilchrist, J.E. Bolton, R.A. Donald, J.H. Livesey, H.K. Roud and M.G. Nicholls, Br. J. Clin. Pharmacol., 16, 465 (1983).
184. A. Roby, J.-C. Willer and B. Bussel, Neuropharmacol., 22, 1121 (1983).
185. G. Nuki, Brit. Med. J., 287, 39 (1983).
186. S.A. Cooper, Ann. Rev. Pharmacol. Toxicol., 23, 617 (1983).
187. H.L. Crossley, S.A. Bergman and R.L. Wynn, J. Am. Dent. Assoc., 106, 61 (1983).
188. R.K. Verbeeck, J.L. Blackburn and G.R. Loewen, Clin. Pharmacokinet., 8, 297 (1983).
189. R.J. Capetola, M.E. Rosenthale, B. Dubinsky and J.L. McGuire, J. Clin. Pharmacol., 23, 545 (1983).
190. R.O. Davies, Pharmacother., 3 (2, pt. 2), 9S (1983).
191. Pharmacol., 27 (Suppl.), pp. 1-96 (1983).
192. A. Sunshine, N.Z. Olson, E.M. Laska, I. Zighelboim, A. deCastro and C. deSarrazin, Clin. Pharmacol. Ther., 34, 254 (1983).
193. D.J.M. Keenan, K. Cave, L. Langdon and R.E. Lea, Brit. Med. J., 287, 1335 (1983).
194. G. Rigamonti, E. Zanella, R. Lampugnani, D. Marrano, O. Campione, G. Bruni, V. Mandelli and G. Sacchetti, Br. J. Anaesth., 55, 513 (1983).
195. A. Pellegrini, B. Massidda, P. Pellegrini, E. Defraia, A. Usai, L. Parrinello, G. Bruni, G. Rigamonti and G. Sacchetti, Int. J. Clin. Pharmacol. Ther. Toxicol., 21, 483 (1983).
196. E.R. Umbenhauer, Pharmacother., 3, 55S (1983).
197. J.A. Forbes, A.L. Kolodny, W.T. Beaver, R.W. Shackelford and V.R. Scarlett, Pharmacother., 3, 47S (1983).
198. A. Sunshine, N.Z. Olson, E.M. Laska, I. Zighelboim, A. DeCastro and C. DeSarrazin, Pharmacother., 3, 177 (1983).
199. P.J. Desjardins, S.A. Cooper, C.M. Ruderman, L.T. Gallegos, D.C. Reynolds and G.O. Kruger, Pharmacother., 3, 52 (1983).
200. E.C. Huskisson and R. Bryans, Curr. Med. Res. Opinion, 8, 350 (1983).
201. N.J. Wenzloff and L. Shimp, Drug Intell. Clin. Pharm., 18, 22 (1984).
202. W.Y. Chan, Ann. Rev. Pharmacol. Toxicol., 23, 131 (1983).
203. A. Marzo, G. Quadro, G. Bruno, E. Treffner, F.L. Reali, G. Borrelli and R. Stradi, Arzneim.-Forsch., 33 (I), 198 (1983).
204. K. Hino, H. Nakamura, Y. Nagai, H. Uno and H. Nishimura, J. Med. Chem., 26, 222 (1983).
205. J.R. Graham, Headache, 21, 243 (1981).
206. E. Skinhoj, Arch. Neurol., 29, 95 (1973).
207. W.K. Amery, Cephalalgia, 2, 83 (1982).
208. M. Botney, Headache, 21, 179 (1981).
209. R.J. McCaffrey and W. Isaac, Headache, 23, 152 (1983).
210. R. Peatfield, Drugs, 26, 364 (1983).
211. J.E. Sutherland, Am. Fam. Physician, 27, 137 (1983).
212. N. Martucci, V. Manna, P. Mattesi, G. Troiani, G. Manzoni, M. Lanfranchi, G. Bono and G. Micieli, Cephalalgia, 3 (Suppl. 1), 151 (1983).
213. B.H. Peters, C.J. Fraim and B.E. Masel, Am. J. Med., 74, 36 (1983).
214. R.C. Peatfield, R.G. Petty and F.C. Rose, Cephalalgia, 3, 129 (1983).
215. S. Diamiond, L. Kudrow, J. Stevens and D. Shapiro, Headache, 22, 268 (1982).
216. R.E. Ryan, Sr., R.E. Ryan, Jr., and A. Sudilovsky, Headache, 23, 26 (1983).
217. P. Louis, Headache, 21, 235 (1981).
218. W.K. Amery, Headache, 23, 70 (1983).
219. J.S. Meyer and J. Hardenberg, Headache, 23, 266 (1983).
220. S.J. Peroutka, Headache, 23, 278 (1983).

Chapter 2. Anti-Anxiety Agents, Anticonvulsants & Sedative-Hypnotics

Joseph P. Yevich, James S. New and Michael S. Eison
Bristol-Myers Research and Development, Evansville, Indiana 47721

Introduction - The prevalence of anxiety and the diverse utility of existing anxiolytic drugs continues to promote a broad spectrum of research in the anxiolytic area. The challenging basic science of anxiolytic drug action and the potential for compounds with profiles different from the traditional antianxiety drugs has focused attention upon more selective structure activity relationships. Several excellent review articles emphasizing medications currently used,[1] approaches to new drug discovery,[2] newer agents and theories,[3] side-effects and hazards,[4-8] and the use of benzodiazepines (BZ) as preoperative medications[9] have been published. Insights into the GABAergic effects of these drugs and their anticonvulsant properties continue to grow.[10-14]

Animal Models - Conflict tests play a prominent role in anxiolytic research, and may be used to demonstrate proconflict, anticonflict, and anxiolytic antagonist actions.[15] A chronic conflict paradigm has been described.[16] The effects of drugs upon deprivation-induced drinking should be considered to avoid false-positive indications.[17] The susceptability of compounds' anticonflict effects to blockade by BZ antagonists appears to distinguish BZ receptor binders from non-displacers.[18] The use of anticonvulsant activity and effects upon BZ binding as indices to dissociate pro-drug from directly acting BZ effects has been described,[19] as has a chronic epilepsy model in rats.[20]

Drug-discrimination techniques have revealed that the diazepam stimulus complex is specific.[21] It has been proposed that drug discrimination techniques can detect anxiogenic BZ withdrawal effects.[22,23] In other behavioral tests, the increase in response rates induced by BZ's to rewarding electrical brain stimulation is facilitated by GABA antagonists,[24] and is observed even in rats trained on a DRL schedule (rewarding low rates of response).[25,26] Conditioned startle,[27] defensive burying,[28] and social interactions[29,30] continue to be used to investigate anxiolytic and BZ antagonist activity. Quantitative EEG activity in animals[31] and man[32] complement cellular electrophysiological investigations of anxiolytic drug effects.[33,34] Effects of BZs upon memory,[35] sleep,[36] and electrically-induced head turning[37] have been investigated, as have effects upon feeding and drinking.[34-40] A protective action of diazepam against gastric ulceration in rats induced by water-immersion or by immobilization stress has been described.[41] Tolerance developing to anxiolytic drugs,[42,43] and models of physical dependence to BZs in rat[44] and dog[45] have been reported.

Benzodiazepine Receptor Neurochemistry and Pharmacology - The concept of heterogeneous benzodiazepine (BZ) receptor sites has stimulated considerable research in the development of drugs which may be more anxioselective than the BZ class. The functional linking of a

ISBN 0-12-040519-9

chloride ionophore, BZ and GABA receptor sites is substantiated by their isolation in a supramolecular complex in which GABA stimulation of [^{3}H]flunitrazepam binding (FLU, 1a) remains intact.[46] Affinity chromatography has also led to the successful isolation of a homogeneous GABA/BZ receptor complex from bovine cerebral cortex.[47] Data continues to accrue indicating GABA, and its agonists, enhance the binding of the BZs to these receptors; diazepam (DZ, 1b) enhances the binding of GABA and muscimol but has no effect on THIP.[48-50] The presence of a single class of saturable [^{3}H] muscimol binding sites (K_D = 23 nM) on a purified BZ receptor is evident.[51] The potentiation of [^{3}H]DZ binding is also elicited by ethylenediamine analogues in a process mediated predominantly by a change in receptor number and not receptor affinity.[52] Both [^{3}H]DZ and [^{3}H]GABA binding are enhanced by the hypnotics (+)etomidate and pentobarbital, presumably through a similar mechanism, which is blocked by GABA antagonists.[53] GABAergic ligands were also observed to protect soluble BZ receptors from heat inactivation.[54]

BZ Type 1 and Type 2 receptors, initially defined on the basis of their different affinities for CL 218,872, have been further characterized in rat cerebellum and cerebral cortex. CL 218,872 preferentially displaced [^{3}H]FLU from Type 1 receptors; Type 1 receptor binding is stimulated in a chloride ion dependent manner by 1mM sodium pentobarbital, while Type 2 sites remain relatively unaffected.[55] Distinction between Type 1 and Type 2 receptor populations is reported to be possible through their differential sensitivity to detergent solubilization techniques.[56]

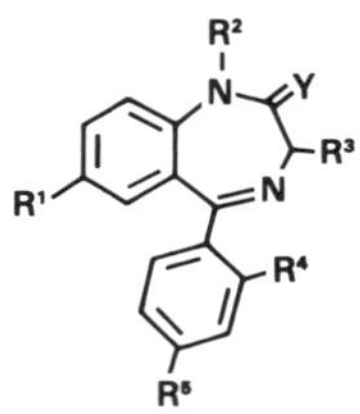

1

	R¹	R²	R³	R⁴	R⁵	Y
1a	NO_2	CH_3	H	F	H	O
1b	Cl	CH_3	H	H	H	O
1c	Cl	CH_2CF_3	H	F	H	S
1d	Cl	CH_3	H	H	Cl	O
1e	Cl	CH_2CF_3	H	H	H	O
1f	Cl	H	OH	Cl	H	O
1g	Cl	H	COOH	H	H	O
1h	Cl	$(CH_2)_2N(C_2H_5)_2$	H	F	H	O
1i	Cl	CH_3	OH	H	H	O
1j	NO_2	H	H	H	H	O

The effects of GABA agonists on [^{3}H]FLU binding to these sites in both cerebellar and hippocampal membranes were found to be broadly similar in one study.[57] Alternatively, the effect of GABA on FLU binding to [^{3}H]FLU and [^{3}H]β-CCP (2a) labelled Type 1 and 2 receptors was found to involve different degrees of GABA regulation dependent on the specific brain region.[58]

SAR studies have examined the β-carbolines as templates in the design of more specific, high affinity ligands for the BZ receptor.[59,60] [^{3}H]DMCM (2b) interacts with high affinity BZ receptors with binding properties different from those of [^{3}H]FLU and β-CCP, which may reflect a subclass selectivity within these receptors.[61] Intraventricular injections of β-CCE (2d) increased the number of [^{3}H]DZ binding sites in a variety of brain regions but, in a different study, decreased the total number of [^{3}H]GABA binding sites in rats habituated to handling procedures.[62,63] β-CCM (2c) has been shown to antagonize the anxiolytic effects of the BZs, possess convulsant properties, and exhibit anxiogenic properties.[64] [^{3}H]β-CCE binds to a homogenous population of recognition sites in rat brain and is completely dissociable by low concentrations of BZs.[65] The β-carboline FG 7142 (2f) was found to induce severe anxiety in man.[66] Pharmacokinetic factors were studied as a parameter controlling the convulsant activity of β-CCM, DMCM, and the nonconvulsant properties of β-CCE in the rat; a slow rate of degradation of these compounds by rat plasma (in vitro) paralleled their potencies as convulsants.[67]

BZ receptor heterogeneity has been further delineated from the differential affinities observed between pairs of ligands for these sites. Subpopulations of [^{3}H]DZ receptor sites are defined by their interactions in bicuculline/GABA or barbiturate/pyrazolopyridine binding studies.[68] Discrimination between these subtypes was also possible with RO 15-1788 (3), but not with CGS 8216.[68] A study of various BZ receptor ligands, with different biological activities, yielded no simple correlation between their biological activity and the observed temperature dependence of their binding.[69]

The non-homogeneity of BZ-receptors has generated several models which attempt to rationalize their binding characteristics with a structurally diverse group of ligands. Kinetic studies suggest these receptors exist as a single population in two distinct conformational states whose interconversion is affected by BZ agonists, GABA, chloride ion, temperature, and the BZ antagonist RO 15-1788.[70] An allosteric model for BZ receptor function suggests the intrinsic activity of an agonist reflects its ability to induce a conformational change in the receptor. This approach predicts that ligands with high affinity for a binding domain on both the ground state and activated conformations of the receptor will be anxioselective agents.[71] A definition of the BZ receptor as a coupling unit between the GABA receptor and chloride channel, is used to explain its interaction with the BZ tranquilizers, the β-carbolines, RO 15-1788, and phenobarbitone. BZ agonists enhance the coupling role of the BZ receptor while β-carbolines block the access of BZ agonists to this coupling site. This actually results in a reduced coupling mechanism in the receptor and elicits opposing pharmacological effects. The β-carbolines, DMCM (2b) and β-CCM (2c), are referred to as "inverse

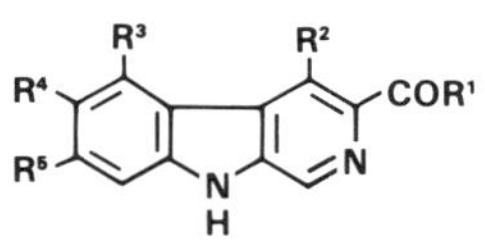

2

	R^1	R^2	R^3	R^4	R^5
2a	OnC_3H_7	H	H	H	H
2b	OCH_3	C_2H_5	H	OCH_3	OCH_3
2c	OCH_3	H	H	H	H
2d	OC_2H_5	H	H	H	H
2e	OC_2H_5	CH_2OCH_3	$OCH_2C_6H_5$	H	H
2f	$NHCH_3$	H	H	H	H

agonists" in this scheme. This distinguishes them from pure antagonists such as RO 15-1788, which blocks the access of both inverse agonists and agonists to the receptor, without inducing a functionally relevant change in it.[72] Ethanol and pentobarbital may modulate the BZ binding component in the chloride ionophore-BZ-GABA receptor complex via the picrotoxinin sensitive site, and a thirsty rat conflict model indicates the anticonflict actions of pentobarbital might be mediated through BZ receptors.[73,74] A peptidoaminobenzophenone, 450088-S(4), was found to effectively inhibit the labeling of BZ

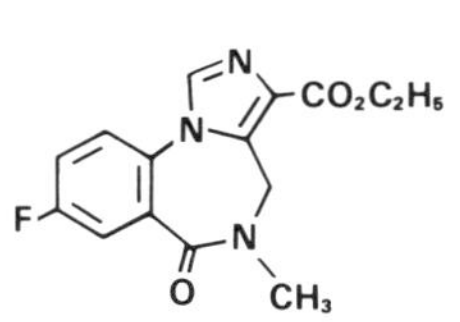

3

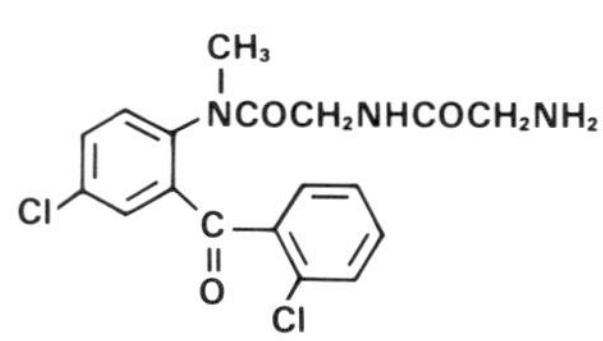

4

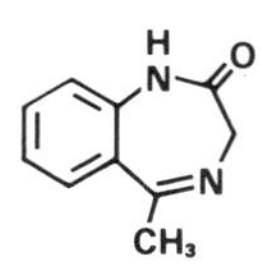

5

receptors by [^{3}H]FLU in vivo despite a lack of in vitro activity, and quazepam (1c) preferentially binds to BZ Type 1 receptors.[75,76]

Several lines of evidence indicate RO 15-1788 may not be a strict BZ antagonist but may also have some partial agonist properties; it abolishes the sedative/ataxic effects of DZ, but slows the development of kindled seizures, suggesting an anticonvulsant action in its pharmacology.[77] Ro 15-1788 is effective in reversing the anticonvulsant effects of DZ and CL 218,872,[64,78] and is shown to block the cognitive, amnesic, psychomotor and subjective effects of DZ in man, without altering the bioavailability of this drug.[79] Psychometric tests sensitive to the effects of BZs showed no intrinsic pharmacological activity associated with RO 15-1788 in man up to 600 mg orally.[80]

The convulsant BZ RO 5-3663 (5) and RO 15-1788, both antagonized the anticonflict effect of chlordiazepoxide, but while RO 5-3663 was distinctly anxiogenic, RO 15-1788 displayed some anticonflict properties.[81] RO 5-4864 (1d), a ligand selective for the BZ low affinity (micromolar) and peripheral receptors, registered significant anxiogenic action in the social interaction test which was reversed by phenytoin, a ligand for the low affinity receptors.[82]

Benzodiazepine Anxiolytic Agents - A critical review examined the concept that multiple BZ receptors underlie the diverse behavioral effects which are observed with this chemical class.[83] A study examining the effects of rapidly increasing doses of halazepam (1e) and DZ on severely anxious hospitalized patients was reported[84] and effects of DZ on short term memory[85] and information processing[86] were examined. DZ was found to potentiate the behavioral effect of neuroleptics in rats,[87] to produce some improvement in neuroleptic resistant schizophrenics,[88] and to alleviate symptoms of tardive dyskinesia.[89] The effects of long term DZ therapy and drug withdrawal effects were also reported.[90,91] Clobazam (6) and DZ were compared in their differential effects on psychomotor performance[92] and anxiolytic responses.[93] Lorazepam (1f) was evaluated for its effects on psychomotor performance vs. clobazam[94] and DZ;[95] clear performance decrements were associated with its use.[96] Several studies indicate alprazolam (7a) has anxiolytic efficacy comparable to DZ but at a lower dose and with a lower incidence of sedation.[97,98] Halazepam and clorazepate (1a) were shown to not be significantly different from each other in treating symptoms of anxiety or tension.[99]

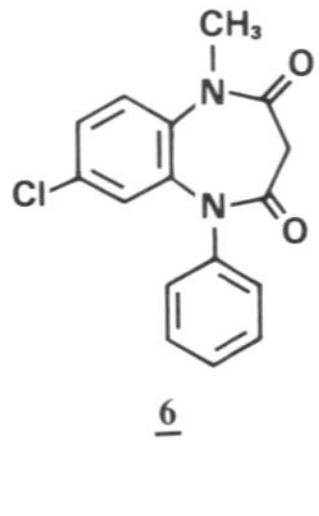

6

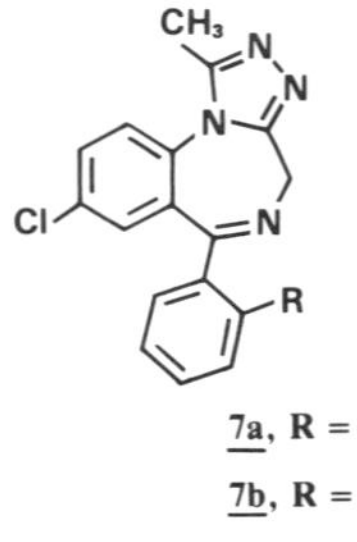

7a, R = H

7b, R = Cl

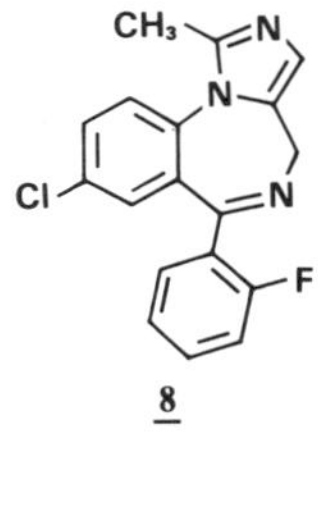

8

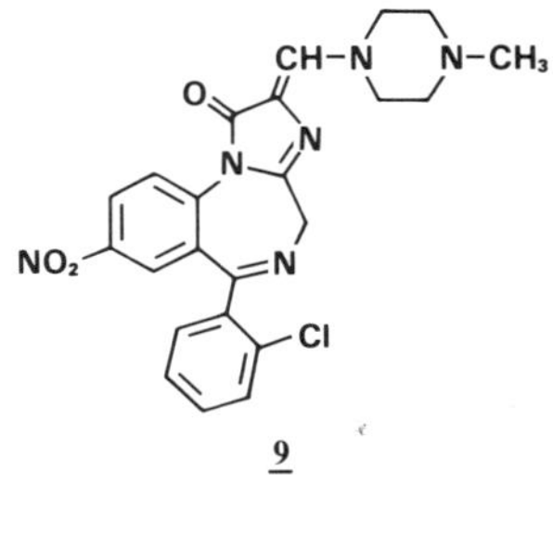

9

Sedative -Hypnotics - Midazolam (8) continues to receive substantial support as an effective hypnotic[100] with neither tolerance effects nor rebound insomnia on drug withdrawal.[101] A series of papers has addressed safety[102] and psychomotor performance aspects of this drug.[103,104] The clinical literature concerning the hypnotic efficacy of flurazepam (1h), temazepam (1i), and triazolam (7b) has been reviewed. Triazolam has been researched as a hypnotic for geriatric patients.[105-107] Considerable differences were reported for the pharmacokinetics of five BZ hypnotics in healthy subjects.[108] There is no distinguishable difference between the nocturnal effects of triazolam or flurazepam, however, the "hangover" effects of these two drugs are still under evaluation.[109,110] Triazolam appeared superior to nitrazepam (1j) in an overall evaluation of sleep.[111] The hypnotic loprazolam (9) was found not to be potentiated by alcohol,[112] but no direct correlation was found between its pharmacokinetic data and various sleep parameters which were scrutinized.[113] Initial indications suggest temazepam may have less hangover potential than nitrazepam,[114] and both nitrazepam and flunitrazepam were well tolerated in elderly patients.[115]

Non-Benzodiazepines

Anxiolytic Agents - Further neurochemical and/or behavioral evaluations have been reported for 10-13a which have previously been shown to exhibit anxiolytic activity in animal models. Electrophysiological and receptor binding studies suggest that etazolate (10a)

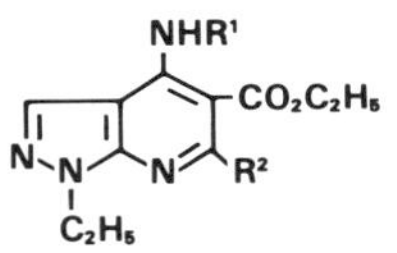

10a, $R' = N{=}C(CH_3)_2$, $R^2 = H$

10b, $R' = n\text{-}C_4H_9$, $R^2 = CH_3$

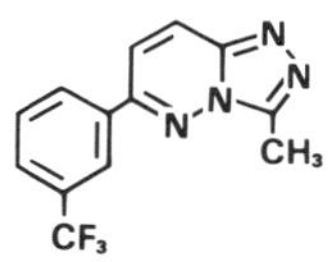

11

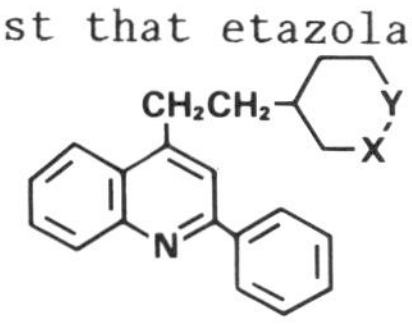

12a, X = CH_2, Y = NH

12b, X = NH, Y = CH_2

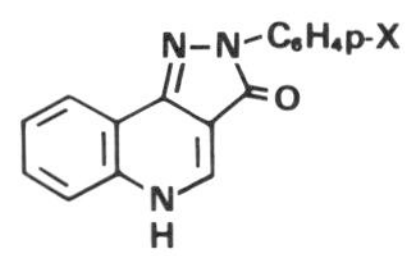

13a, X = Cl

13b, X = H

enhances BZ binding and the effects of GABA by interaction at a site distinct from the GABA and BZ recognition sites.[116] The bicuculline- and picrotoxin-sensitive blockade of electrically-induced head turning in the rat by tracazolate (10b) indicates the drug's GABAmimetic activity.[37] Antagonism of the DZ-induced righting reflex in mice by Cl 218,872 (11) has been interpreted as evidence for its partial agonist activity at the BZ receptor.[117] The quinoline derivatives PK 8165 (12a) and PK 9084 (12b) have also been hypothesized as partial BZ receptor agonists based on their antagonism of the anticonvulsant effects of DZ and their competitive inhibition of [^{3}H]FLU binding.[118] PK 9084 showed a partial anxiolytic profile in a social interaction paradigm in rats, while PK 8165 was ineffective at the doses tested.[29] In a pentylenetetrazol-saline discrimination model, CGS 9896 (13a) exhibited DZ-like activity which was blocked by pretreatment with the BZ antagonist CGS 8216 (13b).[119]

Several clinical studies have shown buspirone (14a) to have anxiolytic efficacy equivalent to that of DZ with significantly less sedation.[120-122] Rats trained to discriminate oxazepam or pentobarbital from vehicle did not generalize to buspirone.[123] At doses above those which are anxiolytically relevant in man, the drug caused a dose-related elevation of plasma prolactin in male subjects, and like the BZ's also increased growth hormone levels.[124]

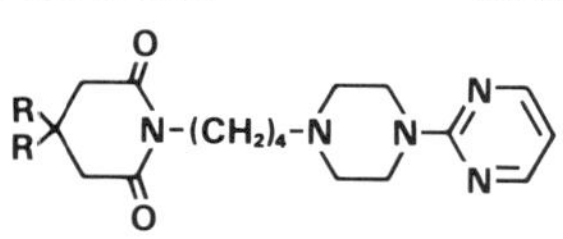

14a, R, R = -$(CH_2)_4$-

14b, R = CH_3

Buspirone elicits a dose-dependent rise in rat striatal dopamine (DA) metabolite levels[125,126] and may do so by selective antagonism of presynaptic DA autoreceptors with minimal postsynaptic effects.[127] Its catalepsy-reversal effects may occur via a DA-independent mechanism.[128] The drug increases rat locus coeruleus noradrenergic neuronal activity,[129] binds to calf hippocampus serotonin type-1 receptors,[130] enhances [^{3}H]FLU binding after oral dosing in rats,[131] causes an increase in the K^+-depolarized uptake of $^{45}Ca^{++}$ into synaptosomes,[132] and affords significant brain levels of 1-(2-pyrimidinyl)piperazine as a metabolite in the rat.[133] The mechanistic implications of these findings are as yet unclear. The analogue MJ 13805 (14b) while equipotent with 14a in conflict testing, has much weaker effects than the latter upon DA neurotransmission, thus suggesting that DAergic effects may not be important to the antianxiety activity of these drugs.[134] The SAR of a series of aryl and heteroarylpiperazine derivatives including 14a and b has been reported.[135]

A single-blind clinical study of the GABA agonist THIP (15a) found it to have weak anxiolytic effects at doses at or near those which caused sedation and other side effects.[136] Members of a series of (ω-piperazinylalkoxy)indanes (16) exhibited BZ-like potency in the fighting mouse test.[137] Various pyrimido[5,4-d]benzazepine derivatives (17) showed potent antimetrazol activity and were more

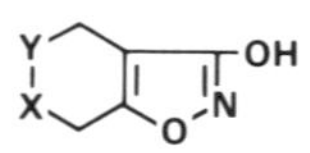

15a, X = NH, Y = CH_2

15b, X = CH_2, Y = NH

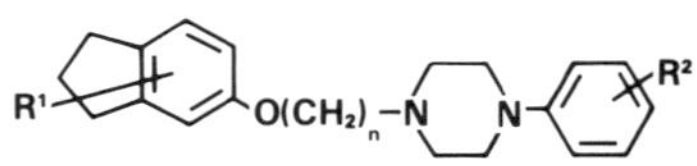

16

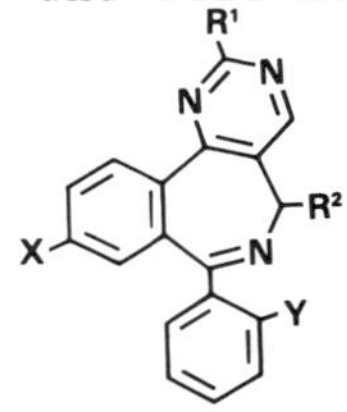

17a, X = Y = Cl, R^1 = R^2 = H

17b, X = Y = Cl, R^1 = H, R^2 = OH

potent in displacing [^{3}H]DZ binding than DZ, with lesser effects than DZ on motor coordination and ethanol interaction.[138] A member of the series (17a) has been selected for anxiolytic evaluation in man and 17b, its major metabolite (in dog and man), was found to be equipotent with the parent drug in pharmacological tests.

Sedative-Hypnotics - The hypnotic and anticonvulsant profile of etomidate (18) has been reviewed.[139] Based on comparative clinical effects on the photopalpebral reflex, zopiclone (19) was slightly more potent than nitrazepam with more rapid onset, shorter duration of action, and fewer side effects.[140] A study in insomniac patients showed that a 7.5 mg dose of zopiclone improved the quality of sleep with a reduction in both sleep onset latency and the number of nocturnal awakenings.[141] Following their microiontophoretic application, the "sleep" peptides, delta sleep-inducing peptide and arginine-vasotocin caused predominantly excitatory actions on neurons in the brain stems of rats and rabbits.[142] Following its microinjection into the brains of rabbits, the sites of action of the sleep-promoting muramyl peptide isolated from human urine were found to be in the forebrain region.[143] While parenteral administration of an adenosine deaminase inhibitor, erythro-9-(2-hydroxy-3-nonyl)adenine, to mice and rats caused profound behavior sedation, the accompanying EEG changes showed greatly enhanced time awake.[144]

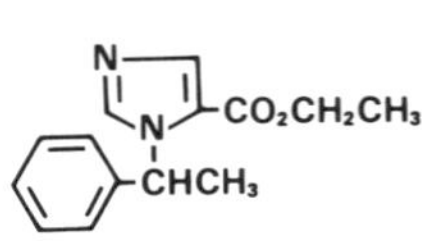

18

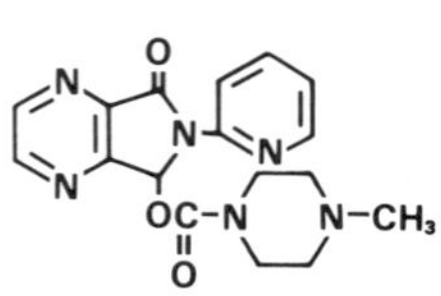

19

Anticonvulsants - The effects of several marketed antiepileptic drugs including carbamazepine, phenytoin, and sodium valproate, upon cognitive function and behavior have been clinically assessed.[145] Adenosine receptor binding studies of carbamazepine indicate that its anticonvulsant action may be due in part to A_1 receptor agonist and A_2 receptor antagonist activities.[146] A computer graphics-based pattern-recognition analysis of several series of 5-ethyl-5-substituted barbiturates has revealed that the region of occupied space for the low energy conformation of the lipophilic side-chain differs between anticonvulsant and convulsant members of the series.[147] N-n-propylnorapomorphine was the most potent of a series of apomorphine-related DA agonists in protecting genetically seizure-prone mice and baboons.[148] Potential anticonvulsant activity of α-adrenergic agonists was indicated by their in vitro inhibition of interictal discharge in rat hippocampal slices; β-adrenergic agonists showed proconvulsant actions.[149] Compared with placebo, the GABA-ergic drug progabide (20) caused a significant decrease in seizures, with mild side effects, in a double-blind clinical trial in "therapy-resistant" epileptic patients.[150] A single-blind evaluation of THIP in epileptic patients failed to show significant differences from placebo with maximal doses of the drug.[151] THPO (15b), a glial-selective GABA uptake inhibitor had anticonvulsant activity in chicks but not in adult mice.[152] Denzimol (21) was shown to have a general anticonvulsant profile similar to that of carbamazepine and phenytoin in animal models.[153] The triazine, LY 81067 (22) protects mice against pentylenetetrazol- and bicuculline-induced convulsions and enhances both [^{3}H]GABA and [^{3}H]FLU binding.[154,155] Other compounds reported to have anticonvulsant activity in various animal models include the β-carboline ZK 91296 (2e),[156] 2-phthalimidoethanesulfonamides (23),[157] eugenol analogues (24),[158] pyrido[3,4-b][1,4]benzothiazines (25)[159] and oxoquinazolines (26).[160]

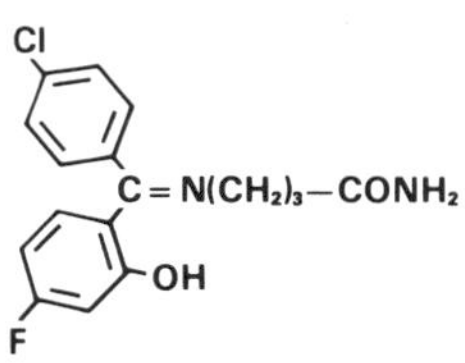

20

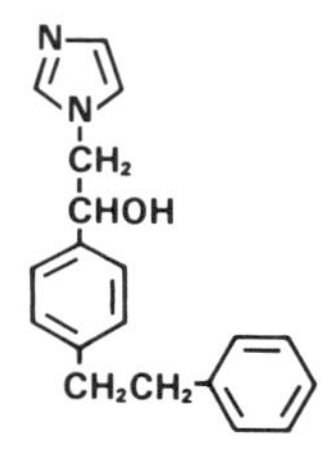

21

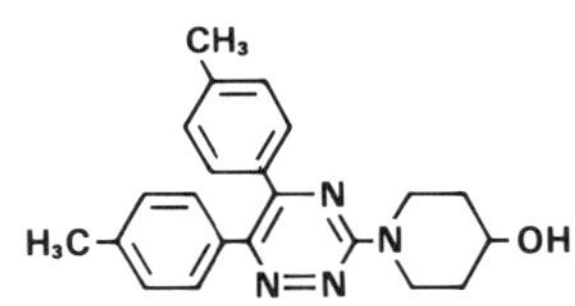

22

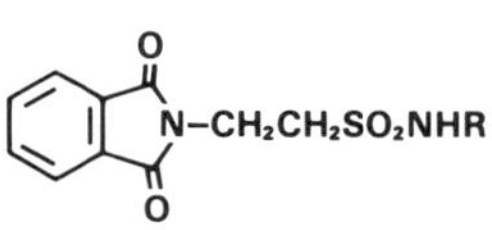

23

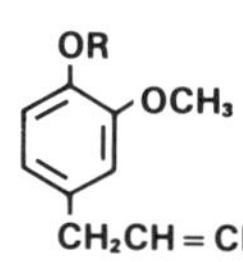

24

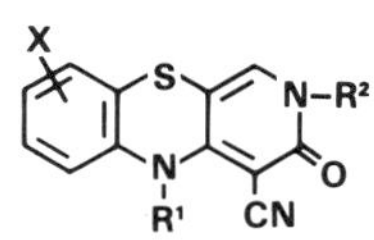

25

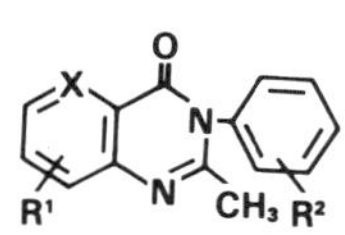

26

X = CH or N

References

1. R.I.Altesman and J.O.Cole, J.Clin.Psychiat., 44, 12 (1983)
2. W.Janke and P.Netter, Neuropsychobiol., 9, 33 (1983)
3. M.Williams, J.Med.Chem., 26, 619 (1983)
4. J.Schopf, Pharmacopsychiat., 16, 1 (1983)
5. D.J.Greenblatt, R.I.Shader, and D.R.Abernethy, The New England J.Med., 309, 410 (1983)
6. T.Roehrs, F.J.Zorick, J.M.Sicklesteel, R.M.Wittig, K.M.Hartse, and T.Roth, J.Clin. Psychopharmacol., 3, 310 (1983)
7. A.Kales, C.R.Soldatos, E.O.Bixler, and J.D.Kales, Pharmacol., 26, 121 (1983)
8. D.Ladewig, Pharmacopsychiat., 16, 103 (1983)
9. R.Miller, The Mt.Sinai J.Med., 50, 289 (1983)

10. K.G.Lloyd, Therapie 1983, 38, 355 (1983)
11. M.J.Croucher, B.S.Meldrum, and P.Krogsgaard-Larsen, Eur.J.Pharmacol., 89, 217 (1983)
12. M.M.Airaksinen, Acta Neurol.Scand., 67, 31 (1983)
13. R.G.Fariello and M.K.Ticku, Life Sci., 33, 1629 (1983)
14. A.Chapman, P.E.Keane, B.S.Meldrum, J.Simiand, and J.C.Vernieres, Prog.Neurobiol., 19, 315 (1982)
15. M.G.Corda, W.D.Blaker, W.B.Mendelson, A.Guidotti, and E.Costa, Proc. Natl.Acad.Sci. USA, 80, 2072 (1983)
16. T.J.McCown, R.A.Vogel, and G.R.Breese, Pharmacol.Biochem.Behav., 18, 277 (1983)
17. J.D.Leander, Drug Dev. Research, 3, 185 (1983)
18. J.B.Patel, C.Martin, and J.B.Malick, Eur.J.Pharmacol., 86, 295 (1983)
19. M.Gerecke, Br.J.Clin.Pharmacol., 16, 11S (1983)
20. S.Banfi, W.Fonio, E.Allievi, and S.Raimond, Pharmacol.Res.Commun., 15, 553 (1983)
21. T.Haug, Eur.J.Pharmacol., 93, 221 (1983)
22. M.W.Emmett-Oglesby, D.G.Spender, M.Lewis,Jr., F.Elmesallamy, and H.Lal, Eur.J. Pharmacol., 92, 127 (1983)
23. M.W.Emmett-Oglesby, D.G.Spender,Jr., F.Elmesallamy, and H.Lal, Life Sci., 33, 161 (1983)
24. L.J.Herberg and S.F.Williams, Pharmacol.Biochem.Behav., 19, 625 (1983)
25. Y.Ichimaru, M.Moriyama, and Y.Gomita, Life Sci., 32, 437 (1983)
26. Y.Gomita, Y.Ichimaru, and M.Moriyama, Jap.J.Pharmacol., 33, 498 (1983)
27. M.Davis, Psychopharmacol.Bull., 19, 457 (1983)
28. N.M.Blampied and R.C.Kirk, Life Sci., 33, 695 (1983)
29. S.E.File and R.G.Lister, Pharmacol.Biochem.Behav., 18, 185 (1983)
30. S.E.File and R.G.Lister, Neurosci.Letters, 39, 91 (1983)
31. M.Mariotti and E.Ongini, Arch.Intl.Pharmacodyn., 264, 203 (1983)
32. Y.Mizuk, M.Hashimoto, T.Tanaka, K.Inanaga, and M.Tanaka, Psychopharmacol., 80, 311 (1983)
33. J.P.Laurent, M.Mangold, U.Humbel, and W.Haefely, Neuropharmacol., 22, 501 (1983)
34. L.Pieri, Br.J.Clin.Pharmacol., 16, 17S (1983)
35. M.C.Cassone, L.Molinengo, and M.Orsetti, Life Sci., 33, 1215 (1983)
36. R.Scherschlicht and J. Marias, Br.J.Clin.Pharmacol., 16, 29S (1983)
37. J.M.Goldstein, L.G.Knobloch, and J.B.Malick, Life Sci., 32, 613 (1983)
38. V.E.Grimm and A.Jancourt, Intl.J.Neurosci., 18, 127 (1983)
39. S.J.Cooper, Life Sci., 32, 1043 (1983)
40. S.J.Cooper, Life Sci., 32, 2453 (1983)
41. K.Yamaguchi, K.Suzuki, T.Niho, M.Shimora, C.Ito, and H.Ohnishi, Can.J.Physiol. Pharmacol., 61, 619 (1983)
42. H.C.Rosenberg, S.Smith, and T.H.Chiu, Life Sci., 32, 279 (1983)
43. E.H.Ellinwood,Jr., M.Linnoila, M.E.Easler, and D.W.Molter, Psychopharmacol., 79, 137 (1983)
44. G.P.Ryan and N.R.Boisse, J.Pharmacol.Exp.Ther., 226, 100 (1983)
45. L.F.McNicholas, W.R.Martin, and S.Cherian, J.Pharmacol.Exp.Ther., 226, 783 (1983)
46. S.T.Mernoff, H.M.Cherwinski, J.W.Becker, and A.L.deBlas, J.Neurochem., 41, 752 (1983)
47. E.Sigel, F.A.Stephenson, C.Mamalaki, and E.A.Barnard, J.Biol.Chem., 258, 6965 (1983)
48. J.H.Skerritt and G.A.R.Johnston, Eur.J.Pharmacol., 89, 193 (1983)
49. J.H.Skerritt and G.A.R.Johnson, Neurosci.Letters, 38, 315 (1983)
50. B.K.Koe, Drug Dev.Res. 3, 421 (1983)
51. C.Martini, T.Rigacci, and A.Lucacchini, J.Neurochem., 41, 1183 (1983)
52. P.F.Morgan and T.W.Stone, Br.J.Pharmacol., 79, 973 (1983)
53. R.Thyagarajan, R.Ramanjaneyulu, and M.K.Ticku, J.Neurochem., 41, 578 (1983)
54. M.Gavish, Life Sci., 33, 1479 (1983)
55. D.L.Niehoff, R.D.Mashal, and M.J.Kuhar, Eur.J.Pharmacol., 92, 131 (1983)
56. M.M.S.Lo, D.L.Niehoff, M.J.Kuhar, and S.H.Snyder, Neurosci.Letters, 39, 37 (1983)
57. R.Mitchell and L.E.Wilson, Neuropharmacol., 22, 935 (1983)
58. K.W.Gee, F.J.Ehlert, and H.I.Yamamura, J.Pharmacol.Exp.Ther., 225, 132 (1983)
59. K.P.Lippke, W.G.Schunack, W.Wenning, and W.E.Müller, J.Med.Chem., 26, 449 (1983)
60. G.Neef, U.Eden, A.Huth, D.Rahtz, R.Schmiechen, and D.Seidelmann, Heterocycles, 20, 1295 (1983)
61. C.Braestrup, M.Nielsen, and T.Honoré, J.Neurochem., 41, 454 (1983)
62. A.Concas, M.Salis, and G.Biggio, Life Sci., 32, 1175 (1983)
63. A.Concas, M.Salis, M.Serra, M.Giuseppa Corda, and G.Biggio, Eur.J.Pharmacol., 89, 179 (1983)
64. L.Prado de Carvalho, G.Grecksch, G.Chapouthier, and J.Rossier, Nature, 301, 64 (1983)
65. I.L.Martin and A.Doble, J.Neurochem., 40, 1613 (1983)
66. R.Dorow, R.Horowski, G.Paschelke, and M.Amin, Lancet, July 9, 98 (1983)
67. M.M.Schweri, J.V.Martin, W.B.Mendelson, J.E.Barrett, S.M.Paul, and P. Skolnick, Life Sci., 33, 1505 (1983)
68. L.M.F.Leeb-Lundberg and R.W.Olsen, Mol.Pharmacol., 23, 315 (1983)
69. A.Doble, J.Neurochem., 40, 1605 (1983)
70. T.H.Chiu and H.C.Rosenberg, Trends in Pharmacol. Sci., 4, 348 (1983)
71. F.J.Ehlert, W.R.Roeske, K.W.Gee, and H.I.Yamamura, Biochem.Pharmacol., 32, 2375 (1983)

72. P.Polc, E.P.Bonetti, R.Schaffner, and W.Haefely, Nauyn.Schiedebergs Arch.Pharmacol., 321, 260 (1982)
73. M.K.Ticku, T.P.Burch, and W.C.Davis, Pharmacol.Biochem.Behav.(Suppl. 1), 18, 15 (1983)
74. W.B.Mendelson, T.Davis, S.M.Paul, and P.Skolnick, Life Sci., 32, 2241 (1983)
75. D.T.Wong and F.P.Bymaster, Drug Dev.Res., 3, 67 (1983)
76. W.Sieghart, Neurosci.Letters, 38, 73 (1983)
77. H.A.Robertson and M.L.Riives, Brain Res., 270, 380 (1983)
78. S.V.Vellucci and R.A.Webster, Eur.J.Pharmacol., 90, 263 (1983)
79. C.O'Boyle, R.Lambe, A.Darragh, W.Taffe, I.Brick, and M.Kenny, Br.J.Anaesth., 55, 349 (1983)
80. A.Darragh, R.Lambe, C.O'Boyle, M.Kenny, and I.Brick, Psychopharmacol., 80, 192 (1983)
81. J.Feldon, T.Lerner, D.Levin, and M.Myslobodsky, Pharmacol.Biochem.Behav., 19, 39 (1983)
82. S.E.File and S.Pellow, Psychopharmacol., 80, 166 (1983)
83. I.L.Martin, C.L.Brown, and A.Doble, Life Sci., 32, 1925 (1983)
84. W.E.Fann, J.Garcia, and B.W.Richman, J.Clin.Pharmacol., 23, 100 (1983)
85. N.Desai, A.Taylor-Davies, and D.B.Barnett, Br.J.Clin.Pharmacol., 15, 197 (1983)
86. Y.D.Lapierre, H.J.Butter, L.K.Oyewumi, Neuropsychobiology, 9, 88 (1983)
87. R.L.Singhal, R.B.Rastogi, and Y.D.Lapierre, J.Neural Transmission, 56, 127 (1983)
88. J.N.Nestoros, N.P.V.Nair, J.R.Pulman, G.Schwartz, and D.Bloom, Psychopharmacol., 81, 42 (1983)
89. M.M.Singh, R.E.Becker, R.K.Pitman, H.A.Nasrallah, and H.Lal, Brain Res.Bull., 11, 179 (1983)
90. K.Rickels, W.G.Case, R.W.Downing, and A.Winokur, J.Am.Med.Assoc., 250, 767 (1983)
91. P.Tyrer, R.Owen, and S.Dawling, Lancet, June 25, 1402 (1983)
92. Y.Kawazu, S.Nakano, N.Ogawa, and K.Taeuber, Drug Dev.Res., 3, 371 (1983)
93. J.John, A.Roy, and A.Verghese, Curr.Ther.Res., 33, 990 (1983)
94. H.Oblowitz and A.H.Robins, Br.J.Clin.Pharmacol., 16, 95 (1983)
95. K.Aranko, M.J.Mattila, and T.Seppälä, Br.J.Clin.Pharmacol., 15, 545 (1983)
96. J.K.Walsh, P.K.Schweitzer and S.Parwatikar, Clin.Pharm.Ther., 34, 496 (1983)
97. K.Davidson, R.G.Farquharson, M.C.Khan, and A.Majid, Psychopharmacol., 80, 308 (1983)
98. K.Rickels, I.Csanalosi, P.Greisman, D.Cohen, J.Werblowsky, H.A.Ross, and H.Harris, Am.J.Psychiat., 140, 82 (1983)
99. T.M.Itil, R.K.Shrivastava, D.M.Collins, S.T.Michael, G.Dayican, and K.Z. Itil, Curr. Ther.Res., 34, 441 (1983)
100. P.Hauri, T.Roth, M.Sateia, and F.Zorick, Br.J.Clin.Pharmacol., 16, 109S (1983)
101. G.W.Vogel and F.Vogel, Br.J.Clin.Pharmacol., 16, 103S (1983)
102. R.Lupolover, U.Ballmer, J.Helcl, J.Escher, and B.Pavletic, Br.J.Clin.Pharmacol., 16, 139S (1983)
103. A.C.Gudgeon and I.Hindmarch, Br.J.Clin.Pharmacol., 16, 121S (1983)
104. Ch.Feldmeier and W.Kapp, Br.J.Clin.Pharmacol., 16, 151S (1983)
105. K.Rickels, J.Clin.Psychopharmacol., 3, 133 (1983)
106. O.Dehlin and G.Björnson, Acta Psychiat.Scand., 67, 290 (1983)
107. O.Dehlin, G.Björnson, L.Börjesson, L.Abrahamsson, and R.B.Smith, Eur.J.Clin.Pharmacol., 25, 91 (1983)
108. R.Jochemsen, C.J.van Boxtel, J.Hermans, and D.D.Breimer, Clin.Pharmacol.Ther., 34, 42 (1983)
109. D.Bliwise, W.Seidel, I.Karacan, M.Mitler, T.Roth, F.Zorick, and W.Dement, Sleep, 6, 156 (1983)
110. C.Gorenstein and V.Gentil, Psychopharmacol., 80, 376 (1983)
111. P.A.Ellingsen, Acta Psychiat.Scand., 67, 154 (1983)
112. I.C.McManus, S.I.Ankier, J.Norfolk, M.Phillips, and R.G.Priest, Br.J.Clin.Pharmacol., 16, 291 (1983)
113. J.Krieger, M.Perianu, C.Bertagna, P.Mangin, and D.Kurtz, Drug Dev.Res., 3, 143 (1983)
114. M.Matejcek, C.Neff, K.Abt, and W.Wehrli, Neuropsychobiology, 9, 52 (1983)
115. M.Viukari, P.Jaatinen, and T.Kylmämaa, Curr.Ther.Res., 33, 828 (1983)
116. D.M.Barnes, W.F.White, and M.A.Dichter, J.Neurosci., 5, 762, (1983)
117. K.W.Gee, R.E.Brinton and H.L.Yamamura, Life Sci., 32, 1037 (1983)
118. K.W.Gee, R.E.Brinton and H.L.Yamamura, Brain Res., 264, 168 (1983)
119. D.G.Spencer and H.Lal, Drug Dev.Res., 3, 365 (1983)
120. K.Rickels, K.Weisman, N.Norstad, M.Singer, D.Stoltz, A.Brown and J.Danton, J.Clin. Psychiatry, 43, 12 (sec. 2), 81 (1982)
121. D.Wheatley, J.Clin.Psychiatry, 43, 12 (Sec. 2), 92 (1982)
122. J.P.Feighner, C.H.Merideth and G.A.Hendrickson, J.Clin.Psychiatry, 43, 12 (Sec. 2), 103 (1982)
123. J.S.Hendry, R.L.Balster and J.A.Rosecrans, Pharmacol.Biochem.Behav., 19, 97 (1983)
124. H.Y.Meltzer, R.Flemming and A.Robertson, Arch.Gen.Psychiat., 40, 1099 (1983)
125. M.Cimino, F.Ponzio, G.Achilli, G.Vantini, C.Perego, S.Algeri, and S.Garattini, Biochem.Pharmacol., 32, 1069 (1983)
126. P.L.Wood, N.P.V.Nair, S.Lal, and P.Etienne, Life Sci., 33, 269 (1983)
127. B.A.McMillen, R.T.Matthews, M.K.Sanghera, P.D.Shepard, and D.C.German, J.Neurosci., 3, 733 (1983)

128. B.A.McMillen and C.C.McDonald, Neuropharmacol., 22, 273 (1983)
129. M.K.Sanghera, B.A.McMillen and D.C.German, Eur.J.Pharmacol., 86, 107 (1983)
130. T.Glaser and J.Traber, Eur.J.Pharmacol., 88, 137 (1983)
131. N.R.Oakley and B.J.Jones, Eur.J.Pharmacol., 87, 499 (1983)
132. S.M.Paul and P.Skolnick, Pharmacol.Biochem.Behav., 17 (Suppl. 1), 37 (1982)
133. S.Caccia, S.Garattini, A.Mancinelli and M.Muglia, J.Chromatography, 252, 310 (1982)
134. B.A.McMillen and L.A.Mattiace, J.Neural Transmission, 57, 255 (1983)
135. J.P.Yevich, D.L.Temple,Jr., J.S.New, D.P.Taylor and L.A.Riblet, J.Med.Chem., 26, 194 (1983)
136. R.Hoehn-Saric, Psychopharmacol., 80, 338 (1983)
137. R.Kikumoto, A.Tobe, H.Fukami and M.Egawa, J.Med.Chem., 26, 246 (1983)
138. E.J.Trybulski, L.E.Benjamin, J.V.Earley, R.I.Fryer, N.W.Gilman, E.Reeder, A.Walser, A.B.Davidson, W.D.Horst, J.Sepinwall, R.A.O'Brien and W.Dairman, J.Med.Chem., 26, 1589 (1983)
139. A.Wauquier, Anaesthesia, 38 (Suppl) 26 (1983)
140. M.Tanaka, Y.Mizuki, H.Isozaki and K.Inanaga, Eur.J.Clin.Pharmacol., 24, 469 (1983)
141. P.J.Chaudoir, N.C.Jarvie and G.J.Wilcox, J.Int.Med.Res., 11, 333 (1983).
142. J.R.Normanton and J.P.Gent, Neuroscience, 8, 107 (1983)
143. J.E.Garcia-Arraras and J.R.Pappenheimer, J.Neurophysiol., 49, 528 (1983)
144. W.B.Mendelson, A.Kuruvilla, T.Watlington, K.Goehl, S.M.Paul and P.Skolnick, Psychopharmacol., 79, 126 (1983)
145. M.R.Trimble and P.J.Thompson, Epilepsia, 24 (Suppl. 1), S55 (1983)
146. J.H.Skerritt, L.P.Davies, and G.A.R.Johnston, Epilepsia, 24, 634 (1983)
147. P.R.Andrews, L.C.Mark, D.A.Winkler and G.P.Jones, J.Med.Chem., 26, 1223 (1983)
148. A.M.Anlezark, D.H.R.Blackwood, B.S.Meldrum, V.J.Ram and J.L.Neumeyer, Psychopharmacol., 81, 135 (1983)
149. A.L.Mueller and T.V.Dunwiddie, Epilepsia, 24, 57 (1983)
150. P.Loiseau, L.Bossi, M.Guyot, B.Orofiamma and P.L.Morselli, Epilepsia, 24, 703 (1983)
151. H.R.Petersen, I.Jensen and M.Dam, Acta Neurol.Scand., 67, 114 (1983)
152. J.D.Wood, D.D.Johnson, P.Krogsgaard-Larsen and A.Schousboe, Neuropharmacol., 22, 139 (1983)
153. G.Graziani, F.Tirone, E.Barbadoro and R.Testa, Arzneim-Forsch., 33 (II), 1155 (1983)
154. D.T.Wong, R.C.Rathbun, F.P.Bymaster and W.B.Lacefield, Life Sci., 33, 917 (1983)
155. D.T.Wong, F.P.Bymaster and W.B.Lacefield, Drug Dev.Res., 3, 433 (1983)
156. B.S.Meldrum, M.C.Evans and C.Braestrup, Eur.J.Pharmacol., 91, 255 (1983)
157. I.B.Linden, G.Gothoni, P.Kontro and S.S.Oja, Neurochem.International, 5, 319 (1983)
158. K.R.D.Zelger, J.L.Zelger and E.A.Carlini, Pharmacol., 27, 40 (1983)
159. R.J.Chorvat, B.N.Desai, S.E.Radak, J.Bloss, J.Hirsch and S.Tenen, J.Med.Chem., 26, 845 (1983)
160. N.A.Vaidya, C.H.Panos, A.Kite, W.B.Iturrian and C.D.Blanton, J.Med.Chem., 26, 1422 (1983)

Chapter 3. Antipsychotic Agents

Tomas de Paulis and Sten Rämsby, Astra Läkemedal AB, Sweden

Introduction - Schizophrenia continues to be one of the most fascinating and challenging research topics for medicinal chemists and neurochemists. Tools such as ^{3}H-spiperone binding assays and an understanding of the biochemical abnormalities associated with schizophrenia may provide the path to future discoveries.[1] Haloperidol-induced dopamine (DA) receptor supersensitivity in schizophrenics appears to be attenuated by concurrent treatment with lithium.[2] Noninvasive radioimaging techniques have brought new insight into the role of the regional distribution and receptor binding of antipsychotic agents.[3] Studies of the blood flow in the brains of schizophrenic patients with positron-emitting Xenon-133 have revealed an over-activation in the left hemisphere during performance of a spatial line orientation test.[4] PET scanning with oxygen-15 (half-life 123 s) in some schizophrenic patients failed to support the hypothesis of reduced blood flow in frontal cortex.[5] Enlarged ventricles and sulcal widening were correlated with type II schizophrenia, which is characterized by predominance of negative symptoms.[6]

The DA hypothesis of schizophrenia has been reviewed in the light of recent findings.[7] Perioral dyskinesias can be induced acutely in rats by the DA antagonists spiroperidol and sulpiride, and by the selective D_1 agonist SKF 38393, but not by the selective D_2 agonist LY 141865.[8] Coadministration of LY 141865 with SKF 38393 did not induce dyskinesias, implying that the perioral movements are mediated by the D_1 receptors and by an imbalance in D_1-D_2 responsivity.[8] It has been hypothesized that the D-3 binding site is the high-affinity agonist state of the D-1 receptor.[9] Accumulated experience with depot neuroleptics, such as clopenthixol decanoate[10] and pipotiazine palmitate,[11] has been reviewed.

Phenothiazines and related rigid compounds - The antihallucinatory effect of chlorpromazine correlates with serum levels of prolactin,[12] consistent with the DA hypothesis of schizophrenia. No regional site-specificity for limbic areas was found with thioridazine or with clozapine.[13] An excellent correlation between daily clinical dose and the dose-dependent increase in cocaine self-administration in rats was found for seven typical and atypical neuroleptics. Clozapine, however, produced a dose-dependent decrease in cocaine intake.[14]

MM2 calculations of the energies of the butaclamol structure have been utilized to suggest that a cisoid *trans* conformer (1) is responsible for the DA antagonist activity of this drug.[15] The previously assumed active conformer was found to have 2.7 kcal/mol higher energy.

Unlike taclamine, the thiophene analog QM 7184 (2) blocks apomorphine or amphetamine induced stereotypy in the rat.[16] It is equipotent with chlorpromazine but less potent than the structurally related compound butaclamol. Binding studies showed that QM 7184 is 30-50 times

ISBN 0-12-040519-9

less active than butaclamol or haloperidol in displacing ^{3}H-spiperone from rat striatum. In contrast to the reference compounds, QM 7184 has a high affinity for α-noradrenergic receptors, as shown in binding studies with ^{3}H-WB-4101.[16].

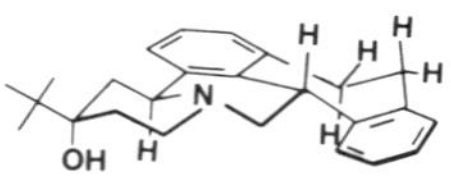

1

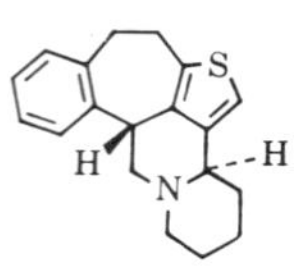

2

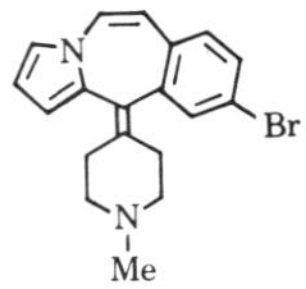

3

Attempts to separate the multitude of effects shown by cyproheptadine have included introduction of nuclear substituents and replacement of a benzene ring by a pyrrole ring.[17] In an extension of this study to include 9-substituted pyrrolobenzapine analogs, compound 3 was found to be 2-3 times more potent than chlorpromazine versus ^{3}H-spiperone binding and also showed high affinity for the serotonin S-2 binding site.[18] The conformational flexibility was higher for these new pyrrolo-compounds than for the corresponding cyproheptadine derivatives as evidenced from the rates of racemization of the atropisomers.

The tetracyclic derivative GP 50302 (4) blocks DA receptors in the same dose range as chlorpromazine, and also shows strong 5-HT blocking properties. Thus in contrast with haloperidol, GP 50302 is much more potent in· vivo in displacing ^{3}H-spiperone from cortical, versus striatal, binding sites.[19] However, in clinical trials GP 50302 had no therapeutic effects in acute schizophrenic patients, but was beneficial in withdrawn and apathetic schizophrenics.[20] The imidazole derivative CGS 10746 B (5), structurally related to clozapine, blocked Sidman avoidance at a dose of 10 mg/kg p.o. in squirrel monkeys.[21] However, 5 did not produce the acute dyskinetic syndrome, characteristic of classical neuroleptic agents. In an open multicenter study in 104 schizophrenic patients, fluperlapine (NB 106-689, 6) showed significant improvement from the fifth day of treatment.[22] Fluperlapine has demonstrated little or no effect in a number of tests, such as catalepsy and inhibition of apomorphine- and amphetamine-induced behavior, which 3re considered specific for antipsychotic agents.[23] It is a muscle relaxant and an uptake inhibitor of norepinephrine.

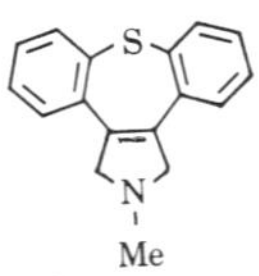

4

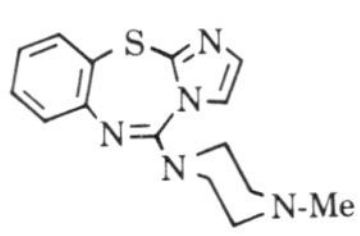

5

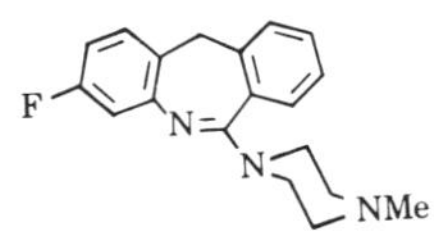

6

Diphenylpiperidines and butyrophenones - Haloperidol decanoate, 100mg once per month, in chronic schizophrenic patients provoked few Parkinson-like side effects.[24] DA receptor supersensitivity induced by chronic pimozide treatment was found to be dependent on the dose and not the duration, if measured as an increase in density of spiperone binding sites. When the induced supersensitivity was measured as increased inhibition of apomorphine-induced stereotypy, it was dependent on the duration and not on the dose. Thus, in the absence of pharmacologically-induced DA receptor stimulation, the functional consequences of neuroleptic-induced supersensitivity at the receptor level remain unknown.[25]

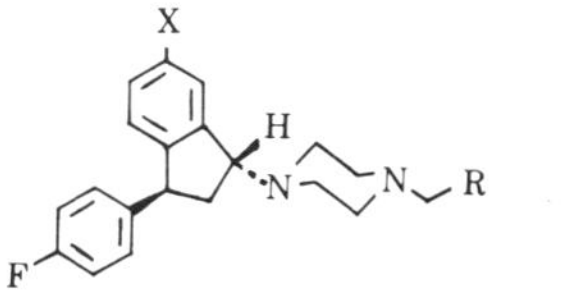

7a X = CF_3 R = CH_2OH

7b X = F R = H

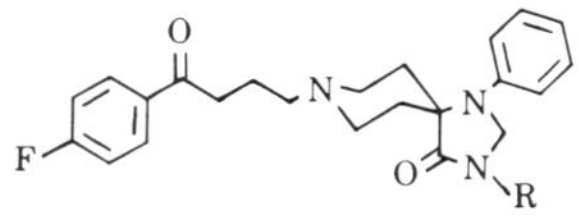

8a R = H

8b R = $^{11}CH_3$

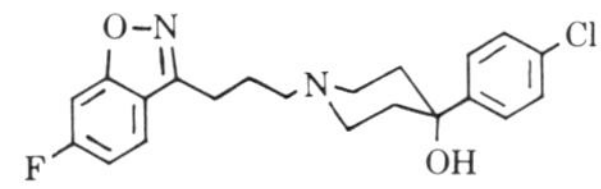

9

Structure-activity studies of 1-piperazino-3-phenylindanes led to the development of the potent and long-lasting neuroleptic, tefludazine (7a). The cis-isomer of tefludazine has 1/1000 the activity of 71 in blocking methylphenidate-induced stereotypy in mice. One member of the tefludazine series, the bisfluoro derivative (7b) was resolved: the antidopaminergic activities resided in the (+)-enantiomer. The (-)-enantiomer, however, proved to be an active uptake inhibitor of DA.[26]

Spiperone (8a) is a highly potent neuroleptic agent. Positron Emission Tomography (PET) imaging with ^{11}C-labelled N-methyl-spiperone (8b) (half-life 20 min) was used to study the distribution of DA receptors in the human brain. There was a high accumulation of activity in the basic ganglia in comparison to the rest of the brain.[27] In another experiment, using ^{18}F-spiperone (half-life 110 min) administered to a baboon 2h before the tomography imaging, it was clearly demonstrated that the ratio of binding in striatum to cerebellum was ten times higher for ^{18}F-spiperone than for ^{18}F-haloperidol.[28]

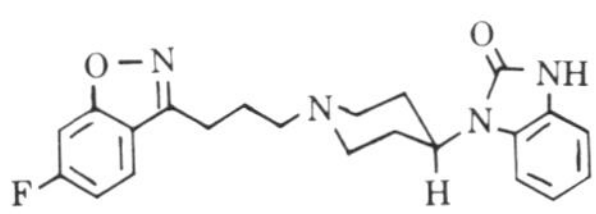

10

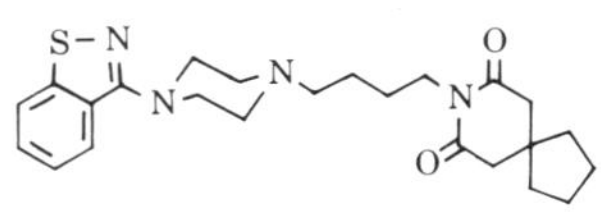

11

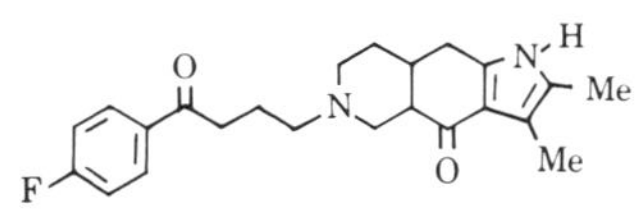

12

Cyclic oxime ethers of butyrophenones exhibit neuroleptic properties. Thus, the benzisoxazole analog of haloperidol, 9, has an ED_{50} of 1.6 µmol kg in blocking apomorphine-induced climbing in mice.[29] The corresponding benzisoxazole analog of benperidol, HRP 913 (10), seems to be a more selective DA antagonist than those presently in clinical use. HRP 913 is twice as potent as haloperidol in displacing spiperone from striatal binding sites and equipotent in the antiapomorphine stereotypy assay.[30] It induces a biphasic atypical catalepsy at a dose 45 times higher than haloperidol, resembling thioridazine in this respect. A similar isosteric arrangement of the aromatic part of the molecule is found in MJ 13859 (11), which in contrast to buspirone exhibits a clear neuroleptic profile.[31]

The molindone-butyrophenone hybrid Ro 22-6600 (12) is both a pre- and postsynaptic DA-antagonist.[32] There are differences in the SAR of the Ro 22-6600 series and typical butyrophenones since hydrogenation of the chain carbonyl group in the former only halves activity in conditioned-avoidance response.[33] Milenperone (R 34009, 13) displayed antiaggressive activity in a group of mentally retarded patients, with no greater incidence of side effects than with placebo.[34] Double blind studies with timiperone (DD-3480, 14) versus clocapramine,[35] sulpiride[35] or haloperidol[36] showed an equal or superior efficacy against schizophrenia. The separation between catalepsy and anti-apomorphine activity for timiperone in the monkey seems to be small.[37] Another new butyrophenone, HR 375 (15), shows weak and short-lasting muscle relaxing properties besides being active on neuroleptic parameters such as block of climbing and conditioned avoidance behavior. However, it neither blocks stereotypy nor does it induce catalepsy in the rat.[38]

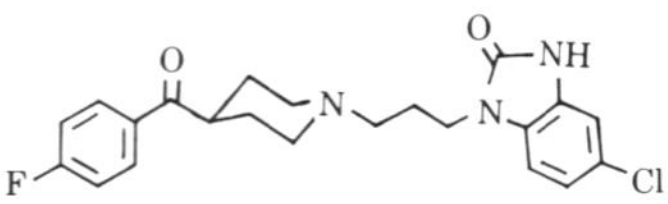

13

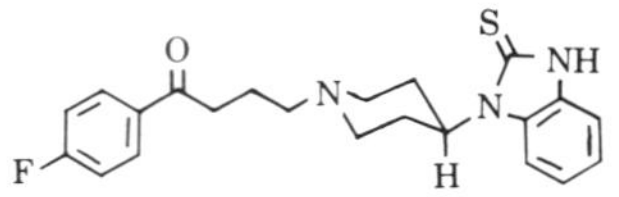

14

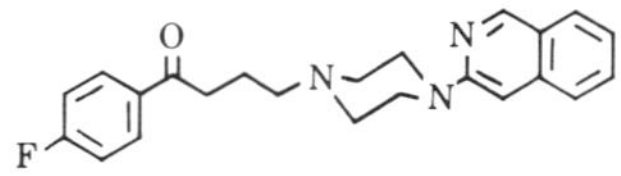

15

Substituted benzamides - The unique behavioral profile in animals of substituted benzamide neuroleptics and their selective action on a subpopulation of D-2 receptors might explain their clinical usefulness in the treatment of schizophrenia.[39] The K_i-value for sulpiride displacement of spiperone binding was approximately 20 times greater than the K_D value for sulpiride binding in the presence of 120 mM Na^+, suggesting that sulpiride and spiperone label different receptor sites.[40] In a comparative study with classical neuroleptics, benzamides were generally 1000 times more potent in stimulating rat prolactin secretion than would have been predicted from their potencies in displacing ^{3}H-spiperone from bovine anterior pituitary membranes.[41]

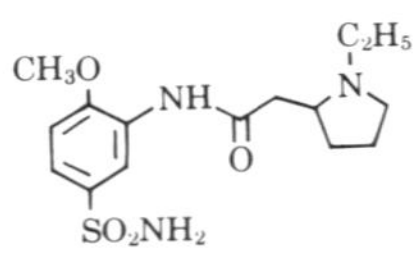

16

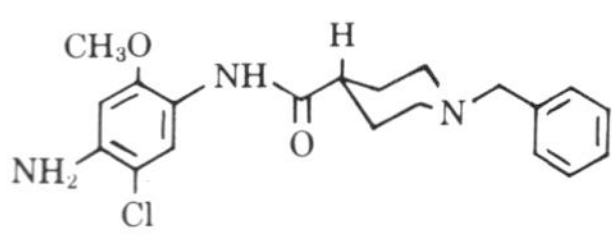

17

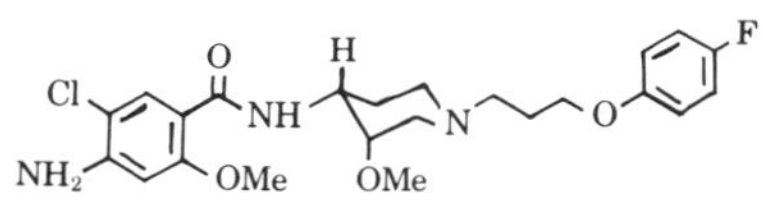

18

In structural analogy with the retro-amide isosulpride[42] (16), the retro-amide of clebopride, BRL-20596 (17), retains the antiapomorphine activity but totally lacks effect on gastric stimulation,[43] thereby showing that separation is possible between the central and peripheral activity. Introduction of a methoxy group in the piperidine ring of clebopride abolishes the central effects. In a series of 3-piperidine substituted benzamides, cisapride (R 51619, 18) is a potent peripheral 5-HT antagonist which shows high potency in increasing gastric promotility.[44] Another compound, BRL 20627 (19), with potent peripheral stimulatory action on the gastric motility of the rat, is only a weak central DA receptor antagonist.[45]

Conformational studies of clebopride and YM-08050 (20a) imply that it is the six-membered pseudo-ring, formed by the hydrogen bond between the amide proton and the ether oxygen atom, rather than the benzene ring, that interacts with the aromatic binding site of the DA receptor.[46] YM 09151-2 (20b) was the most potent and selective D-2 receptor antagonist of a series tested: the K_i value for 20b versus ^{3}H-spiperone binding in rat striatal membranes was 0.1 nM.[47] X-ray studies with YM 09151-2 have revealed that the tertiary amine nitrogen is inaccessible.[48] Thus the active conformer must be different from that in the crystalline state if the lone pair on the tertiary nitrogen interacts with the receptor.

In a double blind study, amisulpride (DAN 2163, 21a) displayed the same antipsychotic efficacy as haloperidol with favorable effects on certain symptoms.[49] At low doses in man, amisulpride behaves like a DA agonist with stimulating properties, but at doses over 600 mg the

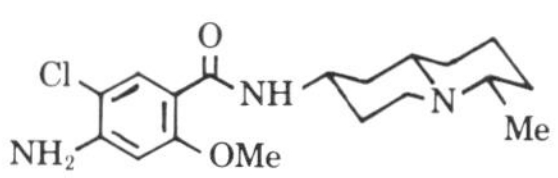

19

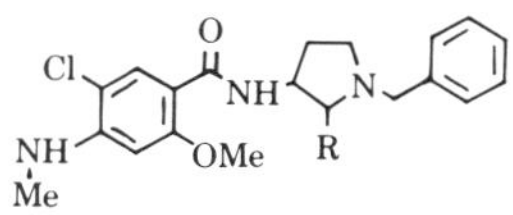

20a R = H

20b R = CH_3

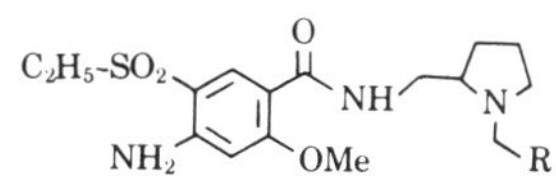

21a R = CH_3

21b R = c-C_3H_6

21c R = C_2H_3

neuroleptic properties predominate.[50] Similar to amisulpride, but in contrast to haloperidol, the N-cyclopropylmethyl analog LUR 2366 (21b) inhibits apomorphine-induced climbing at a 10 times lower dose than that which blocks stereotypy: in fact LUR 2366 potentiates the gnawing behavior in mice.[51] In this series, the n-allyl substituted sulfone derivative alpiropride (RIV 2093, 21c) displays the largest separation between high-affinity apomorphine (D-4) receptor sites and domperidone (D-2) receptor sites, thereby indicating a lack of effect on the pituitary and no consequent release of prolactin.[52]

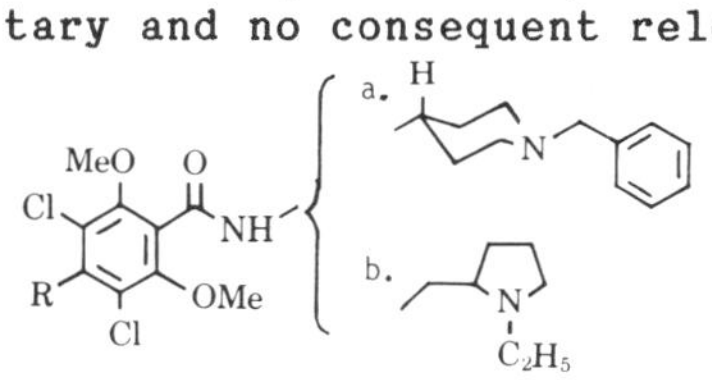

22a R = H

22b R = CH_3

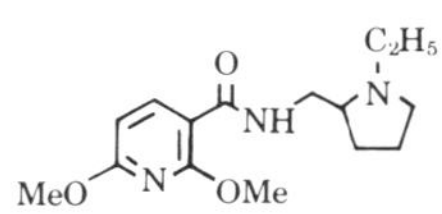

23

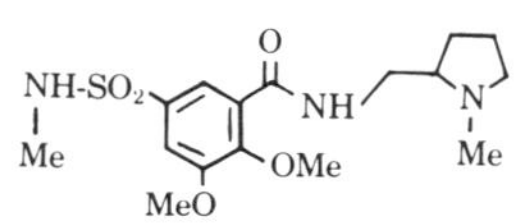

24

A tetrasubstituted benzamide, A 32799 (22a), was equipotent with chlorpromazine in blocking apomorphine-induced stereotypies in the rat, showing that introduction of sterically bulky substituents in the aromatic ring does not abolish antidopaminergic activity.[53] However, the pentasubstituted compound, 22b was only 1/50th as active as its corresponding des-4-methyl substituted analog,[54] indicating the requirement of a free *para* position in this series. The ability of a substituted nicotinamide, CGP 11109 (23), to increase the basal tritium outflow from caudate slices of the rabbit preincubated with ^{3}H-DA, was comparable to that of sulpiride.[55] The metabolic and pharmacokinetic properties of a bis-N-methyl derivative of veralipride, sulverapride (24) were investigated in man. No radioactivity was found in the brain, indicating very poor penetration of sulverapride across the blood-brain barrier.[56]

A new substituted salicylamide, FLB 131 (25a) exhibited potent anti-DA properties in the rat.[57] Selective labeling of the striatum was demonstrated with both FLB 131[58] and the N-methyl analog (25b)[59] in PET imaging experiments in cynomolgus monkeys. In a series of 3-substituted 6-methoxysalicylamides, compound FLA 965 (26) showed optimal activity in displacing ^{3}H-spiperone from rat striatal preparations.[60] A remarkable increase in the antiapomorphine activity through introduction of an alpha-hydroxyethyl group was demonstrated in the substituted benzamide MD 781124 (27). Thus, MD 781124 was 540 times more potent in antagonizing apomorphine-induced stereotypy in the rat than the corresponding 5-sulfonamide analog, sulpiride.[61]

The tropylamine derivative tropapride (MD 790501, 28) shows a neuropharmacological profile equivalent to that of haloperidol but with substantially less sedative, cataleptogenic or dyskinetic properties.[62] In the rat, i.p., tropapride was four times more potent than

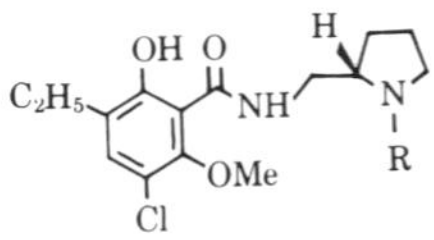

25a R = C_2H_5
25b R = $^{11}CH_3$

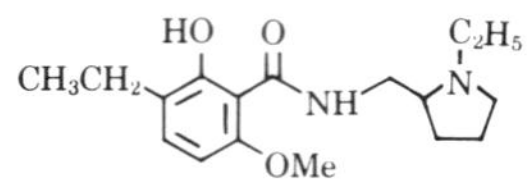

26

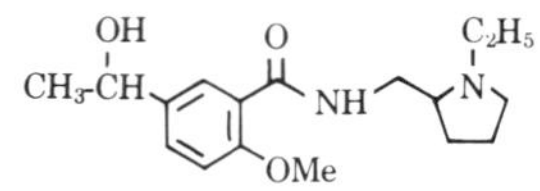

27

haloperidol on antiapomorphine activity. Among a series of azabicyclo-[3.3.1]nonylbenzamide derivatives with neuroleptic properties, compound 29 was one of the most potent agents. Its ED_{50} value in blocking apomorphine-induced climbing in the mouse was 63 times that of clebopride.[63]

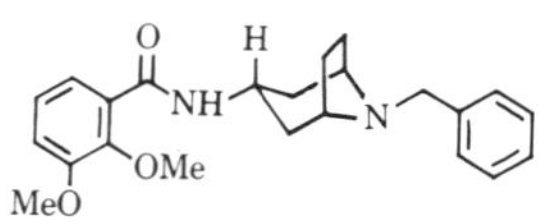

28

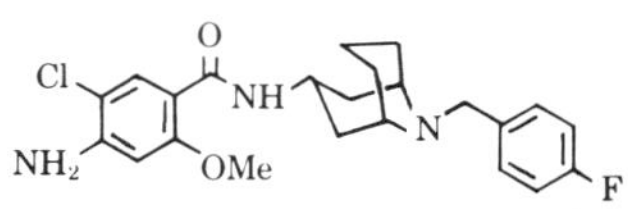

29

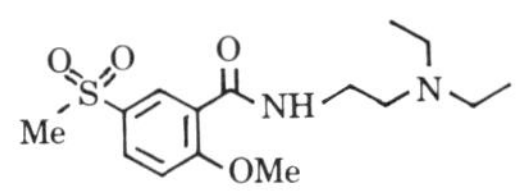

30

The weak blocking properties of tiapride (FLO 1347, 30) at receptors labelled by various neuroleptics could not be explained solely by its poor penetration into the brain.[64] Instead it was suggested to be due to a specific action of tiapride on receptor sites selectively labelled by ^{3}H-sulpiride. Further clinical experience with tiapride in the treatment of tardive dyskinesia has shown that it is an effective agent. In a double-blind comparison with placebo, tiapride significantly improved dyskinesia in 15 patients who had received neuroleptics for 15 years.[65]

Miscellaneous compounds - Similar to dihydro-oxypertine (CL 77328),[66] the structurally related isochroman derivative U-54537 E, (31) displays an atypical neuroleptic profile.[67] Both compounds are weak in displacing ^{3}H-spiperone binding and in antagonizing the behavioral syndrome induced by apomorphine. Yet, U-54537E is three times more potent than thioridazine in decreasing conditioned avoidance response and is equipotent with chlorpromazine in blocking adenylate cyclase coupled D-1 receptors.[67] A thiazoledione analog MJ 13980-1 (32) of buspirone (MJ

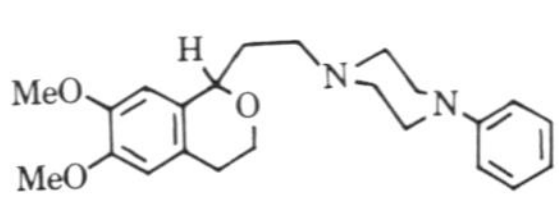

31

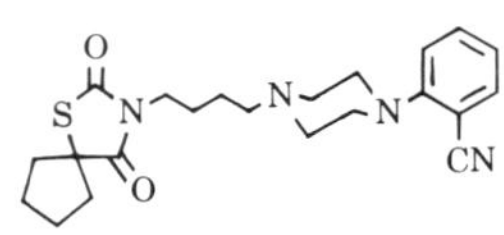

32

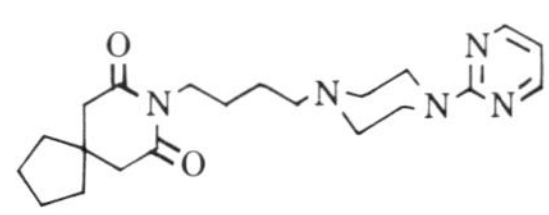

33

9022-1, 33) has shown promise as a potential antipsychotic in that it possesses antidopaminergic activity but does not induce catalepsy. In fact, MJ 13980-1 is potent in reversing phenothiazine-induced catalepsy in the rat.[68] The physicochemical parameters of the aromatic *ortho*-substituent do not correlate with the anti-DA activity in this series.[69]

A double-blind phase I study of zetidoline (DL-308, 34a) revealed a strong sedative effect as judged by self-reported performance scores on logical reasoning and coordination tests.[70] In an open study in 40 schizophrenic patients, doses of 10 to 90 mg daily of zetidoline exerted a good antipsychotic efficacy with akathisia as the most notable side effect.[71] Substitution in the aromatic nucleus with a p-hydroxy group (34b) substantially increases both duration and antiapomorphine activity in rodents.[72]

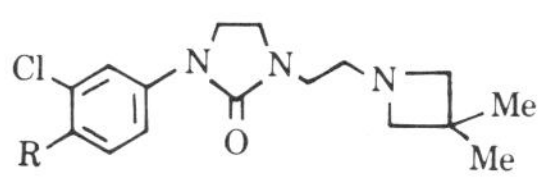

34a R = H

34b R = OH

35

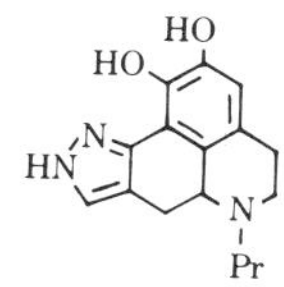

36

2-Bromolisuride (35) is the first ergot derivative with a definite antagonist effect at central DA receptors.[73] In comparison with haloperidol, 35 shows ten times larger separation between drug-induced catalepsy and apomorphine-induced stereotypy in the rat.[74] S(+)-N-propylnorapomorphine shows DA receptor blocking activity[75] as was shown previously for S(+)-apomorphine.[76] A new pyrazole derivative 36 also exhibits DA antagonist properties as evident from its prolactin elevating effect on the pituitary.[77] Both molindone and its rigid derivative Ro 22-1319 (37) were confirmed to be more active as presynaptic DA-antagonists than postsynaptic,[32] thereby indicating an explanation for the relatively weak anti-amphetamine activity of molindone.[78] Ro 22-1319 was equipotent with haloperidol in elevating serum prolactin levels in the rat.[79]

More receptor-specific agonists and antagonists of DA continue to appear. A new benzazepine, SCH 23390 (38) is claimed to be the first selective D-1 antagonist and promises to be a useful neuropharmacological tool.[80] SCH 23390 is equipotent with flupentixol in displacing ^{3}H-piflutixol (D-1) binding, but 1200 times less active in displacing ^{3}H-spiperone (D-2) binding.[81] It is virtually free from blocking effects on apomorphine-induced hypothermia and emesis and does not cause hyperprolactinemia,[82] which suggests that these effects are related to the D-2 receptor.

EEDQ (39) is a potent and irreversible D-2 antagonist which apparently binds covalently to DA receptors in the CNS.[83] EEDQ produces catalepsy and inhibits both amphetamine-induced stereotypy and conditioned avoidance behavior.

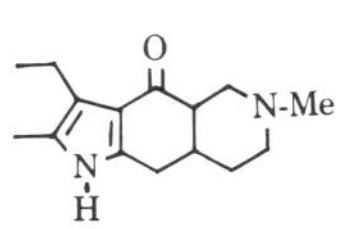

37

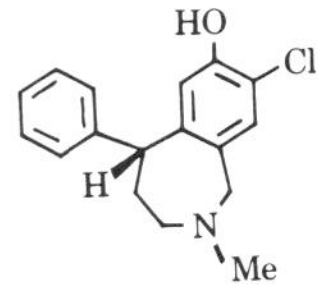

38

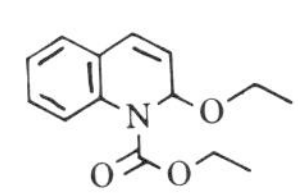

39

A six week study with the GABA agonist progabide (SL 76002, 40), in schizophrenic patients on neuroleptic maintenance therapy, revealed improvement on items such as vigilance and affective involvement.[84] There is evidence that SL 76002 blocks striatal cholinergic neurons through a gabergic mechanism. Thus repeated treatment with SL 76002 in rats prevents tolerance to the cataleptogenic action of haloperidol.[85]

Unlike TL99 (41), neither (±)-3-PPP (42)[86,87] nor its (-)-enantiomer UH 106(-) [87] appear to be presynaptic agonists *in vitro*: thus they do not inhibit evoked release of ^{3}H-DA from rat striatal slices.

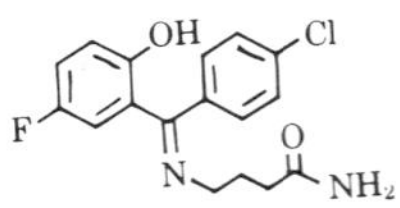

40

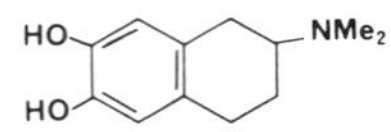

41

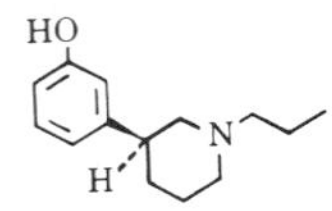

42

Instead, both the racemate[87] and (-)-enantiomer[86,87] behave as post-synaptic D-2 antagonists *in vitro*. For example, the latter prevents the blocking action of the D-2 agonist LY 141865 on potassium-induced release of ^{3}H-acetylcholine from striatum.[88] However, the (-)-enantiomer does behave as a presynaptic DA agonist at low doses *in vivo*.[89]

References

1. R. Rodnight, J. Neurochem., 41, 12 (1983).
2. D. E. Sternberg, M. B. Bowers, Jr., G. R. Heninger and D. S. Charney, Psychiat. Res., 10, 79 (1983).
3. N. D. Heindel and N. Foster, Ann. Rep. Med. Chem., 18, 293 (1983).
4. R. E. Gur, B. E. Skolnick, R. C. Gur, S. Caroff, W. Rieger, W. D. Obrist, D. Younkin and M. Reivich, Arch. Gen. Psychiat., 40, 1250 (1983).
5. G. Sheppard, J. Gruzelier, R. Manchanda, S. R. Hirsch, R. Wise, R. Frackowiak and T. Jones, Lancet 31, 1448 (1983).
6. H. A. Nasrallah, S. Kuperman, C. G. Jacoby, M. McCalley-Whitters and B. Hamra, Psychiat. Res., 10, 237 (1983).
7. Y. Sayed and J. M. Garrison, Psychopharm. Bull., 19, 283 (1983).
8. H. Rosengarten, J. W. Schweitzer and A. J. Friedhoff, Life Sci., 33, 2479 (1983).
9. S. E. Leff and I. Creese, Trends Pharmacol. Sci., 5, 463 (1983).
10. S. Sieberns and H. Spechtmeyer, Int. Pharmacopsychiat., 17, 170 (1982).
11. E. A. Burch, Jr. and F. J. Ayd, Jr., J. Clin. Psychiat., 44, 242 (1983).
12. H. Y. Meltzer and D. Busch, Psychiat. Res., 9, 285 (1983).
13. P. Seeman and C. Ulpian, Eur. J. Pharmacol., 94, 145 (1983).
14. D. C. S. Roberts and G. Vickers, Psychopharmacol., 82 135 (1983).
15. M. Froimowitz and S. Mathysse, Mol. Pharamcol., 24, 243 (1983).
16. E. Arribas, S. Vega, C. Benito, M. P. Fernandez-Tome and J. del Rio, Arzneim.-Forsch, 33, 1417 (1983).
17. D. C. Remy, S. F. Britcher, P. S. Anderson, P. C. Belanger, Y. Girard and B. V. Clineschmidt, J. Med. Chem. 25, 231 (1982).
18. D. C. Remy, S. F. Britcher, S. W. King, P. S. Anderson, C. A. Hunt, W. C. Randall, P. Bélanger, J. G. Atkinson, Y. Girard, C. S. Rooney, J. J. Fuentes, J. A. Totaro, J. L. Robinson, E. A. Risley and M. Williams, J. Med. Chem., 26, 974 (1983).
19. R. Ortman, S. Bischoff, E. Radeke, O. Buech and A. Delini-Stula, Arch. Pharmacol., 321, 265 (1982).
20. A. Delini-Stula, P. Waldmeier, P. Baumann, L. Maitre, H. Blattner and P. Schmidlin, 11th C.I.N.P. Congress (Vienna, July 9-14, 1978) Abst. p. 373.
21. J. M. Liebman, G. P. Quinn, R. A. Lovell, I. Vlattas and T. M. Glenn, Fed. Proc., 42, 1152 (1983). Abst. 5037.
22. K. A. Fisher-Cornelssen, Arzneim-Forsch., 34, 125 (1984).
23. H. R. Bürki, 186th Meet. Am. Chem. Soc. (Washington, Aug. 22-26, 1983) Abst. 765.
24. Y. G. Gelders, A. J. M. Reynjiens, C. W. Ash and T. J. L. Aerts, Int. Pharmacopsychiat., 17, 247 (1982).
25. K. J. Dewey and H. C. Fibiger, Arch. Pharmacol., 322, 261 (1983).
26. K. P. Bögesö, J. Med. Chem., 26, 935 (1983).
27. H. N. Wagner, H. D. Burns, R. F. Dannals, D. F. Wong, B. Langström, T. Duelfer, J. J. Frost, H. T. Ravert, J. W. Links, S. B. Rosenbloom, S. E. Lukas, A. V. Kramer and M. J. Kuhar, Science 221, 1264 (1983).
28. M. J. Welch, M. R. Kilbourn, C. J. Mathias, M. A. Mintun and M. E. Raichle, Life Sci., 33, 1687 (1983).
29. L. Davis, R. C. Effland and J. T. Klein, Eur. Pat. Appl. EP 63799 (1982). Chem. Abst. 98 179353 r (1983).
30. S. Fielding, W. J. Novick, Jr., H. M. Geyer, W. W. Petko, J. C. Wilker, L. Davis, J. T. Klein and M. Cornfeldt, Drug Dev. Res., 3, 233 (1983).

31. J. P. Yevich, D. L. Temple, J. S. New, W. G. Lobeck, J. D. Catt, D. P. Taylor, M. S. Eison and L. A. Riblet, 186th Meet. Am. Chem. Soc., (Washington, Aug. 22-26, 1983) Abst. 69.
32. H. H. Keller, Arch. Pharmacol., 324, Suppl. R 19 (1983) Abst. 74.
33. G. L. Olson, H.-C. Cheung, E. Chiang and L. Berger in "Dopamine Receptors", ACS Symposium Series 224. C. Kaiser and J. W. Kebabian, Eds., Washington D. C., 1983, p. 251.
34. H. De Cuyper, H. M. van Praag and D. Verstraeten, VII World Congr. Psychiat., (Vienna, July 11-16, 1983) Abst. P 714.
35. T. Nakasawa, K. Ohara, Y. Sawa, VII World Congr. Psychiat., (Vienna, July 11-16, 1983) P 722.
36. K. Mori, M. Miyasaka and Y. Shimazono, VII World Congr. Psychiat., (VIenna, July 11-16, 1983) P 72.
37. T. Shibuya, T. Nishimori, H. Matsuda and P. C. Chen, Int. J. Clin. Pharmacol. Ther. Toxicol., 20, 251 (1982).
38. F. J. Hock, H. J. Kruse, H. Gerhards and E. Konz, Pharmacologist, 25, 130 (1983). Abst. 103.
39. P. Jenner and C. D. Marsden, Acta Psychiat. Scand. Suppl. 69, 109 (1984).
40. A. E. Theodorou, P. Jenner and C. D. Marsden, Life Sci., 32, 1243 (1983).
41. H. Y. Melzer, M. Simonovic and R. So, Life Sci., 32, 2877 (1983).
42. G. Le Fur, F. Guilloux and A. Uzan, Biochem. Pharmacol., 29, 267 (1980).
43. F. E. Blaney, M. S. G. Clark, D. V. Gardner, M. S. Hadley, D. Middleton and T. J. White, J. Med. Chem., 26, 1747 (1983).
44. H. J. Cooke and H. V. Carey, Eur. J. Pharmacol., 98, 147 (1984).
45. M. S. Hadley, F. D. King, B. McRitchie, D. H. Turner and E. A. Watts, 2nd SCI-RSC Med. Chem. Symp. (Cambridge, Sept. 12-14, 1983) Abst. p 9.
46. H. van de Waterbeemd and B. Testa, J. Med. Chem., 26, 203 (1983).
47. M. Terai, S. Usuda, I. Kuroiwa, O. Noshiro and H. Maeno, Jap. J. Pharmacol., 33, 749 (1983).
48. T. Furuya, S. Iwanami, A. Takenaka and Y. Sasada, Bull. Chem. Soc. Jap. 55, 2321 (1982).
49. H. E. Klein, D. Dieterle, E. Rüther, E. Eben, N. Nedopil and H. Hippius, VII World Congr. Psychiat. (Vienna, July 11-16, 1983). Abst. S 1090.
50. J. Grunberger, B. Saletu, L. Linzmeyer and H. Stöhr, VII World Congr. Psychiat. (Vienna, July 11-16, 1983). Abst. S 1089.
51. M. P. Martres, P. Sokoloff, M. Delandre, J.-C. Schwartz, P. Protais and J. Costentin, Arch. Pharmacol., 325, 102 (1984).
52. P. Sokoloff, M. Delandre, M. P. Martres, J.-C. Schwartz, P. Laibe and C. Wermuth, 5th Catecholamine Symp. (Gothenburg, June 12-16, 1983). Abst. 459.
53. L. Florvall, M.-L. Persson and S.-O. Ögren, Acta Pharm. Suec., 20, 365 (1983).
54. L. Florvall and S.-O. Ögren, J. Med. Chem., 25, 1280 (1982).
55. K. Starke, L. Späth, J. D. Lang and C. Adelung, Arch. Pharmacol., 323, 298 (1983).
56. A. Benakis, F. R. Sugnaux, C. Plessas, C. Rey, C. Gallay, R. Happe and P. Allais, Pharmacologist, 25, 115 (1983) Abst. 28.
57. L. Florvall, J. Lundström, S. Rämsby and S.-O. Ögren, Eur. Pat. Appl. EP 60235 (1982). Chem. Abst. 98, 53687 j (1983).
58. G. Sedvall, G. Blomqvist, T. de Paulis, E. Ehrin, L. Eriksson, L. Farde, T. Greiz, C. G. Hedström, D. H. Ingvar, J.-E. Litton, J. L. G. Nilsson, S. Stone-Elander, L. Widèn, F. A. Wiesel and G. Wik, Ann. Meet. Am. Coll. Neuropsychopharmacol. (San Juan, Dec. 12-16, 1983).
59. G. Sedvall, G. Blomqvist, T. de Paulis et al., in "Proceedings of VII World Congress of Psychiatry", P. Berner, Ed., Plenum Press, N.Y. 1984.
60. T. de Paulis, U. H. Lindberg, L. Florvall, S.-O. Ögren and H. Hall, 2nd SCI-RSC Med. Chem. Symp. (Cambridge, Sept. 12-14, 1983). Abst. p. 14.
61. M. Langlois, T. Imbert and B. Bucher, Fr. Pat. Appl. 2446274 (1980). Chem. Abst. 94, 208572 m (1981).
62. M. Jalfre, B. Bucher, N. Dorme, G. Mocquet and R. D. Porsolt, Arch. Int. Pharmacodyn., 264, 232 (1983).
63. M. S. Hadley and F. E. Blaney, Eur. Pat. Appl. EP 76592 (1983). Chem. Abst. 99, 70583 k (1983).
64. J. K. Chivers, W. Gommeren, P. Jenner, J. Leysen, C. D. Marsden, C. Reavill and A. Theodorou, Brit. J. Pharmacol., 78, 398 P (1983).
65. VII World Congr. Psychiat. (Vienna, July 11-16, 1983). Abst. S 1085.
66. E. N. Greenblatt, J. Coupet, E. Rauh and V. A. Szucs-Myers, Arch. Int. Pharmacodyn., 248, 105 (1980).
67. P. F. von Voigtlander, R. A. Lahti, A. H. Tang and V. H. Sethy, Arch. Int. Pharmacodyn., 266, 23 (1983).
68. D. P. Taylor, M. S. Eison, D. L. Temple and J. P. Yevich, Fed. Proc., 42, 1153 (1983). Abst. 5045.
69. J. P. Yevich, D. L. Temple, J. S. New, D. P. Taylor and L. A. Riblet, J. Med. Chem., 26, 194 (1983).

70. L. A. Riblet, M. Herbert, P. J. Standen, A. H. Short and A. T. Birmingham, Psychopharmacol., 81, 335 (1983).
71. M. Zapletalek, J. Libiger, I. Tuma and K. Kubicek, VII World Congr. Psychiat. (Vienna, July 11-16, 1983). Abst. F 685.
72. L. Fontanella, E. Martinelli and M. A. Assandri, Ger. Pat. DE 3307172-A (1983).
73. H. Wachtel, Arch. Pharmacol. Suppl., 324, R 90 (1983). Abst. 360.
74. H. Wachtel, W. Kehr and G. Sauer, Life Sci., 33, 2583 (1983).
75. J. L. Neumeyer, D. Reischig, G. W. Arana, A. Campbell, R. J. Baldessarini, N. S. Kula and K. J. Watling, J. Med. Chem., 26, 516 (1983).
76. R. V. Smith, R. E. Wilcox, K. Young, W. H. Riffee and Y. C. Clement-Cormier, Fed. Proc., 41, 105 (1982) Abst. 71.
77. E. B. Smalstig, E. C. Kornfeld and J. A. Clemens, Fed. Proc., 41, 162 (1982). Abst. 118.
78. R. W. Fuller and H. D. Snoddy, Life Sci., 33, 2357 (1983).
79. S. J. Kolis, T. H. Williams, T. F. Mowles, B. Burghardt, V. Toome and M. A. Schwartz, J. Pharmacol. Exp. Ther., 227, 652 (1983).
80. J. Hyttel, Eur. J. Pharmacol., 91, 153 (1983).
81. L. C. Iorio, A. Barnett, F. H. Leitz, V. P. Houser and C. A. Korduba, J. Pharmacol. Exp. Ther., 226, 462 (1983).
82. A. Hilditch, G. M. Drew and R. J. Naylor, Eur. J. Pharmacol., 97, 333 (1984).
83. M. W. Hamblin and I. Creese, Life Sci., 32, 2247 (1983).
84. P. Sevestre, N. Seguier, J. Lagrange, V. Fournier, B. Musch and P. L. Morselli, VII World Congr. Psychiat. (Vienna, July 11-16, 1983) Abst. P 730.
85. G. Bartolini, B. Scatton and B. Zivkovic in "Adv. Biochem. Psychopharmacol.", Vol. 24, B. Cattabeni et al., Eds., Raven, New York, 1980, p. 207.
86. P. Siminia and A. H. Mulder, Eur. J. Pharmacol., 89, 183 (1983).
87. R. Markstein and D. Lahaye, J. Neural Transm., 58, 43 (1983).
88. J. F. Plantje, A. H. Mulder and J. C. Stoof, Pharm. Weekblad Sci. Ed., 5, 265 (1983).
89. S. Hjort, A. Carlsson, D. Clark, K. Svensson, M. Wikström, D. Sanchez, P. Lindberg, U. Hacksell, L.-E. Arvidsson, A. Johansson and J. L. G. Milsson, Psychopharmacol. 81, 89 (1983).

CHAPTER 4. COGNITIVE DISORDERS

Fred M. Hershenson and John G. Marriott
Warner-Lambert/Parke-Davis, Ann Arbor, MI 48105

Introduction - The treatment of cognitive disorders remains one of the great challenges facing medicine and science. In spite of impressive advances in our understanding of the brain, safe and effective treatments for the major disorders of cognition, e.g., mental retardation, learning disabilities, amnestic syndromes, and the dementias, have yet to be demonstrated. To these major disorders of cognition can be added many disorders and diseases of the central nervous system (CNS) which have cognitive sequelae, although they are not primarily cognitive disorders, e.g., epilepsy and schizophrenia.

Research into human disorders of cognition in recent years has focused to a large degree upon Alzheimer's disease or Primary Degenerative Dementia (PDD). Recent clinical evidence,[1-16] along with earlier laboratory studies,[17-21] indicates that the cognitive defects in Alzheimer's disease, the primary cause of dementia in the elderly, are due to a loss of functioning of cholinergic neurons in the brain.[3-15] Since many articles have recently appeared summarizing cholinergic approaches to the treatment of dementia,[22-26] we will only briefly review this active research area and then discuss other chemotherapeutic approaches which show promise in the treatment of cognitive dysfunction.

The Cholinergic System - Following the precedent of ℓ-dopa therapy for Parkinson's disease, many studies have attempted to treat the cholinergic deficit presumed to account for the cognitive impairments in Alzheimer's disease, by the use of precursors to acetylcholine (ACh), such as choline chloride or lecithin. Results from these studies have been almost uniformly negative.[27-33] One recent long-term, six-month study using high doses of lecithin showed beneficial effects in PDD patients on a mental test and a paired-associate learning task.[34]

A report demonstrating greater enhancement of memory in aged rats with a combination of choline chloride and piracetam than with either compound alone has stimulated several clinical studies with combinations of precursor therapy and other drugs.[35] An initial study with piracetam and choline chloride, in patients for whom the diagnosis was mild to moderate PDD, found greater improvement with the combination therapy than with piracetam alone.[36] Recently, the suggestion has been made that precursor therapy be combined with cholinesterase inhibitors in all future trials of the latter in Alzheimer's disease, and in trials of other agents which affect cholinergic function, e.g. naloxone.[37] This suggestion is supported by recent animal studies showing striking reductions in the optimal doses of cholinergic agonists required to produce improved retention on a T-maze task in mice, when given as two or three-drug combinations.[38] However, clinical trials with combinations of anticholinesterases and ACh precursors have shown limited success. Initial positive effects were reported in elderly dementia patients,[39] and two recent studies using oral physostigmine supplemented by lecithin

ISBN 0-12-040519-9

reported positive effects upon cognitive performance in Alzheimer patients.[40,41] Others failed to find positive effects upon cognition in normal aged subjects[42] or in patients with Alzheimer's disease treated in this manner.[43] Many studies have reported beneficial effects in patients treated with anticholinesterases alone,[22,25,44-47] though others have failed to find improvement in cognitive performance in similar patients.[48,49]

4-aminopyridine (4-AP) and 3,4-diaminopyridine (3,4-DAP) have been studied in animal models for potential effects upon cognition. These drugs enhance release of ACh, presumably by blocking potassium channels in nerve cells, thereby promoting calcium flux.[50] 3,4-DAP partially reverses the deficits in ACh metabolism and motor behavior (tight-rope test) produced by hypoxia in mice.[51] It also improved performance on an 8-arm radial maze in aged rats.[52]

Animal Models - Studies of the biochemical and histopathological changes in Alzheimer's patients,[4-16] particularly in the cholinergic system, have opened new approaches to developing animal models for testing hypotheses and new chemical entities. Already several studies have appeared defining the role of ventral pallidal lesions in animals. Acquisition of active and passive avoidance was impaired in rats with unilateral electrolytical lesions in the ventromedial portion of the pallidum, an area which corresponds to the nucleus basalis of Meynert in primates.[53] Responses learned before lesioning were not impaired. Ventral pallidal lesions produced by ibotenic acid did not alter performance of rats on a battery of psychomotor tasks or affect sensitivity to shock.[54] However, severe deficits in retention of a passive avoidance response were found in these lesioned animals. Similar deficits in passive avoidance retention were found in rats lesioned bilaterally in the ventral pallidum using another excitatory neurotoxin, kainic acid.[55] Ethylcholine mustard aziridinium ion (AF64A, 1), a neurotoxic choline analog, has been shown to produce long-lasting hypofunction of central cholinergic systems in mice[56] and a reduction of presynaptic cholinergic markers in rat hippocampus[57] without affecting post-synaptic muscarinic receptor binding. It has been proposed that AF64A-lesions may provide an animal model of Alzheimer's disease, although behavioral evidence to support this view is quite preliminary.[58]

Anticholinergic-induced cognitive deficits have also been used as a model of age-related impairments, with agents being tested for their ability to antagonize or reverse this activity. Systemically administered atropine has been shown to produce increases in running time and working memory errors in mice trained on a six-arm radial maze.[59] In a swim maze, atropine-treated rats were impaired with respect to finding a hidden escape platform.[60] Similar deficits were found in rats with total hippocampalectomy.[61] Atropine was found to disrupt acquisition of a light/dark discrimination and a tone/no tone discrimination in rats.[62] Physostigmine, conversely, facilitated performance on this task. Anticholinergics are also effective in disrupting memory when injected directly into the brain.[63,64] Conversely, cholinergic agents such as arecoline, physostigmine, oxotremorine, and muscarine, all improved retention on an active avoidance task when administered ICV after training and prior to retention testing one week later.[64] 4-(o-Benzylphenoxy)-N-methylbutylamine hydrochloride (MCI-2016, 2), was found to reverse scopolamine-induced impairments of spontaneous alteration of responding in rats similar to the effects of physostigmine, choline and amphetamine.[65]

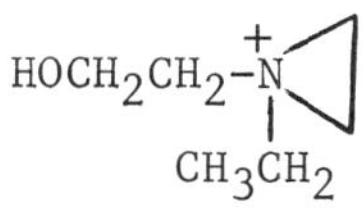

1

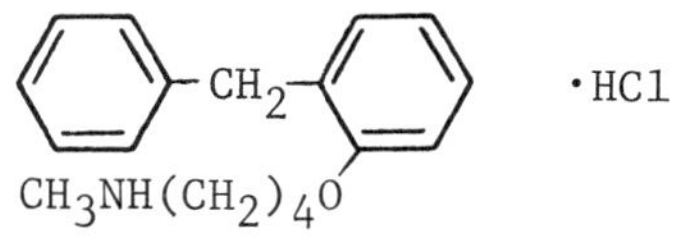

2

Old animals continue to be used extensively as models of age-related cognitive disorders. Old mice, for example, were found to be impaired on passive avoidance compared to young mice.[66] In contrast with clinical data, phosphatidylcholine in the diet enhanced performance of old mice in a shuttle-box avoidance test.[67] Aged rats performed at chance levels after 15 training trials using a 12-arm radial maze, whereas young rats had mastered the task.[68] In this study, positive correlations in aged rats were found between maze performance and hippocampal choline acetyltransferase activity. Aged primates, both cebus and rhesus monkeys, have been employed in studies of age-related memory impairments[69] and drug effects upon memory.[70] Drug trials in monkeys have demonstrated effects with cholinergic agents and neuropeptides similar to those reported in human trials, adding to the validity and utility of these models.[71]

Biogenic Amines - While cholinergic approaches to the treatment of cognitive disorders have been the focus of much clinical work, cumulative laboratory data implies an important role in learning and memory for the adrenergic nervous system.[72,73] In addition, there is a dramatic loss of cells in the locus coeruleus in brains of Alzheimer patients, but not in multiinfarct dementia, suggesting that this disease is not simply a disorder of cholinergic origin.[74,75] There is also a suggestion that cholinergic influences upon memory may involve dopaminergic mechanisms.[76]

Clinical trials of amphetamine, methylphenidate, pentylenetetrazol, and pipradol have generally failed to find significant effects upon cognition in impaired elderly subjects, although mood improvement and reduction in fatigue have been noted.[74] These results contrast with the positive effects upon performance induced by psychostimulant drugs in hyperactive children with attentional deficit disorder.[77,78] Only Hydergine®, which has both alpha-adrenergic and dopaminergic activity, has been shown to have consistent effects in improving mental performance in dementia patients, although the overall degree of improvement is usually small.[79] Medical and psychological improvements have also been reported in normal elderly volunteers treated with Hydergine®.[80] Two recent clinical studies have found impairment of cognitive function in subjects treated with antihypertensive medication.[81,82]

In animals, propranolol attenuated both the memory facilitation and amnesia produced by post-training stimulation of frontal cortex in rats trained using two levels of footshock.[83] Amphetamine attenuated CO_2-induced amnesia for one-trial inhibitory avoidance in young rats, but had no effect in old animals suggesting a change in the neuromodulatory influence of catecholamines on memory processes in senescence.[84] As has been reported for cholinergic drugs, combinations of adrenergic and dopaminergic drugs have produced more improvement in performance than individual drugs alone. The actions of the central stimulants, caffeine, strychnine, or methylphenidate, on retention of shuttle-box avoidance in rats were much more evident when combined with orotic acid.[85] p-Chlorophenylalanine facilitated learning and/or memory when adminstered in conjunction with caffeine, strychnine, or amphetamine.[86] Apomorphine protected against the disturbances in conditioned avoidance responding produced by hypobaric hypoxia.[87] This activity of apomorphine was blocked by adrenalectomy but not by domperidone. Apomorphine injected intrahippocampally immediately after acquisition of a brightness discrimination task improved retention and increased incorporation of L-fucose in the hippocampus of rats.[88]

Neuropeptides - A number of studies have been conducted aimed at correlating abnormal levels of a particular neuropeptide with Alzheimer's disease. The following neuropeptides have been studied recently: somatostatin,[89,90] substance-P,[91] neurotensin,[92] CCK,[93,94] VIP,[93] TRH,[92] glucagon,[94] and beta-endorphin.[95] Only somatostatin levels in CSF and brain appear to be reduced in relation to the severity of the disease.[89,90,96] Angiotensin converting enzyme activity has been reported to be elevated in discrete brain regions from Alzheimer patients when compared to age-matched controls.[97]

The neurophysiological effects[98] of vasopressin and oxytocin and the effects of vasopressin on memory[99] have been recently reviewed. Behavioral studies with these peptides indicate that vasopressin enhances, while oxytocin impairs memory,[100,101] and that these effects are centrally mediated.[102] Arginine vasopressin (3) improved passive avoidance in chicks[103] and lysine vasopressin (4) was found to reverse anisomycin-induced amnesia in mice.[104] Recently, it has been suggested that peripherally administered vasopressin elicits its effect on memory through an aversive property, resulting in an arousal response.[105] Clinical studies with vasopressin in head-injured[106] and Alzheimer patients[107] have failed to demonstrate differences in performance on learning and memory tasks between drug and placebo-control groups. A report that oxytocin produces amnestic effects in normal human subjects has been replicated in a recent study.[108]

A number of clinical investigations have been conducted with the vasopressin analog, 1-desamino-8-D-arginine vasopressin (desmopressin, DDAVP, 5), and have yielded conflicting results. Positive effects on learning and memory were observed in cognitively unimpaired subjects,[109,110] depressed individuals,[110] and patients with diabetes insipidus.[111] A 25 μg intranasal dose of DDAVP, however, failed to reverse the amnesia in patients receiving electroconvulsive shock therapy.[112] This may be due to the single, relatively low dose that was used. Other studies of DDAVP in individuals with memory disorders failed to show any significant improvement.[113,114] One patient developed a paranoid-like psychosis after receiving 30 μg of DDAVP by intranasal administration.[115] The vasopressin analog, desglycinamide arginine vasopressin (DGAVP, 6) improved learning in rats on a food-motivated task.[116] However, DGAVP failed to produce any beneficial effects on memory in patients suffering from Korsakoff's Syndrome.[117]

H-Cys-Tyr-Phe-Gln-Asn-Cys-X   Mpr-Tyr-Phe-Gln-Asn-Cys-Pro-D-Arg-Gly-NH_2

X = Pro-Arg-Gly-NH_2 3 (AVP)   5 (DDAVP)
Pro-Lys-Gly-NH_2 4 (LVP)   (Mpr = Mercapto propionic acid)
Pro-Arg-OH 6 (DGAVP)

Vasopressin may serve as the precursor of smaller molecules with effects on learning and memory. The cyclized derivative of the C-terminal dipeptide of arginine vasopressin (cyclo-(Arg-Gly)) was found to block puromycin-induced amnesia in mice.[118] The C-terminal tripeptide of oxytocin (L-prolyl-L-leucyl-glycineamide or PLG) has been reported to cause an enhancement of memory in chicks.[119]

It has been suggested that while vasopressin is involved in memory processes, the neuropeptides ACTH/MSH affect motivational and attentional processes.[120] However, studies comparing the effects of a number of neuropeptides on short-term memory in young adult primates indicated that α-MSH produced some facilitation of memory.[121] Use of aged primates to assess the effects of neuropeptides on memory suggested that arginine and lysine vasopressin produced the best overall effects followed by $ACTH_{4-10}$.[122] Somatostatin was virtually inactive and oxytocin caused an impairment in memory in half the aged animals tested. $ACTH_{4-10}$ has also been reported to be involved in retrieval processes of memory. In rats treated with antivasopressin serum, which disrupts storage/retrieval processes, $ACTH_{4-10}$ restored the retrieval of memory, possibly by increasing the state of arousal.[123]

The $ACTH_{4-9}$ analog (ORG 2766) has been compared to methylphenidate in hyperkinetic children with attentional deficit disorder.[124] Although anecdotal reports indicated an improvement in the sociability of these children, there did not appear to be an improvement in attention. $ACTH_{4-9}$ has also been administered to patients with severe senile dementia. No improvement was observed after four weeks of drug therapy.[125] There appears to be some relationship between the behavioral effects of both ACTH[126] and the $ACTH_{4-9}$ analog[127] and the opioid system.

The role of enkephalins and endorphins in learning and memory presents a somewhat confusing picture. It has recently been suggested that the apparent conflicts in the literature may be due to the different doses of opiates used in particular studies; high doses appear to facilitate learning while low doses cause a decrement.[128,129] The effects of the endorphins on animal learning have been recently reviewed, and the proposal made that disturbances in the endorphin system may lead to attentional deficit disorders in children.[130]

Actions of opiate antagonists, such as naloxone, on learning and memory have been studied in both laboratory animals and humans. Again, conflicting results have been reported. Naloxone and diprenorphine have been reported to enhance the spatial memory of rats trained on an 8-arm radial maze.[131] Another investigation, however, found that naloxone had a disruptive effect on a learned behavior in rats.[132] Cognitive impairments were evident after infusions of 2 mg/kg naloxone in normal individuals;[133] no changes in cognitive performance were observed at lower doses. Results from a double-blind, placebo-controlled study of naloxone in seven patients with Alzheimer's Disease indicated at least temporary positive effects on cognition,[134] however open-trials conducted at four separate centers did not show any effects.[135]

Nootropics - The term nootropic (noos = thought or reason; trepo = turn or change) describes a drug which selectively affects higher cerebral functions.[136-138] The chemistry and pharmacology of nootropics have been recently reviewed.[139] Most examples of this class are comprised of gamma lactams which are chemically related to the folded conformation of GABA. These include piracetam (7), etiracetam (8), pramiracetam (CI-879, 9), oxiracetam (ISF-2522, 10) and aniracetam (Ro 13-5057, 11). The neuropharmacological[140,141] and neurochemical[142] effects of pramiracetam suggest that the compound is a cognition activator. A clinical study involving computerized quantitative EEG analysis indicated that this drug acts as a vigilance enhancer.[143] Preliminary findings from an open-label study of pramiracetam in Alzheimer patients with mild to moderate cognitive decline indicated an improvement in goal-directed behavior.[144] Oxiracetam has also been studied in elderly patients with organic brain syndrome, and positive effects have been noted.[145] Aniracetam improved impaired learning and memory in rodents,[146] and produced some beneficial effects in aged patients in a geriatric nursing home setting.[147] Finally, cognition activating properties have been described for a series of 3-aryloxypyridines. 3-Phenoxypyridine (CI-844, 12) was found to be the most active member of this series in enhancing learning and/or memory in a single trial passive avoidance task in mice retested one week after training.[148]

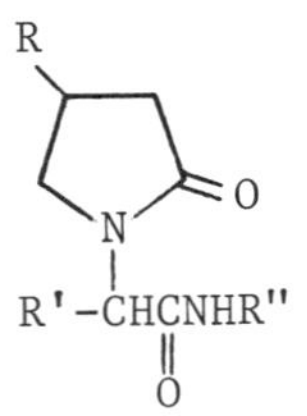

7, R = R' = R'' = H
8, R = R'' = H; R' = Et
9, R = R' = H;
R'' = $CH_2CH_2N(i\text{-}Pr)_2$
10, R = OH; R' = R'' = H

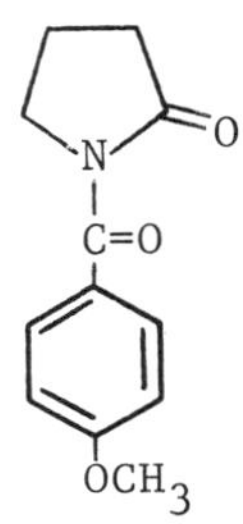

11

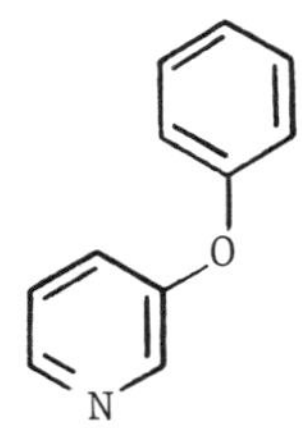

12

Summary and Conclusions - The last few years have seen significant progress made in efforts to understand and treat cognitive disorders. With this progress has come an even greater appreciation of the complexity and richness of the physiological events underlying cognitive function and the difficulties facing the physician and scientist in finding effective treatments for cognitive disorders.

REFERENCES

1. D. A. Drachman in: "Aging of the Brain," D. Samuel, S. Algeri, S. Gershon, V. E. Grimm and G. Toffano, Eds., Raven Press, New York, N.Y., 1983, p. 19.
2. D. M. Bowen, S. J. Allen, J. S. Benton, M. J. Goodhardt, E. A. Haan, A. M. Palmer, N. R. Sims, C. C. T. Smith, J. A. Spillane, M. M. Esiri, D. Neary, J. S. Snowdon, G. K. Wilcock and A. N. Davison, J. Neurochem., 41, 266 (1983).
3. F. Tagliavini and G. Pilleri, Lancet, I, 469 (1983).
4. N. R. Sims, D. M. Bowen, S. J. Allen, C. C. T. Smith, D. Neary, D. J. Thomas and A. N. Davison, J. Neurochem., 40, 503 (1983).
5. J. H. Growdon, Med. Res. Rev., 3, 237 (1983).
6. K. L. Davis, J. Y. K. Hsieh, M. I. Levy, T. B. Horvath, B. M. Davis and R. C. Mohs, Psychopharmacol. Bull., 18, 193 (1982).
7. B. Chipperfield, P. M. Newman and I. C. A. Moyes, Lancet, II, 199 (1981).
8. J. T. Coyle, D. L. Price and M. R. DeLong, Science, 219, 1184 (1983).
9. P. J. Whitehouse, Science, 215, 1237 (1982).
10. P. J. Whitehouse, D. L. Price, A. W. Clark, J. T. Coyle and M. R. DeLong, Ann. Neurol., 10, 122 (1981).
11. D. L. Price, P. J. Whitehouse, R. G. Struble, A. W. Clark, J. T. Coyle, M. R. DeLong and J. C. Hedreen, Neurosci. Comment., 1, 84 (1982).
12. R. G. Struble, Science, 216, 413 (1982).

13. G. K. Wilcock, M. M. Esiri, D. M. Bowen and C. C. T. Smith, J. Neurol. Sci., 57, 407 (1982).
14. E. K. Perry, B. E. Tomlinson, G. Blessed, R. H. Perry, A. Cross and T. T. Crow, Lancet, II, 149 (1981).
15. R. M. Marchbanks, J. Neurochem., 39, 9 (1982).
16. M. N. Rossor, P. C. Emson, C. Q. Mountjoy, M. Roth and L. L. Iversen, Exp. Brain Res., suppl. 5, 153 (1982).
17. H. P. Alpern and J. G. Marriott, Physiol. Behav., 11, 571 (1973).
18. R. T. Bartus, R. L. Dean, B. Beer and A. S. Lippa, Science, 217, 408 (1982).
19. J. A. Deutsch, Science, 174, 788 (1971).
20. T. Cox and N. Tye, Neuropharmacol., 12, 477 (1973).
21. O. Buresova, J. Bures, A. Bohdanechy and T. Weiss, Psychopharmacol., 5, 255 (1964).
22. S. D. Brinkman and S. Gershon, Neurobiol. Aging, 4, 139 (1983).
23. B. S. Greenwald, R. C. Mohs, and K. L. Davis, J. Am. Geriatr. Soc., 31, 310 (1983).
24. B. S. Greenwald and K. L. Davis, Adv. Neurol, 38, 87 (1983).
25. C. A. Johns, B. S. Greenwald, R. C. Mohs and K. L. Davis, Psychopharmacol. Bull., 19, 185 (1983).
26. P. S. Anderson and D. Haubrich, Ann. Rep. Med. Chem., 16, 51 (1981).
27. G. S. Rosenberg and K. L. Davis, Am. J. Clin. Nutr., 36, 709 (1982).
28. P. Etienee, D. Dastoor, S. Gauthier, R. Ludwick and B. Collier, Neurology, 31, 1552 (1981).
29. L. J. Thal, W. Rosen, S. Sharpless and H. Crystal, Neurobiol. Aging, 2, 205 (1981).
30. S. D. Brinkman, N. Pomara, P. J. Goodnick, N. Barnett and E. F. Domino, J. Clin. Psychopharmacol., 2, 281 (1982).
31. S. Weintraub, M. M. Mesulam, R. Auty, R. Baratz, B. N. Cholakos, L. Kapust, B. Ransil, J. G. Tellers, M. S. Albert, S. LoCastro and M. Moss, Arch. Neurol., 40, 527 (1983).
32. N. Pomara, E. F. Domino, H. Yoon, S. Brinkman, C. A. Tamminga and S. Gershon, J. Clin. Psychiatry, 44, 293 (1983).
33. M. Fisman, H. Mersky and E. Helmes, Am. J. Psychiatry, 138, 970 (1981).
34. R. Levy, A. Little, P. Chuaqui and M. Reith, Lancet, I, 987 (1983).
35. R. T. Bartus, R. L. Dean, K. A. Sherman, E. Friedman and B. Beer, Neurobiol. Aging, 2, 105 (1981).
36. B. Reisberg, S. H. Ferris, M. K. Schneck, J. Corwin, P. Mir, E. Friedman, K. A. Sherman, M. McCarthy and R. T. Bartus, Drug Development Res., 2, 475 (1982).
37. R. J. Wurtman, N. Eng. J. Med., 309, 555 (1983).
38. J. F. Flood, G. E. Smith and A. Cherkin, Neurobiol. Aging, 4, 37 (1983).
39. W. H. Kaye, N. Sitaram, H. Weingartner, M. H. Ebert, S. Smallberg and J. C. Gillin, Biol. Psychiatry, 17, 275 (1982).
40. L. J. Thal and P. A. Fuld, N. Engl. J. Med., 308, 720 (1983).
41. L. J. Thal, P. A. Fuld, D. M. Masur and N. S. Sharpless, Ann. Neurol., 13, 491 (1983).
42. D. A. Drachman, G. Glosser, P. Fleming and G. Longenecker, Neurology, 32, 944 (1982).
43. A. Wettstein, Ann. Neurol., 13, 210 (1983).
44. E. Goldberg, L. J. Gerstman, S. Mattis, J. E. Hughes, C. A. Sirio and R. M. Bilder, Jr., J. Clin. Neuropsychol., 4, 219 (1982).
45. E. Goldberg, L. J. Gerstman, S. Mattis, J. E. Hughes, C. A. Sirio and R. M. Bilder, Jr., Arch. Neurol., 39, 581 (1982).
46. K. L. Davis, R. C. Mohs, W. G. Rosen, B. S. Greenwald, M. I. Levy and T. B. Horvath, N. Engl. J. Med., 308, 721 (1983).
47. A. Agnoli, N. Martucci, V. Manna, L. Conti and M. Fioravanti, Clin. Neuropharm., 6, 311 (1983).
48. M. Franceschi, O. Trancredi, G. Savio and S. Smire, Eur. Neurol., 21, 399 (1982).
49. C. Caltagirone, G. Gainotti and C. Masullo, Int. J. Neurosci., 16, 247 (1982).
50. B. Katz and R. Meledi, Proc. R. Soc. (Lond.), 205, 369 (1979).
51. C. Peterson and G. E. Gibson, J. Pharmacol. Exp. Therap. 222, 576 (1982).
52. H. P. Davis, A. Idowu and G. E. Gibson, Exper. Aging Res., 9, 211 (1983).

53. G. LoConte, L. Bartolini, F. Casamenti, I. Marconcini-Pepeu and G. Pepeu, Pharm. Biochem. Behav., 17, 933 (1982).
54. C. Flicker, R. L. Dean, D. L. Watkins, S. K. Fisher and R. T. Bartus, Pharm. Biochem. Behav., 18, 973 (1983).
55. E. Friedman, B. Lerer and J. Kuster, Pharmacol. Biochem. Behav., 19, 309 (1983).
56. A. Fisher, C. R. Mantione, D. J. Abraham and I. Hanin, J. Pharmacol. Exp. Ther., 222, 140 (1982).
57. C. R. Mantione, M. J. Zigmono, A. Fisher and I. Hanin, J. Neurochem., 41, 251 (1983).
58. I. Hanin, W. C. DeGroat, C. R. Mantione, J. T. Coyle and A. Fisher in: "Conference on Biological Aspects of Alzheimers Disease," R. Katzman Ed., Cold Spring Harbor Laboratory, Cold Spring Harbor, N.Y., 1983, p. 243.
59. A. Levy, P. G. Kluge and T. F. Elsmore, Behav. Neural. Biol., 39, 229 (1983).
60. R. J. Sutherland, I. Q. Whishaw and J. C. Regehr, J. Comp. Physiol. Psychol., 96, 563 (1982).
61. R. G. M. Morris, P. Garrud, J. N. P. Rawlins and J. O'Keefe, Nature, 297, 681 (1982).
62. R. T. Rick, K. L. Whittle and S. H. Cross, Neuropharmacol., 20, 747 (1981).
63. D. Blozovski and N. Hannocq, Psychopharmacology, 76, 351 (1982).
64. J. F. Flood, D. W. Landry and M. E. Jarvik, Brain Res., 215, 117 (1981).
65. A. Tobe, M. Egawa and R. Nagai, Jpn. J. Pharmacol., 33, 775 (1983).
66. P. Kubanis, S. F. Zornetzer and G. Freund, Pharmacol. Biochem. Behav., 17, 313 (1982).
67. P. D. Leathwood, E. Heck and J. Mauron, Life Sci., 30, 1065 (1982).
68. D. K. Ingram, E. D. London and C. L. Goodrick, Neurobiol. Aging, 2, 41 (1981).
69. J. G. Marriott and J. S. Abelson, Age, 3, 7 (1980).
70. R. T. Bartus, R. L. Dean and B. Beer, Neurobiol. Aging, 1, 145 (1980).
71. R. T. Bartus, R. L. Dean and B. Beer, Psychopharmacol. Bull., 19, 168 (1983).
72. P. Kubanis and S. F. Zornetzer, Behav. Neural. Biol., 31, 115 (1981).
73. P. E. Gold and S. F. Zornetzer, Behav. Neural. Biol., 38, 151 (1983).
74. B. Reisberg, E. London, S. H. Ferris, R. Anand and M. J. de Leon, Psychopharmacol. Bull., 19 220 (1983).
75. D. M. A. Mann, P. O. Yates and J. Hawkes, J. Neurol. Neurosurg. Psychiatry, 45, 113 (1982).
76. C. M. Baratti, I. B. Introini, P. Huygens and F. Gusovsky, Psychopharmacology, 80, 161 (1983).
77. B. E. Golinko, P. M. Rennick and R. F. Lewis, Prog. Neuropsychopharmacol., 5, 65 (1981).
78. P. T. Ackerman, R. A. Dykman, P. J. Holcomb and D. S. McCray, Psychiatry Res., 7, 199 (1982).
79. P. Cook and I. James, N. Engl. J. Med., 305, 1560 (1981).
80. R. Spiegel, F. Huber and S. Koeberle, J. Am. Geriat. Soc., 31, 549 (1983).
81. L. R. Hartley, S. Ungapen, I. Davie and D. J. Spencer, Br. J. Psychiatry, 142, 512 (1983).
82. S. Solomon, E. Hotchkiss, S. M. Saravay, C. Bayer, P. Ramsey and R. S. Blum, Arch. Gen. Psychiatry, 40, 1109 (1983).
83. D. B. Sternberg, P. E. Gold and J. L. McGaugh, Behav. Neural. Biol., 38, 261 (1983).
84. R. B. Messing, H. Rigter and V. J. Nickolson, Neurobiol. Aging, 3, 133 (1982).
85. D. Yonkov, W. Wetzel, H. Matthies and K. Roussinov, Psychopharmacology, 73, 399 (1981).
86. D. Ionkov, K. Rusinov and Iu Vuglenova, Acta. Physiol. Pharmacol. Bul., 8, 57 (1982).
87. C. Saligaut, N. Moore, P. Chretien, M. Daoust, O. Richard and F. Boismare, Stroke, 13, 859 (1982).
88. R. Jork, G. Grecksch and H. Matties, Pharmacol. Biochem. Behav., 17, 11 (1982).

89. P. L. Wood, P. Etienne, S. Lal, S. Gauthier, S. Cajal and N. P. V. Nair, Life Sci., 31, 2073 (1982).
90. P. Davies and R. D. Terry, Neurobiol. Aging, 2, 9 (1981).
91. H. A. Crystal and P. Davies, J. Neurochem., 38, 1781 (1982).
92. J. A. Biggins, E. K. Perry, J. R. McDermott, A. I. Smith, R. H. Perry and J. A. Edwardson, J. Neurol. Sci., 58, 117 (1983).
93. R. H. Perry, G. J. Dockray, R. Dimaline, E. K. Perry, G. Blessed and B. E. Tomlinson, J. Neurol. Sci., 51, 465 (1981).
94. D. J. Sanders, S. Zahedi-Asl and A. P. Marr, Prog. Brain Res., 55, 465 (1982).
95. H. Kaiya, T. Tanaka, K. Takeuchi, K. Morita, S. Adachi, H. Shirakawa, H. Ueki and M. Namba, Life Sci., 33, 1039 (1983).
96. E. K. Perry, G. Blessed, B. E. Tomlinson, R. H. Perry, T. J. Crow, A. J. Cross, G. J. Dockray, R. Dimaline and A. Arrequi, Neurobiol. Aging, 2, 251 (1981).
97. A. Arrequi, E. K. Perry, M. Rossor and B. E. Tomlinson, J. Neurochem., 38, 1490 (1982).
98. R. E. J. Dyball and A. T. Paterson, Pharmac. Ther., 20, 419 (1983).
99. D. M. Gash and G. J. Thomas, Trends Neurosci., 6, 197 (1983).
100. T. B. van Wimersma Greidanus, J. M. van Ree and D. de Wied, Pharmac. Ther., 20, 437 (1983).
101. G. L. Kovacs and G. Telegdy, Pharmac. Ther., 18, 375 (1982).
102. G. Meisenberg and W. H. Simmons, Neurosci. Biobehav. Rev., 7, 263 (1983).
103. J. L. Davis, R. M. Pico and A. Cherkin, Behav. Neur. Biol., 35, 242 (1982).
104. M. E. Judge and D. Quartermain, Pharmacol. Biochem. Behav., 16, 463 (1982).
105. A. Ettenberg, D. van der Kooy, M. Le Moal, G. F. Koob and F. E. Bloom, Behav. Brain Res., 7, 331 (1983).
106. W. D. Fewtrell, A. O. House, P. F. Jamie, M. R. Oates and J. E. Cooper, Psychol. Med., 12, 423 (1982).
107. R. Durso, P. Fedio, P. Brouwers, C. Cox, A. J. Martin, S. A. Ruggieri, C. A. Tamminga and T. N. Chase, Neurology, 32, 674 (1982).
108. D. J. Kennett, M. C. Devlin and B. M. Ferrier, Life Sci., 31, 273 (1982).
109. B. E. Beckwith, T. Petros, S. Kanaan-Beckwith, D. I. Couk, R. J. Haug, and C. Ryan, Peptides, 3, 627 (1982).
110. H. Weingartner, P. Gold, J. C. Ballenger, S. A. Smallberg, R. Summers, D. R. Rubinow, R. M. Post and F. K. Goodwin, Science, 211, 601 (1981).
111. F. Laczi, J. M. van Ree, A. Wagner, Zs. Valkusz, T. Jardanhazy, G. L. Kovacs, G. Telegdy, J. Szilard, F. A. Laszlo and D. de Wied, Acta. Endocrinol., 102, 205 (1983).
112. B. Lerer, T. Zabow, N. Egnal and R. H. Belmaker, Biol. Psychiat., 18, 821 (1983).
113. J. S. Jenkins, H. M. Mather and A. K. Coughlan, J. Neurol. Neurosurg. Psych., 45, 830 (1982).
114. J. R. Tinklenberg, A. Pfefferbaum and P. A. Berger, Psychopharmacol. Bull., 17, 206 (1981).
115. G. B. Collins, D. J. Marzewski and M. B. Rollins, Lancet, II, (8250) 808 (1981).
116. R. B. Messing and S. B. Sparber, Eur. J. Pharmacol., 89, 43 (1983).
117. F. Laczi, J. M. van Ree, L. Balogh, A. Szasz, T. Jardanhazy, A. Wagner, L. Gaspar, Z. Valkusz, I. Dobranovics, J. Szilard, F. A. Laszlo and D. de Wied, Acta. Endocrinol., 104, 177 (1983).
118. R. Walter, P. L. Hoffman, A. C. Church, J. B. Flexner and L. B. Flexner, Horm. Behav., 16, 234 (1982).
119. J. L. Davis, R. M. Pico and A. Cherkin, Pharmacol. Biochem. Behav., 17, 893 (1982).
120. D. de Wied and J. M. van Ree, Life Sci., 31, 709 (1982).
121. G. A. Olson, R. Roig-Smith, M. D. Mauk, G. J. La Hoste, D. H. Coy, C. W. Hill and R. D. Olson, Peptides, 2, 131 (1981).
122. R. T. Bartus, R. L. Dean and B. Beer, Neurobiol. Aging, 3, 61 (1982).
123. T. B. van Wimersma Greidanus, Peptides, 3, 7 (1982).
124. H. J. Butter, Y. Lapierre, P. Firestone and A. Blank, J. Clin. Psychopharmacol., 3, 226 (1983).

125. J. C. Martin, B. R. Ballinger, L. L. Cockram, F. M. McPherson, R. M. Pigache and D. Tregaskis, Acta. Psychiatr. Scand., 67, 205 (1983).
126. I. Izquierdo and R. D. Dias, Psychoneuroendocrinol., 8, 81 (1983).
127. M. Fekete, F. Drago, J. M. van Ree, B. Bohus, V. M. Wiegant and D. de Wied, Life Sci., 32, 2193 (1983).
128. J. D. Belluzzi and L. Stein, Ann. N.Y. Acad. Sci., 398, 221 (1982).
129. C. Castellano, Behav. Neur. Biol., 36, 189 (1982).
130. S. R. Rose and J. Orlowski, J. Dev. Behav. Pediatr., 4, 131 (1983).
131. M. Gallagher, R. A. King and N. B. Young, Science, 221, 975 (1983).
132. W. C. Lynch and S. Clark, Life Sci., 32, 2801 (1983).
133. M. R. Cohen, R. M. Cohen, D. Pickar, H. Weingartner and D. L. Murphy, Arch. Gen. Psychiat., 40, 613 (1983).
134. B. Reisberg, S. H. Ferris, R. Anand, P. Mir, V. Geibel and M. J. De Leon, N. Engl. J. Med., 308, 721 (1983).
135. J. P. Blass, M. J. Reding, D. Drachman, A. Mitchell, G. Glosser, R. Katzman, L. J. Thal, S. Grenell, J. E. Spar, A. Larue and E. Liston, N. Eng. J. Med., 309, 556 (1983).
136. T. M. Itil, Biol. Psychiatr., 18, 521 (1983).
137. S. Kanowski, Drug Dev. Res., 2, 433 (1982).
138. C. E. Giurgea, Drug Dev. Res., 2, 441 (1982).
139. B. J. R. Nicolaus, Drug Dev. Res., 2, 463 (1982).
140. T. A. Pugsley, B. P. H. Poschel, D. A. Downs, Y. H. Shih and M. I. Gluckman, Psychopharmacol. Bull., 19, 721 (1983).
141. B. P. H. Poschel, J. G. Marriott, and M. I. Gluckman, Psychopharmacol. Bull., 19, 720 (1983).
142. T. A. Pugsley, Y. H. Shih, L. Coughenour and S. F. Stewart, Drug Dev. Res., 3, 407 (1983).
143. T. M. Itil, S. Mukherjee, G. Dayican, D. M. Shapiro, A. M. Freedman and L. A. Borgen, Psychopharmacol. Bull., 19, 708 (1983).
144. R. J. Branconnier, J. O. Cole, E. C. Dessain, K. F. Spera, S. Ghazvinian and D. DeVitt, Psychopharmacol. Bull., 19, 726 (1983).
145. T. M. Itil, G. N. Menon, M. Bozak and A. Songar, Drug. Dev. Res., 2, 447 (1982).
146. R. Cumin, E. F. Bandle, E. Gamzu and W. E. Haefely, Psychopharmacol., 78, 104 (1982).
147. P. Foltyn, P. W. Lucker and J. Schnitker, Arzneim-Forsch., 33, 865 (1983).
148. D. E. Butler, B. P. H. Poschel and J. G. Marriott, J. Med. Chem., 24, 346 (1981).

Chapter 5. Adaptive Changes in Central Nervous System Receptor Systems

B. Kenneth Koe and Fredric J. Vinick
Pfizer Central Research, Groton, CT 06340

Introduction – The phenomenon that repeated (subacute, chronic) administration of drugs can induce pharmacological effects that become progressively attenuated or accentuated compared to the original effects of a single (acute) dose is well known. In the past decade, the molecular events involved in the rapid loss of the ability of the norepinephrine (NE) receptor-coupled adenylate cyclase system of intact cells to generate cyclic AMP (cAMP) following stimulation of β-adrenoceptors by β-agonists have been well studied.[1,2] In such cells, isoproterenol stimulates the formation of cAMP, but continual exposure to this agonist induces a refractoriness, characterized by a rapid desensitization of agonist-stimulated adenylate cyclase and a slower loss of β-adrenoceptors. The reduction in sensitivity is measured as a change in maximal response of the adenylate cyclase, while the loss in β-adrenoceptors (down-regulation) is ascertained as a decrease in receptor density (determined by Scatchard analysis of saturation binding of antagonists). The $K_{0.5}$ values for adenylate cyclase activity and the K_D values (affinity^{-1}) for antagonist binding are unchanged in desensitization and down-regulation, respectively. Prior to the irreversible disappearance of β-adrenoceptors, an uncoupling process occurs; although the number of receptors remains essentially the same, the sensitivity of these receptors to isoproterenol is less than that of untreated cells. A rapid and reversible desensitization and down-regulation of the β-receptors of rat cerebral cortex is observed on incubation of tissue slices with isoproterenol[3] or desipramine (DMI), an NE uptake blocker.[4] Similar adaptive changes in the receptors of the NE system also occur *in vivo* in the brain when NE levels are altered. Prolonged NE depletion can be effected by drugs (reserpine) or selective lesions (6-hydroxydopamine [6-HDA] or electrolytic), whereas prolonged elevation can be achieved by preventing metabolism (MAO inhibitors) or increasing synaptic concentrations through inhibition of reuptake.[5,6] This chapter examines the recent literature on adaptive changes in central NE, serotonin (5-HT) and dopamine (DA) receptor systems and concentrates on drug-induced effects which could relate to therapeutic modes of action in human diseases. Reviews of related topics have appeared.[7-10]

Norepinephrine Receptors – Depression is a CNS disease believed to be associated with a perturbation of central NE transmission. The first effective drugs in depression were MAO inhibitors and tricyclic uptake blockers which increase the concentration of neuronal or synaptic NE, respectively. With the recognition that the onset of action of antidepressants was several weeks, attention turned to studying the effects of subacute and chronic administration of antidepressant drugs. In rats, multiple treatments with antidepressant drugs[11-14] and electroconvulsive shock (ECS)[11,15-17] desensitized the NE receptor-coupled adenylate cyclase system in limbic forebrain slices and decreased the number of β-adrenoceptors measured by antagonist binding. In contrast to the subsensitivity observed after these chronic treatments, acute administration showed no effects. In general, antidepressants,[6,11,12,18-34] MAO inhibitors,[5,9,21,23,37,38] lithium[35,36] and REM sleep deprivation[39] have all been shown to induce desensitization and/or down-regulation of NE receptors.

Administration of multiple daily doses of drug produces down-regulation faster than single daily dosing and allows a lower dose per injection.[6,29] In a time course study in rats, DMI (10 mg/kg b.i.d.) elicited maximal desensitization of cAMP accumulation by isoproterenol in cortical slices in 5 to 7 days and down-regulation of β-adrenoceptors in cortical membranes after 7 days. After DMI withdrawal, the subsensitive NE receptors returned to normal within a week.[6] After multiple ECS treatments, β-adrenoceptors remain down-regulated for at least a

ISBN 0-12-040519-9

week.[15] These studies indicate that faster onset of therapeutic effects might be achieved by rapidly increasing the dose of antidepressants. Since side effects may preclude such a strategy, other ways to facilitate the desensitization/down-regulation have been investigated. The co-administration of α_2-adrenergic antagonists (yohimbine, dihydroergotamine or phenoxybenzamine) with DMI and other antidepressants increased the rate of down-regulation.[40-42] In contrast, the α_1-antagonist, prazosin, was ineffective in accelerating down-regulation.[42] Rapid desensitization/down-regulation of β-adrenoceptors by co-administration of iprindole and amphetamine occurred at doses of each which were ineffective alone.[43,44] On the other hand, co-administration of chlorpromazine with imipramine failed to enhance the down-regulation of β-receptors observed with imipramine alone, even though this combination is reportedly more effective in treating delusional depression.[45] Unlike most studies, some recent reports claim down-regulation of β-adrenoceptors following a single ECS treatment or single doses of DMI on days 1 and 17.[46]

One explanation for the facilitation of down-regulation of β-adrenoceptors by α_2-adrenergic antagonists is as follows: the expected synaptic build-up of NE after DMI is attenuated by concurrent stimulation of α_2-autoreceptors, which inhibits NE release. Only when these α_2-autoreceptors become down-regulated will synaptic levels of NE increase, thus accounting in part for the slow onset of therapeutic defects.[47] The notion that repeated administration of antidepressants or ECS can induce subsensitivity of α_2-autoreceptors is based on the tolerance to various actions of low doses of clonidine in mice or rats induced by such regimens.[48,49] Chronic administration of clorgyline desensitized α_2-autoreceptors and down-regulated β-adrenoceptors before a decrease in the cAMP response to NE could be observed.[38] The α_2-adrenoceptor adaptation elicited by clorgyline reverted to normal very gradually.[50] One potential disadvantage of using α_2-antagonists is that autoreceptors may quickly become supersensitive. Reduced antagonism of responses to clonidine after long-term administration of mianserin, an α_2-antagonist, may be the result of the development of supersensitive autoreceptors.[51,52] Some authors propose that depression may be caused in part by supersensitivity of α_2-adrenoceptors and report evidence that chronic antidepressant treatments elicit down-regulation of these receptors.[53,54] In other studies, however, an up-regulation of α_2-adrenoceptors has been observed.[55,56] In contrast, α_1-adrenoceptor binding is not affected by chronic administration of antidepressants.[21]

The subsensitive β-adrenergic system resulting from chronic administration of antidepressants implies that postsynaptic responsiveness to NE is diminished, e.g., long-term administration of DMI, clomipramine, maprotiline or tranylcypromine reduced the sensitivity of cingulate cortical neurons to iontophoretically applied NE.[57] Similarly, chronic DMI markedly decreased the responsiveness of Purkinje cells in cerebellum to applied NE.[58] In contrast, other studies have shown that chronic antidepressant treatments enhanced postsynaptic NE and 5-HT responses in the facial motor neurons.[59] Possibly related to the enhanced NE responses is the finding that chronic antidepressant treatment can apparently facilitate the interaction of the guanine nucleotide binding protein with the catalytic protein.[60] The subgroup of antidepressants comprising the selective 5-HT uptake blockers, zimelidine,[24,31] fluvoxamine[33] and sertraline[34] also down-regulate β-adrenoreceptors. Such agents may induce this subsensitivity by the removal of an inhibitory 5-HT constraint on NE neurons through inhibition of firing of raphe 5-HT neurons[34] or enhancement of 5-HT reuptake.[61] The notable exception is fluoxetine.[21,25,26,30]

In rats deprived of a 5-HT input, via lesions with 5,7-dihydroxytryptamine (5,7-DHT) or by p-chlorophenylalanine (PCPA), chronic administration of DMI failed to down-regulate β-adrenoceptors, although the usual desensitization of NE sensitive adenylate cyclase was still observed.[62,63] When NE receptors were desensitized and reduced in number by the DMI treatment, PCPA reversed the decrease of β-receptors without affecting the cyclase.[14] These persistent β-receptors may be "uncoupled", since their affinity for isoproterenol is reduced.[64] Possibly, the step from uncoupled receptors to actual disappearance of NE receptors is interrupted by the loss of 5-HT input. An interesting corollary is the observation that depressed patients benefiting from either imipramine or tranylcypromine showed reemergence of their depressive symptoms when PCPA was administered along with the antidepressant.[65,66]

Since drugs for the treatment of depression appear to increase NE activity, β-agonists may be therapeutically effective in this disorder. Antidepressant activity has been reported following administration of the β-agonist, salbutamol, with a suggestion of more rapid onset of action.[67,68] The time course of clinical improvement with oral salbutamol paralleled the induction of subsensitivity of a plasma NE-sensitive adenylate cyclase.[68] Subacute infusion of the β-agonist, clenbuterol, was found to desensitize isoproterenol-stimulated adenylate cyclase and down-regulate β-adrenoceptors in the rat cerebellum.[69] In contrast, the β-antagonist, propranolol, not only blocked the effects of chronic DMI but also elicited up-regulation of β-receptors and enhanced sensitivity of isoproterenol-sensitive adenylate cyclase.[6] Subacute administration of phenoxybenzamine or propranolol also increased the cAMP response to NE in mouse limbic forebrain.[70] Interestingly, major depressive episodes have been reported for several patients on propranolol therapy.[71] Unlike the rat experiment described above,[6] propranolol did not attenuate the therapeutic effects of an antidepressant in a human study.[72] Pharmacological treatments which deplete neurons of NE,[70,73] and 6-HDA[74] or DSP-4 induced lesions,[75] increased the cAMP response to NE or isoproterenol, and β-adrenoceptor density. Unilateral electrolytic lesions of the locus ceruleus prevented the desensitizing effects of DMI or iprindole on the lesioned side only.[76] The β-adrenoceptors involved in down-regulation or up-regulation in rat cortex belong mainly to the β_1 subtype.[77] A subpopulation of receptors with neither α nor β characteristics is also coupled to the adenylate cyclase of rat limbic forebrain.[78]

Desensitization of the β-receptor system by repeated administration of antidepressants also occurs in the pineal gland, which is innervated by postganglionic sympathetic fibers. The stimulation of the β-receptor/adenylate cyclase system increases the synthesis of serotonin-N-acetyl transferase, a critical enzyme for melatonin production. Thus, the subsensitivity of pineal β-adrenoceptors found after chronic DMI or nialamide was accompanied by a fall in melatonin formation stimulated by isoproterenol or darkness.[79] Chronic administration of amitriptyline, maprotiline and clenbuterol also attenuated the increase of pineal melatonin stimulated by isoproterenol, but iprindole, mianserin, fluoxetine and repeated ECS had no effect.[80] Attempts to detect subsensitivity of the β-receptors in patients treated with DMI by monitoring night-time plasma melatonin levels failed.[81] The exclusive β-adrenergic regulation of melatonin synthesis seen in rats may not extend to man.[82]

Stress applied to rats (electric foot shock) induced desensitization of NE receptor-coupled adenylate cyclase in cortical slices.[83] This finding suggests that a similar desensitization could be a natural response to prolonged emotional stress. Thus, the NE receptors in depressed patients may be unable to down-regulate in response to stress, necessitating antidepressant therapy.[84] In another study, however, the number of β-adrenoceptors in rats was not altered by chronic foot shock stress.[15] Most studies on adaptive changes in the β-adrenergic system focus on cerebral cortex, frontal cortex, limbic forebrain and hippocampus. Although ECS did not down-regulate β-adrenoceptors in striatum, hypothalamus or cerebellum,[15] chronic drug treatments do cause a decrease in the density of β-receptors in these brain regions.[85,86,69]

Adaptation of NE receptors is also observed after long-term administration of other types of drugs. For example, chronic chlorpromazine desensitized cortical NE-sensitive adenylate cyclase in rats[18] but caused supersensitivity in mouse limbic forebrain.[70] Chronic haloperidol did not desensitize cortical NE-stimulated adenylate cyclase[73] but apparently induced up-regulation of α_2-adrenoceptors (enhanced locomotor response to clonidine) in mice.[87] The methylxanthines (caffeine,[88] pentoxyfylline[89]), which increase turnover of brain NE, caused rapid down-regulation of β-adrenoceptors.

<u>Serotonin Receptors</u> – The relationships between 5-HT receptors and the actions of drugs are much less clear than those for NE or DA receptors. Numerous conflicting findings result from the inherent difficulties of relating behavioral and electrophysiological effects. Also, the aforementioned effects do not correlate well with drug-induced changes in binding sites that

are poorly defined functionally. Serotonin receptors are termed 5-HT_1 (based on ^{3}[H]5-HT binding[90]) or 5-HT_2 (based on ^{3}[H]spiperone binding in frontal cortical membranes).[21,91] The latter sites are also labeled by ^{3}[H]ketanserin[92] or ^{3}[H]mianserin.[93,94] Serotonin agonists exhibit nanomolar affinities for 5-HT_1 receptors but only micromolar affinities for 5-HT_2 binding sites. A 5-HT sensitive adenylate cyclase has been characterized, but its connection with 5-HT_1 or 5-HT_2 receptors is not clear.[91,95,96] Increases in hippocampal[90] and cortical[21] 5-HT_1 receptors are found in rats lesioned with 5,7-DHT. Similar lesions increased ^{3}[H]mianserin binding sites[61] but not ^{3}[H]spiperone[97] or ^{3}[H]ketanserin binding.[98] Chronic PCPA treatment effected increases in 5-HT_1[99] and 5-HT_2 receptors (^{3}[H]spiperone binding)[100] and ^{3}[H]mianserin binding sites.[61] The binding of ^{3}[H]imipramine to brain membranes is supposedly associated with 5-HT uptake sites, but there is a question whether ^{3}[H]imipramine labels a physiologically relevant site.[101] ^{3}[H]Imipramine binding sites are decreased in rats with 5,7-DHT lesions but remain unchanged in rats following chronic treatment with chronic PCPA.[61]

The role of serotonergic changes in depression and antidepressant therapy has been a subject of intense study. For example, brains of suicide victims exhibited an increase in cortical 5-HT_2 binding sites[102,103] and either an increase[103] or a decrease[104] in ^{3}[H]imipramine binding. Furthermore, cortisol secretion stimulated by 5-hydroxytryptophan was found to be greater for depressed and manic patients than normal controls, suggesting the involvement of supersensitive 5-HT receptors.[105] Thus, an increased risk to suicide and depression may be characterized by a decrease in serotonergic function.[106] If up-regulation of 5-HT receptors is associated with suicide and depression, then the question arises as to whether antidepressant therapy normalizes 5-HT receptor density. Several investigators have reported that chronic administration of MAO inhibitors results in a decrease in the number of 5-HT_1 receptors (K_D unchanged) in rat cortex or brain stem,[99,107] inhibits the "5-HT syndrome" induced by 5-HT agonists[108] and attenuates the response of cortical neurons to microiontophoretically applied 5-HT.[109] However, variable results have been obtained with tricyclic antidepressants. Many studies have indicated that chronic administration of these drugs does not down-regulate 5-HT_1 receptors,[21-23,107,111,112] but there are a few exceptions.[19,112,113] One group reported a decrease in the number of ^{3}[H]5-HT binding sites after long-ter DMI, imipramine, amitriptyline or dimetacrine treatments.[113] Another group found that chronic treatment with imipramine or DMI produced high and low affinity 5-HT_1 receptors in rat dorsal cortex.[112] Even 5-HT specific uptake blockers have led to conflicting findings: chronic clomipramine had no effect on ^{3}[H]5-HT binding,[110] whereas D-fenfluramine led to a decrease in ^{3}[H]5-HT binding (lowered K_D and B_{max}).[114,115] Fluoxetine was found by several groups not to alter 5-HT_1 receptors;[19,21,111,116] however, others have reported either a reduction in the number of high affinity 5-HT_1 receptors in frontal cortex[117] or a decrease in the affinity of 5-HT_1 receptors in dorsal cortex.[112] Similarly, one chronic study with zimelidine failed to show down-regulation of 5-HT_1 receptors in rat cortex,[118] while a different group observed an induction of high and low affinity ^{3}[H]5-HT binding sites, amounting to a "selective down-regulation" of 5-HT_1 receptors.[112] Chronic lithium treatment can also alter 5-HT_1 receptors as evidenced by the fact that receptor density decreased in the rat hippocampus; no change was observed in the cortex.[119] The 5-HT autoreceptor blocker, methiothepin, facilitated the down-regulation of 5-HT_1 receptors induced by imipramine or mianserin, presumably by increasing 5-HT release.[120]

Chronic administration of imipramine, DMI, amitriptyline, iprindole or pargyline down-regulated ^{3}[H]spiperone-labeled 5-HT_2 receptors in rat brain.[21,97,121] Pretreatment with the serotonergic neurotoxin, p-chloroamphetamine (PCA), did not lessen the effectiveness of amitriptyline, which suggests that the drug itself may be interacting with postsynaptic receptors. Interestingly, the PCA treatment, while reducing 5-HT uptake drastically, had no effect on 5-HT_2 receptors.[97] Chronic administration of mianserin decreased the number of 5-HT_2 sites (^{3}[H]spiperone binding) in rat cortex.[122] Even a single dose of this drug lowered 5-HT_2 binding.[100,122] Similarly, amoxapine and loxapine were able to reduce the density of 5-HT_2 receptors after both acute and chronic treatment.[100] Clozapine blocked the binding of ^{3}[H]spiperone to frontal cortex in intact mice as effectively as amitriptyline;[123] long-term administration of clozapine decreased the number of 5-HT_2 sites labeled by ^{3}[H]ketanserin.[124] Rats with lesioned 5-HT neurons did not show alterations in 5-HT_2 receptor density as measured by ^{3}[H]spiperone,[97] ^{3}[H]LSD[125] or ^{3}[H]ketanserin binding.[98] In contrast, chronic PCPA treatment increased 5-HT$_2$ receptor binding.[100] Chronic administration of imipramine decreased ^{3}[H]ketanserin binding and ^{3}[H]spiperone binding[21] without affecting the binding of ^{3}[H]mianserin,[98] which also labels 5-HT$_2$ receptors.[94] Repeated administration of mianserin itself did not down-regulate ^{3}[H]mianserin binding[61] nor did lesions of presynaptic 5-HT neurons.[93] In fact, up-regulation of these binding sites has been reported after either lesions or PCPA treatment.[61] Electroconvulsive shock, unlike antidepressant drugs, caused an increase in the number of 5-HT$_2$ binding sites labeled by ^{3}[H]spiperone in rat cortex.[126,127] Receptor affinity was unchanged as was 5-HT$_1$ binding.[128] Amitriptyline administered together with the α_2-antagonist, yohimbine, decreased 5-HT$_2$ receptor binding after 4 days compared to 10 days for amitriptyline alone.[129] Phenoxybenzamine and dihydroergotamine were also effective, whereas the specific α_1-antagonist, prazosin, was not. Very recently, it has been shown that subacute co-administration of imipramine and chlorpromazine produced a greater reduction of 5-HT$_2$ binding in rat cortex than either drug alone;[130] in contrast, this same drug combination had no effect on α_2- and β-receptor densities.[45]

Repeated administration of imipramine diminished ^{3}[H]imipramine binding in rat hippocampus but not in cerebellum or cortex; in the same animals, down-regulation of β-receptors was induced in the latter regions but not in the hippocampus.[131] The down-regulation of ^{3}[H]imipramine binding sites by chronic imipramine was accompanied by an increase in 5-HT reuptake in hippocampal slices.[61] It was proposed that chronic administration of imipramine and DMI may effect the desensitization/down-regulation of β-adrenoceptors by eventually facilitating 5-HT reuptake, thereby attenuating the inhibitory effects of 5-HT on NE neurons.[61] This diminished 5-HT function could also lead to the enhanced responses to 5-HT observed after chronic tricyclics.[132] The down-regulation of ^{3}[H]imipramine binding sites is supposedly a reduction in the effector sites for an "endogenous modulator" (inhibitory) of 5-HT reuptake.[61] It is notable that either deprivation of REM sleep or chronic ECS treatment can down-regulate ^{3}[H]imipramine sites as well as β-receptors.[39] However, iprindole and mianserin, which produce subsensitivity of NE receptors, do not down-regulate ^{3}[H]imipramine sites.[61,131]

Chronic tricyclic antidepressants, but not fluoxetine, femoxetine, or chlorpromazine, increased sensitivity of postsynaptic forebrain neurons to microionotophoretically-applied 5-HT.[132] Similar treatments did not affect the sensitivity of raphe 5-HT autoreceptors, as assessed by the effect of LSD on neuronal firing rates.[133] Electrophysiological studies following chronic zimelidine treatment revealed enhanced 5-HT neurotransmission on electrical stimulation of rat hippocampus but not an enhanced responsivity of hippocampal neurons to applied 5-HT.[134] These results are at variance with the notion that long-term zimelidine may attenuate 5-HT transmission[112] and with the observation that chronic antidepressants increase postsynaptic functional sensitivity.[59,132] Acute zimelidine inhibited the firing of raphe 5-HT neurons, but the effect became progressively less with continued drug administration. After 14 days of zimelidine, the firing rate returned to normal, but the response of raphe 5-HT autoreceptors to intravenous LSD was much less.[134] These findings suggest that the autoreceptors mediating the firing rate of 5-HT neurons become desensitized by continual exposure to this 5-HT uptake blocker. The postsynaptic 5-HT functional supersensitivity elicited by chronic clomipramine was not blocked by chronic lithium administration.[135]

Acute administration of amitriptyline, imipramine, trazodone, mianserin or viloxazine (but not DMI or iprindole) antagonized the "head-twitch" response to 5-methoxy-N,N-dimethyltryptamine (5-MeODMT) in mice. Chronic administration of the same agents, including DMI and iprindole but excluding mianserin or viloxazine, elicited enhanced responsiveness to this 5-HT agonist.[136] Recently, the head twitch response has been ascribed to 5-HT$_2$ receptor stimulation, whereas the 5-HT syndrome may involve activation of 5-HT$_1$ receptors.[137] Antagonism of the head twitch response by acute tricyclic drugs but enhanced responses after chronic treatments suggest that "receptors" blocked after one dose should be up-regulated by multiple doses. Instead, the effects of chronic tricyclic antidepressants on 5-HT$_2$ receptors (down-regulation) and the behavior presumably mediated by these receptors are opposite. As noted earlier, mianserin can down-regulate 5-HT$_2$ receptors after acute or chronic administration; both treatments also antagonized the head twitch response in mice.[122,136] The up-regulation of $5\text{-}HT_2$ receptors by multiple ECS treatments is accompanied by enhanced behavioral responses to 5-HT agonists.[126] Thus, the effects on $5\text{-}HT_2$ receptors seem to parallel a behavioral response for mianserin and ECS but not other antidepressant drugs. However, a recent investigation showed parallel increases in the 5-HT head twitch response and $5\text{-}HT_2$ receptor density following chronic DMI treatment.[138] Finally, another study reported that $5\text{-}HT_1$ receptor binding was unaltered, whereas $5\text{-}HT_2$ binding was decreased after chronic administration of clomipramine or fluoxetine. The 5-HT syndrome in the clomipramine and fluoxetine-treated rats from 5-MeODMT was also diminished.[111] This result supports other findings that postsynaptic activity appears to be attenuated following long-term administration of 5-HT uptake blockers,[112] but the behavioral subsensitivity to 5-HT agonists is in contrast to the supersensitivity to microiontophoretically-applied 5-HT in chronic clomipramine treated rats.[135] Amitriptyline, trazodone and other agents, in fact, may function as antagonists of 5-HT at the postsynaptic receptor.[139-141] Chronic treatment with zimelidine induces a lower affinity $5\text{-}HT_1$ site in both rat cortex and hypothalamus, in concert with the persistent reduction of 5-HT synthesis. These effects might result in reduced presynaptic and postsynaptic activity.[112,142] A similar decrease in 5-HT transmission after chronic 5-HT uptake blockers could be responsible for their ability to desensitize and down-regulate β-adrenoceptors.[34] Other investigators have proposed that antidepressant drugs alter the structure of the postsynaptic 5-HT receptor to give a higher affinity 5-HT binding form which is inactive, i.e., no longer coupled to a 5-HT sensitive adenylate cyclase.[143] The relevance of these findings to the action of antidepressant drugs is obscure, since MAO inhibitors were ineffective.

Dopamine Receptors – All antipsychotic drugs (neuroleptics) are known to block DA receptors. The number of receptor subtypes claimed for DA, based on binding studies, include "D-1, D-2, D-3 and D-4".[144,145] However, only the postsynaptic D-2 receptors (labeled by ^{3}[H]butyrophenones) are most consistently relatable to the effects of drugs. The D-1 subtype, a DA-stimulated adenylate cyclase site, has no clearcut function in brain.[146] Recent studies on agonist binding disclosed that ^{3}[H]DA binds to two sites, both postsynaptic; one is related to D-2 sites and the other to D-1 sites.[147]

Changes in DA receptors are of particular relevance to three major types of CNS disorders, schizophrenia, tardive dyskinesia and Parkinson's disease. Strong evidence points to the involvement of the DA system in many of the clinical manifestations of schizophrenia. Several studies have been conducted to determine whether this disease ultimately causes alterations in DA receptors. In one investigation, ^{3}[H]spiperone binding in 19 postmortem brains from schizophrenic patients showed an increase in the density of postsynaptic DA receptors (D-2) versus a control group.[148] A concurrent increase in K_D of ^{3}[H]spiperone binding was attributed to residual neuroleptics. Two patients had never received antipsychotic drugs, and five others had been drug-free for at least a year prior to death (this latter group had normal K_D values). Thus, up-regulation of postsynaptic receptors in the caudate nucleus, putamen and nucleus accumbens may be an aspect of the disease process. Another study gave similar values, but it was difficult to rule out supersensitivity induced by prior neuroleptic treatment.[149] In Parkinson's disease, DA receptor supersensitivity is postulated as a compensatory response to the degeneration of nigrostriatal DA neurons. Examination of postmortem brains of Parkinson patients revealed an increase in ^{3}[H]haloperidol binding in the putamen and a similar trend in the caudate.[150] In the brains of such patients on long term L-dopa therapy, ^{3}[H]haloperidol binding was normal. L-Dopa thus normalizes the DA system. It was proposed that deliberate up-regulation by decreasing or withdrawing L-dopa treatment may be beneficial to patients who become refractory to this drug due to the development of subsensitivity.[150,151] The more disabled Parkinson patients showed a decreased number of striatal D-2 receptors and loss of therapeutic response to L-dopa.[152] In Huntington's disease, postmortem brain tissue from patients showed a significant decrease in the number of ^{3}[H]spiperone D-2 sites (unchanged K_D) in the putamen, caudate nucleus, and frontal cortex.[153] This diminished dopaminergic function in the basal ganglia and frontal cortex may be a cause of the motor and mental disorders, respectively, in Huntington's disease.

Many animal studies have been carried out to determine the effects of antipsychotic drugs on DA receptors.[7,144,145] Rats treated chronically with haloperidol exhibited increases in ^{3}[H]haloperidol binding in the striatum and mesolimbic region.[154] Also, ^{3}[H]apomorphine binding was increased in the aforementioned brain areas. Enhanced ^{3}[H]haloperidol binding after chronic haloperidol, fluphenazine and reserpine was reflected by an increase in receptor density, not in affinity.[155] In accord with the greater number of D-2 receptors, an augmented behavioral response to apomorphine,[156,157] amphetamine[157] or tranylcypromine + L-dopa[158] was observed. In general, chronic administration of neuroleptics alters the number of D-2 receptors; except for thioxanthenes, the number of D-1 sites, i.e., DA sensitive adenylate cyclase activity, is unchanged.[156-158] The thioxanthene, cis-flupenthixol, which exhibits similar affinities for D-1 and D-2 receptors, up-regulated D-1 receptors after chronic administration.[156,159] Longer treatment (18 months) increased the number of D-2 sites without affecting that of D-1 sites (labeled by ^{3}[H]piflutixol) despite increased DA-sensitive adenylate cyclase activity.[160,161] Receptor plasticity was demonstrated by the fact that withdrawal after 18 months of cis-flupenthixol treatment resulted in the return of enhanced apomorphine-induced stereotypy to normal and elevated B_{max} for ^{3}[H]spiperone binding to control values after 1 month and 3 months, respectively. However, DA-stimulated adenylate cyclase activity remained elevated for 6 months, returning to normal activity 1 year following drug withdrawal.[161] Similar effects were found upon continual administration of trifluoperazine (1 year). This drug initially acted as a DA antagonist, but behavioral supersensitivity to high dose apomorphine, increased density of D-2 receptors and enhanced DA-stimulated adenylate cyclase activity was observed at 6 and 12 months and persisted for 3 months after drug withdrawal.[160] These receptor sensitivity changes appeared to parallel the development of drug-induced human tardive dyskinesia. Administration of haloperidol, trifluoperazine, fluphenazine or pifluthixol for 6-9 months resulted in an enduring functional blockade of DA receptors, as measured by low dose apomorphine elicited responses; D-2 binding was elevated.[162] Other investigators have reported that neuroleptic-induced DA receptor supersensitivity can be reversed by chronic L-dopa administration, presumably by increasing the concentration of DA, which in turn down-regulates the receptors.[151,163] Concurrent chronic treatment of rats with haloperidol and lithium did not lead to the development of DA receptor supersensitivity.[164] This prophylactic effect of lithium in preventing up-regulation of the DA receptors may be involved in its antimanic properties.

Studies of receptor regulation have also been useful in attempts to understand the mechanism of action of antipsychotic drugs. In monkeys, it was found that after chronic haloperidol treatment, DA receptor sensitivity to a challenge dose of haloperidol is maintained in the frontal and cingulate cortex, whereas tolerance is seen in the putamen and caudate.[165,166] Thus, these cortical regions may represent the site of therapeutic action of antipsychotic drugs. Furthermore, the unique effects of neuroleptics on the mesocortical DA neurons may be caused by the lack of autoreceptors in this brain region.[167,168] In electrophysiological studies, DA autoreceptors (substantia nigra) were found to be 6 to 10 times more sensitive to DA agonists than postsynaptic receptors (caudate nucleus); therefore, low doses of agonists may act preferentially at autoreceptors to inhibit DA release.[169] Thus, selective autoreceptor agonists could be effective antipsychotic agents; the development of tolerance may be a limiting factor.[170,171] Conflicting results have been reported regarding the effects of chronic antidepressant administration on DA receptors: imipramine and iprindole induced subsensitivity of autoreceptors,[172] but efforts to repeat these findings have failed.[173-175] Also, some investigators have observed a decrease in D-2 binding following tricyclic antidepressant treatment,[20] whereas others have claimed no change.[21]

Conclusion – Recent years have witnessed an increased interest in the study of receptor plasticity in the CNS following drug administration or in relation to disease processes. Generally, stimulation of receptors with agonists causes desensitization/down-regulation; antagonists produce the opposite effects. The NE system seems to exhibit the best correlation between therapeutic drug response and receptor alteration. Many contradictions in the literature remain to be explained. A better understanding of the functional significance of receptor subtypes would help answer many questions regarding adaptive mechanisms.

References

1. Y.-F. Su, T. K. Harden and J. P. Perkins, *J. Biol. Chem.*, **255**, 7410 (1980).
2. V. Homburger, M. Lucas, B. Cantu, J. Barabe, J. Perrit and J. Bockaert, *J. Biol. Chem.*, **255**, 10436 (1980).
3. M. D. Dibner and P. B. Molinoff, *J. Pharmacol. Exp. Ther.*, **210**, 433 (1979).
4. D. C. U'Prichard and S. J. Enna, *Eur. J. Pharmacol.*, **59**, 297 (1979).
5. J. Vetulani, R. J. Stawarz and F. Sulser, *J. Neurochem.*, **27**, 661 (1976).
6. B. B. Wolfe, T. K. Harden, J. R. Sporn and P. B. Molinoff, *J. Pharmacol. Exp. Ther.*, **207**, 446 (1978).
7. I. Creese and D. R. Sibley, *Ann. Rev. Pharmacol. Toxicol.*, **21**, 357 (1981).
8. M. F. Sugrue, *Pharmacol. Ther.*, **21**, 1 (1983).
9. G. N. Pandley and J. M. Davis, *Drug Dev. Res.*, **3**, 393 (1983).
10. J. L. Anderson, *Life Sci.*, **32**, 1791 (1983).
11. J. Vetulani, R. J. Stawarz, J. V. Dingell and F. Sulser, *Naunyn-Schmiedeberg's Arch. Pharmacol.*, **293**, 109 (1976).
12. S. P. Banerjee, L. S. Kung, S. J. Riggi and S. K. Chandra, *Nature*, **268**, 455 (1977).
13. F. Sulser, A. J. Janowsky, F. Okada, D. H. Manier and P. L. Mobley, *Neuropharmacology*, **22**, 425 (1983).
14. F. Sulser, *Psychopharmacol. Bull.*, **19**, 300 (1983).
15. K. J. Kellar, C. S. Cascio, D. A. Bergstrom, J. A. Butler and P. Iadarola, *J. Neurochem.*, **37**, 830 (1981).
16. G. N. Pandley, W. J. Heinze, B. D. Brown and J. M. Davis, *Nature*, **280**, 234 (1979).
17. D. D. Gillespie, D. H. Manier and F. Sulser, *Commun. Psychopharmacol.*, **3**, 191 (1979).
18. J. Schultz, *Nature*, **261**, 417 (1976).
19. A. Maggi, D. C. U'Prichard and S. J. Enna, *Eur. J. Pharmacol.*, **61**, **91** (1980).
20. T. Koide and H. Matsushita, *Neuropharmacology*, **20**, 285 (1981).
21. S. Peroutka and S. H. Snyder, *Science*, **210**, 88 (1980).
22. D. A. Bergstrom and K. J. Kellar, *J. Pharmacol. Exp. Ther.*, **209**, 256 (1979).
23. M. M. Sellinger-Barnette, J. Mendels and A. Frazer, *Neuropharmacology*, **19**, 447 (1980).
24. V. H. Sethy and D. W. Harris, *Eur. J. Pharmacol.*, **75**, 53 (1981).
25. R. Mishra, N. J. Leith, L. Steranka and F. Sulser, *Naunyn-Schmiedeberg's Arch. Pharmacol.*, **316**, 218 (1981).
26. R. Mishra and F. Sulser, *Commun. Psychopharmacol.*, **2**, 365 (1978).
27. C. Kopanski, M. Türck and J. E. Schultz, *Neurochem. Int.*, **5**, 649 (1983).
28. S. Clements-Jewery, *Neuropharmacology*, **17**, 779 (1978).
29. R. Mishra, D. D. Gillespie, R. A. Lovell, R. D. Robson and F. Sulser, *Life Sci.*, **30**, 1747 (1982).
30. R. Mishra, A. Janowsky and F. Sulser, *Eur. J. Pharmacol.*, **60**, 279 (1979).
31. R. Mishra, A. Janowsky and F. Sulser, *Neuropharmacology*, **19**, 983 (1980).
32. O. Gandolfi, M. L. Barbaccia, D. M. Chuang and E. Costa, *Neuropharmacology*, **22**, 927 (1983).
33. V. Claasen, *Brit. J. Clin. Pharmacol.*, **15**, 349S (1983).
34. B. K. Koe, A. Weissman, W. M. Welch and R. G. Browne, *J. Pharmacol. Exp. Ther.*, **226**, 686 (1983).
35. J. E. Rosenblatt, C. B. Pert, J. F. Tallman, A. Pert and W. E. Bunney, *Brain Res.*, **160**, 186 (1979).
36. S. Treiser and K. J. Kellar, *Eur. J. Pharmacol.*, **58**, 85 (1979).
37. R. Mishra, D. D. Gillespie, M. Youdim and F. Sulser, *Fed. Proc.*, **41**, 1066 (1982).
38. R. M. Cohen, R. P. Ebstein, J. W. Daly and D. L. Murphy, *J. Neurosci.*, **2**, 1588 (1982).
39. E. Mogilnicka, S. Arbilla, H. Depoortere and S. Z. Langer, *Eur. J. Pharmacol.*, **65**, 289 (1980).
40. N. L. Wiech and R. C. Ursillo, *Commun. Psychopharmacol.*, **4**, 95 (1980).
41. F. T. Crews, S. M. Paul and F. K. Goodwin, *Nature*, **290**, 787 (1981).
42. J. A. Scott and F. T. Crews, *J. Pharmacol. Exp. Ther.*, **224**, 640 (1983).
43. D. H. Manier, D. D. Gillespie and F. Sulser, *Naunyn-Schmiedeberg's Arch. Pharmacol.*, **313**, 113 (1980).
44. T. D. Reisine, D. C. U'Prichard, N. L. Wiech, R. C. Ursillo and H. I. Yamamura, *Brain Res.*, **188**, 587 (1980).
45. M. Mikuni, D. M. Stoff and H. Y. Meltzer, *Eur. J. Pharmacol.*, **89**, 313 (1983).
46. J. W. Lace and S. M. Antelman, *Brain Res.*, **278**, 359 (1983).
47. B. A. McMillen, W. Warnack, D. C. German and P. A. Shore, *Eur. J. Pharmacol.*, **61**, 239 (1980).
48. P. F. Von Voightlander, H. J. Triezenberg and E. G. Losey, *Neuropharmacology*, **17**, 375 (1978).
49. D. J. Heal, H. Akagi, J. M. Bowdler and A. R. Green, *Eur. J. Pharmacol.*, **75**, 231 (1981).
50. C. S. Aulakh, R. M. Cohen, C. McLellan and D. L. Murphy, *Brit. J. Pharmacol.*, **80**, 10 (1983).
51. F. Cerrito and M. Raiteri, *Eur. J. Pharmacol.*, **70**, 425 (1981).
52. M. F. Sugrue, *Eur. J. Pharmacol.*, **68**, 377 (1980).
53. L. J. Siever, R. M. Cohen and D. L. Murphy, *Am. J. Psychiatry*, **138**, 681 (1981).
54. C. B. Smith, J. A. Garcia-Sevilla and P. J. Hollingsworth, *Brain Res.*, **210**, 413 (1981).
55. R. W. Johnson, T. Reisine, S. Spotnitz, N. Wiech, R. Ursillo and H. I. Yamamura, *Eur. J. Pharmacol.*, **67**, 123 (1980).
56. M. Asakura, T. Tsukamoto and K. Hasegawa, *Brain Res.*, **235**, 192 (1982).
57. H.-R. Olpe and A. Schellenberg, *Eur. J. Pharmacol.*, **63**, 7 (1980).
58. J. E. Schultz, G. R. Siggins, F. W. Schocker, M. Türek and F. E. Bloom, *J. Pharmacol. Exp. Ther.*, **216**, 28 (1981).

59. D. B. Menkes, G. K. Aghajanian and R. B. McCall, *Life Sci.*, **27**, 45 (1980).
60. D. B. Menkes, M. M. Rasenick, M. A. Wheeler and M. W. Bitensky, *Science*, **219**, 65 (1983).
61. M. L. Barbaccia, N. Brunello, D. M. Chuang and E. Costa, *Neuropharmacology*, **22**, 373 (1983).
62. A. Janowsky, F. Okada, D. H. Manier, C. D. Applegate and F. Sulser, *Science*, **218**, 900 (1982).
63. N. Brunello, M. L. Barbaccia, D. M. Chuang and E. Costa, *Neuropharmacology*, **21**, 1145 (1982).
64. D. H. Manier, F.Okada, A. J. Janowsky, L. R. Steranka and F. Sulser, *Eur. J. Pharmacol.*, **86**, 137 (1983).
65. B. Shopsin, S. Gershon, M. Goldstein, E. Friedman and S. Wilk, *Psychopharmacol. Commun.*, **1**, 239 (1975).
66. B. Shopsin, E. Friedman and S. Gershon, *Arch. Gen. Psychiatry*, **33**, 811 (1976).
67. Y. Lecrubier, A. J. Puech, R. Jouvent, P. Simon and D. Widlocher, *Br. J. Psychiatry*, **136**, 354 (1980).
68. B. Lehrer, R. P. Ebstein and R. H. Belmaker, *Psychopharmacology*, **75**, 169 (1981).
69. D. J. Dooley and K. L. Hauser, *Neurosci. Lett.*, **36**, 93 (1983).
70. A. Dolphin, M. C. B. Sawaya, P. Jenner and C. D. Marsden, *Naunyn-Schmiedeberg's Arch. Pharmacol.*, **299**, 167 (1977).
71. W. M. Petrie, R. J. Maffucci and R. L. Woosley, *Am. J. Psychiatry*, **139**, 92 (1982).
72. C. B. Nemeroff and D. L. Evans, *Biol. Psychiatry*, **18**, 237 (1983).
73. G. C. Palmer and H. R. Wagner, *Res. Commun. Psychol. Psychiat. Behav.*, **1**, 567 (1976).
74. J. R. Sporn, B. B. Wolfe, T. K. Harden and P. B. Molinoff, *Mol. Pharmacol.*, **13**, 1170 (1977).
75. D. J. Dooley, H. Bittiger, K. L. Hauser, S. F. Bischoff and P. C. Waldmeier, *Neuroscience*, **9**, 889 (1983).
76. A. J. Janowsky, L. R. Steranka, D. D. Gillespie and F. Sulser, *J. Neurochem.*, **39**, 290 (1982).
77. K. P. Minneman, M. D. Dibner, B. B. Wolfe and P. B. Molinoff, *Science*, **204**, 866 (1979).
78. P. L. Mobley and F. Sulser, *Eur. J. Pharmacol.*, **60**, 221 (1979).
79. W. E. Heydorn, D. J. Brunswick and A. Frazer, *J. Pharmacol. Exp. Ther.*, **222**, 534 (1982).
80. P. J. Cowen, S. Fraser, D. G. Grahame-Smith, A. R. Greene and C. Stanford, *Br. J. Pharmacol.*, **78**, 89 (1983).
81. C. Thompson, S. A. Checkley, T. Corn, C. Franey and J. Arendt, *The Lancet*, 1101 (1983).
82. G. M. Vaughan, R. W. Pelham, S. F. Pang, L. L. Loughlin, K. M. Wilson, K. L. Sandrock, M. K. Vaughan, S. H. Koslow and R. J. Heiter, *J. Clin. Endocrinol. Metab.*, **42**, 752 (1976).
83. E. A. Stone, *J. Neurochem.*, **32**, 1335 (1979).
84. E. A. Stone, *Res. Commun. Psychol. Psychiat. Behav.*, **4**, 241 (1979).
85. K. P. Minneman, B. B. Wolfe and P. B. Molinoff, *Brain Res.*, **252**, 309 (1982).
86. C. Stanford, D. J. Nutt and P. J. Cowen, *Neuroscience*, **8**, 1161 (1983).
87. R. Dunstan and D. M. Jackson, *Psychopharmacology*, **48**, 105 (1976).
88. M. R. Goldberg, P. W. Curatolo, C.-S. Tung and D. Robertson, *Neurosci. Lett.*, **31**, 47 (1982).
89. P. R. Lowenstein, M. I. Vacas and D. P. Cardinali, *Neuropharmacology*, **21**, 243 (1982).
90. D. L. Nelson, A. Herbet, S. Bourgoin, J. Glowinski and M. Hamon, *Mol. Pharmacol.*, **14**, 983 (1978).
91. J. E. Leysen and J. P. Tollenaere, *Ann. Rep. Med. Chem.*, **17**, 1 (1982).
92. J. E. Leysen, C. J. E. Niemegeers, J. M. Van Neuten and P. Laduron, *Mol. Pharmacol.*, **21**, 301 (1982).
93. A. Dumbrille-Ross, S. W. Tang and D. V. Coscina, *Life Sci.*, **29**, 2049 (1981).
94. S. J. Peroutka and S. H. Snyder, *J. Pharmacol. Exp. Ther.*, **216**, 142 (1981).
95. D. L. Nelson, A. Herbet, A. Enjalbert, J. Bockaert and M. Hamon, *Biochem. Pharmacol.*, **29**, 2445 (1980).
96. M. L. Barbaccia, N. Brunello, D.-M. Chuang and E. Costa, *J. Neurochem.*, **40**, 1671 (1983).
97. S. Clements-Jewery and P. A. Robson, *Neuropharmacology*, **21**, 725 (1982).
98. M. L. Barbaccia, O. Gandolfi, D.-M. Chuang and E. Costa, *Neuropharmacology*, **22**, 123 (1983).
99. D. D. Savage, J. Mendels and A. Frazer, *Neuropharmacology*, **19**, 1063 (1980).
100. D. M. Helmeste and S. W. Tang, *Life Sci.*, **33**, 2527 (1983).
101. S. Z. Langer and P. M. Laduron, *Trends Pharmacol. Sci.*, **5**, 51 (1984).
102. Stanley and J. J. Mann, *The Lancet*, 214 (1983).
103. L. R. Meyerson, L. P. Wennogle, M. S. Abel, J. Coupet, A. S. Lippa, C. E. Rauh and B. Beer, *Pharmacol. Biochem. Behav.*, **17**, 159 (1982).
104. M. Stanley, V. J. Virgilo and S. Gershon, *Science*, **216**, 1337 (1982).
105. H. Y. Meltzer, B. Uberkoman-Wiita, A. Robertson, B. J. Tricou and M. Lowy, *Life Sci.*, **33**, 2541 (1983).
106. F. K. Goodwin and R. M. Post, *Br. J. Clin. Pharmacol.*, **15**, 393S (1983).
107. D. D. Savage, J. Mendels and A. Frazer, *J. Pharmacol. Exp. Ther.*, **212**, 259 (1980).
108. I. Lucki and A. Frazer, *Psychopharmacology*, **77**, 205 (1982).
109. H.-R. Olpe, *Eur. J. Pharmacol.*, **69**, 375 (1981).
110. A. Wirz-Justice, K. Krauchi, M. Lichtsteiner and H. Feer, *Life Sci.*, **23**, 1249 (1978).
111. J. F. Stolz, C. A. Marsden and D. N. Middlemiss, *Psychopharmacology*, **80**, 150 (1983).
112. K. Fuxe, S.-O. Ögren, L. F. Agnati, F. Benfenati, B. Fredholm, K. Andersson, I. Zini and P. Eneroth, *Neuropharmacology*, **22**, 389 (1983).
113. T. Segawa, T. Mizuta and Y. Nomura, *Eur. J. Pharmacol.*, **58**, 75 (1979).
114. R. Samanin, T. Mennini, A. Ferraris, C. Bendotti and F. Borsini, *Brain Res.*, **189**, 449 (1980).
115. R. Samanin, T. Mennini, A. Ferraris, C. Bendotti and R. Borsini, *Eur. J. Pharmacol.*, **61**, 203 (1980).

116. E. C. Hwang, J. Magnussen and M. H. Van Woert, *Res. Comm. Chem. Pathol. Pharmacol.*, **29**, 79 (1980).
117. D. T. Wong and F. P. Bymaster, *Res. Commun. Chem. Pathol. Pharmacol.*, **32**, 41 (1981).
118. S. B. Ross, H. Hall, A. L. Renyi and D. Westerlund, *Psychopharmacology*, **72**, 219 (1981).
119. S. Treiser and K. J. Kellar, *Eur. J. Pharmacol.*, **64**, 183 (1980).
120. T. Segawa and M. Uehara, *Life Sci.*, **30**, 809 (1982)
121. S. J. Peroutka and S. H. Snyder, *J. Pharmacol. Exp. Ther.*, **215**, 582 (1980).
122. M. A. Blackshear and E. Sanders-Bush, *J. Pharmacol. Exp. Ther.*, **221**, 303 (1982).
123. S. Clements-Jewery and P. A. Robson, *Neuropharmacology*, **19**, 657 (1980).
124. G. P. Reynolds, N. J. Garrett, N. Rupniak, P. Jenner and C. D. Marsden, *Eur. J. Pharmacol.*, **89**, 325 (1983).
125. M. A. Blackshear, L. R. Steranka and E. Sanders-Bush, *Eur. J. Pharmacol.*, **76**, 325 (1981).
126. K. J. Kellar, C. S. Cascio, J. A. Butler and R. N. Kurtzke, *Eur. J. Pharmacol.*, **69**, 515 (1981).
137. J. Vetulani, U. Lebrecht and A. Pilc, *Eur. J. Pharmacol.*, **76**, 81 (1981).
128. K. J. Kellar and D. A. Bergstrom, *Neuropharmacology*, **22**, 401 (1983).
129. F. T. Crews, J. A. Scott and N. H. Shorstein, *Neuropharmacology*, **22**, 1203 (1983).
130. M. Mikuni and H. Y. Meltzer, *Life Sci.*, **34**, 87 (1984).
131. W. J. Kinnier, D.-M. Chuang and E. Costa, *Eur. J. Pharmacol.*, **67**, 289 (1980).
132. C. De Montigny and G. K. Aghajanian, *Science*, **202**, 1303 (1978).
133. P. Blier and C. De Montigny, *Naunyn-Schmiedeberg's Arch. Pharmacol.*, **314**, 123 (1980).
134. P. Blier and C. De Montigny, *J. Neurosci.*, **3**, 1270 (1983).
135. D. W. Gallager and W. E. Bunney, Jr., *Naunyn-Schmiedeberg's Arch. Pharmacol.*, **307**, 129 (1979).
136. E. Friedman, T. B. Cooper and A. Dallob, *Eur. J. Pharmacol.*, **89**, 69 (1983).
137. I. Lucki, M. S. Nobler and A. Frazer, *J. Pharmacol. Exp. Ther.*, **228**, 133 (1984).
138. A. R. Greene, D. J. Heal, P. Johnson, B. E. Lawrence and V. L. Nimgaonkar, *Br. J. Pharmacol.*, **80**, 377 (1983).
139. K. Fuxe, S.-O. Ögren, L. Agnati, J. Å. Gustafsson and G. Jonsson, *Neurosci. Lett.*, **6**, 339 (1977).
140. H. Nagayama, J. N. Hingtgen and M. H. Aprison, *Pharmacol. Biochem. Behav.*, **13**, 575 (1980).
141. J. N. Hingtgen, H. C. Hendric and M. H. Aprison, *ibid.*, in press (1984).
142. K. Fuxe, S.-O. Ögren and L. F. Agnati, *Neurosci. Lett.*, **13**, 307 (1979).
143. G. Fillion and M. P. Fillion, *Nature*, **292**, 349 (1981).
144. P. Seeman, *Pharmacol. Rev.*, **32**, 229 (1980).
145. I. Creese, M. W. Hamblin, S. E. Leff and D. R. Sibley, in "Handbook of Psychopharmacology", Vol. 17, L. L. Iversen, S. D. Iversen and S. H. Snyder, Ed., Plenum Press, New York, NY, 1983, p. 81.
146. B. Costall and R. J. Naylor, *Life Sci.*, **28**, 215 (1981).
147. N. G. Bacopoulos, *Life Sci.*, **34**, 307 (1984).
148. F. Owen, T. J. Crow, M. Poulter, A. J. Cross, A. Longden and G. J. Riley, *The Lancet*, 223 (1978).
149. T. Lee, P. Seeman, W. W. Tourtello, I. J. Farley and O. Hornykeiwicz, *Nature*, **274**, 897 (1978).
150. T. Lee, P. Seeman, A. Rajput, I. J. Farley and O. Hornykeiwicz, *Nature*, **273**, 59 (1978).
151. A. J. Friedhof and J. C. Miller, *Ann. Rev. Neurosci.*, **6**, 121 (1983).
152. U. K. Rinne, J. O. Rinne, J. K. Rinne, K. Laakso, A. Laihinen and P. Lonnberg, *J. Neural Transm. (Suppl.)*, *18*, 279 (1983).
153. T. D. Reisine, J. Z. Fields, L. Z. Stern, P. C. Johnson, E. D. Bird and H. I. Yamamura, *Life Sci.*, **21**, 1123 (1977).
154. P. Muller and P. Seeman, *Life Sci.*, **21**, 1751 (1977).
155. D. R. Burt, I. Creese and S. H. Snyder, *Science*, **196**, 326 (1977).
156. S. Fleminger, N. M. J. Rupniak, M. D. Hall, P. Jenner and C. D. Marsden, *Biochem. Pharmacol.*, **32**, 2921 (1983).
157. P. von Voigtlander, E. G. Losey and H. Triezenberg, *J. Pharmacol. Exp. Ther.*, **193**, 88 (1975).
158. D. J. Heal, A. R. Green, D. J. Boullin and D. G. Grahame-Smith, *Psychopharmacology*, **49**, 287 (1976).
159. J. Hyttel and A. V. Christensen, *J. Neural Transm. (Suppl.)*, **18**, 157 (1983).
160. N. M. J. Rupniak, P. Jenner and C. D. Marsden, *Life Sci.*, **32**, 2289 (1983).
161. K. Murugaiah, S. Fleminger, A. Theodorou, P. Jenner and C. D. Marsden, *Biochem. Pharmacol.*, **32**, 2495 (1983).
162. J. L. Waddington, A. J. Cross, S. J. Gamble and R. C. Bourne, *Adv. Biochem. Psychopharmacol.*, **37**, 299 (1983).
163. A. J. Friedhoff, K. Bonnet and H. Rosengarten, *Res. Commun. Chem. Pathol. Pharmacol.*, **16**, 411 (1977).
164. A. Pert, J. E. Rosenblatt, C. Sivit, C. B. Pert and W. E. Bunney, Jr., *Science*, **201**, 171 (1978).
165. N. G. Bacopoulos, G. Bustos, D. E. Redmond, J. Baulu and R. H. Roth, *Brain Res.*, **157**, 396 (1978).
166. N. G. Bacopoulos, D. E. Redmond, J. Baulu and R. H. Roth, *J. Pharmacol. Exp. Ther.*, **212**, 1 (1980).
167. M. J. Bannon, J. F. Reinhard, Jr., E. B. Bunney and R. H. Roth, *Nature*, **296**, 44 (1982).
168. M. J. Bannon, R. L. Michaud and R. H. Roth, *Mol. Pharmacol.*, **19**, 270 (1980).
169. L. R. Skirboll, A. A. Grace and B. S. Bunney, *Science*, **206**, 80 (1979).
170. H. Y. Meltzer, *Pharmacol. Biochem. Behav.*, **17**, Suppl. 1, 1 (1982).
171. H. Y. Meltzer, T. Kolakowska, A. Robertson and B. J. Tricou, *Psychopharmacology*, **81**, 37 (1983).
172. L. A. Chiodo and S. M. Antelman, *Nature*, **287**, 451 (1980).
173. J. Welch, H. Kim, S. Fallon, J. Liebman, D. A. McNeill and M. Gower, *Nature*, **298**, 301 (1982).
174. H. H. Holcomb, M. J. Bannon and R. H. Roth, *Eur. J. Pharmacol.*, **82**, 173 (1982).
175. I. Creese, R. Kuczenski and D. Segal, *Pharmacol. Biochem. Behav.*, **17**, 375 (1982).

Chapter 6. CNS Autoreceptors as Targets for Drug Design

Pieter B. M. W. M. Timmermans, Department of Pharmacy,
Division of Pharmacotherapy, University of Amsterdam,
Plantage Muidergracht 24, 1018 TV Amsterdam, The Netherlands

<u>Introduction</u> - Several lines of evidence strongly suggest that presynaptic receptors are of physiological importance. In contrast to somadendritic receptors, which primarily modulate impulse generation, presynaptic receptors primarily modulate synthesis and release of transmitter. Presynaptic receptors are located on, in or near the axon terminals of many neurons. Activation results in facilitation or inhibition of transmitter release/synthesis, whereas blockade generally induces the opposite of activation.[1-3] In view of the ubiquity of presynaptic receptors at central and peripheral sites,[3] this Chapter focusses on the subset of presynaptic receptors termed autoreceptors, i.e. presynaptic receptors on a particular neuron which are activated by its own neurotransmitter. CNS autoreceptors have been postulated for, <u>inter alia</u>, norepinephrine (NE)/epinephrine, dopamine (DA), serotonin (5HT), acetylcholine (ACh), histamine (H) and γ-aminobutyric acid (GABA).

<u>CNS Autoreceptors for Norepinephrine (NE)</u> - Though the concept of autoinhibition of NE release is not universally accepted,[4,5] the consensus is that release of NE from sympathetic nerves is modulated through inhibitory α_2-autoreceptors. By contrast, somadendritic α_2-adrenoceptors of central noradrenergic neurons inhibit firing.[2,3,6] Release of ^{3}H-amezinium from rat cortical noradrenergic axons has been advanced as a model for the study of the α_2-autoreceptor hypothesis.[7] There are a number of mechanisms possible for the link between presynaptic α_2-adrenoceptor activation and transmitter release, in which a pivotal role for Ca^{2+} has been established.[3,8] More recent data implicates inhibition of adenylate cyclase with receptor activation and subsequent attenuation of transmitter release.[9,10]

The α-adrenoceptor agonists B-HT 920 (<u>1</u>) and 933 (<u>2</u>),[11,12] UK-14,304 (<u>3</u>),[11,13] M-7 (<u>4</u>),[14] TL-99 (<u>5</u>),[15] DP-6,7-ADTN (<u>6</u>),[16] ASL-7022 (<u>9</u>),[17] and some <u>exo</u>-isomers of 2-amino-6,7-dihydroxybenzonorbornene (<u>10</u>)[18] preferentially stimulate this type of α-adrenoceptor. The antagonists yohimbine and rauwolscine,[8] RS 21361 (11),[19] idazoxan (RX 781094 <u>12</u>)[20] and derivatives,[21] BDF 6143 (<u>13</u>),[22] Wy 26392 (<u>14</u>)

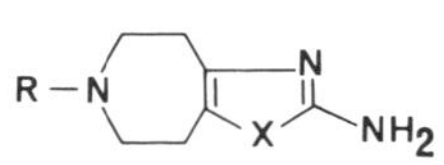

<u>1</u> R = CH_2=CH-CH_2-; X=S
<u>2</u> R = CH_3-CH_2-; X=O

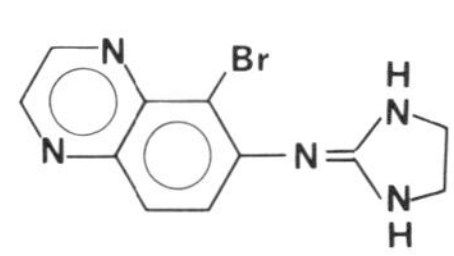

<u>3</u>

and Wy 26703 (<u>15</u>)[23] as well as SK&F 86466 (<u>16</u>)[24] block α_2-adrenoceptors selectively. The stereochemical requirements of α_2-adrenoceptors have been described in detail.[25].

ISBN 0-12-040519-9

Presynaptic α_2-adrenoceptors possibly represent only a minor fraction of the total α_2-adrenoceptor population in the CNS.[26] Although the general characteristics of presynaptic (auto) and postsynaptic α_2-adrenoceptors with respect to drug selectivity are very similar,[8] some findings suggest that in rat cerebral cortex the α_2-adrenoceptors for NE differ from the α_2-adrenoceptors located on serotonergic terminals.[27] Surprisingly, in rat frontal cortex α_1-adrenoceptors seem to modulate 5HT release.[28] The suggestion that different recognition sites on the presynaptic α_2-autoreceptor exist in

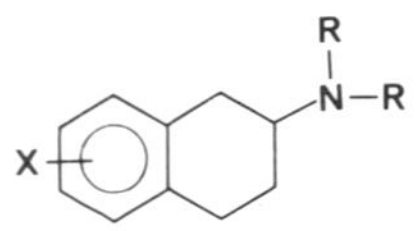

4 X = 5,6-di-OH; R = CH_3
5 X = 6,7-di-OH; R = CH_3
6 X = 6,7-di-OH; R = n-C_3H_7
7 X = 7-OH; R = n-C_3H_7
8 X = 8-OH; R = n-C_3H_7

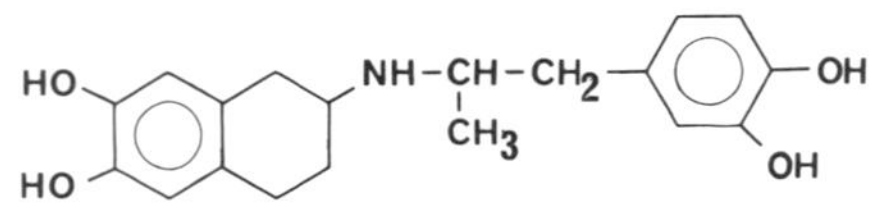

9

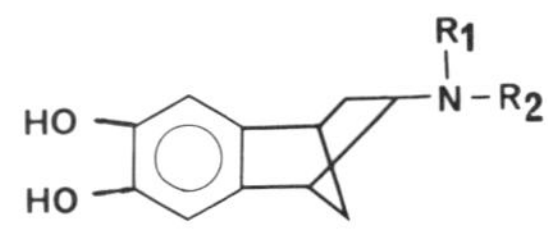

10 R_1R_2=H orCH_3

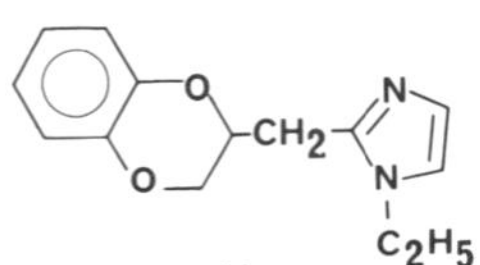

11

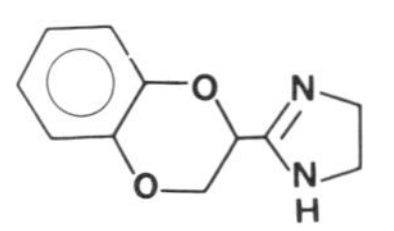

12

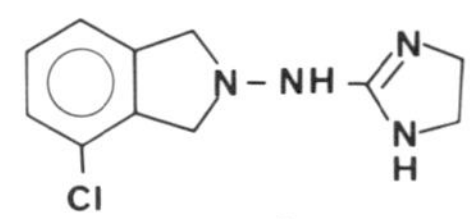

13

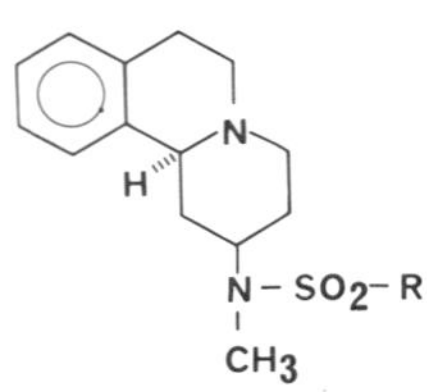

14 R = n-propyl

15 R = i-butyl

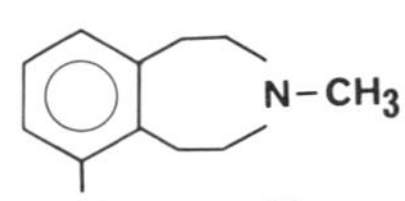

16

the CNS for imidazoli(di)nes and catecholamines,[29] has not found support.[30] Direct labeling of central α_2-autoreceptors has been unsuccessful.[31,32] The autoreceptors regulating NE release undergo sensitivity changes after long-term drug-induced activation or blockade: activation typically leads to a decrease in receptor number, whereas blockade produces the converse.[33]

<u>CNS Autoreceptors for Dopamine (DA)</u> - DA autoreceptors, probably located on dopaminergic nerve terminals, modulate calcium-dependent stimulation evoked release of DA *via* a negative feed-back mechanism,[2,34,35] as has been reported for the field-stimulated release of ^{3}H-DA from slices of the striatum/caudate nucleus of rats,[36] rabbits[37] and cats.[36,38] Presynaptic DA autoreceptors have also been implicated in the regulation of the synthesis of DA,[39] whereas somadendritic DA receptors inhibit the firing of the dopaminergic neurons.[40] In the rat, autoreceptor regulation of DA synthesis appears more significant in the mesolimbic as compared with the nigrostriatal DA pathway, while autoreceptors may be absent in the median eminence.[41]

The controversies in the classification of DA receptors have been reviewed.[42] The majority of current receptor binding data supports two major DA receptor classes for the CNS, D-1 and D-2, with the DA autoreceptor probably belonging to the latter.[42] However, the data is not sufficiently quantitative to assess conclusively whether more than one distinct DA autoreceptor is present on DA neurons.[42] Furthermore, in vivo studies with behavioral models of postsynaptic and neurochemical models of presynaptic dopaminergic activity have demonstrated that DA autoreceptors and postsynaptic DA receptors are diffierent.[38,43,44] The suggestion that the so-called D-3 binding site corresponds to presynaptic DA autoreceptors[45,46] has been refuted both on the basis of pharmacological and neurochemical evidence.[47,48]

Some agents have been proposed as selective presynaptic DA agonists, e.g. 3-PPP (17)[49,50], RSD-127 (18),[51] N,N-dipropyl-2-amino-7-hydroxytetralin (7)[52] and the selective α_2-adrenoceptor stimulants TL-99[53] and B-HT 920.[54] TL-99 and 7 inhibit the release of ^{3}H-DA from brain slices, but 3-PPP is inactive in this in vitro model for presynaptic DA activity.[47,55] The apparent DA autoreceptor selectivity of TL-99 as assessed in vivo may be due at least partially to its α_2-agonist activity.[56] It has been hypothesized that the presynaptic DA receptors mediating biochemical effects and those mediating effects on release are different.[48,52] Both isomers of 3-PPP are selective agonists for the autoreceptor at low doses. At higher doses postsynaptic activity is also observed, with the (+)- and (-)-enantiomers behaving as an agonist and antagonist, respectively.[57-60] For the more rigid analogs of 3-PPP, c7-OBQ (19) and t7-OBQ (20), the pre/post selectivity is lost for the trans compound (20), whereas the cis compound (19) is inactive.[61] Inhibition by d-amphetamine of the electrically-evoked release of ^{3}H-DA does not involve the activation of presynaptic inhibitory DA autoreceptors.[62,63]

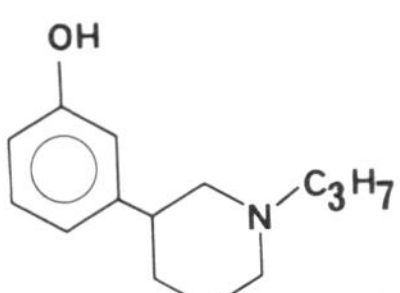

17

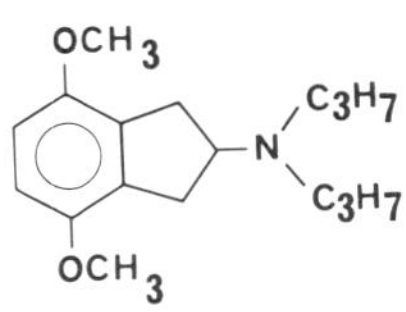

18

19 cis

20 trans

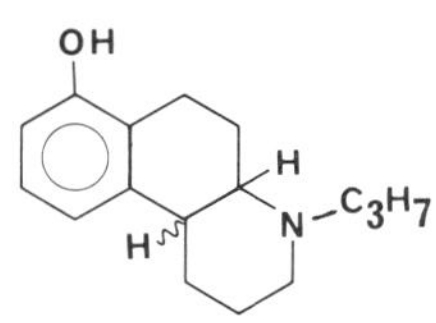

Effects of chronic neuroleptic administration on cerebral DA receptor function have been reviewed.[64] Following chronic haloperidol treatment, the DA autoreceptors modulating ^{3}H-DA release in the rabbit caudate develop supersensitivity to apomorphine[65] or to DA itself.[66]

CNS Autoreceptors for Serotonin (5-HT) - Various experimental findings imply that presynaptic autoreceptors modulate the release of 5HT from the corresponding neurons in rat brain.[67] In rat brain slices, or synaptosomes preincubated with ^{3}H-5HT, the Ca^{2+}-dependent stimulation-evoked release of the amine is inhibited by 5HT,[68,69] 5HT analogs,[70,71] and numerous 5HT receptor agonists.[72] Metitepine, metergoline and quipazine are competitive antagonists at the 5HT autoreceptor.[68-73] The 5HT autoreceptors in rat cortical tissue may be different from the presynaptic excitatory 5HT M-receptors[74] on sympathetic nerve endings of the rabbit heart,[70] and from the somadendritic 5HT receptors in rat raphe nuclei[75,76] which mediate an inhibition of neuronal-activity.[70] Rank orders of 5HT agonists indicate that 5HT autoreceptors, inhibitory presynaptic 5HT receptors on sympathetic nerve endings of canine saphenous vein, and the binding sites of ^{3}H-5HT in rat brain, designated as

$5HT_1$ sites, are similar,[70,71,77] but relative blocking potencies of antagonists show that they are not identical.[69,70,72,78] In fact, $5HT_1$ binding sites in rat brain cortex are not homogenous, since various compounds produce biphasic or shallow displacement curves.[71,79,80] This may be one of the reasons why correlations between binding affinity for $5HT_1$ sites and functional activity are frequently difficult to interpret.[81] Two subtypes of the ^{3}H-5HT recognition site, designated $5HT_{1A}$ (high affinity site) and $5HT_{1B}$ (low affinity site), have been recognized from displacement studies with spiperone, which exhibits a 3000-fold difference in affinity between the two sites.[80] Both sites are differentially distributed in rat brain.[82] A subset of the ^{3}H-5HT binding sites (low affinity, rather than high affinity sites, for antagonists) may be localized on the serotonergic neurons and may correspond to the 5HT autoreceptors.[71] The 5HT agonist PAT (8) potently and selectively displaces ^{3}H-5HT from the purported $5HT_{1A}$ binding site in rat frontal cortex.[83] However, ^{3}H-PAT has been reported to bind to pre and postsynaptic 5HT sites in rat brain.[84] In the hippocampus, ^{3}H-PAT binding sites have been associated with (postsynaptic) $5HT_1$ sites, whereas in the striatum they exhibit some properties of 5HT autoreceptors.[84] The agonist activity of PAT on the 5HT autoreceptor in vitro appears rather limited,[85,86] and so the potent central effects of PAT on 5HT turnover are unlikely to be mediated by 5HT autoreceptors.[87] Additional selective $5HT_1$ (A or B) agonists/antagonists may provide further clarification of several discrepancies. The selective $5HT_1$ agonist RU 24969 (21) may be a possible candidate.[88]

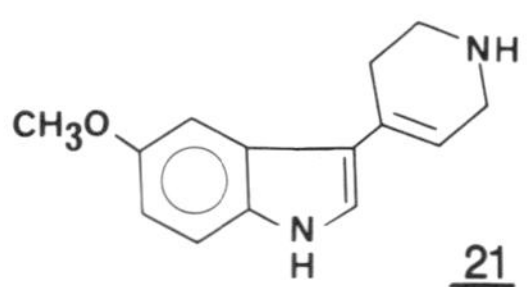

21

Citalopram, a selective inhibitor of 5HT uptake, does not affect per se the stimulation-evoked overflow of ^{3}H-5HT from rat hypothalamic slices, but antagonizes the inhibitory effect of LSD[89] and 5-methoxytryptamine[90] on ^{3}H-5HT release, indicating an interaction between the neuronal uptake mechanism and 5HT autoreceptors. On the other hand, stimulation or blockade of 5HT autoreceptors does not modulate the 5HT uptake process.[91] The autoreceptor regulating 5HT release undergoes sensitivity changes after chronic pharmacological activation or blockade.[33] However, long-term treatment with metitepine does not induce changes in the sensitivity or function of 5HT autoreceptors in rat hypothalamus.[92]

CNS Autoreceptors for Acetylcholine (ACh) - Proposals have been made for the existence of an inhibitory muscarinic feedback regulation of ACh in the CNS, and at certain locations in the periphery.[1,93,94] Evidence for a negative feedback mechanism involving ACh autoreceptors has been obtained in vitro in both brain slices and synaptosomes.[94-96] Cyclic GMP has been found to act as a second messenger for the muscarinic autoreceptor regulating ACh release.[97,98] Rat hippocampal muscarinic autoreceptors become super or subsensitive following chronic in vivo treatment with the muscarinic antagonist, scopolamine, or the cholinesterase inhibitor, paraoxon, respectively.[99] The inhibitory muscarinic autoreceptors in synaptosomes of rat striatal nerve terminals have a markedly lower release modulating capacity than autoreceptors in cortical or hippocampal nerve endings.[100] Since autoregulation of ACh release through muscarinic receptors occurs as effectively in the striatum as in the cortex or hippocampus, it has been suggested that in the former area the autoregulation of ACh release may not necessarily require the activation of autoreceptors on cholinergic nerve terminals.[100] Unlike the periphery,[101] the autoreceptors for ACh in the CNS seem to be different from postsynaptic muscarinic receptors.[94]

CNS Autoreceptors for Histamine (H) - The first experimental evidence for histamine autoreceptors in rat brain has been presented recently.[102] ^{3}H-Histamine can be released in a calcium-dependent manner with K^+ or veratridine from pre-labeled pools in slices or synaptosomes from rat cerebral cortex. Histamine, as well as its N-methyl and N,N'-dimethyl congeners, produces up to 60% inhibition. Other H_1- or H_2-agonists, like 2-methyl and 4-methylhistamine and dimaprit, were ineffective. The partial H_2-receptor agonist impromidine also lacked effect, but competitively antagonized the action of histamine. The H_1-receptor antagonists mepyramine, cyclizine and chlorpheniramine were found inactive. The H_2-receptor blocker burimamide was active, but tiotidine was without effect. The relative potencies of agonists and antagonists led the authors to suggest that autoinhibition of histamine release in rat brain is mediated by an as yet unidentified class of histamine receptor, for which they propose the designation H_3.[102]

CNS Autoreceptors for GABA - Studies *in vitro* have demonstrated that the release of GABA is subject to negative feedback control through GABA autoreceptors on GABAergic terminals. In rat cortical synaptosomes, GABA and muscimol each inhibit evoked release of ^{3}H-GABA, and are blocked by bicuculline.[103] The K^+-evoked release of GABA from rat frontal cortex is reduced by GABA, muscimol and 3-amino-propane sulphonic acid; the effect of muscimol being antagonized by bicuculline and picrotoxin.[104] The GABA agonist δ-aminolaevulinic acid inhibits K^+-induced release of GABA from preloaded synaptosomes from rat cerebral cortex, and again this effect may be blocked by bicuculline and picrotoxin.[105] Inhibition of the K^+-evoked, Ca^{2+}-dependent release of ^{3}H-GABA from rat substantia nigra has been demonstrated with exogenous GABA and muscimol, and antagonism quantified for picrotoxin.[106]

Bicuculline-sensitive postsynaptic GABA receptors are called $GABA_A$ receptors, whereas baclofen-sensitive GABA receptors present on peripheral sympathetic nerve terminals are termed $GABA_B$ receptors.[107-108] The $GABA_B$ site has also been detected in the CNS, but its relationship to the presynaptic receptor which modulates transmitter release remains uncertain.[109]

Physiological Role of Autoreceptors in the CNS; Targets for Drugs - Though autoreceptors play a key role in modulating transmitter release, it should be emphasized that many other types of prejunctional receptors can influence transmitter synthesis, release, or neuron firing.[1-3] For example, NE release can be inhibited by stimulation of muscarinic, dopamine, serotonin, histamine, GABA, opiate, prostaglandin and adenosine receptors located at noradrenergic nerve endings. Furthermore, in many cases the net effect of an agonist or antagonist will be the sum of presynaptic and postsynaptic actions, since the majority of drugs cannot discriminate between both receptor populations. Thus it is frequently extremely difficult to interpret the (patho)physiological role of CNS autoreceptors or their interactions with drugs.

Functional and biochemical studies suggest that endogenous depression may be related to impaired central noradrenergic transmission.[110,111] Drug-induced desensitization of central α_2-autoreceptors, stemming from chronic inhibition of neuronal NE uptake, and hence elevated synaptic NE levels, has been postulated to play a key role in the mechanism of action of many antidepressants.[111] This would lead to yet further enhanced neurotransmission, with subsequent down regulation of central ß-adrenoceptors[111] - possibly a pivotal event in the induction of the desired therapeutic effect.[112] However, the situation is far from clear with, on the one hand, conflicting reports as to whether chronic desipramine and imipramine do,[113] or do not,[114,115] down regulate presynaptic α_2-adrenoceptors in rat brain; and on the other, the finding that mianserin treatment results in supersensitive α_2-adrenoceptors.[116] It has been suggested recently that the induction of subsensitive central presynaptic α_2-adrenoceptors cannot represent the fundamental mode of action of antidepressants.[111]

Irrespective of the mechanism of action of tricyclic antidepressants, antagonists of α_2-autoreceptors per se might have a beneficial effect in depressive illness, since they facilitate noradrenergic neurotransmission. Phenoxybenzamine, dihydroergotamine and yohimbine, but not the selective α_1-adrenoceptor antagonist prazosin, have been shown to be synergistic with several antidepressants in decreasing ß-adrenoceptor density in rat cerebral cortex.[117-119] Thus combined administration of an α_2-adrenoceptor blocker with other antidepressants may provide a rapid onset therapy.[119] RS 21361 produces a more rapid onset of ß-adrenoceptor desensitization by itself than desipramine.[120]

Chronically administered antidepressants also down-regulate the number of $5HT_2$ recognition sites in rat brain.[121-123] Inasmuch as $5HT_2$ receptors are also implicated in the therapeutic action of antidepressants, blockade of central presynaptic α_2-(auto)receptors accelerates the down regulation of $5HT_2$ receptors by antidepressants.[124]

Desensitization of presynaptic DA receptors following chronic administration of antidepressants has been observed in some,[125,126] but not in other studies.[127-128] However, central DA turnover is unaltered by such regimens[111] and it is too soon to assess the relevance of these findings to the mechanisms underlying the clinical efficacy of these drugs.

Clonidine and some related drugs suppress narcotic withdrawal signs in the rat, monkey and in man[129,130] possibly via stimulation of presynaptic α_2-adrenoceptors, thereby inhibiting the firing of the locus ceruleus through reduction of NE release.[131] Clonidine has also been found effective in the treatment of alcohol withdrawal states.[132]

Elevated central NE activity (e.g. in the locus ceruleus) is an important correlate of human anxiety, though not the sole one.[133,134] The beneficial effects of clonidine as an antianxiety agent in humans could result from its inhibitory action on NE activity by stimulation of α_2-autoreceptors.[135,136] The observation that yohimbine is anxiogenic in humans is in accord with this hypothesis.[137]

The ideal agent for facilitating central cholinergic transmisson would act simultaneously as an antagonist at ACh autoreceptors and as an agonist at postsynaptic ACh-receptors.[138] It has been suggested[138] that compounds, with this profile would be of therapeutic value in the treatment of e.g. Huntington's chorea[139] and Alzheimer-type dementia.[140] BM-5 (22), an oxotremorine analog, is a presynaptic antagonist and a postsynaptic agonist at peripheral and central muscarinic synapses *in vitro*.[138]

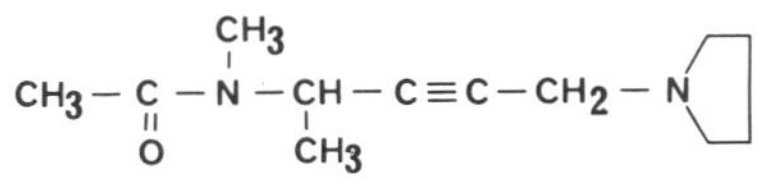

22

The question of a physiological role for GABA autoreceptors remains open.[3] GABA is involved in many CNS functions. Impaired GABA-mediated inhibition may contribute to the pathophysiology and phenomenology of epileptic seizures.[141] Consonant with this hypothesis, the purported GABA autoreceptor agonist δ-aminolaevulinic acid is a convulsant.[105] "GABA replenishment" treatment has been advocated for seizures, psychosis, movement disorders, pain, hypertension and sleep disturbances.[141] Presynaptic autoreceptors for GABA have yet to be investigated as targets for potential drugs (e.g. GABA autoreceptor antagonists).

The clinical significance of DA autoreceptor stimulation has been discussed.[142] Selective DA autoreceptor agonists, which reduce DA synthesis and release, represent a novel pharmacological approach to the treatment of various neurological disorders, including schizophrenia and Huntington's chorea, where an overactivity of central dopaminergic neurons is believed to be etiologic. Such drugs might lack some of the side-effects, e.g. tardive dyskinesia, associated with the currently prescribed neuroleptic agents, which are postsynaptic DA antagonists.[143-145] Apomorphine, N-propylnorapomorphine, and bromocryptine have been studied in the clinic.[146-148] There is evidence for an antipsychotic effect following acute administration, particularly for the two former agents. However, preliminary evidence suggests that the effect tolerates rapidly on repeated administration.[149]

References

1. E. S. Vizi, Progr. Neurobiol., 12, 181 (1979).
2. S. Z. Langer, Pharmacol. Rev., 32, 337 (1981).
3. K. Starke, Ann. Rev. Pharmacol. Toxicol., 21, 7 (1981).
4. M. J. Rand, M. W. McCulloch and D. F. Story, Trends Pharmacol. Sci., 3, 8 (1982).
5. S. Kalsner, Trends Pharmacol. Sci., 3, 11 (1982).
6. G. K. Aghajanian and M. A. Rogawski, Trends Pharmacol. Sci., 4, 315 (1983).
7. L. Hedler, K. Starke and A. Steppeler, Brit. J. Pharmacol., 78, 645 (1983).
8. P. B. M. W. M. Timmermans and P. A. van Zwieten, J. Med. Chem., 25, 1389 (1982).
9. A. N. M. Schoffelmeer and A. H. Mulder, Naunyn-Schmiedeb. Arch. Pharmacol., 323, 188 (1983).
10. A. N. M. Schoffelmeer and A. H. Mulder, Eur. J. Pharmacol., 87, 449 (1983).
11. J. C. A. van Meel, A. de Jonge, P. B. M. W. M. Timmermans and P. A. van Zwieten, J. Pharmacol. Exp. Ther., 291, 760 (1981).
12. W. Kobinger and L. Pichler, Eur. J. Pharmacol., 73, 313 (1981).
13. D. Cambridge, Eur. J. Pharmacol., 72, 413 (1981).
14. P. B. M. W. M. Timmermans, B. Wilffert, D. Davidesko, M. J. Mathy, D. Dijkstra, A. S. Horn and P. A. van Zwieten, J. Pharmacol. Exp. Ther., 225, 407 (1983).
15. P. E. Hicks and C. Waldron, Brit. J. Pharmacol., 74, 254P (1980).
16. P. B. M. W. M. Timmermans, M. J. Mathy, B. Wilffert, H. O. Kalkman, G. Smit, D. Dijkstra, A. S. Horn and P. A. van Zwieten, Eur. J. Pharmacol., 97, 55 (1984).
17. R. R. Ruffolo, Jr. and E. L. Yaden, Eur. J. Pharmacol., 93, 117 (1983).
18. P. E. Hicks, C. Waldron, P. Burn and P. A. Crooks, J. Pharm. Pharmacol., 35, 94 (1983).
19. J. M. Caroon, R. D. Clark, A. F. Kluge, R. Olah, D. B. Repke, S. H. Unger, A. D. Michel and R. L. Whiting, J. Med. Chem., 25, 666 (1982).
20. J. C. Doxey, A. G. Roach and C. F. C. Smith, Brit. J. Pharmacol., 78, 489 (1983).

21. J. C. Doxey, A. G. Roach, D. A. Strachan and N. Virdee, Brit. J. Pharmacol., 79, 311P (1983).
22. I. B. Armah, D. R. Ferry and H. Glossmann, Brit. J. Pharmacol., 78, 151P (1983).
23. V. Pierce and J. F. Waterfall, Brit. J. Pharmacol., 76 263P (1982).
24. J. M. Roesler, L. J. Kopaciewicz and J. P. Hieble, The Pharmacologist, 25(3), 242 (1983).
25. R. R. Ruffolo, Jr., in "Stereochemistry and Biological Activity of Drugs", E. J. Ariëns, W. Soudijn and P. B. M. W. M. Timmermans, Eds., Blackwell, Oxford, 1983, p. 103.
26. J. P. Dausse, K. H. Le Quan-Bui and P. Meyer, Eur. J. Pharmacol., 78, 15 (1982).
27. M. Raiteri, G. Maura and P. Versache, J. Pharmacol. Exp. Ther., 224, 679 (1983).
28. C. Ennis, J. Pharm. Pharmacol. 35, 750 (1983).
29. F. Pelayo, M. L. Dubocovich and S. Z. Langer, Eur. J. Pharmacol., 64, 143 (1980).
30. E. Schlicker, M. Göthert, F. Köstermann and R. Clausing, Naunyn-Schmiedeb. Arch. Pharmacol., 323, 106 (1983).
31. D. C. U'Prichard, T. D. Reisine, S. T. Mason, H. C. Fibiger and H. T. Yamamura, Brain Res., 187, 143 (1980).
32. B. B. Hoffmann and R. J. Lefkowitz, Ann. Rev. Pharmacol., 20, 581 (1980).
33. M. Raiteri, M. Marchi, G. Maura, P. Paudice and P. Versache, in "Basic and Clinical Aspects of Molecular Neurobiology", A. M. Giuffrida Stella, G. Gombos, G. Benzi and H. S. Bachelard, Eds., Fondazione Internationale Menarini, Milan, 1982, p. 207.
34. S. Hjorth, A. Carlsson, D. Clark, K. Svensson, H. Wikström, D. Sanchez, P. Lindberg, U. Hacksell, L.-E. Arvidsson, A. Johansson and J. L. G. Nilsson, Acta Pharm. Suec. (Suppl) 1, 130 (1983).
35. S. Z. Langer, S. Arbilla, L. Kamal and R. Cantrill, Acta Pharm. Suec. (Suppl) 1, 108 (1983).
36. J. Lehmann, S. Arbilla and S. Z. Langer, Naunyn-Schmiedeb. Arch. Pharmacol., 317, 31 (1981).
37. L. X. Cubeddu and I. S. Hoffmann, J. Neurochem., 41, 91 (1983).
38. J. Lehmann, R. V. Smith and S. Z. Langer, Eur. J. Pharmacol., 88, 81 (1983).
39. A. H. Tissari, L. Atzori and M. T. Galdieri, Naunyn-Schmiedeb. Arch. Pharmacol., 322, 89 (1983).
40. G. K. Aghajanian and B. S. Bunney, Naunyn-Schmiedeb. Arch. Pharmacol., 297, 1 (1977).
41. T. C. Westfall, L. Naes and C. Paul, J. Pharmacol. Exp. Ther., 224, 199 (1983).
42. S. E. Leff and I. Creese, Trends Pharmacol. Sci., 4, 463 (1983).
43. E. Eriksson, K. Modigh, A. Carlsson and H. Wikström, Eur. J. Pharmacol., 96, 29 (1983).
44. K. Starke, L. Späth, J. D. Lang and C. Adelung, Naunyn-Schmiedeb. Arch. Pharmacol., 323 298 (1983).
45. P. Seeman, Pharmacol. Rev., 32, 229 (1980).
46. P. Sokoloff, M.-P-Martres and J.-C. Schwartz, Naunyn-Schmiedeb. Arch. Pharmacol., 315, 89 (1980).
47. J. Lehmann, M. Briley and S. Z. Langer, Eur. J. Pharmacol., 88, 11 (1983).
48. S. E. Leff and I. Creese, Nature, 306, 586 (1983).
49. J. L. G. Nilsson and A. Carlsson, Trends Pharmacol. Sci., 3, 322 (1982).
50. J. F. Plantjé, A. H. Mulder and J. C. Stoof, Eur. J. Pharmacol., 91, 317 (1983).
51. J. P. Arneric, J. P. Long, M. Williams, D. B. Goodale, G. M. Lakoski and G. F. Gebhart, J. Pharmacol. Exp. Ther., 224, 161 (1983).
52. M. G. P. Feenstra, C. Summers, J. H. Goedemoed, J. B. de Vries, H. Rollema and A. S. Horn, Naunyn-Schmiedeb. Arch. Pharmacol., 324, 108 (1983).
53. D. B. Goodale, D. B. Rusterholz, J. P. Long, J. R. Flynn, B. Walsh, J. G. Cannon and T. Lee, Science, 210, 1141 (1980).
54. N. E. Andén, H. Nilsson and U. Thornström, Acta Pharmacol. Toxicol., 52, 51 (1983).
55. P. Sminia and A. H. Mulder, Eur. J. Pharmacol., 89, 183 (1983).
56. M. Williams, G. E. Martin, D. E. McClure, J. J. Baldwin and K. J. Watling, Naunyn-Schmiedeb. Arch. Pharmacol., 324, 275 (1983).
57. S. Hjorth, A. Carlsson, D. Clark, K. Svensson, H. Wikström, D. Sanchez, P. Lindberg, U. Hacksell, L.-E. Arvidsson, A. Johansson and J. L. G. Nilsson, Acta Pharm. Suec. (Suppl) 1, 130 (1983).
58. S. W. Koch, B. K. Koe and N. Bacopoulos, Eur. J. Pharmacol., 92, 279 (1983).
59. K. J. Watling, Brit. J. Pharmacol., 80, 646P (1983).
60. S. Arbilla and S. Z. Langer, Brit. J. Pharmacol., 79, 193P (1983).
61. J. C. van Oene, H. A. Houwing, D. Dijkstra and A. S. Horn, Eur. J. Pharmacol., 87, 491 (1983).
62. L. Kamal, S. Arbilla, A.-M. Galzin and S. Z. Langer, J. Pharmacol. Exp. Ther., 227, 446 (1983).
63. S. Arbilla, P. Baud and S. Z. Langer, Brit. J. Pharmacol., 80 472P (1983).
64. N. M. J. Rupniak, P. Jenner and C. D. Marsden, Life Sci., 32, 2289 (1983).
65. J. Z. Novak, S. Arbilla, A.-M. Galzin and S. Z. Langer, J. Pharmacol. Exp. Ther., 226, 558 (1983).
66. S. Arbilla, S. Z. Langer and J. Z. Novak, Brit. J. Pharmacol., 78, 16P (1983).
67. M. Göthert, Trends Pharmacol. Sci., 3, 437 (1982).

68. M. Göthert, Naunyn-Schmiedeb. Arch. Pharmacol., 314, 223 (1980).
69. I. Mounsey, K. A. Brady, J. Carroll, R. Fischer and D. N. Middlemiss, Biochem. Pharmacol., 31, 49 (1982).
70. M. Göthert and E. Schlicker, Life Sci., 32, 1183 (1983).
71. G. Engel, M. Göthert, E. Müller-Schweinitzer, E. Schlicker, L. Sistonen and P. A. Stadler, Naunyn-Schmiedeb. Arch. Pharmacol., 324, 119 (1983).
72. C. Ennis and B. Cox, Neuropharmacol., 21, 41 (1982).
73. L. L. Martin and E. Sanders-Bush, Neuropharmacol., 21, 445 (1982).
74. J. R. Fozard and G. M. P. Mwaluko, Brit. J. Pharmacol., 57, 115 (1976).
75. M. A. Rogawski and G. K. Aghajanian, J. Neurosci., 1, 1148 (1981).
76. P. Blier and C. de Montigny, Neuropharmacol., 22, 495 (1983).
77. L. L. Martin and E. Sanders-Bush, Naunyn-Schmiedeb. Arch. Pharmacol., 321, 165 (1982).
78. B. Cox and C. Ennis, J. Pharm. Pharmacol., 34, 438 (1982).
79. D. L. Nelson, N. W. Pedigo and H. I. Yamamura, J. Physiol. (Paris), 77, 369 (1980).
80. N. W. Pedigo. H. I. Yamamura and D. L. Nelson, J. Neurochem., 36, 220 (1981).
81. J. R. Fozard, Trends Pharmacol. Sci., 4, 288 (1983).
82. P. P. Deshmukh, D. L. Nelson and H. I. Yamamura, Fed. Proc., 41, 6238 (1982).
83. D. N. Middlemiss and J. R. Fozard, Eur. J. Pharmacol., 90, 151 (1983).
84. H. Gozlan, S. El Mestikawy, L. Pichat, J. Glowinski and M. Hamon, Nature, 305, 140 (1983).
85. H. Gozlan, S. El Mestikawy, S. Bourgoin, M. Hall, L. Pichat, J. Glowinski and M. Hamon, Naunyn-Schmiedeb, Arch. Pharmacol., 324 (Suppl), R19 (1983).
86. D. N. Middlemiss, Naunyn-Schmiedeb. Arch. Pharmacol., 324 (Suppl), R19 (1983).
87. S. Hjorth, A. Carlsson, P. Lindberg, S. Sanchez, H. Wikström, L.-E. Arvidsson, U. Hacksell and J. L. G. Nilson, J. Nueral Transm., 55, 169 (1982).
88. J. R. Fozard and M. D. Tricklebank, Naunyn-Schmiedeb. Arch. Pharmacol., 324 (Suppl), R20 (1983).
89. S. Z. Langer and C. Moret, J. Pharmacol. Exp. Ther., 222, 220 (1982).
90. A.-M. Galzin, S. Z. Langer and B. Verzier, Brit. J. Pharmacol., 80, 718P (1983).
91. M. Briley and C. Moret, Brit. J. Pharmacol., 80, 673P (1983).
92. R. M. Hagan and I. E. Hughes, Brit. J. Pharmacol., 80, 513P (1983).
93. J. C. Szerb, in "Presynaptic Receptors", S. Z. Langer, K. Starke and M. L. Dubocovich, Eds., Pergamon, Oxford, 1979, p. 293.
94. P. C. Molenaar and R. L. Polak, Progr. Pharmacol., 3(4), 39 (1980).
95. Ö. Nordström and T. Bartfai, Acta Physiol. Scand., 108, 347 (1980).
96. M. Marchi, P. Paudice and M. Raiteri, Eur. J. Pharmacol., 73, 75 (1981).
97. Ö. Nordström and T. Bartfai, Brain Res., 213, 467 (1981).
98. P. Alberts and L. Stjärne, Acta Physiol. Scand., 115, 269 (1982).
99. M. Raiteri, M. Marchi and P. Paudice, Eur. J. Pharmacol., 74, 109 (1981).
100. M. Marchi, A. Caviglia, P. Paudice and M. Raiteri, Eur. J. Pharmacol., 91, 63 (1983).
101. S. Halim, H. Kilbinger and I. Wessler, Scand. J. Gastroenterol. (Suppl), 17, 87 (1982).
102. J. M. Arrang, M. Garbarg and J.-C. Schwartz, Nature, 302, 832 (1983).
103. S. R. Snodgrass, Nature, 273, 392 (1978).
104. P. R. Mitchell and I. L. Martin, Nature, 274, 905 (1978).
105. M. J. W. Brennan and R. C. Cantrill, Nature, 280, 514 (1979).
106. S. Arbilla, L. Kamal and S. Z. Langer, Eur. J. Pharmacol., 57, 211 (1979).
107. N. G. Bowery, A. Doble, D. R. Hill, A. L. Hudson, J. S. Shaw, M. J. Turnbull and R. Warrington, Eur. J. Pharmacol., 71, 53 (1981).
108. M. S. Muhyaddin, P. J. Roberts and G. N. Woodruff, Eur. J. Pharmacol., 92, 9 (1983).
109. N. G. Bowery, D. R. Hill and A. L. Hudson, Brit. J. Pharmacol., 78, 191 (1983).
110. C. B. Smith and J. A. Garcia-Sevilla, in "New Vistas in Depression", S. Z. Langer, R. Takanaski, T. Segawa and M. Briley, Eds., Pergamon, Oxford, 1982, p. 99.
111. M. F. Sugrue, Biochem. Pharmacol., 32, 1811 (1983).
112. F. Sulser, Trends Pharmacol. Sci., 1, 92 (1979).
113. M. F. Sugrue, Life Sci., 28, 377 (1981).
114. A. N. M. Schoffelmeer and A. H. Mulder, Naunyn-Schmiedeb. Arch. Pharmacol., 318, 173 (1982).
115. E. Schlicker, M. Göthert and R. Clausing, Naunyn-Schmiedeb. Arch. Pharmacol., 320, 38 (1982).
116. F. Cerrito and M. Raiteri, Eur. J. Pharmacol., 70, 425 (1981).
117. F. T. Crews, S. M. Paul and F. K. Goodwin, Nature, 290, 787 (1981).
118. T. Reisine, R. Johnson, N. Wiech, R. Ursillo and H. I. Yamamura, Adv. Biochem. Psychopharmacol., 31, 63 (1982).
119. J. A. Scott, and F. T. Crews, J. Pharmacol. Exp. Ther., 224, 640 (1983).
120. A. Dye, C. M. Page and R. L. Whiting, Brit. J. Pharmacol., 80, 665P (1983).
121. S. J. Peroutka and S. H. Snyder, Science, 210, 88 (1980).
122. K. J. Kellar, C. S. Cascio, D. A. Bergstrom, J. A. Butler and P. Iadarola, J. Neurochem., 37, 830 (1981).
123. D. A. Kendall, R. Duman, J. Slopis and S. J. Enna, J. Pharmacol. Exp. Ther., 222, 566 (1982).

124. F. T. Crews, J. A. Scott and N. H. Shorstein, Neuropharmacology, 22, 1203 (1983).
125. G. Serra, A. Argiolas, V. Klimek, F. Fadda and G. L. Gessa, Life Sci., 25, 415 (1979).
126. G. Serra, A. Argiolas, F. Fadda, M. R. Melis and G. L. Gessa, Psychopharmacology, 73, 194 (1981).
127. C. Spyraki and H. C. Fibiger, Eur. J. Pharmacol., 74, 195 (1981).
128. H. H. Holcomb, M. J. Bannon and R. H. Roth, Eur. J. Pharmacol., 82, 173 (1982).
129. M. S. Gold, D. E. Redmond and H. D. Kleber, Lancet, 2, 599 (1978).
130. H. Lal and G. T. Shearman, Progr. Clin. Biol. Res., 71, 99 (1981).
131. H. Lal and S. Fielding, Trends Pharmacol. Sci., 4, 70 (1983).
132. J. A. Steiner, A. J. Wilkins and W. Jenkins, Brit. J. Clin. Pharmacol., 15, 583P (1983).
133. D. E. Redmond and Y. H. Huang, Life Sci., 25, 2149 (1979).
134. R. Hoehnsaric, A. F. Merchant, M. L. Keyser and V. K. Smith, Arch. Gen. Psychiatry, 38, 1278 (1981).
135. D. E. Redmond, Trends Pharmacol. Sci., 3, 477 (1982).
136. D. S. Charney, G. R. Heninger, D. E. Sternberg, K. M. Hafstad, S. Giddings and D. H. Landis, Arch. Gen. Psychiatry, 39, 290 (1982).
137. D. S. Charney, G. R. Heninger and D. E. Redmond, Life Sci., 33, 19 (1983).
138. Ö. Nordström, P. Alberts, A. Westlind, A. Undén and T. Bartfai, Mol. Pharmacol., 24, 1 (1983).
139. E. G. S. Spokes, Trends Neurochem. Sci., 4, 115 (1981).
140. S. Corkin, Trends Neurochem. Sci., 4, 287 (1981).
141. R. G. Fariello and M. K. Ticku, Life Sci., 33, 1629 (1983).
142. H. Y. Meltzer, Dopamine Receptors and Their Behavioral Correlates Pharmacology Biochemistry & Behavior, Vol. 17 (Suppl 1), 1 (1982).
143. H. Meltzer, Schizophrenia Bull., 6, 456 (1980).
144. R. W. Fuller, Trends Pharmacol. Sci., 4, 394 (1983).
145. K. J. Watling, Trends Pharmacol. Sci., 3, 232 (1982).
146. H. Y. Meltzer, Schizophr. Bull., 6, 456 (1980).
147. C. A. Tamminga, M. D. Gotts and M. R. Miller, Acta Pharm. Suec. (Suppl) 2, 153 (1983).
148. H. Y. Meltzer, T. Kolakowska, A. Robertson and B. J. Tricou, Psychopharmacology, 81, 37 (1983).
149. C. A. Tamminga, 22nd Ann. Meeting Am. College Pharmacol., San Juan, Puerto Rico, December 12-16, 1983.

Section II - Pharmacodynamic Agents

Editor: William T. Comer, Bristol-Myers Research and Development
345 Park Avenue, New York, New York 10154

Chapter 7. Antihypertensive Agents

Peter Sprague and James R. Powell, The Squibb Institute for Medical Research, Princeton, New Jersey 08540

Introduction - Arterial hypertension continues as a major risk factor in the development of cardiovascular disease. Currently it is not possible to prevent essential hypertension and this presents a challenge to the clinician, pharmacologist, and medicinal chemist to discover adequate therapies. The beneficial effects of drug therapy in the management of hypertension have been recently reviewed.[1]

A number of recent reviews describe topics related to hypertension. The significance of catecholamine levels in hypertension has been discussed and a cautionary note offered that levels in mixed venous blood are unreliable indicators of sympathetic neural activity in hypertension.[2,3] The role of the sympathoadrenal axis in hypertension has been reviewed.[4] The need for intact vascular endothelium for vascular responses to a number of agents has been described.[5] A recent review examines the role of renin substrate in hypertension.[6] Lastly, based on the unimodal distribution of blood pressures, the existence of hypertension as an entity has been questioned.[7]

Renin Inhibitors - A review of progress made in defining the structure of mouse submaxillary gland renin has appeared.[8] The presence of two aspartyl, two tyrosyl and one arginyl residues at the active site has been demonstrated.

Current approaches to renin inhibition have been reviewed.[9] Monoclonal antibodies with high affinity and specificity for human renin have been developed and several new peptides modeled after the minimal substrate octapeptide (1) have been prepared.

1 His-Pro-Phe-His-Leu-Leu-Val-Tyr
2 His-Pro-Phe-His-Leu-^{R}-Leu-Val-Tyr
3 Iva-His-Pro-Phe-His-Sta-Ile-Phe-NH_2
4 BOC-Phe-His-Sta-Leu-(4-amido-1-benzylpiperidine)

H-77 (2), in which the Leu-Leu amide is reduced to an amine, suppresses the pressor effect in dogs of injected dog renin (ID_{50}=0.06 mg/kg/hr).[10,11] Heptapeptide 3, in which Leu_{10}-Leu_{11} in 1 is replaced by statine (Sta), is one of the most potent human renin inhibitors *in vitro* yet discovered (I_{50}= 2 nM).[12] Further refinement of this class gave 4, a protected tetrapeptide amide with a 60 picomolar K_i against human renin.[13]

Angiotensin Converting Enzyme Inhibitors - An excellent review on the mechanisms by which inhibitors of angiotensin converting enzyme (CEI) reduce blood pressure has appeared.[14] Current evidence supports the hypothesis that CEI exert their hypotensive effects primarily by inhibition of circu-

ISBN 0-12-040519-9

lating and tissue-bound converting enzyme.

Several new CEI modeled after either captopril or enalapril have been reported. Pivalopril (RHC 3659, 5) has been shown equivalent to captopril in vitro but only 1/4 as active in man.[15-17] Zofenopril (6) represents a second generation captopril showing 10-fold greater potency and a longer duration of action.[18] The dihydroindole analog 7 (WY 44,221) has activity similar to 6.[19-21] Tetrahydroisoquinoline variation 8 is equivalent in activity to captopril.[22] SA 446 (9), given orally, inhibits angiotensin I (i.c.v.)-induced hypertension more effectively than captopril, implying delivery across the blood-brain barrier.[23] The side-chain benzoyl analog 10 is more potent than captopril with a longer duration of activity.[24] Lactam 11, 2-fold less active than captopril, was developed using a computer-assisted molecular modeling approach.[25]

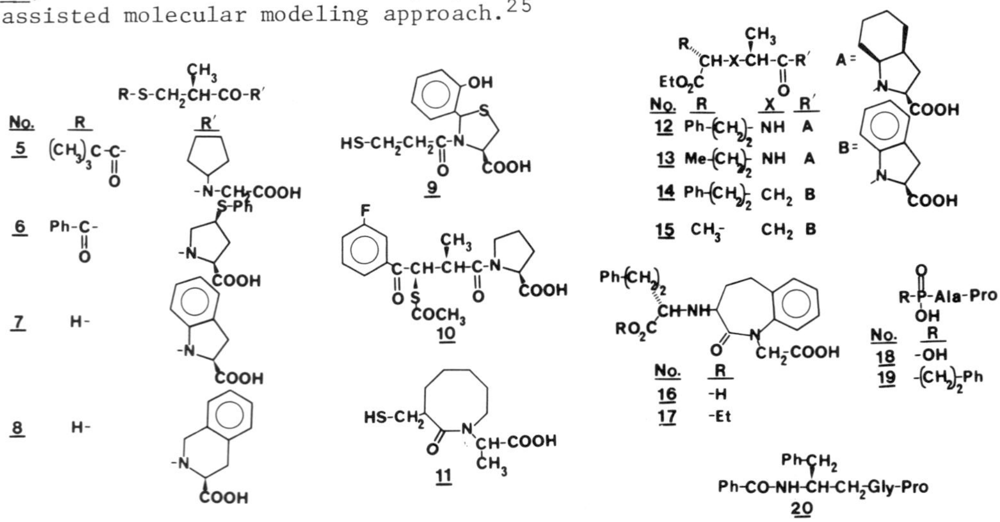

A summary of the pharmacology of enalapril (MK-421) has appeared along with an account of clinical studies done with lisinopril (MK-521) which suggest this drug will be suitable for once-a-day antihypertensive therapy.[26-28] Analogs of enalapril which have appeared include the perhydroindole derivative 12 (SCH 31,846). This compound has a quicker onset of action, better absorption, and greater biliary excretion than enalapril but is otherwise indistinguishable in vivo.[29-31] Synthesis of the closely related 13 has been reported.[32] Glutaric acid analog 14, missing a side-chain nitrogen, is intermediate in activity in vitro between captopril and enalapril.[33] A related simplified derivative 15 (CGS 13,945), is much less potent than captopril in man.[34] The internal lactam stratagy has also been employed in the enalapril series. Benzazepine 16 (CGS 14,831) is potent in vitro (I_{50}=3 nM), and its ester 17 (CGS 14,824A) has a more rapid onset of action plus greater potency than enalapril in dogs.[35,36] The phenethylphosphinamidate 19 refinement of the phosphoramidate CEI 18 is 3-fold more potent in vitro.[37] Compound 20, a derivative of benzoyl-Phe-Gly-Pro in which the scissile amide bond is reduced to an amine, is a poorly active CEI.[38]

Vasodilators - The cellular mechanisms by which vasodilators lower blood pressure have been reviewed.[39] One of these involves the arachidonic acid cascade in which both the cyclooxygenase and lipoxygenase pathways play a role. Evidence for increased synthesis of prostaglandins (PG) by the renal glomerulus of SHR relative to normotensive rats has been presented.[40] The use of synthetic analogs of natural PG as antihypertensive agents remains an active area of research. The main obstacles to overcome are manifold side effects, short duration of action and poor oral effectiveness.

CL 115,347, a 16-vinyl-16-hydroxy-PGE_2 analog, is a more potent vasodilator than PGE_2 with less emesis and diarrhea potential.[41,42] The short duration of action after oral administration is overcome by giving the drug topically. Thiazolidinone 21 increases renal blood flow maximally 5 hr after oral dosing with potency comparable to PGE_2.[43] The 11,12-seco-PG analog 22, increased renal blood flow after oral administration without concomitant increases in heart rate or blood pressure.[44] Carbacyclin analog ZK 36,374 (23) is 5-fold more potent than PGI_2 *in vitro*, and stabilized enolether 24 (CG 4203) also mimics PGI_2 in animal tests.[45,46] Both compounds have a longer duration of action than PGI_2. BW 245C (25) mimics PGD_2 in its effects on platelets and *in vivo* has both PGD_2 and PGI_2-like activities.[47] When given to man by constant infusion, 25 shows activity and side effects similar to PGI_2 with a greater duration of action.[48] However, this compound is stable only at high pH and is not active orally.

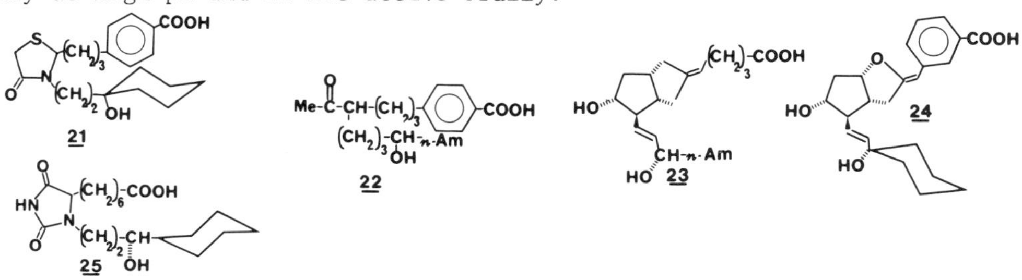

MDL 899 (26) is a direct acting arteriolar vasodilator in man, but coadministration of a beta-blocker is necessary to prevent reflex-related side effects.[49] A clinical study with cadralazine (ISF 2469, 27) shows this compound similar to hydralazine with fewer side effects.[50] Two other vasodilators, budralazine (28) and pinacidil (29) may owe some of their activity to calcium channel blockade.[51,52] Anthranilamide 30 (WIN 48,049) lowers blood pressure in monkeys without accompanying tachycardia.[53] This compound acts primarily as a direct vasodilator but also has sympatholytic and dopaminergic activities.[54] Compounds 31 and 32 also lower blood pressure by apparently direct vasodilator mechanisms.[55,56]

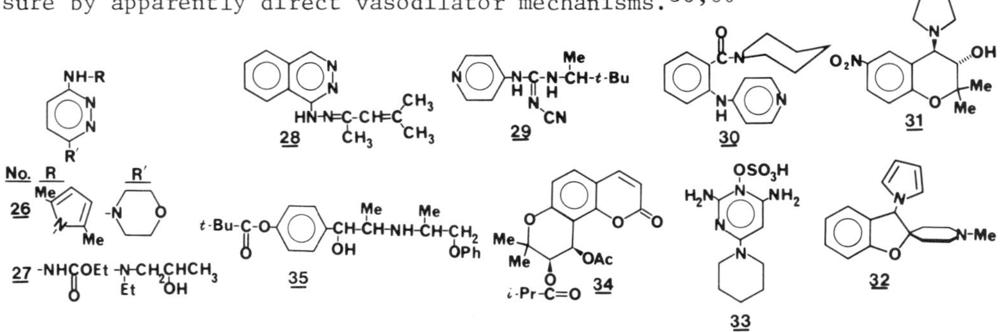

The O-sulfate 33 of minoxidil has been proposed as an important active metabolite.[57] Unlike minoxidil, 33 is active *in vitro* and is more potent with a faster onset of action *in vivo*. Evidence suggests the dihydropyrano and dihydrofuranocoumarin vasodilators (e.g. 34) exert their effect by cAMP phosphodiesterase inhibition.[58] Finally, the pivalic acid ester of isoxsuprine 35 has a slower onset and longer duration of action than the parent phenol.[59]

Calcium Entry Blockers - In the U.S. nifedipine, verapamil, and diltiazem are marketed for antianginal or antiarrhythmic use, though none has FDA approval yet for use in hypertension.

The potential therapeutic use of calcium entry blockers (CEB) in the treatment of atherosclerosis has recently been reviewed.[60] Verapamil and nifedipine have been shown to suppress atherogenesis in cholesterol-fed rabbits without altering serum cholesterol levels.[61,62]

New evidence suggests certain CEB affect neural regulation of blood vessels at several sites. Studies in anesthetized dogs indicate that nifedipine increases and verapamil decreases the sensitivity of carotid sinus baroreceptors apparently by interference with a calcium-dependent and a sodium-dependent mechanism, respectively.[63] A similar increase in baroreflex sensitivity was found in man treated chronically with nifedipine, which may explain why there is evidence of sympathetic activation during initial therapy.[64] Tachycardia and increased levels of plasma norepinephrine are found during acute but not chronic phases of nifedipine treatment.[65]

During activation of postganglionic sympathetic nerves there is an exocytotic release of norepinephrine which is dependent upon the movement of calcium across the neural cell membrane. The effects of CEB on stimulus-secretion coupling in sympathetic nerves has been studied in several different preparations. Verapamil does not affect the neural release of norepinephrine in either isolated cat heart or in the pithed rat.[66,67] In the latter model nifedipine blunted increases in blood pressure due to activation of sympathetic nerves but did not affect cardiovascular responses to exogenous norepinephrine suggesting a prejunctional site of inhibition. A similar prejunctional CEB effect was found for diltiazem in guinea pig mesenteric and dog basilar arterial preparations.[68] Cinnarizine and flunarizine were shown more potent than verapamil, diltiazem or nifedipine in reducing the potassium-induced uptake of calcium in a synaptosomal preparation.[69]

New studies have examined the interaction of various CEB with α-adrenoceptors and their effect on processes mediated by α-adrenoceptors.[70] The (+) and (-) antipodes of verapamil, nicardipine and (±) D-600 showed similar inhibition of radioligand binding to α_1-adrenoceptors, whereas diltiazem and nifedipine showed a very low level of activity.[71] A different pattern was seen in regard to α_2-adrenoceptors. (-) Verapamil was the most potent inhibitor of binding with (+) verapamil, (±) D-600, diltiazem, and nicardipine showing intermediate activity. Nifedipine was inactive. Verapamil, but not diltiazem, impairs the α_2-adrenoceptors present in rabbit hypothalmus which modulate the release of norepinephrine.[72] The dihydropyridines nifedipine (36), nimodipine (37), or PY-108-068 (38), as well as verapamil and diltiazem selectively inhibit vasoconstriction mediated by α_2-adrenoceptors and not that mediated by α_1-adrenoceptors.[73-79] Thus, vasodilatation (and presumably antihypertensive effects) caused by CEB may be due to impairment of α_2-adrenoceptor mediated vascular tone. There may be some impairment of vascular α_1-adrenoceptor processes as well.[78] Nifedipine decreases vasopressor responses to norepinephrine in man.[80]

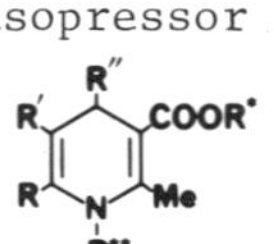

No.	R	R′	R″	R*	R**
36	Me	CO_2Me	B	Me	H
37	Me	CO_2-i-Pr	A	$-(CH_2)_2-OMe$	H
38	Me	CO_2Et	D	Et	H
39	Me	CO_2Et	C	Et	$-(CH_2)_2-N$ 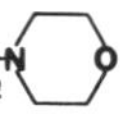
40	CN	CO_2Me	A	i-Pr	H
41	Me	NO_2	C	Me	H

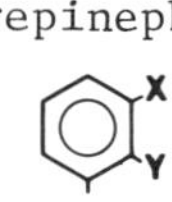

R″	Y	X
A	H	NO_2
B	NO_2	H
C	CF_3	H

D= 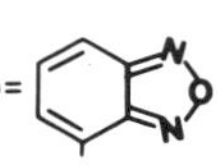

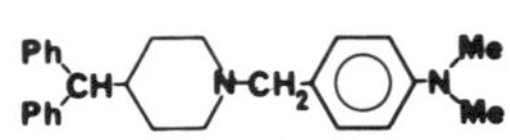

42

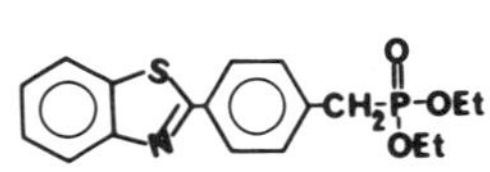

43

The dihydropyridine CEB flordipine (39) has pharmacological properties unlike other dihydropyridines.[81] Flordipine and nifedipine showed similar potency in vitro in relaxing potassium-depolarized canine veins but flordipine was >100-fold less potent in causing relaxation of arterial strips. Flordipine was 1000-fold less potent than nifedipine in causing depression of feline cardiac tissue. Thus, flordipine is a "vascular selective" CEB. In SHR, a 30 mg/kg dose of flordipine produced blood pressure lowering which persisted for 24 hrs and, in dogs, oral doses from 0.03 to 3.0 mg/kg lowered blood pressure due to reduced peripheral vascular resistance.[82] The lowered blood pressure was associated with tachycardia, presumably of reflex origin. Oral doses of 200 or 300 mg lowered blood pressure in normotensive human subjects and caused reflex tachycardia.[83] In hypertensive subjects, 60 mg of flordipine lowered blood pressure without any accompanying changes in cardiac rate. In addition to calcium entry blockade, flordipine has been found to inhibit cyclic nucleotide phosphodiesterase.[84,85]

Another vascular selective dihydropyridine, FR34235, (40) is about four times more potent in vitro than nifedipine in causing relaxation in potassium-depolarized rabbit aortic strips but 50 times less potent in causing depression of rabbit cardiac tissue.[86] FR34235 also displayed selectivity for relaxation of arterial strips from various vascular beds.[87]

Given i.v. to anesthetized dogs, BAY K 8644 (41) caused increases in myocardial contractility, blood pressure, and peripheral vascular resistance.[88,89] These changes were reversed by nifedipine but not by α- or β-adrenoceptor blockade. BAY K 8644 increased cardiac rate and contractility and also caused coronary vasoconstriction in an isolated cardiac preparation. This compound may be useful as a tool to ascertain the function of CEB in tissues.

SC-30552 (42) given orally to SHR (spontaneously hypertensive rats) lowered blood pressure and caused bradycardia.[90] Tachycardia caused by isoproterenol was not affected by SC-30552, whereas isoproterenol-induced decreases in blood pressure were attenuated. Ganglionic blockade caused by hexamethonium did not affect vasodilatation caused by this CEB. A pA_2 value against calcium-induced contractions of rabbit aortic strips of 7.42 was calculated. These effects suggest 42 possesses some β_2- but not β_1- or α-adrenoceptor activity and may depress the SA node.

KB-944 (43), a new type of CEB, given to normotensive rats orally at 10-50 mg/kg produced graded reductions in blood pressure and increased cardiac rate.[91] The increased heart rate was apparently due to the baroreceptor reflex as it was blunted by concomitant propranolol treatment. KB-944 was intermediate in potency (nifedipine>KB-944>diltiazem) as an antihypertensive when given to SHR, DOCA-NaCl hypertensive rats, and renal hypertensive rats, and had a longer duration of effect. KB-944 given intravenously to anesthetized dogs produced negative dromotropic effects reflected by increased AV nodal conduction time and 3° heart block in some animals and thus KB-944 may be useful as an antiarrhythmic.[92]

Serotonin Receptor Antagonists - Ketanserin (44) reduced blood pressure in experimental and clinical hypertension acting by specific blockade of serotonin ($5\text{-}HT_2$) receptors.[93] However, ketanserin has been found to possess potent α-adrenoceptor blocking properties in vitro and the antihypertensive effects of this compound more closely correlate with α-adrenoceptor rather than serotonin antagonism.[94] LY53857 (45) is a highly specific serotonin receptor antagonist which does not lower blood pressure in SHR.[95] Thus, it appears that antihypertensive effects of serotonin receptor antagonists may be due, at least in part, to α-adrenoceptor blockade.

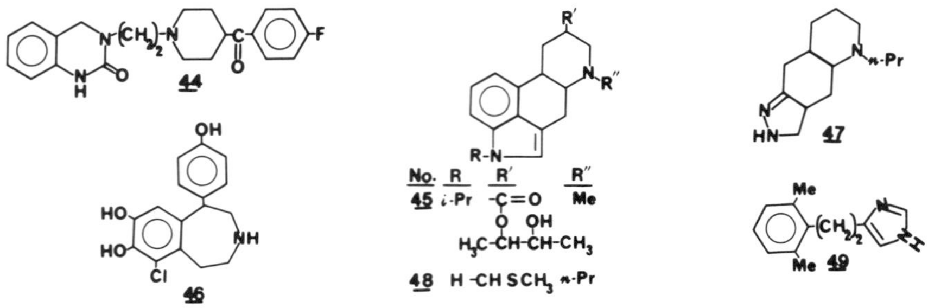

Dopamine Receptor Agonists - Compounds which mimic dopamine can lower blood pressure by pre- and postjunctional as well as central neural mechanisms. The R- and S-enantiomers of SKF 82526 (46) are both renal vasodilators, but the R-enantiomer operates in hypertensive rats by a postjunctional mechanism whereas the S-enantiomer does not.[96] LY141865 (47) is a pergolide (48) derivative which lowers blood pressure after oral administration to SHR.[97] LY141865 in low doses reduced the hypermotility observed in SHR but not in their corresponding normotensive controls.[98] Pergolide given i.v. lowers blood pressure and reduces peripheral sympathetic nerve activity suggesting a central nervous system site of action.[99]

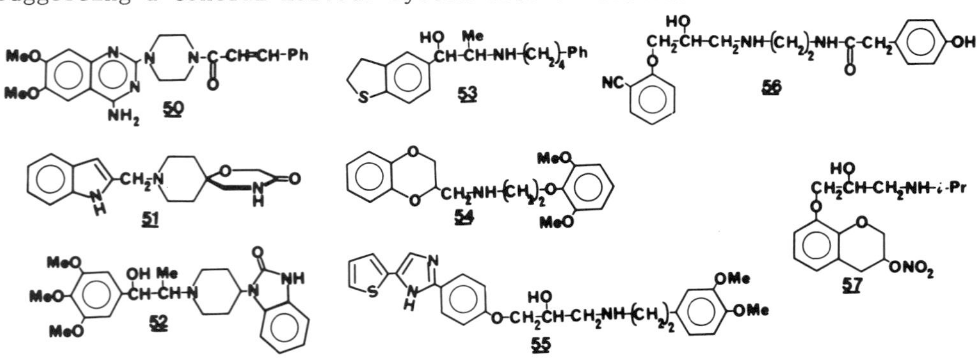

α-Adrenoceptor Agonists and Antagonists - MPV295 (49) lowers blood pressure by stimulation of α-adrenoceptors in the brainstem.[100] MPV295 stimulates only cardiovascular α_2-adrenoceptors whereas clonidine stimulates both α_1- and α_2-adrenoceptors.

Interest continues in the area of selective α_1-adrenoceptor blockade as a mechanism to lower blood pressure. Compound 50 was less than 1/10 as potent as prazosin *in vitro* as an α-adrenoceptor blocker but was nearly equivalent in lowering blood pressure in renal hypertensive rats.[101] Unlike prazosin, trimazosin lowered blood pressure by some unknown mechanism in addition to blockade of α_1-adrenoceptors.[102] Compound 51 showed potency as an α_1-adrenoceptor blocker *in vitro* less than that found with prazosin.[103] The compound lowered blood pressure in SHR and may have central antihypertensive effects due to blockade of brain α_1-adrenoceptors. Like prazosin, KF-4942 (52) lowers blood pressure due to blockade of vascular α_1- adrenoceptors.[104] Tibalosine (53) is a selective α_1-adrenoceptor antagonist with both central and peripheral actions.[105] Tibalosine is about 1000 times less potent than prazosin in causing blockade of vascular α_1-adrenoceptors and displays anxiolytic activity in animals and a tranquilizing effect in man.[106] The S-enantiomer of WB-4101 (54) was more potent than the R-enantiomer in blocking α_1- and α_2-adrenoceptors and was also more selective toward α_1-receptors in pithed rats.[107]

β-Adrenoceptor Blocking Agents - Compound 55 displayed nearly a 9000-fold greater affinity for β_1- than for β_2-adrenoceptors.[108] ICI 141,292 (56) was also selective for β_1-adrenoceptor mediated cardiovascular and metabolic responses. Intrinsic sympathomimetic activity was found toward cardiovascular, but not metabolic β_1-adrenoceptors.[109] K-351 (57), an antihypertensive agent in SHR, showed non-selective β-adrenoceptor blockade, α-adrenoceptor blockade, and direct vascular relaxation *in vitro*.[110]

Miscellaneous Agents - Atriopeptins I and II (AtI and AtII), two peptides of atrial tissue origin, have been isolated and characterized.[111] These substances are diuretic, natriuretic and relax intestinal (AtI and AtII) and vascular (AtII) tissue.[112] The discovery of these compounds suggests a new endogenous control mechanism involving the heart and kidneys that may be important for maintenance of normal blood pressure.

The ratio of CNS to antihypertensive activities, typically from 1 to 10 for hypotensive tetrahydrocannabinol analogs, has been increased to 100 in (58).[113] However, tachyphylaxis was observed for the antihypertensive activity.

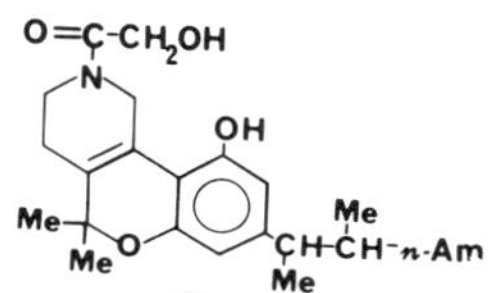

58

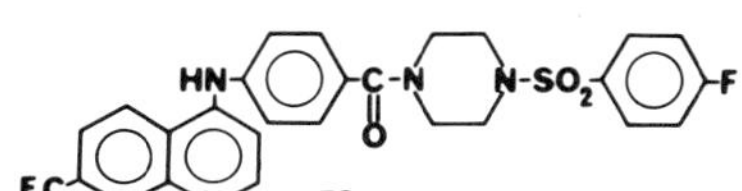

59

U-54,669F (59) lowered blood pressure in hypertensive and normotensive rats and monkeys.[114] The compound inhibited sympathetic neuronal function and depleted catecholamines in cardiac and other tissues. However, there were no effects on postjunctional organ responsiveness to catecholamines nor effects on blood pressure during postural changes.

References

1. M. Moser, Hypertension, 5, 808 (1983).
2. D.S. Goldstein, Hypertension, 5, 86 (1983).
3. B. Folkow, G.F. DiBona, P. Hjemdahl, P.H. Thoren and B.G. Walin, Hypertension, 5, 339 (1983).
4. B.G. Zimmerman, Circ. Res., 53, 121 (1983).
5. R.F. Furchott, Circ. Res., 53, 557 (1983).
6. D.B. Gordon, Hypertension, 5, 353 (1983).
7. M.F. O'Roarke, Aust. NZ J. Med., 13, 84 (1983).
8. T. Inagami, J.J. Chang, C.W. Dykes, Y. Takii, M. Kisaragi and K.S. Misono, Fed. Proc. 47, 2729-2734 (1983).
9. E. Haber, Clin. and Exper. Hyper.-Theory and Practice, A5, 1193-1205 (1983).
10. M. Szelke, B. Leckie, A. Hallett, D.M. Jones, J. Sueiras, B. Atrash and A.F. Lever, Nature, 299, 555-557 (1982).
11. M.J.A. Arnstein, H.B. Bell, D.P. Clough, J.S. Major, A.A. Oldham and J. Saunders, Br. J. Pharmacol., 79, 253P (1983).
12. J. Boger, N.S. Lohr, E.H. Ulm, M. Poe, E.H. Blaine, G.M. Fanelli, T.Y. Linn, L.S. Paine, T.W. Schorn, B.I. LaMont, T.C. Vassil, I.I. Stabilito, D.F. Veber, D.H. Rich and A.S. Bopari, Nature, 303, 81-84 (1983).
13. European Patent Application 82111359.6.
14. T. Unger, D. Ganten and R.E. Lang, Clin. and Exper. Hyper.-Theory and Practice, A5, 1333-1354 (1983).
15. A. Schwab, I. Weinryb, R. Marcerata, W. Rogers, J. Suh and A. Khandwala, Biochem. Pharm., 32, 1957-1960 (1983).
16. M. Burnier, G.A. Turini, H.R. Brunner, M. Porchet, D. Kruithof, R.A. Vukovich and H. Gavras, Br. J. Clin. Pharmacol., 12, 893-899 (1981).
17. T.A. Solomon, F.S. Caruso and R.A. Vukovich, Clin. Pharmacol. Ther., 33, 231 abstr. C28 (1983).
18. J.R. Powell, B. Rubin, C.W. Cushman, J. Krapcho and M.J. Antonaccio, Abstr. Eur. Meet. Hypertens. 1st Milan 355 (1983).

19. D.H. Kim, C.J. Guinosso, G.C. Buzby, Jr., D.R. Herbst, R.J. McCaully, T.C. Wicks and R.L. Wendt, J. Med. Chem., 26, 394-403 (1983).
20. J.L. Dinish, N.K. Metz, R.W. Lappe, D.H. Kim and R.L. Wendt, Pharmacologist, 25, 102 abstr. 8 (1983).
21. J.L. Stanton, N. Gruenfeld, J.E. Babiarz, M.H. Ackerman and R.C. Friedmann, J. Med. Chem., 26, 1267-1277 (1983).
22. K. Hayashi, Y. Ozaki, K. Nunami, T. Uchida, J. Kato, K. Kinashi and N. Yoneda, Chem. Pharm. Bull., 31, 570-576 (1983).
23. T. Unger, T. Yukimura, M. MarinGrez, R.E. Lang, W. Rascher and D. Ganten, Europ. J. Pharmacol., 78, 411-420 (1982).
24. F.J. McEvoy, F.M. Lai and J.D. Albright, J. Med. Chem., 26, 381-394 (1983).
25. E.D. Thorsett, E.E. Harris, S. Aster, E.R. Peterson, D. Taub, A.A. Patchett, E.H. Ulm and T.C. Vassil, Biochem. Biophys. Res. Comm., 111, 166-171 (1983).
26. C.S. Sweet, Fed. Proc. 42, 167-170 (1983).
27. H.H. Rotmensch, P.H. Vlasses, B.N. Swanson, R.O. Davies, D.D. Merill and R.K. Ferguson, Clin. Pharmacol. Ther., 33, 210 abstr. A24 (1983).
28. H.H. Rotmensch, P.H. Vlasses, B.N. Swanson, J.D. Irvin, K.E. Harris, D.G. Merrill and R.K. Ferguson, Am. J. Card., 53, 116-119 (1984).
29. E.J. Sybertz, T. Baum, H.S. Ahn, S. Nelson, E. Eynon, D.M. Desiderio, K. Pula, F. Becker, C. Sabin, R. Moran, G.V. Vliet, B. Kastner and E. Smith, J. Cardiovasc. Pharmacol., 5, 643-654 (1983).
30. M.L. Powell, K. Stabins and J.E. Patrick, Fed. Proc., 42, 378 abstr. 499 (1983).
31. T. Baum, E.J. Sybertz, R.W. Watkins, H.S. Ahn, S. Nelson, E. Eynon, G.V. Vliet, K.K. Pula, C. Sabin, D.M. Desiderio, F.T. Becker and S. Vermulapalli, J. Cardiovasc. Pharmacol., 5, 655-667 (1983).
32. M. Vincent, G. Remond, B. Portevin, B. Serkiz and M. Laubie, Tet. Lett., 23, 1677-1680 (1982).
33. N. Gruenfeld, J.L. Stanton, A.M. Yuan, F.H. Ebetino, L.J. Browne, C. Gude and C.F. Huebner, J. Med. Chem., 26, 1277-1282 (1983).
34. B. Jacot des Combes, G.A. Turini, H.R. Brunner, M. Porchet, D.S. Chen and Supradip (Ben) Sen, J. Cardiovasc. Pharmacol., 5, 511-516 (1983).
35. W. H. Parsons, J.L. Davidson, D. Taub, S.D. Aster, E.D. Thorsett, A.A. Patchett, E.H. Ulm and B.I. Lamont, Biochem. Biophys. Res. Comm., 117, 108-113 (1983).
36. D.S. Chen, R.A. Dotsen, R.D. Burrell and B.E. Watkins, Pharmacologist, 25, 240 abstr. 715 (1983).
37. R.E. Galardy, V. Kontoyiannidou-Ostrem and Z.P. Kortylewicz, Biochem., 22, 1990-1995 (1983).
38. R.G. Almquist, P.H. Christe, W. Chao and H.L. Johnson, J. Pharm. Sci., 72, 63-67 (1983).
39. R. Zelis, Amer. J. Med., 74(6B), 3-12 (1983).
40. M. Konieczkowski, M.J. Dunn, J.E. Stork and A. Hassid, Hypertension, 5, 446-452 (1983).
41. P. Cervoni, P.S. Chan, F.M. Lai and J.E. Birnbaum, Fed. Proc., 42, 157-161 (1983).
42. P.S. Chan, P. Cervoni, M.A. Ronsberg, R.C. Accomando, G.J. Quirk, P.A. Scully and L.M. Lipchuck, J. Pharmacol. Exp. Ther., 226, 727-732 (1983).
43. J.B. Bicking, M.G. Bock, E.J. Cragoe, Jr., R.M. DiPardo, N.P. Gould, W.J. Holtz, T.J. Lee, C.M. Robb, R.L. Smith, J.P. Springer and E.H. Blaine, J. Med. Chem., 26, 342-348 (1983).
44. J.B. Bicking, C.M. Robb, E.J. Cragoe, Jr., E.H. Blaine, L.S. Watson and M.C. Dunlay, J. Med. Chem., 26, 335-341 (1983).
45. J. Casals-Stenzel, M. Buse and W. Losert, Prostaglandins Leukotrienes Med., 10, 197-212 (1983).
46. B. Muller, J. Schneider, K. Wilsmann, W. Lintz and L. Flohe, Prostaglandins Leukotrienes Med., 11, 361-372 (1983).
47. B.J. Whittle, S. Moncada, K. Mullane and J.R. Vane, Prostaglandins, 25, 205-223 (1983).
48. M.A. Orchard, J.M. Ritter, G.L. Shepherd and P.J. Lewis, Br. J. Clin. Pharmacol., 15, 509-511 (1983).
49. C.W. Howden, H.L. Elliot, C.B. Lawrie and J.L. Reid, J. Cardiovasc. Pharmacol., 5, 552-556 (1983).
50. M. Catalano, J. Parini and A. Libretti, Eur. J. Clin. Pharmacol., 24 , 157-161 (1983).
51. T. Chiba, M. Hirohashi, I. Suzuki, S. Tanaka, K. Watanabe and A. Akashi, Arzneim.-Forsch., 33(1), 112-116 (1983).
52. M.J.M.C. Thoulen, J.C.A. van Meel, B. Wilffert, P.B.M.W.M. Timmermans and P.A. van-Zweiten, Pharmacology, 27, 245-254 (1983).
53. H. Lape, A. DeFelice and D. Bailey, Pharmacologist, 25, 101 abstr. 1 (1983).
54. A. DeFelice and H. Lape, Pharmacologist, 25, 101 abstr. 2 (1983).
55. J.M. Evans, C.S. Fake, T.C. Hamilton, R.H. Poyser and E.A. Watts, J. Med. Chem., 26, 1582-1589 (1983).
56. L. Davis, M.N. Agnew, R.C. Effland, J.T. Klein, J.M. Kitzen and M.A. Schwenkler, J. Med. Chem., 26, 1505-1510 (1983).
57. J.M. McCall, J.W. Aiken, C.G. Chidester, D.W. DuCharme and M.G. Wendling, J. Med. Chem., 26, 1791-1793 (1983).
58. O. Thastrup, B. Fjalland and J. Lemmich, Acta Pharmacol. et Toxicol., 52, 246-253 (1983).

59. A. Salimbeni, E. Manghisi, G. Ferni and G.B. Fregnan, II Farmaco.-Ed. Sci., 38, 904-910 (1983).
60. F.V. DeFeudis, Life Sci., 557 (1983).
61. J.L. Rouleau, W.A. Parmley, J. Stevens, I. Wikman-Coffelt, R. Sievers, R.W. Mahley and R.J. Havel, J. Am. Coll. Cardiol., 1, 1453 (1983).
62. P.D. Henry and K.I. Bentley, J. Clin. Invest., 68, 1366 (1981).
63. C.M. Heesch, B.M. Miller, M.D. Thames, and F.M. Abboud, Am. J. Physiol., 245, H653 (1983).
64. W. Kiowski, O. Bertel, P. Erne, P. Bolli, U.L. Hulthen, R. Ritz and F.R. Buhler, Hypertension, 5 (Suppl I), I-70 (1983).
65. R.A.B. McLeay, T.J. Stollard, R.D.S. Watson, and W.A. Littler, Circulation, 67, 1084 (1983).
66. G. Haeusler, J. Pharmacol. Exp. Therap., 180, 672 (1972).
67. J.R. Powell and L.J. Kerwin, Fed. Proc. 42, 842 (1983).
68. H. Kuriyama, Y. Ito, H. Suzuki, K. Kitamura, T. Itoh, M. Kajwara and S. Fujiwara, Circ. Res., 52 (Suppl I), 92 (1983).
69. M. Wibo, I. Delfosse and T. Godfraind, Arch. int. Pharmacodyn., 263, 333 (1983).
70. P.A. van Zweiten and P.B.M.W.M. Timmermans, J. Mol. Cell. Cardiol., 15, 717 (1983).
71. W.G. Nayler, J.E. Thompson and B. Jarrott, J. Mol. Cell. Cardiol., 14, 185 (1982).
72. A-M. Galzin and S.Z. Langer, Br. J. Pharmacol., 78, 571 (1983).
73. J.C.A. van Meel, B. Wilffert, K. de Zoeten, P.B.M.W.M. Timmermans and P.A. van Zweiten, Arch. int. Pharmacodyn., 260, 206 (1982).
74. P.B.M.W.M. Timmermans, A. de Jonge, J.C.A. van Meel, M.J. Mathy and P.A. van Zweiten, J. Cardiovas. Pharmacol., 5, 1 (1983).
75. J.C. van Meel, P.B.M.W.M. Timmermans and P.A. van Zweiten, Eur. J. Pharmacol., 92, 27 (1983).
76. P.A. van Zweiten, J.C.A. van Meel and P.B.M.W.M. Timmermans, Circ. Res., 52, (Suppl I), 77 (1983).
77. M. Saeed, J. Holtz, D. Elsner and E. Bassenge, Eur. J. Pharmacol., 94, 149 (1983).
78. P.B.M.W.M. Timmermans, M.J. Mathy, B. Wilffert, H.O. Kalkman, M.J.M.C. Thoolen, A. de Jonge, J.C.A. van Meel and P.A. van Zweiten, Naunyn-Schmiedebergs Arch. Pharmacol., 324, 239 (1983).
79. I. Cavero, N. Shepperson, F. Lefevre-Borg and S.Z. Langer, Circ. Res. 52, (Suppl I), 69 (1983).
80. H. Vierhapper and W. Waldhousl, Eur. J. Clin. Med., 12, 263 (1982).
81. R.D. Smith, D.V. Romano, B. Loev, T.P. Pruss and P.S. Wolf, Pharmacologist, 24, 137 (1982).
82. W.S. Mann, P.S. Wolf, R.D. Smith and B. Loev, Pharmacologist, 24, 137 (1982).
83. S.M. Ringel, A. Darragh, J. Rosenthal, F.S. Caruso and R.A. Vukovich, Clin. Pharmacol. Therap. 33, 231 (1983).
84. R.G. Van Inwegen, L. Baukh, P. Feeney, B. Loev, J. Schroff and A. Khandwala, Fed. Proc. 42. 843 (1983).
85. P.M. Thorpe, Drugs Future, 8, 770 (1983).
86. T. Furuta, S. Shibata, I. Kodama and K. Yamada, J. Cardiovas. Pharmacol., 5, 836 (1983).
87. M. Ohtusuka, T. Ono, J. Hiroi, K. Esumi and S. Kamada, J. Cardiovas. Pharmacol., 5, 1074 (1983).
88. M. Schramm, G. Thomas, R. Towart and G. Franckowiak, Nature (London), 303, 535 (1983).
89. M. Schramm, G. Thomas, R. Toward and G. Franckowiak, Arzneim-Forsch., 33, 1268 (1983).
90. J.F. Lee, L.F. Rozek, J.H. Sanner and F.M. Radzialowski, Meth. Find. Exptl. Clin. Pharmacol., 5, 235 (1983).
91. T. Morita, T. Kanazawa, K. Ito and T. Nose, Arzneim-Forsch., 32, 1047 (1982).
92. J.J. Lynch, E. Patterson, D.G. Montgomery and B.R. Lucchesi, Pharmacologist, 25, 247 (1983.
93. P.M. Van Houtte, J.M. van Neuten, J. Symoens and P.A.J. Janssen, Fed. Proc. 42, 182 (1983).
94. M.L. Cohen, R.W. Fuller and K.D. Kurz, Hypertension, 5, 676 (1983).
95. M.L. Cohen, R.W. Fuller and K.D. Kurz, J. Pharmacol. Exp. Therap., 227, 327 (1983).
96. D.M. Ackerman, A.L. Blumberg, J.P. McCafferty, S.S. Sherman, J. Weinstock, C. Kaiser and B. Berkowitz, Fed. Proc., 42, 186 (1983).
97. R.A. Hahn, B.R. MacDonald and M.A. Martin, J. Pharmacol. Exp. Therap. 224, 206 (1983).
98. R.W. Fuller, S.K. Hemrick-Luecke, D.T. Wong, D. Pearson, P.G. Threlkeld and M.D. Hynes, J. Pharmacol. Exp. Therap., 227, 354 (1983).
99. A.L. Jadhav, R.N. Willet, H.N. Sapru and M.F. Lockhandwala, Naunyn-Schmiedebergs Arch. Pharmacol., 324, 281 (1983).
100. H. Ruskoaho, J-M, Savoala, S. Kaipiainen, J. Puurunen and N. Karki, Nauyn-Schmiedebergs Arch. Pharmacol., 322, 279 (1983).
101. T. Sekia, H. Hiranuma, S. Hata, S. Miyogama, M. Hanazuka and S. Yamada, J. Med. Chem., 26, 411 (1983).
102. J.W. Constantine, W. Lebel and R. Weeks, J. Cardiovas. Pharmacol., 6, 142 (1983).
103. R.D. Clark, J.M. Caroon, A.F. Kluge, D.B. Repke, A.P. Roszkowski, A.M. Strosberg, S.M. Bitter and M.D. Okada, J. Med. Chem., 26, 657 (1983).

104. A. Karawawa, K. Shuto, K. Kubo, Y. Kasya, M. Hashikomi and K. Shigenobu, Arch. int. Pharmacodyn., 261, 278 (1983).
105. J.Q. Qian, M.J. Mathy, M.J.M.C. Thoolen, P.B.M.W.M. Timmermans and P.A. van Zweiten, Arch. int. Pharmacodyn., 266, 264 (1983).
106. J. Staessen, R. Fagard, R. Grauwels, P. Lijnen, L. Verschueren and A. Amery, Clin. Pharmacol. Therap., 33, 556 (1983).
107. R.R. Ruffolo, W.L. Nelson and E.L. Yoden, Naunyn-Schmiedebergs Arch. Pharmacol., 322, 93 (1983).
108. J.J. Baldwin, G.H. Denny, R. Hirschmann, M.B. Freedman, G.S. Ponticello, D.M. Gross and C.S. Sweet, J. Med. Chem., 26, 950 (1983).
109. H.J. Smith, S.E. Halliday, D.C.N. Earl and D. Stribling, J. Pharmacol. Exp. Therap., 226, 211 (1983).
110. Y. Uchida, M. Nakamura, S. Shimizu, Y. Shirasawa and M. Fujii, Arch. int. Pharmacodyn., 262, 132 (1983).
111. M.G. Currie, D.M. Geller, B.R. Cole, J.G. Boylan, W.Y. Sheng, S.W. Holmberg and P. Needleman, Science, 221, 71-73 (1983).
112. M.G. Currie, D.M. Geller, B.R. Cole, N.R. Siegel, K.F. Fok, S.P. Adams, S.R. Eubanks, G.R. Galluppi and P. Needleman, Science, 223, 67-69 (1984).
113. H.E. Zaugg and J. Kyncel, J. Med. Chem., 26, 214-217 (1983).
114. M.G. Wendling, D.W. DuCharme, G.A. Johnson, R.B. McCall and D.T. Pals, Fed. Proc. 42, 162 (1983).

Chapter 8. Cardiotonic Agents

Bernd Wetzel and Norbert Hauel
Dr. Karl Thomae GmbH, D-7950 Biberach, W.-Germany

Introduction - Heart failure is an abnormality in cardiac function that results in the inability of the heart to pump blood to the peripheral sites, with the consequence that the metabolic requirements of the body remain unsatisfied. Factors which contribute to heart failure include myocardial ischemia, hypertension, non-obstructive cardiomyopathy, and congenital heart disease. Therapy of heart failure has focused on three approaches: (1) Reduction of volume overload by diuretic therapy, (2) improvement of depressed cardiac function by stimulation of myocardial contractility, (3) reduction of vascular resistance by means of systemic vasodilators. The principles of treating heart failure have been the subject of several review articles in recent years.[1-7] This chapter is primarily concerned with the major developments of new inotropic agents over the last three years. Compounds which are not reviewed are glucagon,[8] anthopleurin,[9] ionophores,[10] the cardiotonic principles of ginger,[11] a newly synthesized analog of PGI_2,[12] and SKF 86466, a selective α_2-adrenoceptor agonist.[13]

Cardiac glycosides - Although their clinical usefulness is unquestionable, therapy with cardiac glycosides is, even today, associated with numerous side effects.[14,15] Three types of intoxication are known: arrhythmias, gastrointestinal disorders and CNS effects. There is still a debate on the mode of action of cardiac glycosides. The positive inotropic activity of these agents seems to be due to the inhibition of myocardial cell membrane Na^+-K^+-ATPase (Na^+-pump), on the other hand there is evidence that this may not be the only mechanism of action.[16-18] Pengitoxin (1), an orally active analog of gitoxin, produced a dose-dependent shortening of systolic time intervals in healthy volunteers. Fatigue and nausea were observed, but no arrhythmias.[19] Clinical studies indicated that it is as effective as digitoxin with therapeutic plasma concentrations of 17.5 - 37.5 µg/ml; at higher levels toxic side effects developed.[20] The structurally related gitoformate (2) has a calculated therapeutic index (toxicity threshold / ED_{50}) of 4.6.[21,22] In vivo gitoformate is hydrolyzed to gitoxin.[23] In patients with clinical signs of latent cardiac insufficiency (stages II and II-III) p.o. administration of gitoformate led to complete recompensation.[24] No side effects occurred.[25] ASI-222 (3), a semisynthetic 4-aminosugar cardiac glycoside, demonstrated a greater inotropic potency and therapeutic index in dogs than its neutral ß-D-galactose analog, digoxin or ouabain.[26-28] On the other hand, it is reported that 3 causes neural activation and does not offer an improved therapeutic-to-toxic ratio, despite the fact that it is a highly polar compound which does not cross the blood-brain barrier.[29] Substitution of the ß-lactone moiety of digoxin by a pyridazine ring to provide SC-4453 (4), did not show much difference in interaction with the inhibitory sites of Na^+- K^+-ATPase.[30] The bioavailability of 4 is high, the cardioactive properties are similar to those of digoxin, while the pharmacokinetic parameters in guinea pigs are different.[31,32] Meproscillarin (5) showed an unusually rapid onset of action following i.v. administra-

ISBN 0-12-040519-9

tion in healthy volunteers. Oral 5 is well absorbed.[33] Since 90% is excreted in the bile, dosage adjustment is not necessary for renal function impairment.[34,35] Acrihellin (D-12.316) (6) is an acylated derivative of hellebrigenin.[36] Recent results from electrophysiological studies on the Purkinje fiber suggest a significantly smaller arrhythmogenic action than other known glycosides.[37]

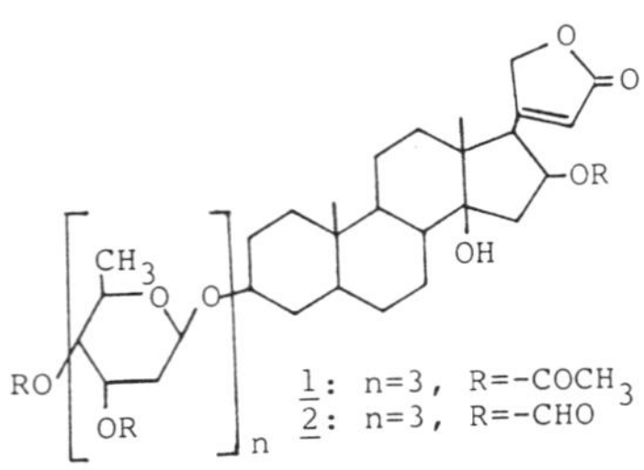

1: n=3, R=-$COCH_3$
2: n=3, R=-CHO

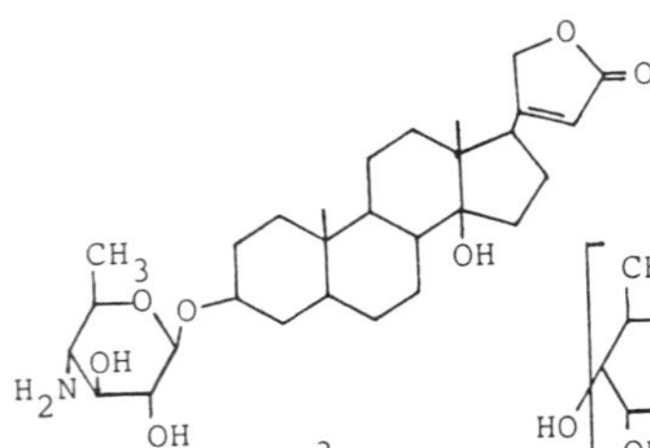

3

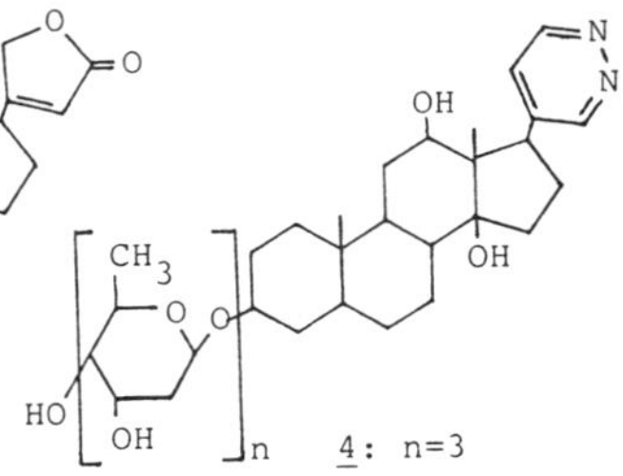

4: n=3

In healthy volunteers significant inotropic effects developed after 0.015 mg/kg p.o. The maximum tolerable single oral dose was found to be 0.017 mg/kg.[38] Proscillaridin and its 14,15-ß-oxido analog, HOE-040 (7), are reported to have equipotent positive inotropic effects in dogs. The lethal dose of 7 and the dose causing arrhythmia are two to five times higher.[39] The compound is well absorbed in dogs. The rate of decline of plasma levels is 100% in 24 hours.[40]

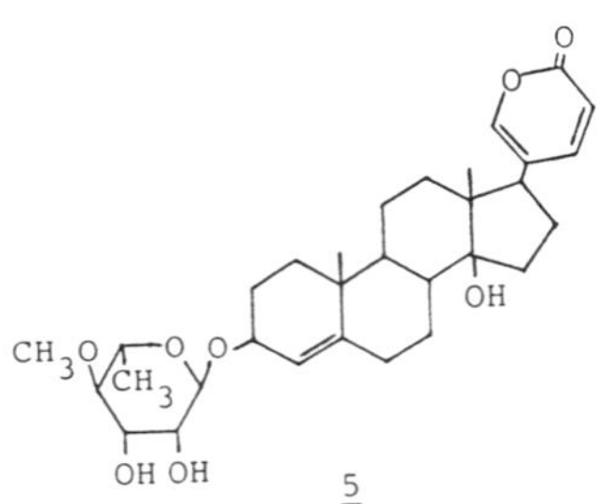

5

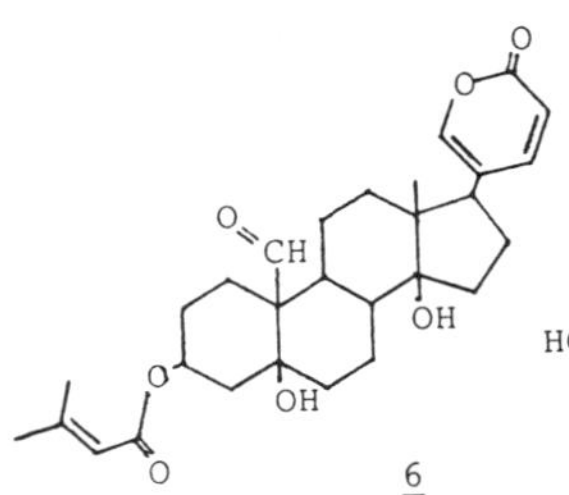

6

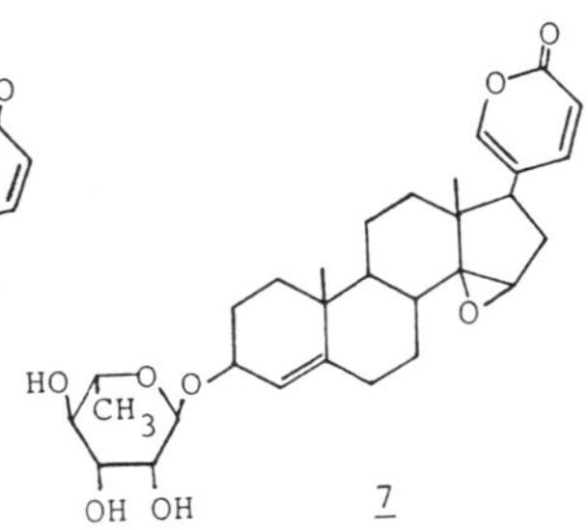

7

Besides the well characterized compounds 1 - 7, the chemical and pharmacological properties of a variety of novel natural and semisynthetic cardiac glycosides were reported.[41-46] Recently an endogenous factor, "cardiodigin", with digitalis-like activity was discovered.[47] It was recovered from the postnuclear particulate material of bovine, guinea pig, and rat heart homogenates. This material inhibited Na^+-K^+-ATPase in a dose-dependent manner, and had an affinity for the digitalis receptor 10-100 times higher than that of digoxin.

ß-Adrenoceptor Stimulants - Catecholamines increase contractility by stimulating postsynaptic cardiac adrenoceptors, which have been recently reviewed.[48] The clinical usefulness of the standard amines, such as epinephrine, norepinephrine, and isoproterenol, is limited by their relative chronotropic and peripheral receptor effects. It is the goal of current research to reduce these unwanted side-effects and to overcome the lack in oral absorption. Studies in the ventricular myocardium of kittens were performed with 23 different adrenoceptor agonists.[49] It was demonstrated that the inotropic potency is determined by both the intrinsic activity for stimulating adenylate cyclase and the affinity for the ß-adrenoceptors. In a short communication the role of ß-agonists in heart failure was discussed with respect to their receptor subtype specificity.[50]

Dopamine and dobutamine (8) have both been approved for the i.v.-

treatment of congestive heart failure. Their clinical use has been reviewed.[51] In a crossover study in patients with cardiac failures both drugs were shown to have comparable beneficial effects. Maximal doses were limited by frequent occurrence of premature ventricular beats and angina pectoris.[52] The stereoisomers of 8 were analyzed with regard to their α- and ß-adrenergic activity *in vitro* and their vascular effects in pithed rats.[53,54] Ibopamine (9) is the orally effective diisobutyric ester of N-methyldopamine.[55] A number of clinical evaluations have now been published. In one study the invasive assessment of hemodynamic parameters in patients with severe congestive heart failure (150 mg p.o) was reported.[56] In another study, 50 mg twice a day for seven days was administered to eight patients, who all showed symptomatic improvement during therapy.[57] Butopamine (10) is an orally active analog of dobutamine, which has no catecholamine-like structure.[58] After i.v. administration to volunteers, 10 manifested substantial positive inotropic effects at 0.04 - 0.08 mcg/kg/min with positive chronotropy at higher doses.[59] The structure-activity relationships of several aminotetralin derivatives of A-5,6-DIN and A-6,7-DIN (11), two rigid forms of dopamine, were investigated.[60,61] One of them, ASL-7022 (12), was found to be more inotropically selective than dopamine or dobutamine.[62] TA-064 (13) increased dp/dt in conscious dogs for longer than seven hours after an oral dose of 0,4 mg/kg.[63,64] In normal and diseased human hearts, 13 improved contractile function at a dose that did not affect heart rate and blood pressure.[65]

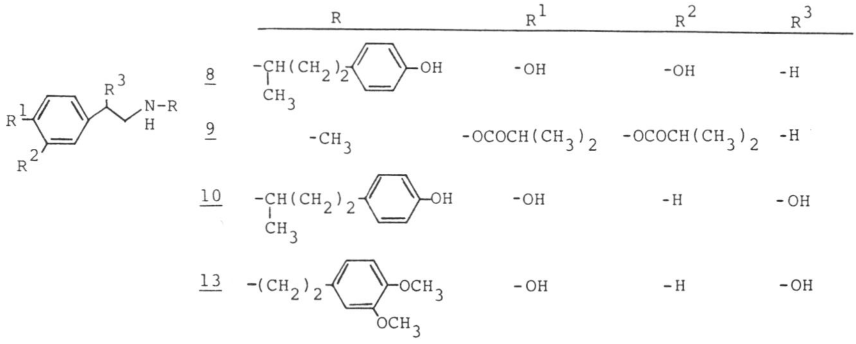

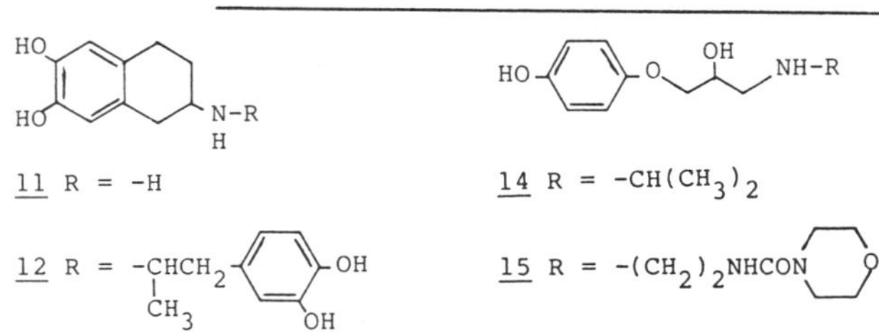

Prenalterol (14) was introduced as an orally active ß-adrenoceptor agonist for the treatment of mild to moderate heart failure.[51] Several recent clinical studies were reported.[66,67] Its effect was no longer significant after three to six months of treatment.[68] Recent studies have dealt with the mechanism of action of 14 and its receptor subtype selectivity.[69] The agent was characterized as a partial $ß_1$-agonist.[70] However, another study suggested that the $ß_1/ß_2$-selectivity of 14 is tissue dependent but not receptor subtype dependent.[71] From a series of arylethanolamines and aryloxypropanolamines the orally active xamoterol (ICI 118,587) (15) was selected for clinical development because of its partial agonist properties and its high cardioselectivity.[72,73] Xamoterol improved systolic and diastolic left ventricular function in patients with coronary artery disease without inducing myocardial ischaemia.[74] Results both in dog and in man showed that the drug is able to act as an agonist when sympathetic activity is low, and as an antagonist reducing heart rate when activity is high, such as during exercise.[75] It appears likely that partial β_1-agonists like 14 or 15 will be used primarily in patients with mild to moderate heart failure (class II - III of NYHA).[76] A review of the structure-activity relationships in ß-adrenoceptor-aryloxypropanolamines dealing with the question of optimal substitution pattern for cardiac β_1-selectivity was presented.[77] A dibenzoxepine derivative, doxaminol (16), is an orally active ß-sympathomimetic agent, having positive inotropic activity without obvious effects on heart rate or blood pressure in animals, healthy volunteers, and heart failure patients.[78-81]

Pirbuterol (17), albuterol (18), and terbutaline (19) have been widely used in the management of asthma because of their well known beneficial bronchodilatory effects. These β_2-adrenoceptor agonists relax vascular smooth muscles and increase myocardial contractility. Oral and i.v. activity of 17 have been evaluated clinically by several groups of investigators.[51,82-84] The drug appears to be well tolerated. In one study a marked decrease in lymphocyte ß-receptors was observed after a four weeks treatment with accompanying tolerance to the hemodynamic effects.[85] Pirbuterol was launched in the UK with the claim that it is a new selective bronchodilator with beneficial effects on the heart.[86]

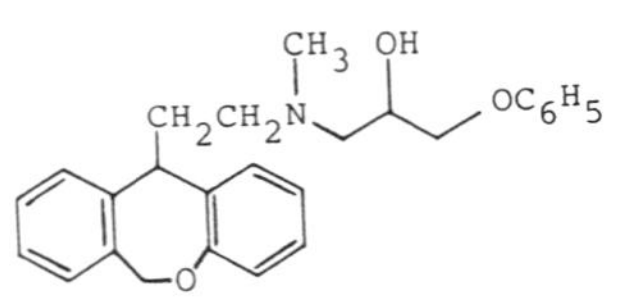

16

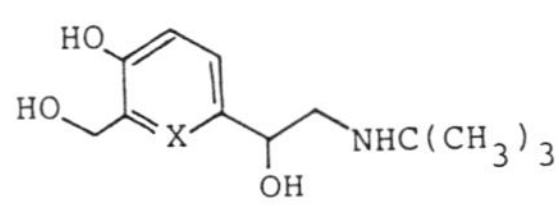

17 X = N

18 X = CH

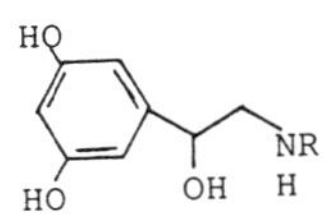

19 R = $C(CH_3)_3$

20 R = 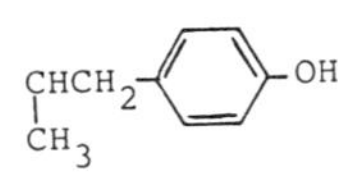

Albuterol has been used in acute myocardial infarction and in severe chronic heart failure. In a recent study a sustained improvement of cardiac function was found after a single oral administration of 4-8 mg.[87] The hemodynamic effects of 19 were measured in patients with severe heart failure.[88] Fenoterol (20), another β_2-adrenoceptor agonist, showed significant improvement in pump function in seven patients with severe congestive heart failure (NYHA IV) after intravenous administration.[89] All these β_2-adrenoceptor agonists can be classified as vasodilators which exert their effect primarily due to a reduction in afterload and possess some positive inotropic effect.[90] In summary long term studies with chronic administration have still to be performed for all ß-adrenoceptor agonists before any of them can be advocated for routine maintenance therapy.

Inhibitors of phosphodiesterase increase the intracellular level of cyclic AMP, which is associated with increased myocardial contractility. One of the first PDE-inhibitors so studied is theophylline (21), an important drug in the treatment of asthma. Its actions have been extensively reviewed.[91] Several recent studies have indicated that it might stimulate contractility also by adenosine receptor antagonism.[92] In patients with chronic obstructive pulmonary disease, theophylline (14 mg/kg, p.o.) improved cardiac performance.[93,94] Severe adverse side effects were observed.[95,96] Many derivatives of theophylline and other purine analogs were prepared and tested as PDE-inhibitors and cardiac stimulants, some of them being several times more active than theophylline *in vitro* and *in vivo*.[97-99]

Buquineran (UK 14.275) (22) showed PDE-inhibiting activity 20 times as potent as theophylline and a dose-related increase in cardiac output and left ventricular dp/dt in anesthetized dogs.[100] In patients, 22 increased cardiac output and reduced systemic and pulmonary vascular resistance.[101] Since it showed a potentially arrhythmogenic effect (prolongation of the QT interval), no further clinical development has occurred.[93] Carbazeran (UK 31.557) (23) is a potent inotropic compound in experimental animal models. In volunteers it showed cardiotonic and positive chronotropic activity without significant prolongation of the QT interval.[93,102] UK 36.327 (24) produced dose-related increases in dp/dt max. in anesthetized dogs, accompanied by transient falls in blood pressure and small increases in heart rate.[103] After oral administration an increase in contractility was observed in conscious dogs.[104]

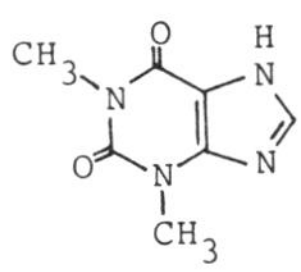

21

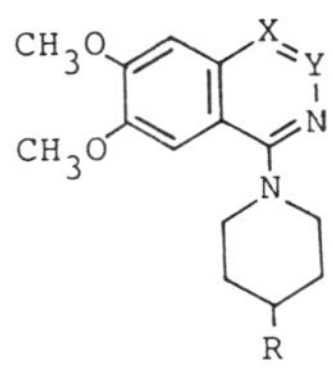

22: X = N; Y = CH
R = NHCONH-n-C_4H_9
23: X = CH; Y = N
R = $OCONHC_2H_5$
24: X = N; Y = CH
R = $CH(CH_3)CH_2N(CH_3)COCH_3$

A variety of 4,5-dihydro-6-phenyl-3(2H)-pyridazinone derivatives are PDE-inhibitors. CI-914 (25) produced cardiotonic effects in anesthetized dogs as well as in conscious dogs after oral administration, accompanied by only slight decreases in blood pressure and moderate increases in heart rate.[105-107] Oral absorption was complete, and the inotropic activity correlated well with plasma levels.[108] PD-112.548 (26), an analog of CI-914, is 5-10 times more potent in conscious dogs.

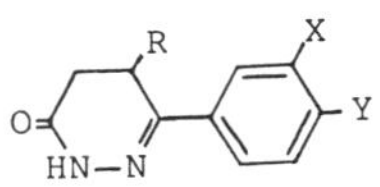

25: R = H; X = H;
Y = 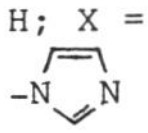
26: R = CH_3; X = H
Y = 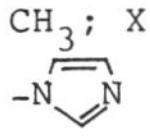
27: R = CH_3
X,Y = 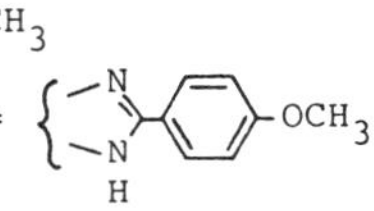

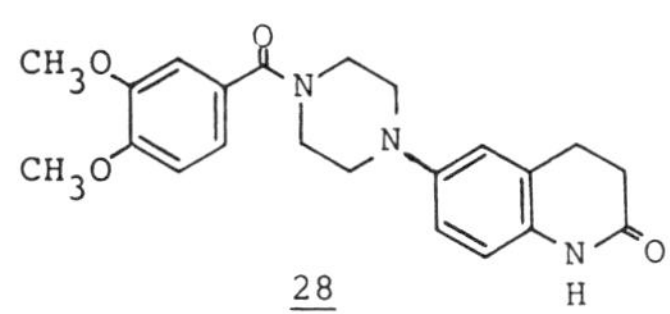

28

Pimobendan (UD-CG 115) (27) is another pyridazinone which produces cardiotonic activity by elevating intracellular cAMP levels.[109] By the i.v.-route, it increased cardiac contractility in anesthetized cats up to 110%, heart rate rose only up to 21%, and arterial blood pressure decreased by 34 mm Hg. Oral administration to conscious dogs enhanced dp/dt max. up to 97% for 6 to 10 hr with little or no changes in heart rate and blood pressure. Cardiac oxygen consumption was not significantly changed, but coronary blood flow increased slightly.[110] The cardiovascular properties of OPC-8212 (28) were examined in various isolated heart preparations. Intravenous and oral administration produced positive inotropic effects in conscious dogs, without significant effects on heart rate and blood pressure.[111]

Forskolin (29), a natural diterpene, has become an important tool in cell physiology due to its unique mechanism of action.[112-114] It exerts its activity either by direct action on the catalytic subunit of adenylate cyclase or by indirect action via a previously unrecognized cyclase component.[113,115-117] In a recent study, structural analogs of 29 were synthesized and investigated in a guinea pig atrium model.[118] None was more potent than forskolin (EC_{50} for the force of contraction: 2.48 x 10^{-8} M), which is now scheduled for clinical trials.

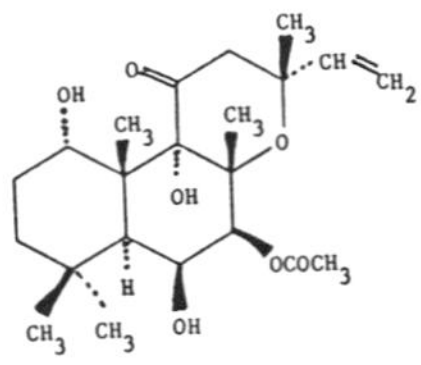

29

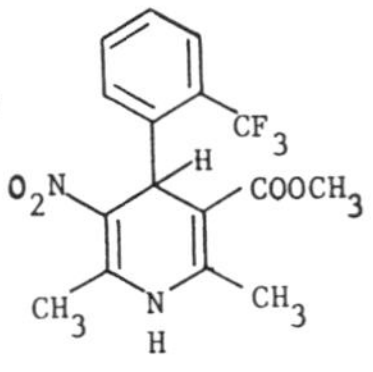

30

Bay K 8644 (30) is a novel 1,4-dihydropyridine derivative.[119,120] In contrast to the nifedipine-like dihydropyridines which have calcium antagonistic actions, 30 promotes the influx of calcium ions and therefore produces positive inotropic and vasoconstrictor effects. In the isolated guinea pig heart, the positive inotropic effect begins at 10^{-9} mol/l and reaches a plateau at 10^{-7} mol/l. In various models the pharmacological effects of 30 were found to be competitively antagonized by nifedipine, but not by other calcium antagonists such as verapamil or diltiazem.

Agents Acting by Novel or Incompletely Defined Mechanisms

The pharmacological and clinical effects of amrinone (31), a bipyridyl analog, have been reviewed.[121-123] Following intravenous (0,1 -1,0 mg/kg) and oral (2-10 mg/kg) administration, 31 produced positive inotropic effects in experimental animals. In dogs with acute left ventricular failure it significantly improved hemodynamic parameters. These effects are possibly due to an increase in the Ca^{++} inward current, a slight histaminergic action, and an increase in intracellular cAMP-levels. Clinical effects were studied in patients with moderate to severe congestive heart failure. Oral 31 (100 mg) increased cardiac index (+29%), decreased pulmonary capillary wedge pressure (-31%), and decreased right atrial pressure (-25%). The drug seems to have significant beneficial effects in the acute treatment of congestive heart failure. Since a variety of adverse reactions such as gastrointestinal disorders and thrombocytopenia were observed, the long term benefits of the drug remain uncertain.[121,124] Milrinone (WIN-47.203) (32), an analog of amrinone, is about 30 times more potent. I.v. injection (0,01-0,1 mg/kg) and oral doses (0,1-1,0 mg/kg) increased contractile force in dogs by 35 to 99%, with a maximum increase in heart rate of 40% and no significant change in blood pressure.[125] Although 32 is a potent cAMP inhibitor, this effect seems unlikely to be responsible for the initiation of the inotropic action. The clinical effects were studied in patients with severe chronic congestive heart failure.[126-128] Intravenous infusion and oral administration resulted in a significant improvement of hemodynamic parameters. Since long-term administration remained well tolerated, this drug appears to be very promising for treatment of chronic cardiac failure. APP 201-533 (33) is another analog of

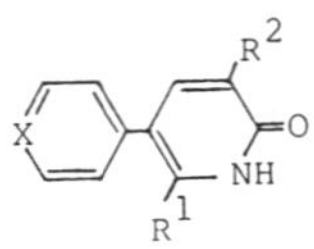

31: X = N; R^1 = H; R^2 = NH_2

32: X = N; R^1 = CH_3; R^2 = CN

33: X = CH; R^1 = CH_3; R^2 = NH_2

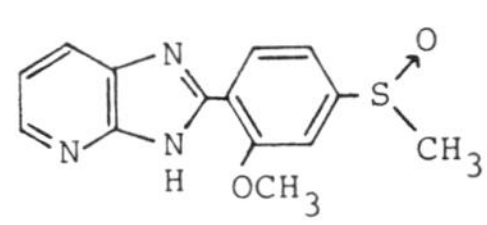

37

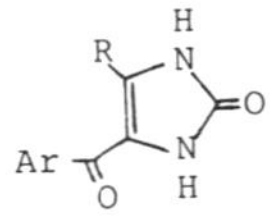

34: R = CH_3; Ar = 4-methoxyphenyl

35: R = CH_3; Ar = 4-methylthiophenyl

36: R = C_2H_5; Ar = 4-pyridyl

amrinone which is reported to produce positive inotropic actions in experimental animals following intravenous and oral administration.[129,130] Compared to amrinone it seems to be less toxic and to have a broader therapeutic range.

A series of 4-aroyl-1,3-dihydro-2H-imidazol-2-ones were synthesized and evaluated for pharmacological activity in the anesthetized dog.[131] Most of these compounds produced dose-related increases in cardiac contractile force and heart rate, and decreases in systemic arterial blood pressure. Three (34, 35 and 36) were studied in detail, two of them are currently undergoing clinical trials. RMI-82,249 (34) increased force and rate of cardiac contraction and decreased blood pressure in anesthetized dogs.[132,133] Oral activity was demonstrated in conscious dogs.[134] These effects were not antagonized by propranolol but by verapamil, suggesting a dependence on calcium flux via the "slow channels". MDL-17,043 (35) produced positive inotropic activity in anesthetized dogs and in conscious dogs.[135] The effects were not altered by reserpine pretreatment or by α- or ß-blockade.[136] The drug did not alter uptake, binding or release of Ca^{++} from isolated sarcoplasmic reticulum, but an inhibition of dog heart cAMP-phosphodiesterase was observed. I.v. administration of 35 produced a positive inotropic effect in patients with reduced left ventricular performance and severe congestive heart failure.[137,138] The effects lasted for up to six hours and were not associated with any adverse side-effects. MDL-19,205 (36) is reported to be equipotent with 35, while producing a greater increase in heart rate in anesthetized dogs. It selectively inhibited the low Km cAMP-phosphodiesterase from dog heart *in vitro*. Oral administration in conscious dogs produced dose-dependent increases in dp/dt.[139,140] The drug is completely and rapidly absorbed, and renal clearance is the major mechanism of elimination.[141]

Sulmazole (AR-L 115) (37), an imidazopyridine derivative, increased dp/dt max. by 58% and heart rate by 12 beats/min in anesthetized cats after 1 mg/kg i.v. Systolic blood pressure was decreased by 13 mm Hg.[142] Oral administration augmented dp/dt max. up to 92% for 5 to 12 hours in conscious dogs.[143] Despite similar PDE-inhibitory activity of 37 and theophylline, it seems likely that effects other than PDE-inhibition contribute to the positive inotropic action.[144] Measurement of the effect of 37 on relations between free Ca^{++}, bound Ca^{++}, and ATPase activity of dog cardiac myofibrils indicates that the positive inotropic actions may involve direct activation of myofibrils by an increased affinity of thin filament receptors for Ca^{++}.[145] Several clinical studies in heart failure showed that in patients with congestive cardiomyopathy and with coronary artery disease, 37 improved hemodynamics and regional wall motion via decrease in pre- and afterload and increase in contractility, myocardial oxygen consumption and coronary sinus flow, with no myocardial ischemia.[146-149]

References:

1. H. Scholz, Eur. Heart J., 4, (Suppl. A), 161 (1983)
2. J.A. Bristol and D.B. Evans, Med. Res. Rev., 3, 259 (1983)
3. D.B. Evans, R.E. Weishaar and H.R. Kaplan, Pharmac. Ther. 16, 303 (1982)
4. P. Jerie, Schweiz. Rundsch. Med. Prax., 72, 15 (1983)
5. S.H. Taylor, B. Silke and G.I.C. Nelson, Eur. Heart J., 3, (Suppl. D), 19 (1982)
6. F. Sestier, J.Y. Pagny and M. Sestier, Ann. Cardiol. Angiol. 31, 7 (1982)
7. A.B. Schwartz and K. Chatterjee, Drugs, 26, 148 (1983)
8. L.F. Yao, K.M. MacLeod and J.H. McNeill, Eur. J. Pharmacol., 79, 147 (1982)
9. T.R. Norton, Fed. Proc., 40 (1), 21 (1981)
10. J.A. Bristol and D.B. Evans, Ann. Rep. Med. Chem., 16, 93 (1981)
11. N. Shoji, A. Iwasa, T. Takemoto, Y. Ishida and Y. Ohizumi, J. Pharm. Sci, 71, 1174 (1982)
12. G. Fassina, F. Tessari and P. Dorgio, Pharmacol. Res. Commun., 15, 735 (1983)
13. R.M. De Marinis, J.P. Hieble and W.D. Matthews, J. Med. Chem., 26, 1213 (1983)
14. K.-O. Haustein, in Pharmacology and Therapeutics, Vol. 18, W.C. Bowman, A.M. Breckenridge and A.S. Sartorelli, Pergamon Press, 1982
15. L. Storstein and S.H. Taylor, Herz, 8, 1 (1983)
16. T. Godfraind, F. Noel and M. Finet, J.Mol.Cell. Cardiol. 15, (Suppl. 2), 43 (1983)

17. H. Luellmann, T. Peters and J. Preuner, Eur. Heart J., 3 (Suppl. D), 45 (1982)
18. L.S. Cook, R.W. Caldwell, C.B. Nash, J.E. Doherty and K.D. Straub, J. Pharmacol. Exp. Ther., 223, 761 (1982)
19. H. Hentschel and K.-O. Haustein, Zentralbl. Pharm. Pharmakother. Laboratoriumsdiagn., 122, 225 (1983)
20. K.-O. Haustein and M. Wesser, Farmakol. Toksikol., 45, 67 (1982)
21. E. Schellberg and U. Fricke, Arch. Pharmacol., 322 (Suppl. R 34), (1983)
22. A. De Pover and T. Godfraind, Arch. Pharmacol., 321, 135 (1982)
23. L. Dodion, Drugs Exp. Clin. Res., 6, 189 (1980)
24. R.C. Mueller Haake and H.C. Kuemmel, Therapiewoche 31, 5417, 5423 (1981)
25. L. Dei Cas, Curr. Ther. Res., 29, 685 (1981)
26. R.W. Caldwell and C.B. Nash, J. Pharmacol. Exp. Ther., 197, 19 (1976)
27. L.S. Cook, R.W. Caldwell and C.B. Nash, J. Pharmacol. Exp. Ther., 215, 198 (1980)
28. R.W. Caldwell, Drugs of the Future VII (10), 719 (1982)
29. J.C. Somberg, J.E. Kuhlmann and T.W. Smith, Circulation Res., 49, 226 1981)
30. T. Godfraind and N.T. Lutete, Eur. J. Pharmacol., 60, 329 (1979)
31. A. Marzo, P. Ghirardi, C. Casagrande and R.W. Jelliffe, Drugs Exp. Clin. Res., IX, 207 (1983)
32. F. Randimbivololona and M. Lesne, J. Pharmacol., 14, 9 (1983)
33. U. Peters, Eur. Heart J., 3 (Suppl. D), 65 (1982)
34. F. Salomon and E. Quelhorst, Therapiewoche, 31, 5789 (1981)
35. H. Brass, Therapiewoche, 31, 1918, 1923 (1981)
36. K. Eiser, H. Mattern, W. Runkel and G. Fricke, Z. Kardiol., 71, 601 (1982)
37. R. Ziskoven, J. Wiemer and C. Achenbach, Drug. Res., 33, 1106 (1983)
38. J. Engel, O. Isaac, A. v. Schlichtegroll and K. Thiemer, Drugs of the Future, 6 459 (1981)
39. E. Lindner, G.V. Reitzenstein and H.H. Schoene, Arzneim.-Forsch., 29, 221 (1979)
40. U. Stache, K. Radscheit, W. Haede and W. Fritsch, Ann. Chem., 1982, 342
41. Y. Gomita, M. Moriyama, Y. Ichimaru, A. Uchikado, T. Nohava and T. Kawasaki, J. Pharm. Soc. Japan, 102, 495 (1982)
42. K. Shimada, T. Kyuno, T. Nambara and J. Uchida, Chem. Pharm. Bull., 30, 4075 (1982)
B. Swynghedauw, F.X. Jarreau, A. Nittenberg, C. Mouas, M. Preteseille and L. Lelievre, J.Mol.Cell. Cardiol. 15, (Suppl. 2), 55 (1983)
43. H. Tawfik, A.A.L. Fox and K. Greeff, Z. Kardiol., 72 (Suppl. 1), 104 (1983)
J. Engel, O. Isaac, K. Posselt, K. Thiemer and H. Uthemann, Drug Res. 33(II), 1215 (1983)
44. G. Pastelin and R. Mendez, Life Sci., 32, 1905 (1983)
45. H. Luellmann, T. Peters and J. Preuner, Eur. Heart J., 3 (Suppl. D), 45 (1982)
46. H. Jin, T.Y.R. Tsai and K. Wiesner, Can. J. Chem. 61, 2442 (1983)
47. T. Godfraind, A. De Pover, G. Castaneda Hernandez and M. Fagoo, Arch. Int. Pharmacodyn., 258, 165 (1982)
48. K.J. Broadley, J. Anton. Pharmacol. 2, 119 (1982)
49. A.J. Kaumann, Naunyn Schmiedeberg's Arch. Pharmacol., 317, 13 (1981)
50. A.J. Kaumann, Lancet, 5, 1063 (1983)
51. A.P. Rae, A.C. Tweddel, I. Hutton, Herz, 8, 23 (1983)
K. Chatterjee, R. Bendersky, W.W. Parmley, Eur. Heart. J., 3 (Suppl. D), 107 (1982)
52. B. Knippa, M.S. Verani, A.C. Montero, R.R. Miller, M.A. Quinones, Circulation, 66, 278 (1982)
53. R.R. Ruffolo, E.L. Yaden, J. Pharmacol. Exp. Ther., 224, 46 (1983)
54. R.R. Ruffolo, T.A. Spradlin, G.D. Pollock, J.E. Waddell, P.J. Murphy, J. Pharmacol. Exp. Ther.,219,447 (1981)
55. Drugs of the Future, VI, 681 (1981)
56. L. Dei Cas, R. Bolognesi, F. Cucchini, A. Fappani, S. Riva, O. Visioli, J. Cardiovas. Pharmacol. 5, 249(1983)
57. L. Dei Cas, C. Manca, B. Bernardini, G. Vasini, O. Visioli, J. Cardiovasc. Pharmacol. 4, 436 (1982)
58. Drugs of the Future, VI, 340 (1981)
59. S. Nelson, C.V. Leider, Current Therap. Research, Vol. 30, 405 (1981)
60. R.J. Gorczynski, W.G. Anderson, D.M. Stout, J. Med. Chem. 24, 835 (1981)
61. D.M. Stout, R.J. Gorczynski, J. Med. Chem. 25, 326 (1982)
62. R.J. Gorczynski, J. Pharmacol. Exp. Ther., 223, 7 (1982)
63. Drugs of the Future, VIII, 568 (1983)
64. T. Ikeo, T. Nagao, T. Suzuki, H. Yabana, H. Nakajima, Jap. J. Pharmacol. 31 (Suppl.), 168 (1982)
65. M. Kino, Y. Hirota, S. Yamamoto, K. Sawada, M. Moriguchi, M. Kotaka, S. Kubo and K. Kawamura, Am.J. Cardiology, 51, 802 (1983)
66. V. Dahlstrom, C. Graffner, U. Jonsson, K.J. Hoffmann, E. Karlsson, P.O. Lagerstrom, Eur. J. Clin. Pharmacol., 24, 495 (1983)
67. C. Graffner, K.J. Hoffmann, G. Johnsson, P. Lundborg, O. Rönn, Eur. J. Clin. Pharmacol., 20, 91 (1981)
68. H. Lambertz, J. Meyer, R. Erbel, S. Effert, Circulation 66, 20 (1982)
H. Lambertz, J. Meyer and R. Erbel, Circulation 69, 298 (1984)
69. A. Hedberg, E. Carlsson, E. Fellenius, B. Lundgren, Naunyn-Schmiedeberg's Arch. Pharmacol., 318, 185 (1982)
70. A. Wirtzfeld, G. Klein, H.V. Bibra, E. Sauer, G. Baumann, Z. Kardiol. 71, 619 (1982)
71. G.H. Apperley, G.M. Drew, A.T. Sullivan, Eur. J. Pharmacol., 81, 659 (1982)
72. Drugs of the Future, VII, 571 (1982)
SCRIP 867, 19 (1984)
73. J.G. Barlow, B.G. Main, H.M. Snow, J. Med. Chem., 24, 315 (1981)
74. H. Pouleur, M.F. Rousseau, P. Mengeot et al., Eur. Heart. J., 3 (Suppl. D), 123 (1982)
75. H.M. Snow, A. Nuttall, H.F. Marlow, P.M. Dawes, Clin. Sci., 64, 4P-5P, (1983)
76. H. Pouleur, J.M. Detry, G. Mancia, Eur. J. Clin. Invest., 12, 441 (1982) and references cited
77. B.G. Main, J. Chem. Tech. Biotechnol., 32, 617 (1982)

78. Drugs of the Future, VII, 552 (1982)
79. G. Sponer, K. Strein, W. Bartsch, B. Mueller-Beckmann, Z. Kardiol., 71, 171 (1982)
80. B. Whiting, A.W. Kelman, D.J. Sumner, W.S. Hillis, H. Ledermann, Brit. J. Clin. Pharmacol., 13, 529 (1982)
81. K. Bödigheimer, T. Ischinger, F.G. Nowak, W. Delius, Z. Kardiol., 72, 48 (1983)
82. Drugs of the Future, VIII, 65 (1983)
83. B. Sharma, I. Hobach, G.S. Francis, M. Hodges, R.W. Asinger, J.N. Cohn, C.R. Taylor, Am. Heart J., 102, 533 (1981)
84. J. Bayliss, M. Norell, R. Canepa-Anson, P.A. Poole-Wilson, S. Reuben, G.C. Sutton, Am. J. Cardiol., 49, 1039 (1982)
F.X. Pamelia, M. Georghiade, G.A. Beller, H.L. Bishop, A.Y. Olikotun, C.R. Taylor, D.D. Watson, A.M. Grunwald, J. Sirowatka and B.A. Carabello, Am. Heart Journal 106, 1369 (1983)
85. W.S. Colucci, R.W. Alexander, G.H. Williams, R.E. Lude, B.L. Holman, M.A. Konstam, J. Wynne, G.H. Mudge, E. Braunwald, N. Eng. J. Med., 305, 185 (1981)
86. Scrip No. 837, 1983, p. 18
87. J. Mifune, K. Kuramoto, K. Ueda, S. Matsushita, I. Kuwajima, M. Sakai, T. Iwasaki, N. Moroki and M. Murakami, Am. Heart J. 104, 1011 (1982)
88. R.Y.C. Wang, T.F. Tse, D.Y.C. Yu, P.K. Lee, M.S.S. Chow, Am. Heart J., 104, 1016 (1982)
89. M. Irmer, H. Wollschläger, H. Just, Klin. Wochenschr., 59, 639 (1981)
90. R. Canepa-Anson, J.R. Dawson, W.S. Frankl, P. Kuan, G.C. Sutton, S. Reuben and P.A. Poole-Wilson, Eur. Heart. J., 3, (Suppl. D), 129 (1982)
91. J.A. Beavo, N.L. Rogers, O.B. Crofford, J.G. Hartmann, E.W. Sutherland and E.V. Neuman, Mol. Pharmacol., 6, 597 (1970)
92. R. Buenger, F.J. Haddy and E. Gerlich, Pfluegers Arch., 358, 213 (1975),
L. Belardinelli, F.L. Belloni, R. Rubio and R.M. Berne, Circ. Res., 47, 684 (1980)
G. Burnstock, Circ. Res., 46 (Suppl. I), 175 (1980)
93. W.S. Hillis and M. Been, Eur. Heart J., 3 (Suppl. D), 97 (1982)
94. R.A. Matthay, H.J. Berger, R. Davies, J. Loke, A. Gottschalk and B.L. Zaret, Am. Heart J., 104, 1022 (1982)
95. D. Reinhardt, O. Richter, M. Schaefers and B. Becker, Internist, 23, 728 (1982)
96. D.F. LeGatt and A. Edmonton, Drug Intell. Clin. Pharm., 17, 59 (1983)
97. M. Gorczyca, M. Pawlowski, and B. Lucka Sobstel, Acta Pol. Pharm., 39, 315 (1982)
98. T. Novinson, R.H. Springer, D.E. O'Brien, M.B. Scholten, J.P. Miller and R.K. Robins, J. Med. Chem., 25, 420 (1982)
99. T. Okabe, B. Bhooshan and T. Novinson, J. Heterocycl. Chem., 20, 735 (1983)
100. P.G. Jackson, G. Jackson, D. Kitson and D.E. Jewitt, Br. J. Clin. Pharmacol., 5, 7 (1978)
101. J. Hutton, W.S. Hillis, C.E. Langan, J.M. Conely and T.D.V. Lawrie, Br. J. Clin. Pharmacol., 4, 513 (1977)
102. C.R. Taylor, J.R.C. Baird, K.J. Blackburn, D. Cambridge, J.W. Constantine, M.S. Ghaly, M.L. Hayden, H.M. McIlhenny, P.F. Moore, A.Y. Olukotun, L.G. Pullman, D.S. Salsburg, C.A.P.D. Saxton and S. Shevde, Am. Heart J., 1981, 515
103. P. Ellis and J.R. Joice, Br. J. Pharmacol., 78 (3), Suppl., 125P (1983)
104. C.T. Albaster and C.G. Henderson, Br. J. Pharmacol., 78 (3), Suppl., 124P (1983)
105. D.B. Evans, W.E. Burmeister, C.M. Eldon, R.W. McNish, R.E. Potoczak, J.A. Schenden, R.P. Steffen and H.R. Kaplan, Pharmacologist, 25, 207 (1983)
106. R.E. Weishaar, M.M. Quade, J.A. Schenden, D.K. Boyd and D.B. Evans, Pharmacologist, 25, 207 (1983)
107. D.B. Evans, W.E. Burmeister, C.M. Eldon, R.W. McNish, R.E. Potoczak, J.A. Schenden, Pharmacologist 25, 207 (1983)
108. T. Chang, S. Olson, A. Hayes, R. Potoczak, C. Eldon, D. Evans and R. Steffen, Pharmacologist, 25, 249 (1983)
109. A. Heiss, M. Schaefer Kortin and P. Honerjaeger, Arch. Pharmacol., 321, Suppl., R36 (1982)
110. W. Diederen, J. Daemmgen and R. Kadatz, Arch. Pharmacol., 321, Suppl., R36 (1982)
111. M. Inoue, M. Hori, K. Tsujioka, Y. Ishida, S. Fukunami and T. Kagiya, Circulation 68, 62 (1983)
New positive inotropic agent, OPC-8212, Arzneim.-Forsch./Drug Res. 34 (I), Nr. 3a (1984)
112. K.B. Seamon, J.W. Daly, J. Cycl. Nucleotide Res., 7 (4), 201 (1981)
113. K.B. Seamon, J.W. Daly, Trends Pharmacol., 1983
114. N.J. de Sonza, A.N. Dohawalla, J. Reden, Med. Res. Reviews 3, 201 (1983)
F. Späh, J. Cardiovasc. Pharmacol. 6, 99 (1984)
115. G. Brooker, C. Pedone, K. Barowsky, Science, 220, 1169 (1983)
116. H. Metzger, E. Lindner, Arzneim.-Forsch./Drug Res. 31, 1248 (1981)
117. E. Lindner, H. Metzger, Arzneim.-Forsch./Drug Res. 33, 1436 (1983)
118. S.V. Bhat, A.N. Dohadwalla, B.S. Baywa, N. Dadkar, H. Dornauer, N.J. de Sonza, J. Med. Chem., 26, 486 (1983)
119. M. Schramm, G. Thomas, R. Towart, G. Franckowiak, Nature, 309, 535 (1983)
120. M. Schramm, G. Thomas, R. Towart, G. Franckowiak, Arzneim.-Forsch./Drug Res. 33, 1268 (1983)
121. T.H. Le Jemtel and E.H. Sonnenblick in: Recent Developments in Cardiovascular Drugs. J. Coltart and D. Jewitt (Eds.). Churchill Livingstone, Edingburgh-London-Melbourne-New York 1982; 200-212
122. A.A. Alousi and H.P. Dobreck in: New Drugs Annual: Cardiovascular Drugs. A. Scriabine (Ed.). Raven Press, New York 1983; 259
123. A. Ward, R.N. Brogden, R.C. Heel, T.M. Speight and G.S. Avery, Drugs 26, 468 (1983)
124. P.T. Wilmshurst and M.M. Webb-Peploe, Br. Heart J., 49, 447 (1983)
SCRIP 866, 16 (1984)
125. A.A. Alousi, J.M. Canter, M.J. Montenaro, D.J. Fort and R.A. Ferrari, J. Cardiovasc. Pharmacol., 5, 792(1983)
126. D.S. Baim, A.V. McDowell, J. Cherniles, E.S. Monrad, J.A. Parker, J. Edelson, E. Braunwald and W. Grossman, The New England J. Medicine, 309, 748 (1983)

127. L.S. Sinoway, C.S. Maskin, B. Chadwick, R. Forman, E.H. Sonnenblick and T.H. Le Jemtel, J. Am. Coll. Cardiol, 2, 327 (1983)
128. C.S. Maskin, L.S. Sinoway, B. Chadwick, E.H. Sonnenblick and T.H. Le Jemtel, Circulation, 67, 1065 (1983)
129. R. Salzmann, G. Bormann and G. Scholtysik, Naunyn-Schmiedeberg's Arch. Pharmacol., 324, 138 (1983)
130. J.W. Herzig, G. Bormann, L. Botelho, E. Erdmann, R. Salzmann, and R.J. Solaro, J. Molec. Cell. Cardiol., 15 (Suppl. 1), abstr. 244
131. R.A. Schnettler, R.C. Dage and J.M. Grisar, J. Med. Chem. 25, 1477 (1982)
132. R.C. Dage, C.P. Hsieh and N.L. Wiech, Fed. Proc., 39 (3), No. 4418 (1980)
133. C.P. Hsieh and R.C. Dage, Fed. Proc., 39 (3), No. 4419 (1980)
134. L.E. Roebel, H.C. Cheng, R.W. Lucas, R.J. Hodgeman, O.K. Reavis and J.K. Woodward, Pharmacologist, 22, 287 (1980)
135. L.E. Roebel, R.C. Dage, H.C. Cheng, J.K. Woodward, J. Cardiovasc. Pharmacol., 4, 721 (1982)
136. R.C. Dage, L.E. Roebel, C.P. Hshieh, D.L. Weines and J.K. Woodward, J. Cardiovasc. Pharmacol., 4, 500 (1982)
137. M.H. Crawford, S.G. Sorensen, K.L. Richards and M. Sodums, Clin. Res., 30, 866A (1982)
138. B.F. Uretsky, T. Generalovich, P.S. Reddy, R.B. Spangenberg and W.P. Follansbee, Circulation, 67, 823 (1983)
J.L. Martin, M.J. Likoff, J.S. Janicki, W. Laskey, J.W. Hirshfeld and K.T. Weber, Clin.Res. 31, 667A (1983)
139. T. Kariya, L.J. Wille and R.C. Dage, J. Cardiovasc. Pharmacol. 6, 50 (1984)
140. R.C. Dage, L.E. Roebel, C.P. Hsieh and J.K. Woodward, J. Cardiovasc. Pharmacol. 6, 35 (1984)
L.E. Roebel, R.C. Dage, H.C. Cheng and J.K. Woodward, J. Cardiovasc. Pharmacol. 6, 43 (1984)
141. R.A. Okerholm, F.J. Keeley, D.L. Weiner and R.B. Spangenberg, Fed. Proc., 42 (4), 1131 (1983)
142. Drugs of the Future, 8, 657 (1983)
143. W. Diederen and R. Kadatz, Arzneim.-Forsch., 31 (1a), 146 (1981)
144. W. Diederen and H. Weisenberger, Arzneim.-Forsch., 31 (1a), 177 (1981)
145. R.J. Solaro and J.C. Ruegg, Circ. Res., 51, 290 (1982)
146. E. Maurer, W. Klein and D. Brandt, Arzneim.-Forsch., 33 (5), 760 (1983)
147. M.L. Simoons, G. Muskens and P.G. Hugenholtz, Herz, 8 (1), 34 (1983)
M. Stauch, H. Sigel, H. Seibold and W. Nechwatal, Klin.Wochenschr. 61, 759 (1983)
148. J. Thormann, M. Schlepper and W. Kramer, Clin. Exp. Pharmacol. Physiol., 9, 235 (1982)
149. J. Thormann, M. Schlepper, W. Kramer and M. Gottwik, Z. Kardiol., 71, 736 (1982)

Chapter 9. Agents for the Treatment of Peptic Ulcer Disease

David E. Bays and Roger Stables
Glaxo Group Research Ltd., Ware, Hertfordshire, England

Introduction - The natural history of duodenal ulcer and its etiology have been reviewed, and strategies for the prevention of the recurrence of peptic ulcer described.[1-3] The consequences of selective gastric and parietal cell vagotomy in 685 patients have been assessed.[4] Antisecretory agents have been reviewed including a discussion of experimental methods used to assess H_2-antagonists.[5-7] Parietal cell function, receptors and H_2-antagonists are some of the subjects discussed in a recent book.[8] In a review of the use of anticholinergics as inhibitors of acid secretion, a tentative classification of cholinergic receptors is presented.[9] The mechanisms by which prostaglandins protect the gastric mucosa have been comprehensively reviewed.[10] The potential role of intragastric nitrosamide formation in the etiology of gastric cancer is discussed in a review.[11]

Histamine H_2-Antagonists - The clinical development of cimetidine 1 over the last 6 years has been briefly reviewed,[12] and its pharmacokinetics in subjects ranging from healthy volunteers to elderly patients have been surveyed.[13] Drug interactions with cimetidine continue to be reviewed.[14] Post-marketing surveillance for 12 months of 9928 patients produced no evidence of any fatal adverse effects.[15] Nitrosocimetidine, unlike nitrosomethylnitroguanidine, was not carcinogenic in rats when given in drinking water for more than 2 years.[16]

Reviews of ranitidine 2 continue to substantiate its clinical efficacy and selectivity of action.[17-19] A multicenter trial in America has been assessed.[20] Drug interactions, pharmacodynamics and pharmacokinetics of ranitidine and cimetidine have been compared.[21,22] A broad review of the pharmacokinetic basis for H_2-antagonist drug interactions includes a discussion of structural features present in cimetidine, but not ranitidine, that contribute to binding to the cytochrome P450 linked mixed function oxidase.[23]

Ranitidine contracted the rat isolated lower esophageal sphincter at high concentrations (3×10^{-6} - 10^{-3}M), but had no effect on lower esophageal sphincter pressure in man.[24,25] Cimetidine[26] and ranitidine[27] can inhibit human cholinesterases *in vitro* but ranitidine had no such effect *in vivo* nor have cholinergic side effects been reported in patients.[28] In the conscious dog using bethanechol as stimulant, neither ranitidine nor cimetidine inhibited pepsin secretion and ranitidine was more potent than cimetidine in inhibiting acid suggesting a significant involvement of histamine in this pathway.[29] Cimetidine and ranitidine inhibited cell division in colonic tumors in the rat and the growth of xenografts in immune deprived mice.[30]

ISBN 0-12-040519-9

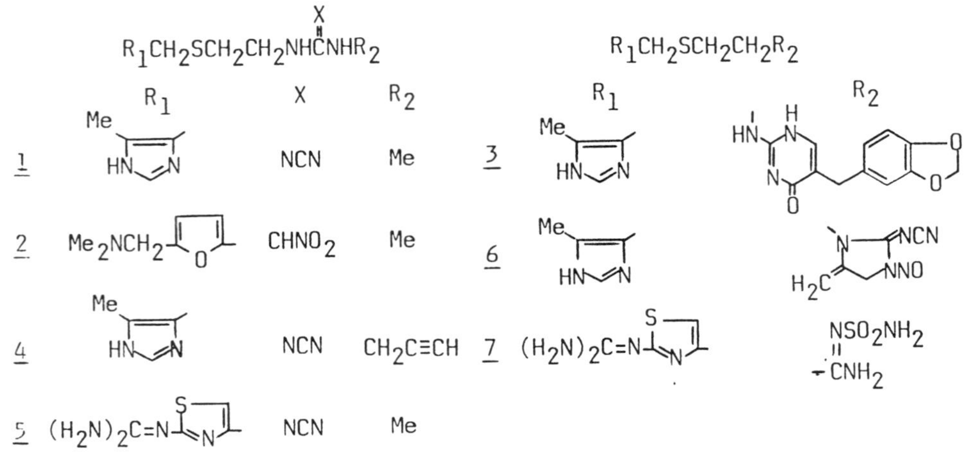

Clinical results obtained with ranitidine and cimetidine have encouraged discussion of the choice of treatment of ulcer disease.[31,32] Ranitidine (150mg bid) suppressed 24h gastric secretion more effectively than cimetidine (400mg bid) in healthy subjects and ulcer patients.[33,34] Single doses of ranitidine (300 mg nocte) and cimetidine (800mg nocte) have also been compared in ulcer patients. A single night time dose was as effective as a bid regime for both compounds (150mg and 400mg respectively) in reducing acid secretion. This paper underlined the importance of controlling nocturnal secretion and showed that nocte ranitidine reduced acid concentration by 85% whereas cimetidine decreased it by 56%.[35]

In ulcer patients oxmetidine 3 was effective at 400mg bid.[36] High doses of oxmetidine given intraarterially did not cause prolactin release in the rat.[37] Trials were suspended when a patient on maintenance therapy developed jaundice and had elevated levels of liver enzymes.[38]

Etintidine 4 was slightly more potent than cimetidine in inhibiting meal stimulated acid secretion in man.[39] In the dog, the order of potency as inhibitors of acid secretion was cimetidine $<$ etintidine $<$ ranitidine $<$ tiotidine 5.[40] Nitrosation of etintidine and cimetidine have been compared, and the formation of the imidazoline 6 is discussed.[41] The development chemistry on tiotidine has been described.[42] ^{3}H-Tiotidine is the most suitable ligand for H_2-receptor binding studies.[43] The Yamanouchi group found that in the pouch dog i.v. famotidine 7, another guanidinothiazole, and cimetidine competitively antagonised dimaprit induced acid secretion, but that against pentagastrin the antagonism was neither simply competitive nor non-competitive.[44] Famotidine and cimetidine competitively inhibited a histamine stimulated guinea pig adenylate cyclase with IC_{50}'s of 5.9×10^{-7}M and 1.4×10^{-5}M.[45] The Merck group described famotidine (MK-208) as a slowly dissociable unsurmountable H_2-antagonist in isolated guinea pig atria. In the chronic fistula dog it was about 7 times more potent than ranitidine and when given at 2.1mg/kg p.o. (70 x ED_{50}) it still inhibited acid output by 59% after 24h.[46] In man against pentagastrin stimulation, 5mg MK-208 was equivalent to 300mg cimetidine.[47] In another study, single night time doses of cimetidine (800mg), ranitidine (300mg) and famotidine (40mg) inhibited acid secretion during the following day by 10, 55 and 32% respectively.[48,49] Confusion over the structure of famotidine gave the Bristol-Myer's group an opportunity to comment upon the potential advantages of a long acting compound BMY-25260, 8.[50] A new long acting compound from this group is the cyclobutenedione (BMY-25368, 9).[51]

The Merck group also described 8 (L-643,441) as long acting. In the fistula dog using histamine as the stimulant, it is a little more potent than ranitidine, but at 15mg/kg p.o. (~50xED_{50}) it inhibited acid secretion by 80% after 24h. At this dose ranitidine's effect had disappeared after this time.[52] The relatively high p.o. to i.v. ED_{50} ratio for L-643,441 was attributed to incomplete absorption and rapid first-pass metabolism.[53]

The triazole, loxtidine (AH23844, 10), is like lamtidine (AH22216, 11) long acting in the pouch dog.[54,55] Given orally they were about four and eight times as potent as ranitidine against histamine induced acid secretion, and inhibition of about 20% and 60% was seen 18h after 0.1mg/kg p.o. (2 to 4 x ED_{50}). On the guinea pig atria loxtidine was an unsurmountable antagonist, and this may be related to the dissociation energy of the drug-receptor complex.[54] The interaction of lamtidine with guinea pig isolated parietal cells was either non-competitive or competitive depending on the mathematical treatment of the results.[56] It interacted non-competitively with H_2-receptors on human gastric cancer cells.[57] A radioimmunoassay for loxtidine and lamtidine has been developed[58] and the metabolism of loxtidine has been investigated in animals.[59]

Some formamidines have been described as H_2-antagonists 20 to 50 times as potent as cimetidine in the rat perfused stomach preparation.[60] One, DA4577, (12), showed little cholinergic effect on isolated tissues.[61]

The rigid phenylene analogue 13 is more potent in the dog than the flexible molecules cimetidine and tiotidine given i.v., but is similar to cimetidine p.o. Structure-activity relationships are described.[62] Bioisosteric design and restricted conformation provided the rationale for the synthesis of the triazole 14 that was somewhat more potent than cimetidine.[63]

The combined H_1/H_2-antagonist, SK&F 93319 (15), has similar activity at both receptors having pA_2s of 7.77 and 7.49 in the guinea pig ileum and atria. Given i.v. it was more potent as an inhibitor of gastric secretion than cimetidine in the rat and dog. It was less active than mepyramine against histamine induced bronchoconstriction in the guinea pig, and behaved as an antagonist at both H_1 and H_2-receptors on peripheral vasculature.[64,65]

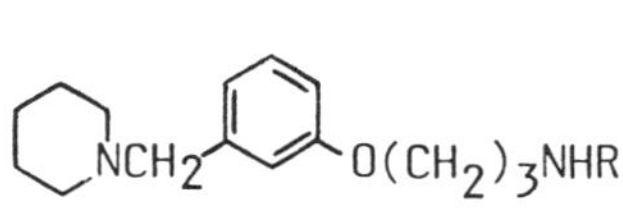

12 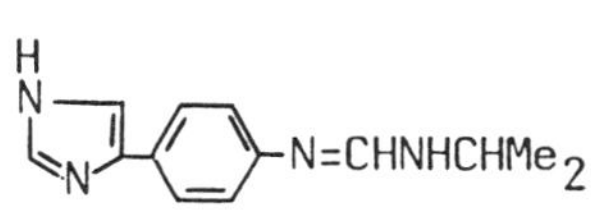

8 R= 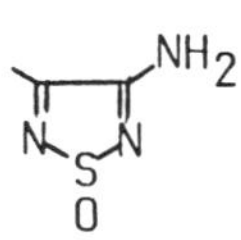

9 R= 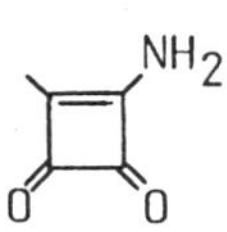

10 R= 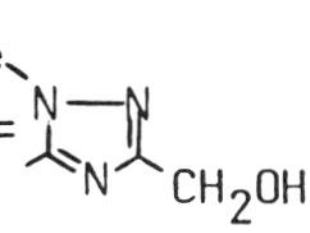

11 R= 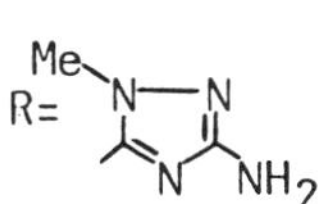

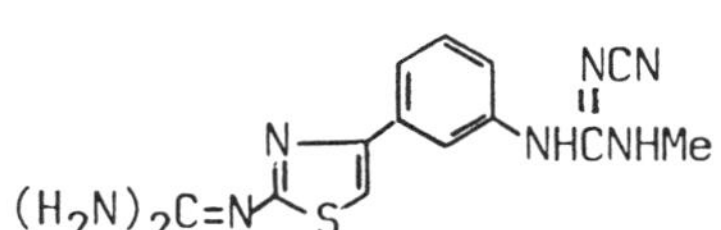

13

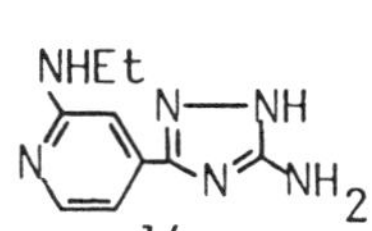

14

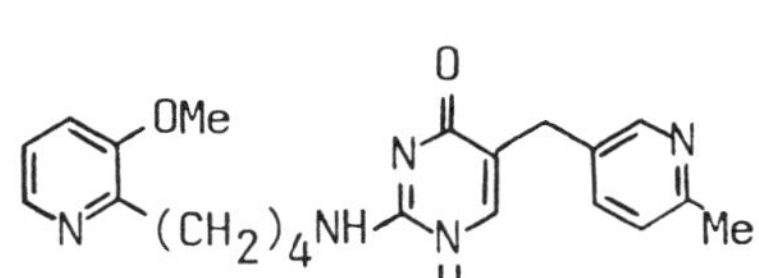

15

Inhibitors of H^+/K^+-ATPase - The development of omeprazole (H 168/68, 16) has continued. In the Heidenhain pouch dog, orally it was twice as potent as cimetidine with a bioavailability of about 15%. Intraduodenally the bioavailability was around 70%, suggesting that omeprazole is partially degraded by gastric juice. The antisecretory effect after a dose of 2.5μmol/kg i.d. was still observable after 3-4 days. It also has a long duration of action in the rat.[66] Infusion of omeprazole directly into a Heidenhain pouch produced a dose related reduction of acid in the pouch but not in the main stomach, suggesting a local action on the oxyntic glands.[67] Given intragastrically, but not parenterally, omeprazole prevents mucosal damage caused by aspirin, and at a higher dose, that induced by ethanol.[68] It was also effective orally.[69]

The mechanism and site of action of omeprazole has been investigated further. Sixteen hours after an i.v. injection of labelled drug into mice, high levels of radioactivity remained only in the gastric mucosa suggesting that the gastric H^+/K^+-ATPase resides in the parietal cells.[70] The related imidazole picoprazole (H 149/94, 17) inhibited gastric H^+/K^+-ATPase but not a related Na^+/K^+-ATPase, even at 10^{-4}M.[71] Pepsinogen release from rabbit chief cells was not affected by omeprazole, but it did affect parietal cell secretion.[72] In vitro studies on the action of omeprazole, cimetidine and thiocyanate ion provided further evidence for its selective action.[73] In guinea pig parietal cells the inhibitory action of omeprazole was non-competitive but reversible.[74] The different mechanism of action of omeprazole and ranitidine is reflected in the way each drug modifies the ultrastructure of canine parietal cells.[75]

Omeprazole is effective in man against acid secretion induced by pentagastrin, insulin, or a meal in oral doses in the range 20-90mg depending on the trial.[76-78] In the Hassle study a single dose of 40mg reduced secretion for three days, despite a short drug half-life.[76] In duodenal ulcer patients intragastric omeprazole inhibited acid secretion induced by sham feeding, a cholinergic pathway, without affecting serum gastrin.[79] Intragastric acidity over 24 hours was almost eliminated (95%) in ulcer patients by omeprazole at 30mg p.o., whereas cimetidine (1g/day) and ranitidine (300mg/day) decreased it by 48 and 69%.[80] Very good healing rates in ulcer patients have been obtained with oral doses of 20-60mg given once daily over 28 days.[81,82] Omeprazole, 60mg daily, was very effective in seven patients with Zollinger-Ellison syndrome.[83]

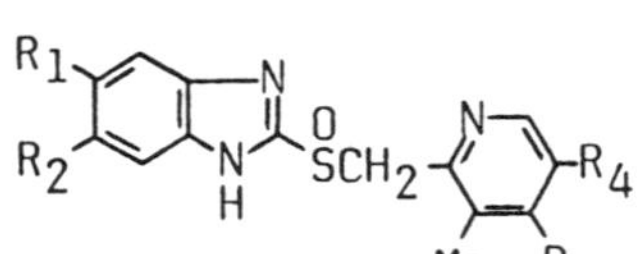

	R_1	R_2	R_3	R_4
16	OMe	H	OMe	Me
17	CO_2Me	Me	H	H

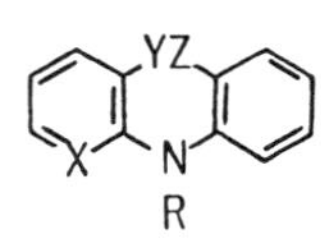

18 X = CH, YZ = CH_2CH_2

R = $CH_2CH(Me)CH_2NMe_2$

19 X = N, YZ = NHCO

R = $COCH_2N$ NMe

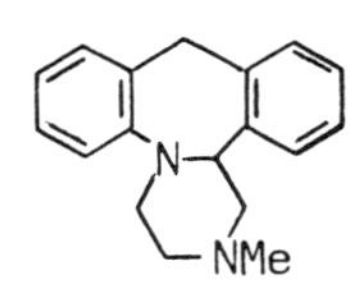

20

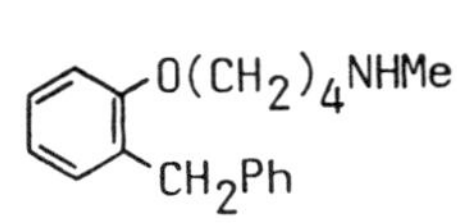

21

Tricyclic compounds - The therapeutic use of tricyclic compounds such as trimipramine 18 and pirenzepine 19 has been reviewed.[84] Trimipramine (50mg nocte) produced similar healing rates to cimetidine in duodenal and gastric ulcers, and was effective in maintenance treatment.[85,86] Tiredness and dry mouth were side effects. The related antidepressant mianserin 20 inhibited acid secretion in volunteers,[87] but MCI 2016, 21, a non-tricyclic compound with antidepressant-like properties, prevented experimental ulcers in rats without inhibiting gastric secretion.[88] The mechanism of action of the selective antimuscarinic agent, pirenzepine, has been discussed.[9] A relatively weak inhibitor of carbachol-stimulated acid secretion in rat isolated parietal cells, it was only 5-7 times less active than atropine in reducing both acid and pepsin output elicited by bethanechol or 2-deoxy-D-glucose in the dog, but was 200-300 times less potent than atropine in causing tachycardia.[89,90] In man pirenzepine (10mg i.v.) reduced salivation and lacrimation as well as gastric secretion.[91] Clinically pirenzepine 50-75 mg bid. seems as effective as cimetidine in healing duodenal ulcers.[84] Doses causing moderate inhibition of gastric secretion did not increase nitrite or bacteria in gastric aspirates, but enhanced gastric mucosal blood flow.[92,93]

Other structures - An imidazopyridine, SCH 28080 (22) was 7-10 times more potent than cimetidine as an antisecretory agent in the pylorus ligated rat, but was only 0.3 times as active orally in the dog.[94] It differed from cimetidine and atropine in its antisecretory mechanism of action being effective against histamine, methacholine and cyclic AMP in the guinea pig gastric mucosa.[95] SCH 28080 prevented ethanol lesions in the rat (ED_{50} 3.0mg/kg p.o.).[94] This cytoprotective effect was not blocked by indomethacin, and could result from stimulation of mucus and bicarbonate secretion.[95] The development of SCH 28080 was suspended because of unspecified problems in toxicology.[94]

Some analogues of fenoctimine (McN-4097, 23) and its antisecretory effect in volunteers have been described.[96,97] In the gastric fistula dog the mast cell stabilising agent, FPL 52694 (24), given topically, in addition to reducing output of acid, increased the output of gastric $NaHCO_3$.[98] It also reduced lesions induced by cysteamine in the rat.[99] The antisecretory effect of WHR-1370 (25) in the rat was probably due to an action on α_2-receptors since it could be blocked by yohimbine.[100] The pyridazinones 26 and MUN-114 (27), inhibit acid secretion in the Shay rat and prevent stress induced ulcers. They are as potent as cimetidine, but are neither H_2-antagonists nor anticholinergics.[101] The urea 28 was the most active of a series of pyridine derivatives. In the dog it was at least as potent as cimetidine against secretion induced by gastrin tetrapeptide but not histamine. It decreased the number of parietal cells in dogs when given at 2.5 to 22.5mg/kg p.o. over three months.[102] The quaternary derivative 29 was one of a series that inhibited acid secretion by an unknown mechanism.[103]

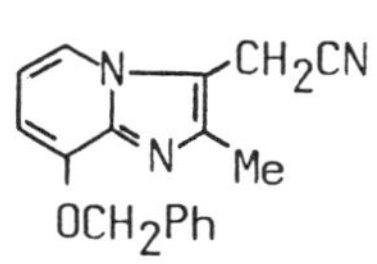

22

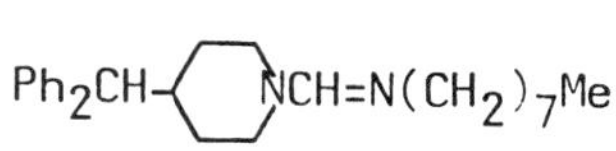

23

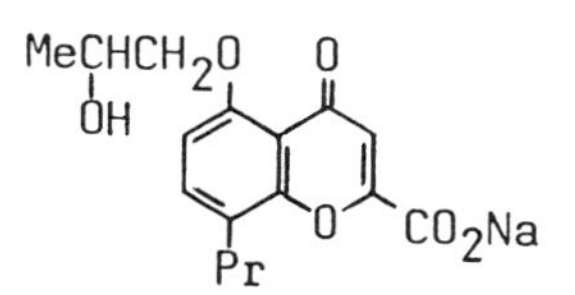

24

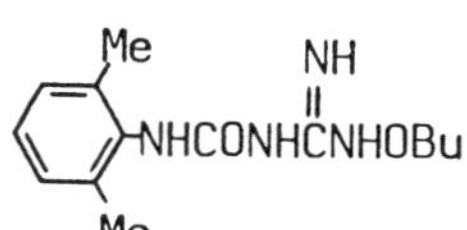

25

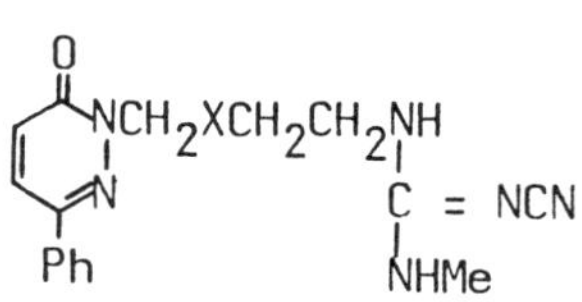

26 X = S 27 X = CH_2

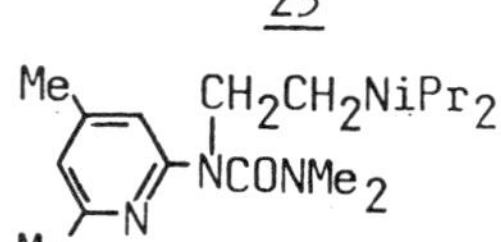

28

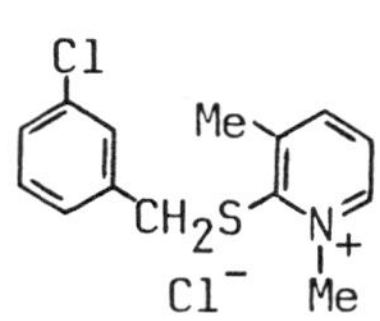

29

Prostaglandins - The PGE_1 analogue misoprostil (SC29333,30) is a potent, but relatively short acting inhibitor of gastric acid secretion in man.[104,105] After 2 weeks treatment with 100μg p.o. (qid), pentagastrin stimulated acid output was only reduced up to 2h after dosing.[104] Food-stimulated acid secretion was significantly inhibited for 0.5 - 2h, but not longer than 4h, after a single dose of 200μg p.o.[105] Lower doses, 25 and 50μg qid, reduced aspirin-induced gastric bleeding.[106] Misoprostil contains four stereoisomers in approximately equal proportions, but studies in the dog suggested that the 16-S-16-hydroxy isomer SC 30249 (31) was mainly responsible for antisecretory activity.[107] Another PGE_1 analogue, CL 115,574 (32), inhibited gastric acid secretion in man at doses of 500-1000μg p.o., and enhanced mucus output.[108] MDL 646 (33) inhibited gastric acid secretion when given intragastrically at doses of 15-50μg/kg in rats, cats and dogs, but caused salivation and vomiting in dogs at 100μg/kg i.g.[109]

The PGE_2 analogue trimoprostil, RO 216,937 (34), inhibited gastric acid secretion and protected against gastric mucosal damage in animals and man. Antisecretory dose levels in man were 2.5 - 10μg/kg, but even at these doses side effects such as epigastric distress, headache, dizziness and nausea were noted.[110] In volunteer studies basal acid output was reduced by 80% over 6h after 1500μg p.o.[111] Doses of 250-3000μg p.o. inhibited pentagastrin stimulated gastric acid secretion and increased HCO_3^- secretion.[112] The stable PGI_2 analogue nileprost (35) inhibited gastric secretion in man by 28% at a dose of 50μg i.g., and at 10μg i.g. reduced by 46% the fall in transmucosal potential difference caused by aspirin.[113]

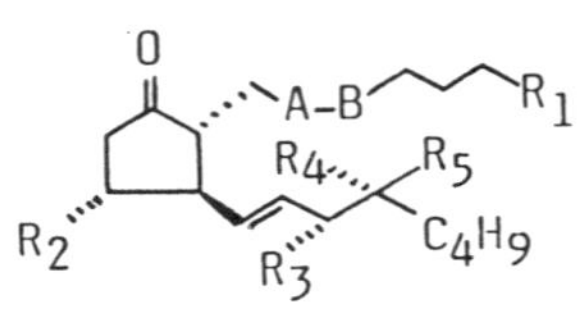

	R^1	R^2	R^3	R^4	R^5	A - B
30	CO_2CH_3	OH	H	CH_3/OH		CH_2-CH_2
31	CO_2CH_3	OH	H	OH	CH_3	CH_2-CH_2
32	$COCH_2OH$	OH	H	CH_3/OH		CH_2-CH_2
33	CO_2CH_3	OH	OH	OCH_3	CH_3	CH_2-CH_2
34	CO_2H	CH_3	OH	CH_3	CH_3	CH=CH

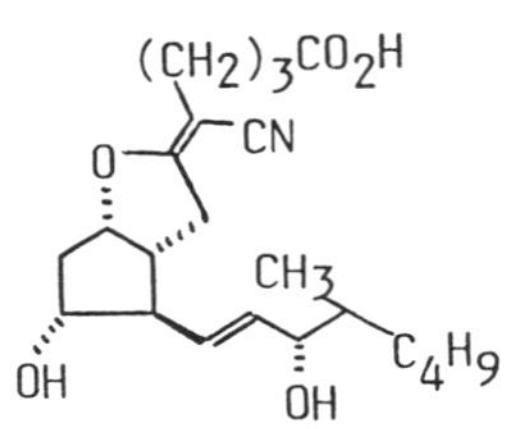

35

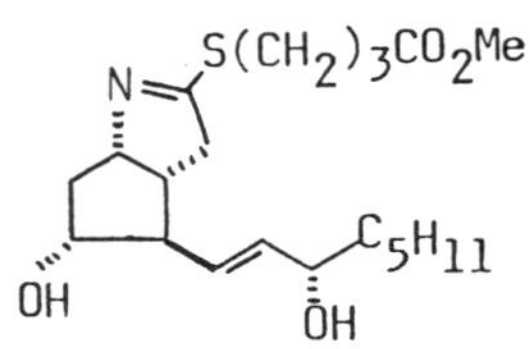

36

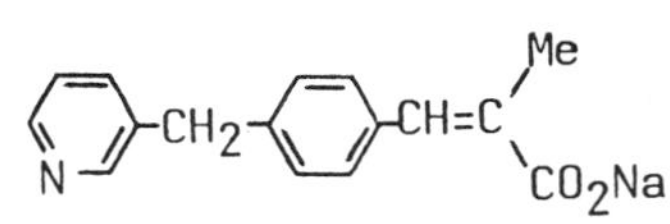

37

The ability of prostanoids to protect the gastrointestinal mucosa from experimental damage (termed "cytoprotection") continues to be much studied. 16,16-Dimethyl PGE_2 reduced aspirin-induced gastric lesions and indomethacin-induced intestinal lesions in pigs, but failed to modify gastric mucosal damage in volunteers dosed with 10mM taurocholate.[114,115] The type of damage that prostanoids can inhibit has been debated.[116] In rats PGE_2 or the prostaglandin precursor, arachidonic acid, prevented deep necrotic lesions produced in the gastric mucosa by ethanol, but failed to alter the initial desquamation of the surface epithelium.[117,118] However, in arachidonic acid treated animals a continuous layer of surface epithelium had reappeared by 3h, compared to 15h in controls.[118]

Mechanisms of cytoprotection have been the subject of a symposium and a review.[10,119] Interest has centred on the "mucus-bicarbonate barrier" as prostaglandins have been shown to stimulate bicarbonate secretion in both stomach and duodenum of rats, cats and dogs.[120-123] In the dog PGE_2, 16,16-dimethyl PGE_2, $PGF_{2\alpha}$ and a stable prostacyclin analogue, HOE 892 (36), stimulated bicarbonate secretion when applied topically to pouches of fundic, antral or duodenal mucosa.[123] In dog fundus this response may involve a cholinergic mechanism since it was blocked by atropine or tetrodotoxin.[124] The importance of bicarbonate secretion as a cytoprotective mechanism has been disputed.[120,125] The carbonic anhydrase inhibitor, acetazolamide, prevented the stimulation of bicarbonate secretion by 16,16-dimethyl PGE_2 without modifying its cytoprotective effect against ethanol damage in rats.[126] It was also suggested that high luminal bicarbonate might enhance mucosal damage.[127] Nevertheless, most papers favour a role for bicarbonate in cytoprotection.

The gastroduodenal mucosa is covered by a mucus gel layer which not only reduces H^+ diffusion but traps secreted bicarbonate, resulting in a pH gradient from lumen to mucosa.[128] For example, in rat stomach and duodenum when luminal contents were at pH2, a pH of 7 was maintained at the mucus/mucosa interface.[121,129] Following administration of PGE_2 or 16,16-dimethyl PGE_2, this intramucus pH became alkaline and mucus gel thickness was enhanced.[121,129-131] $PGF_{2\beta}$ and cysteamine stimulated release of mucus glycoprotein at doses that protected against ethanol-induced gastric mucosal damage in the rat, but the two effects could be dissociated.[132]

Other mechanisms which have been suggested are increased stability of lysosomes and cell membranes and release of phospholipids. 16,16-Dimethyl PGE_2 (0.2μg/kg/day) prevented the decrease in mucosal lysosome stability and the acute gastric erosions following bacterial peritonitis in dogs.[133] PGE_2 reduced the fall in DNA synthesis in human mucosal biopsy samples caused by incubation with ethanol.[134] However, 16,16-dimethyl PGE_2 did not reduce hemolysis of rat erythrocytes by 280mM sodium acetylsalicylate or 14% ethanol.[135] The cytoprotective effect of prostaglandins may be due to their ability to increase mucosal levels of hydrophobic phospholipids. Dog fundic mucosa was more hydrophobic than other areas of the gastrointestinal tract, and damaging agents markedly reduced this hydrophobicity.[136] In rats a cytoprotective dose of 16,16-dimethyl PGE_2 (0.5μg/kg) increased gastric mucosal phospholipids, and administration of a phospholipid suspension reduced the damaging effects of 0.6N HCl on the gastric mucosa.[137]

Endogenous prostanoids may have a role in mucosal protection. The ability of aluminum antacids to prevent taurocholate induced lesions in rats was inhibited by indomethacin, as was the cytoprotective effect of low doses of ethanol, HCl or NaOH against damage caused by high doses of these agents.[138,139] Increased formation of prostanoids in gastric mucosa was detected following antacids or low doses of NaOH.[138,139] In the dog, indomethacin prevented the progressive decrease in gastric damage on repeated exposure to ethanol.[140] In rats the thromboxane synthesis inhibitor OKY 1581 (37) reduced gastric mucosal damage by taurocholate and increased mucosal generation of PGE_2 and PGI_2.[141] Accumulation of PGE_2 and stable metabolites of prostacyclin and thromboxane A_2 was lower in cultured gastric mucosal tissue from duodenal ulcer patients than from normal subjects.[142]

Mucosal protectants - Sucralfate was the subject of a symposium, and its efficacy against experimental gastric erosions but its lack of antisecretory activity was confirmed.[143,144] In a double blind trial, tripotassium-dicitratobismuthate (DeNol) enhanced healing of gastric but not duodenal ulcers, and its ultrastructural localisation in the gastrointestinal tract was described.[145,146] Lozilurea (38) inhibited experimental gastric lesions in the rat. It was devoid of antisecretory activity but had some sedative properties and increased gastric levels of hexosamines and mucoproteins.[147]

Gut peptides - The inhibitory effect of somatostatin in the dog on secretion elicited by tetragastrin or 2-deoxy-D-glucose was not due to changes in mucosal blood flow or circulating gut hormones.[148] In the rat somatostatin inhibited bethanechol-induced secretion possibly via endogenous prostaglandins since it potentiated their release and was blocked by indomethacin.[149] In isolated gastric glands and cells from rabbit or rat, somatostatin was inactive against carbachol but appeared to inhibit secretion to gastrin indirectly, and to histamine by a direct effect on the parietal cell.[150,151]

The presence of opiate receptors on parietal cells was suggested by binding studies.[152] However, endogenous opioids did not seem to be involved in gastric secretion or emptying in rhesus monkeys since these processes were unaffected by naloxone, although they were inhibited by metenkephalin.[153] The enkephalin analogue (D-Ala^2,MePhe^4,Met(O)^{5}ol) enkephalin (FK33-824), given intracerebroventricularly in rats, reduced stress ulcers and gastric secretion, possibly via prostaglandin release.[154]

Interest in gastrin antagonists continues, and C-2-phenylthiotryptophanyl^2-pentagastrin and C-γ^4-pentagastrinyl-histamine inhibited pentagastrin-induced secretion in rats.[155] In rabbit isolated parietal cells gastrin binding and secretion stimulated by gastrin were both inhibited by proglumide 39 and benzotript 40.[156] Proglumide (50mg/kg + 50mg/kg/h i.v.) reduced acid hypersecretion in Zollinger-Ellison patients, but was less effective than cimetidine at 2mg/kg.[157] A new technique for studying gut hormone release from isolated glands was described.[158]

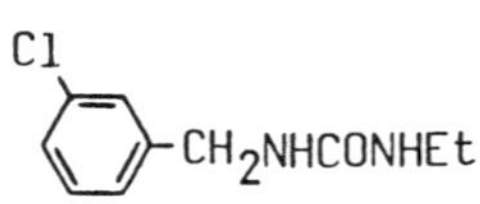

38

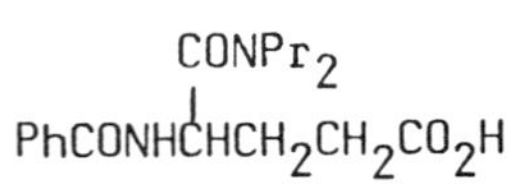

39

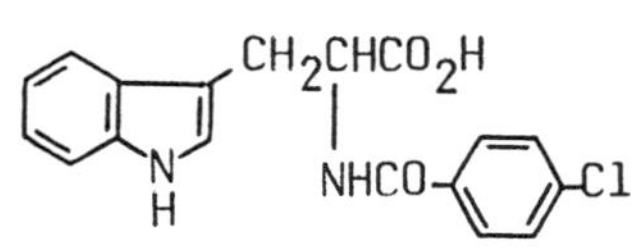

40

References

1. B.I. Hirschowitz, Gastroenterol., 85, 967 (1983).
2 K.G. Wormsley, Gut, 24, 775 (1983).
3 D.W. Piper, Drugs, 26, 439 (1983).
4. J. Ornsholt, E. Amdrup, D. Andersen and H. Hostrup, Scand. J. Gastrenterol., 18, 455 and 465 (1983).
5. K.W. Somerville and M.J.S. Langman, Drugs, 25, 315 (1983).
6. D.E. Bays and R. Stables, Pharm. Int., 4, 199 (1983).
7. M.J. Daly and B.J. Price in "Progress in Medicinal Chemistry", Vol. 20, G.P. Ellis and G.B. West, Ed., Elsevier Science Publishers, B.V., 1983, p.337.
8. "Receptors and the Upper Gl Tract", B.I. Hirschowitz and J.G. Spenney, Ed., Advanced Therapeutics Communications, Inc., N.Y., 1983.
9. M. Feldman, Gastroenterol., 86, 361 (1984).
10. T.A. Miller, Am. J. Physiol., 245, G601 (1983).
11. S.S. Mirvish, J. Natl. Cancer Inst., 71, 631 (1983).

12. T. Gledhill, Canadian J. Surgery, 26, 312 (1983).
13. A. Somogyi and R. Gugler, Clin. Pharmacokinet., 8, 463 (1983).
14. E.M. Sorkin and D.L. Darvey, Drug Intell. Clin. Pharm., 17, 110 (1983).
15. D.G. Colin-Jones, M.J.S. Langman, D.H. Lawson and M.P. Vessey, Br. Med. J., 286, 1713 (1983).
16. W. Lijinsky and M.D. Reuber, Cancer Res., 44, 447 (1984).
17. J.H. Bauman and T.S. Gaginella, Drug Intelligence Clin. Pharm., 17, 873 (1983).
18. R. Cockel, J. Dawson, G.T. Dixon, D.A. Richards and R. Stables, J. Clin. Hospital Pharm., 8, 1 (1983).
19. J.E. McGuigan, Clinics in Gastroenterol., 12, 819 (1983).
20. B.I. Hirschowitz, J. Clin. Gastroenterol, 5 (Suppl. 1), 115 (1983).
21. P.D. Hansten, Drug Interactions Newsletter, 3, 31 (1983).
22. D.A. Richards, J. Clin. Gastroenterol., 5 (Suppl. 1), 81 (1983).
23. J.R. Powell and K.H. Donn, J. Clin. Gastroenterol., 5 (Suppl. 1), 95 (1983).
24. J.M. Humphray, J.J. Reeves and R. Stables, Br. J. Pharmacol., 79, 384P (1983).
25. J.M. Thomas, I.F. Trotman and J.J. Misiewicz, Lancet, 1 (8321), 418 (1983).
26. W.E. Hansen and S. Bertl, Arzneim. Forsch., 33, 161 (1983).
27. W.E. Hansen and S. Bertl, Lancet, 1 (8318), 235 (1983).
28. R.T. Brittain, D. Jack and L.E. Martin, Lancet, 1 (8321), 418 (1983).
29. B.I. Hirschowitz and E. Molina, J. Pharmacol. Exp. Therap., 224, 341 (1983).
30. P.J.M. Tutton and D.H. Barkla, Anticancer Res., 3, 7 (1983).
31. J. Feely and K.G. Wormsley, Br. Med. J., 286 (6366), 695 (1983).
32. T. Gledhill and R.H. Hunt, Br. Med. J., 286 (6373), 1283 (1983).
33. H.G. Dammann, W. Friedl, P. Muller and B. Simon, Br. J. Clin. Pharmacol., 16, 461 (1983).
34. R. Corinaldesi, V. Stanghellini, T. Sacco, C. Raiti, A. Galassi and L. Barbara, Curr. Ther. Res., 34, 92 (1983).
35. T. Gledhill, O.M. Howard, M. Buck, A. Paul and R.H. Hunt, Gut, 24, 904 (1983).
36. L. Loof, H.O. Adami, S. Gustavsson, I. Kagevi, A. Nyberg and O. Nyren, Scand. J. Gastroenterol., 18, 839 (1983).
37. C. Netti, F. Guidobono, V.R. Olgiati, V. Sibilia and A. Pecile, Endocrinology, 113, 412 (1983).
38. Scrip, 830, 23 (1983).
39. D.C. Brater, W.M. Meyers Jr., K.A. Dandekar, K.A. Pittman and W. Peterson, Eur. J. Clin. Pharmacol., 23, 495 (1982).
40. R.L. Cavanagh, J.J. Usakewicz and J.P. Buyniski, J. Pharmacol. Exp. Therap., 224, 171 (1983).
41. T.A. Montzka, P.F. Juby, J.D. Matiskella, H.M. Holava and R.R. Crenshaw, Can. J. Chem., 61, 1771 (1983).
42. G.E. Robinson, Chem. and Ind., 349 (1983).
43. G.A. Gajtkowski, D.B. Norris, T.J. Rising and T.P. Wood, Nature, 304, 65 (1983).
44. M. Takeda, T. Takagi and H. Maeno, Eur. J. Pharmacol., 91, 371 (1983).
45. M. Harada, M. Terai and H. Maeno, Biochem. Pharmacol., 32, 1635 (1983).
46. C. Chung, B.V. Clineschmidt, P. Cook, R.G. Pendleton, M.L. Torchiana and S. Wiese, Arch. int. Pharmacodyn., 266, 4 (1983).
47. A.N. Chremos, M. Gamal, D.Y. Graham and J.L. Smith, Clin. Pharmacol. Ther., 33, 216 (B65) (1983).
48. H.G. Dammann, P. Muller and B. Simon, Lancet, II (8358), 1078 (1983).
49. H.G. Dammann, P. Muller and B. Simon, Lancet, I (8367), 54 (1984).
50. J.P. Buyniski, R.R. Crenshaw, P.F. Juby and S.E. Schwartz, Lancet, I, 53 (1984).
51. A.A. Algieri, G.M. Luke, M. Brown, R.T. Standridge, R.A. Partyka and R.R. Crenshaw, 186th ACS Meeting, Washington, Aug.-Sept. 1983, Section A, 63.
52. M.L. Torchiana, R.G. Pendleton, P.G. Cook, C.A. Hanson and B.V. Clineschmidt, J. Pharm. Expl. Therap., 224, 514 (1983).
53. D.E. Duggan and K.F. Hooke, Fed. Proc., 41, 1555 (1982).
54. R.T. Brittain and D. Jack, J. Clin. Gastroenterol., 5 (Suppl. 1), 71 (1983).
55. R. Stables, M.J. Daly and J.M. Humphray, Agents and Actions, 13, 166 (1983).
56. K.-Fr. Sewing and H. Hannemann, Hepato-gastroenterol., 30, 73 (1983).
57. I. Menez, C. Gespach, S. Emami and G. Rosselin, Biochem. Biophys. Res. Comm., 116, 251 (1983).
58. C. Harrison, W.N. Jenner, L.E. Martin and S.N. Young, Biochem. Soc. Trans., 11, 713 (1983).
59. J.A. Bell, A. Bradbury, W.N. Jenner, G.R. Manchee and L.E. Martin, Biochem. Soc. Trans., 11, 715 (1983).
60. A. Donetti, E. Cereda, A. Gallazzi, A. Giachetti and P.C. Vanoni, 186th ACS Meeting, Washington, Aug-Sept. 1983, Section A,62.
61. G. Bertaccini, G. Coruzzi and E.S. Vizi, Int. J. Tiss. Reac., V, 257 (1983).
62. J.M. Hoffman, A.M. Pietruszkiewicz, C.N. Habecker, B.T. Phillips, W.A. Bolhofer, E.J. Cragoe Jr., M.L. Torchiana, W.C. Lumma Jr. and J.J. Baldwin, J. Med. Chem., 26, 140 (1983).

63. C.A. Lipinski, J. Med. Chem., 26, 1 (1983).
64. R.C. Blakemore, T.H. Brown, D.G. Cooper, G.J. Durant, C.R. Ganellin, R.J. Ife, M.E. Parsons, A.C. Rasmussen and G.S. Sach, Br. J. Pharmacol., 80 (Suppl.), 437P (1983).
65. C.A. Harvey and D.A.A. Owen, Br. J. Pharmacol., 80 (Suppl.), 438P (1983).
66. H. Larsson, E. Carlsson, U. Junggren, L. Olbe, S.E. Sjostrand, I. Skanberg and G. Sundell, Gastroenterol., 85, 900 (1983).
67. S.J. Konturek, M. Cieszkowski, N. Kwiecien, J. Konturek, J. Tasler and J. Bilski, Gastroenterol., 86, 71 (1984).
68. S.J. Konturek, T. Brzozowski and T. Radecki, Digestion, 27, 159 (1983).
69. H. Mattsson, K. Andersson and H. Larsson, Eur. J. Pharmacol., 91, 111 (1983).
70. H.F. Helander, A. Smolka, C.H. Ramsay, G. Sachs and E. Fellenius, Hepato-gastroenterol., 30, 74 (1983).
71. B. Wallmark, G. Sachs, S. Mardh and E. Fellenius, Biochim et Biophysica. Acta., 728, 31 (1983).
72. J. Fryklund, B. Wallmark, H. Larsson and H.F. Helander, Biochem. Pharmacol., 33, 273 (1984).
73. B. Wallmark, B.-M. Jaresten, H. Larsson, B. Ryberg, A. Brandstrom and E. Fellenius, Am. J. Physiol., 245, G64 (1983).
74. K.-Fr. Sewing, P. Harms, G. Schulz and H. Hannemann, Gut, 24, 557 (1983).
75. J. Stachura, S.J. Konturek, M. Cieszkowsky, W. Dabros, J. Zakreuska and J. Konturek, Hepato-gastroenterol., 30, 205 (1983).
76. T. Lind, C. Cederberg, G. Ekenved, U. Haglund and L. Olbe, Gut, 24, 270 (1983).
77. R.J. Utley and D.C. Carter, Br. J. Surg., 70, 295 (1983).
78. W. Londong, V. Londong, C. Cederberg and H. Steffen, Gastroenterol., 85, 1373 (1983).
79. S.J. Konturek, N. Kweicien, W. Obtulowicz, B. Kopp and J. Oleksy, Gut, 25, 14 (1984).
80. R.P. Walt, M.de F.A. Gomes, E.C. Wood, L.H. Logan and R.E. Pounder, Br. Med. J., 287, 12 (1983).
81. M.J. Daly, A.L. Howe, J.J. Misiewicz, I.F. Trotman, J.A. Wilson, K.G. Wormsley, R.E. Pounder, B. Sharma, J. Spencer and J.H. Baron, Gut, 24, 1002 (1983).
82. S. Gustavsson, H.-O. Adami, L. Loof, A. Nyberg and O. Nyren, Lancet, II, 124 (1983).
83. T. Lind, J.B.M.J. Jansen, C.B.H. Lamers, S. Moberg and L. Olbe, Scand. J. Gastroenterol., 18, (Suppl. 86) 45 (1983).
84. R.R. Berardi and N.B. Caplan, Clin. Pharmacy, 2, 425 (1983).
85. K. Valnes, S. Wetterhus, J. Myren, P. Arva and S. Larsen, Scand. J. Gastroenterol., 18, 33 (1983).
86. K. Valnes, J. Myren, S. Wetterhus, S. Larsen, S. Dyb, E.F. Blekjaer, L. Malvorsen and N. Hovdenak, Scand. J. Gastroenterol., 17, 1003 (1982).
87. J.A. Wilson, J.R.M. Read, E.J.S. Boyd and K.G. Wormsley, Br. J. clin. Pharmac., 15, 329S (1983).
88. A. Tobe, Y. Yoshida and M. Egawa, Arzneim.-Forsch., 33, 397 (1983).
89. G.C. Rosenfeld, Eur. J. Pharmacol., 86, 99 (1983).
90. B.I. Hirschowitz, J. Fong and E. Molina, J. Pharmacol. Exp. Ther., 225, 263 (1983).
91. G. Dobrilla, L. Lucchin, M. Comberlato, G. Coruzzi and G. Bertaccini, Hepato-gastroenterol., 30, 246 (1983).
92. R. Stockbrugger, S. Seeberg, L. Hellner, B. Jaup and G. Dotevall, Scand. J. Gastroenterol., 18 (Suppl. 86), 84 (1983).
93. S. Abate, A. Fresini, G.P. Ferulano, P. Valentini and G. Califano, Clin. Trials J. 20, 255 (1983).
94. J.F. Long, P.J.S. Chiu, M.J. Derelanko and M. Steinberg, J. Pharmacol. Exp. Ther., 226, 114 (1983).
95. P.J.S. Chiu, C. Casciano, G. Tetzloff, J.F. Long and A. Barnett, J. Pharmacol. Exp. Ther., 226, 121 (1983).
96. M.K. Scott, H.I. Jacoby, J.E. Mills, A.C. Bonfilio and C.R. Rasmussen, J. Med. Chem., 26, 535 (1983).
97. J.G. Williams, R.J. Robertson and G.J. Milton-Thompson, Br. J. clin. Pharmacol., 15, 673 (1983).
98. S.P. Canfield and B.P. Curwain, Br. J. Pharmacol., 80, 451 (1983).
99. J.A. Smith, Eur. J. Pharmacol., 88, 215 (1983).
100. J.F. Dijoseph, B.L. Hughes and G.N. Mir, Fed. Proc., 42, 1368 (1983).
101. T. Yamada, H. Shimamura, Y. Tsukamoto, A. Yamaguchi and M. Ohki, J. Med. Chem., 26, 1144 (1983).
102. W.A. Bolhofer, A.A. Deana, C.H. Habecker, J.M. Hoffman, N.P. Gould, A.M. Pietruszkiewicz, J.D. Prugh, M.L. Torchiana, E.J. Cragoe Jr. and R. Hirschmann, J. Med. Chem., 26, 538 (1983).
103. D.E. Beattie, R. Crossley, K.H. Dickinson and G.M. Dover, Eur. J. Med. Chem., 18, 277 (1983).
104. M.D. Ene, T.K. Daneshmend and C.J.C. Roberts, Br. J. Clin. Pharmac., 15, 611P (1983).
105. J.K. Ramage, A. Denton and J.G. Williams, Gut, 24, (1983)
106. J.N. Hunt, J.L. Smith, C.L. Jiang and L. Kessler, Dig. Dis. Sci., 28, 897 (1983).
107. E.Z. Dajani, D.R. Driskill, R.G. Bianchi, E.L. Phillips, E.M. Woods, D.G. Colton, P.W. Collins and R. Pappo, Drug Dev. Res., 3, 339 (1983).

108. H. Kaymakcalan, E. Ramsamooj and D.E. Wilson, Am. J. Gastroenterol, 77, 688 (1982).
109. C. Scarpignato, G. Spina, G.C. Signorini and G. Bertaccini, Chem. Pathol. Pharmacol., 39, 211 (1983).
110. G. Friedman, Am. J. Gastroenterol, 78, 387 (1983).
111. H.G. Damman, P. Muller and B. Simon, Arzneim-Forsch., 33, 905 (1983).
112. M. Feldman, J. Clin. Invest., 72, 295 (1983).
113. W.F. Caspary, W. Seifert and H. Schäfer-Pöss, Acta Endocrinol., 103 (suppl. 256), 174 (1983).
114. K. Tabata and S. Okabe, Digestion, 26, 61 (1983).
115. St. A. Muller-Lissner, C.J. Fimmel and A.L. Blum, Hepato-gastroenterol, 30, 107 (1983).
116. P. Zeitoun, A. Duchateau, E.R. Lacy and S. Ito, Gastroenterol, 84, 1641 (1983).
117. J.L. Wallace, G.P. Morris, E.J. Krausse and S.E. Greaves, Can. J. Physiol. Pharmacol., 60, 1686 (1982).
118. A. Tarnawski, D. Hollander, J. Stachura and W.J. Krause, J. Lab. Clin. Med., 102, 340 (1983).
119. "Mechanisms of Mucosal Protection in the Upper Gastrointestinal Tract", Ed. A. Allen, G. Flemstrom, A. Garner, W. Silen and L.A. Turnberg, Raven Press, New York, 1984.
120. A.A. van Kolfschoten, F. Hagelen, F.C. Hillen, L.P. Jager, P. Zandberg and J. van Noordivijk, Dig. Dis. Sci., 28, 1127 (1983).
121. G. Flemstrom and E. Kivilaakso, Gastroenterol., 84, 787 (1983).
122. L.A. Smeaton, B.H. Hirst, A. Allen and A. Garner, Am. J. Physiol., 245, G751 (1983).
123. S.J. Konturek, J. Tasler, J. Bilski and J. Kania, Am. J. Physiol., 245, G539 (1983).
124. T.A. Miller, J.M. Henagan, L.A. Watkins and T.M. Loy, J. Surg. Res., 35, 105 (1983).
125. W.D.W. Rees, L.C. Gibbons and L.A. Turnberg, Gut, 24, 784 (1983).
126. B.J. Reichstein and M.M. Cohen, Surg. Forum, 33, 171 (1983).
127. P. Vadgama and K.G.M.M. Alberti, Digestion, 27, 203 (1983).
128. J. Sarosiek, A. Slomiany and B.L. Slomiany, Biochem. Biophys. Res. Comm., 115, 1053 (1983).
129. I.N. Ross and L.A. Turnberg, Gut, 24, 1030 (1983).
130. S. McQueen, D. Hutton, A. Allen and A. Garner, Am. J. Physiol., 245, G388 (1983).
131. K. Takeuchi, D. Magee, J. Critchlow, J. Matthews and W. Silen, Gastroenterol., 84, 331 (1983).
132. J.T. Lamont, A.S. Ventola, E.A. Maull and S. Szabo, Gastroenterol., 84, 306 (1983).
133. H.S. Himal and C. Mowat, Can. J. Surg., 26, 142 (1983).
134. S. Domschke, A. Dembinski and W. Domschke, Scand. J. Gastroenterol., 18, 113 (1983).
135. P.H. Guth, G. Paulsen and K. Hirabayashi, Dig. Dis. Sci., 28, 903 (1983).
136. B.A. Hills, B.D. Butler and L.M. Lichtenberger, Am. J. Physiol., 244, G561 (1983).
137. L.M. Lichtenberger, L.A. Graziani, E.J. Dial, B.D. Butler and B.A. Hills, Science, 219, 1327 (1983).
138. I. Szelenyi, S. Postius and H. Engler, Eur. J. Pharmacol., 88, 403 (1983).
139. A. Robert, J.E. Nezamis, C. Lancaster, J.P. Davis, S.O. Field and A.J. Hanchar, Am. J. Physiol., 245, G113 (1983).
140. T.A. Miller and J.M. Henagan, Surg. Forum, 33, 156 (1983).
141. S.J. Konturek, T. Brzozowski, I. Piastucki, T. Radecki and A. Dembinska-kiec, Dig. Dis. Sci., 28, 154 (1983).
142. P. Sharon, F. Cohen, A. Zifroni, F. Karmeli, M. Ligumsky and D. Rachmilevitz, Scand. J. Gastroenterol., 18, 1045 (1983).
143. I.N. Marks, Scand. J. Gastroenterol., 18 (Suppl. 83), (1983).
144. S. Okabe, K. Takeuchi, H. Kunimi, M. Kanno and M. Kawashima, Dig. Dis. Sci., 28, 1034 (1983).
145. S.C. Glover, J.S. Cantlay, J. Weir and N.A.G. Mowat, Dig. Dis. Sci., 28, 13 (1983).
146. S.B. Coghill, D. Hopwood, S. McPherson and S. Hislop, J. Pathol., 139, 105 (1983).
147. L. Bruseghini, N. Basi, J. Gras, J. Llenas, G. de Miguel, B. Rodes, C. Sanfeliu, J. Vilageliu, J. Zapatero, A. Giraldez, J. Freixes and E. Pacei, Arzneim. Forsch., 33, 1655 (1983).
148. T.M. Lin, D.C. Evans, C.J. Shaar and M.A. Root, Am. J. Physiol., 244, G40 (1983).
149. M. Ligumsky, Y. Goto, H. Debas and T. Yamada, Science, 219, 301 (1983).
150. C.S.Chew, Am. J. Physiol., 245, G221 (1983).
151. W. Schepp, H-K. Heim and H.-J. Ruoff, Agents and Actions, 13, 200 (1983).
152. W. Kromer, B. Skowronek, H. Stark and S. Netz, Pharmacol., 27, 298 (1983).
153. P.T. Shea-Donohue, N. Adams, J. Arnold and A. Dubois, Am. J. Physiol., 245, G196 (1983).
154. S. Ferri, R. Arrigo-Reina, S. Candaletti, G. Costa, G. Murari, E. Speroni and G. Scoto, Pharmacol. Res. Comm., 15, 409 (1983).
155. M.H. Charon, P. Pham Van Chuong, M. Dubrasquet and P. Fromageot, C.R. Acad. Sci. Ser. II, 296, 895 (1983).
156. R. Magous and J.P. Bali, Reg. Peptides, 7, 233 (1983).
157. C.B.H.W. Lamers and J.B.M.J. Jansen, J. Clin. Gastroenterol., 5, 21 (1983).
158. B. Richelson, J.F. Rehfeld and L.I. Larsson, Am. J. Physiol., 245, 463 (1983).

Chapter 10. Pulmonary and Antiallergy Agents

John H. Musser, Anthony F. Kreft and Alan J. Lewis
Wyeth Laboratories, Inc., Philadelphia, PA 19101

Introduction - Reviews dealing with the history of treatment,[1] pharmacology,[2] pathogenesis[3] and inflammatory mediators[4] of asthma have appeared. Although no new breakthroughs in pulmonary and antiallergic drug development were reported in 1983, the biochemical and pharmacologic groundwork for the discovery of new agents based on novel mechanisms of action became more apparent. Research areas of significant interest included: the role of leukotrienes (LTs) in the pathophysiology of asthma, the interactions of LTs with other inflammatory mediators and various theories of airway hyperreactivity.[5] Of particular interest, was the finding that LTs can alter the *in vitro* reactivity of airways smooth muscle to other bronchoconstrictor agents;[6] also, increased lysophosphatidyl choline may cause various biochemical changes associated with airway hyperreactivity.[7] An *in vivo* model of human allergic disease in which nasal challenge with antigen leads to physiologic changes together with increased release of inflammatory mediators into nasal secretions was described. This may provide a valuable means to evaluate antiallergic drugs in man.[8]

Leukotrienes and other Mediators - Comprehensive reviews on the chemistry,[9-11] biosynthesis and metabolism,[12,13] and pharmacology[14-16] of LTs have appeared (see also Chapter 24). Chiral starting materials have been employed in the enantiospecific synthesis of LTs C_4, D_4 and E_4.[17] LTF_4 and LTF_4 sulfone have been synthesized and their biological activities determined in the guinea pig (GP).[18] Total syntheses of LTB_4 (non-carbohydrate based), LTB_5 and 14,15-LTA_4 have been reported.[19-21] In the biosynthesis of LTs in human PMNs, it was shown that the 7-H (pro-S) hydrogen is removed from arachidonic acid (AA) to form 5-hydroperoxy-6E, 8Z, 11Z, 14Z-eicosatetraenoic acid (5-HPETE) and that the 10-D (pro-R)-hydrogen is eliminated in the conversion of 5-HPETE into LTA_4.[22,23]

The availability of synthetic LTs and their corresponding radio-labeled analogs has made the study of both receptor and binding site interactions possible. Schild analysis of FPL 55712 antagonism provided evidence for two distinct receptors for LTD_4 in the GP trachea.[24] Separate receptors for LTD_4 and LTB_4 and LTC_4 and LTD_4 in GP lung parenchymal strip and a specific receptor for LTB_4 on human PMNs were also described.[25-27]

Evidence for the role of LTs in the pathophysiology of asthma is accumulating. The production of LTs in neutrophils and macrophages of asthmatics and eosinophils of individuals with hypereosinophilic syndrome is greatly above normal.[28-30] A clinical evaluation of the effects of LTD_4 on the airways of asthmatics has provided evidence which may explain asthmatic hyperresponsiveness.[31] Allergen challenge of lung tissue from asthmatics elicits bronchial contraction that correlates with release of LTC_4, D_4 and E_4.[32] LTC_4 and D_4 increase glycoprotein and lysozyme secretion by human bronchial mucosa.[33] When injected

ISBN 0-12-040519-9

intracutaneously into humans, LTC_4, D_4 and E_4 elicit erythema and wheal formation whereas LTB_4 produces neutrophil infiltration.[34,35] The pharmacology and pathophysiology of LTB_4 was recently reviewed.[36] It is a potent chemotaxin but a weak secretagogue for human PMN.[37] Because of the possibility of impure preparations being used in earlier experiments, LTB_4 may be a more potent chemotactic agent than previously reported.[38] The chemotactic effect of LTB_4 on human lung phagocytes has been demonstrated in vivo by measuring the increase in total bronchoalveolar lavage cells.[39]

LT interaction with other mediators was described in smooth muscle preparations and cell systems. Both LTC_4 and LTB_4 induced contractions in GP lung parenchymal strip and LTD_4 mediated bronchoconstriction in the GP are attenuated by inhibiting thromboxane A_2 (TxA_2) production, thus indicating a role for TxA_2 in LT induced smooth muscle contractions in this species.[40-41] In human endothelial cells, LTD_4 promotes prostacyclin (PGI_2) synthesis.[42] Prostaglandins E_1 and E_2 (PGE_1 and PGE_2) inhibit LTB_4 release from activated human PMN.[43] Both LTB_4 and platelet activating factor (PAF) induce the aggregation of rat PMN; however, PAF does not induce neutrophil aggregation by releasing LTB_4.[44]

The role of PAF in asthma has recently been reviewed.[45] Acetylation of 1-0-alkyl-2-glyceryl-3-phosphorylcholine is an important step in the biosynthesis of PAF both in human monocytes and rat alveolar macrophages.[46,47] Specific receptor sites for PAF on rabbit platelet and GP ileum have been identified.[48] PAF contracts smooth muscle of GP lung parenchyma independently of endogenous histamine, AA metabolites or platelets trapped within the pulmonary vasculature.[49] PAF causes inflammation and edema in the lungs of rabbits and dogs, broncho-constriction in the baboon and smooth muscle contraction and plasma exudation in the hamster cheek pouch.[50-55] In addition to platelets and neutrophils, macrophages now must be included as known target cells for the action of PAF since it induces an oxidative burst in elicited GP peritoneal macrophage.[56] PAF evokes release of PGE_2 and TxB_2 from albumin-elicited GP macrophages in vitro and TxB_2 release from rabbits in vivo.[57,58] A synergistic action has been noted between PAF and PGE_2 with respect to neutrophil migration in rabbit lung and the wheal and flare response in humans.[59,60] PAF and 5-L-hydroxyeicosatetraenoic acid (5-HETE) selectively interact to induce degranulation of human neutrophils.[61]

Inhibitors of LT Biosynthesis and Release - Reviews on the inhibition of phospholipase A_2 (PLA_2), the modulation of LT biosynthesis and the inhibition of lipoxygenase have appeared.[62-64] Recent results suggest that human platelet PLA_2 activity reported in the literature may have been underestimated, apparently due to the presence of an endogenous inhibitor.[65] Activity of microsomal PLA_2 in rat lung can be inhibited by the β-adrenergic drug, fenoterol.[66] Certain protease inhibitors, such as ethyl-4-[6-guanidinohexanoyloxy]-benzoate methansulfonate inhibit PLA_2 from rabbit PMN's.[67] However, selective inhibitors of PLA_2 have yet to be described.

FPL 55712 was found to inhibit the formation of lipoxygenase (LO) products (IC_{50} of 20.6 μM for 5-HETE) but had no effect on synthesis of cyclooxygenase (CO) products in a cell-free homogenate of rat basophilic leukemic (RBL) cells.[68] This finding is significant because other SRSA antagonists may also be specific 5-LO inhibitors and vice versa. Aryl AA analog 1 was shown to inhibit the formation of 5-HETE and LTB_4 from intact human PMN's.[69] Two dimethyl AAs 2 inhibited the formation of 5-LO products 5-HETE and 5,12-diHETE in RBL cells.[70] An irreversible inhibition of the LT pathway was observed with the allenic 4,5-dehydro AA 3 in RBL cells.[71]

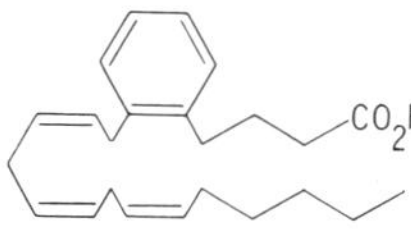

1

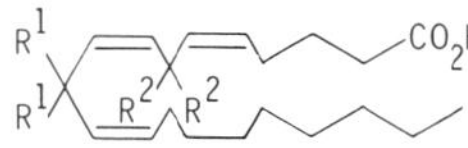

2 a) $R^2 = CH_3$, $R^1 = H$
b) $R^1 = CH_3$, $R^2 = H$

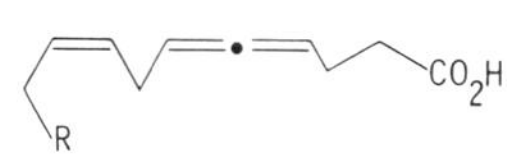

3 R = 

4 R = $C_{10}H_{21}$

The 4,5,8-eicosatrienoic acid 4 was found to have the same inhibitory activity as ETYA toward the formation of LTB_4 and was twice as potent as ETYA toward the formation of 5-HETE using intact human PMN's.[72] Of ten eicosanoids prepared and tested in GP PMNs for 5-LO inhibitory activity, the 5,6-methano LTA_4 5 was the most potent with an IC_{50} of 3 µM.[73] In the same study, LTA_4, 5-HETE and 4,12-diHETE all had IC_{50}s of 2 µM. Using a cultured mastocytoma P-815 cell line, compound 5 had a 5-LO inhibitory activity of 44 µM, whereas the methyl ester of 14-E-5,6-methano LTA_4 6 had an IC_{50} of 6 µM.[74] The secoleukotriene A_4 7 causes a decrease in the formation of LTB_4 in human PMN's.[75] Two members (8 E and Z) of a series of unsaturated sulfur substituted fatty acids

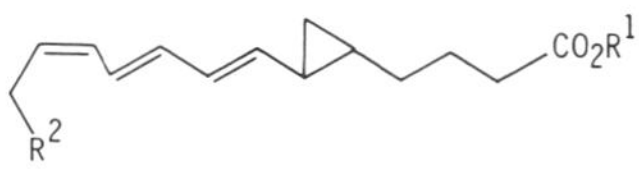

5 R^2 = 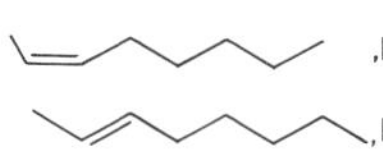 , $R^1 = H$

6 R^2 = , $R^1 = CH_3$

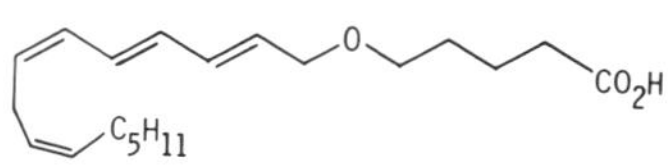

7

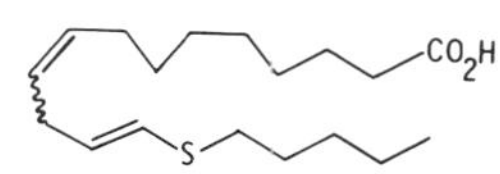

8

inhibit (IC_{50} 3 µM) soybean lipoxygenase.[76] U-60257 9 and its methyl ester block LTB_4 synthesis in human PMN's (IC_{50} 1.8 and 0.42 µM, respectively).[77] The phenolic natural products quercetin 10, eupatilin 11 and esculetin 12 are inhibitors of 5-LO.[78-81] Acetylenic phenone 13 (AA - 861) inhibited the biosynthesis of LTD_4, LTB_4 and 5-HETE in RBL-1 extracts at a concentration of 1 µM (70-80% inh.).[82] Bay 0 8278 (14) inhibits the formation of LO products 5-HETE, 12-HETE, 15-HETE, LTB_4 and its isomer in rabbit and human PMN's.[83] LC-6 (15) has the ability to inhibit AA 5-LO activity.[84] A correlation was observed between known antipsoriatic drugs (resorcinol, salicyclic acid, anthralin, AA, ETYA, mycophenolic acid, trans-retinoic acid and etretinate) and their inhibitory activity with soybean LO.[85] Calmodulin (CM) antagonists trifluoroperamine, chlorpromazine and diphenylbutlamine inhibited 5-LO from RBL-1 cells with potencies (IC_{50} 5.6, 9.6 and 8.3 µM, respectively) similar to those reported for their CM antagonism.[86] 13-*cis* retinoic acid (Ro 4-3780) is a selective inhibitor of 5-LO with an IC_{50} of 9 µM.[87]

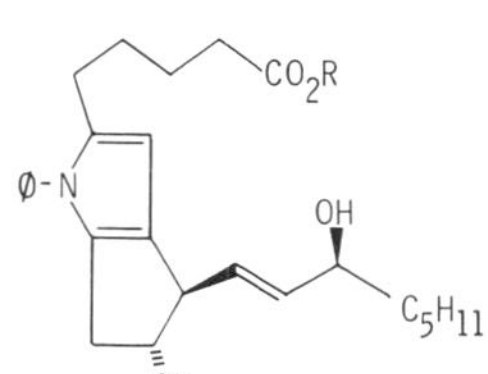

9

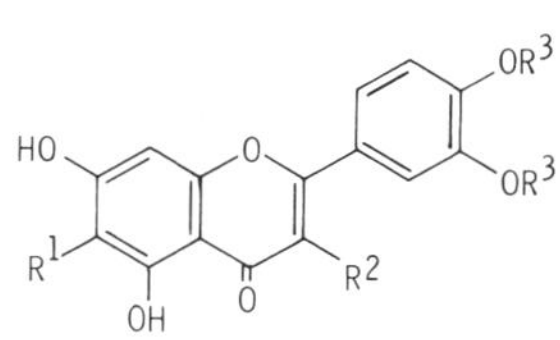

10 $R^1 = R^3 = H$, $R^2 = OH$

11 $R^1 = OCH_3$, $R^2 = H$, $R^3 = CH_3$

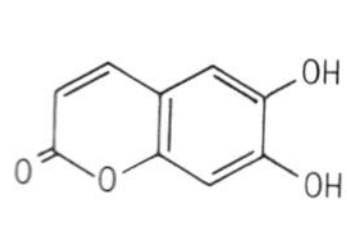

12

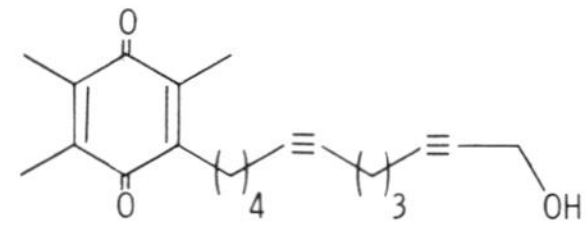

13

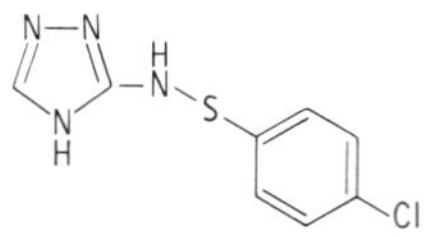

14

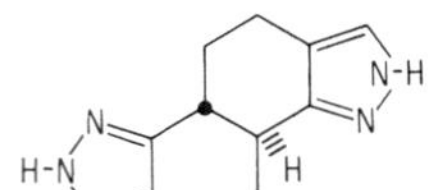

15

Leukotriene Antagonists - FPL 55712 (16) competitively antagonized LTD_4 (pA_2 = 7.35) using GP ileum while the longer acting analog FPL 59257 (17) was a non-competitive antagonist.[88] In contrast to its short biological half-life when administered i.v. to anesthetized GPs, FPL 55712 possesses a surprisingly long biological half-life when given via the aerosol route (120 min vs LTD_4).[89] Several compounds of novel structure were reported to antagonize the effects of LTs (in the form of SRS-A or as the pure substances). KC-404 (18) in anesthetized GPs inhibited SRS-A-induced bronchoconstriction when administered either i.v. or i.d. (ED_{50} = 1.4 x 10^{-3} and 6.5 x 10^{-3} mg/kg, respectively).[90]

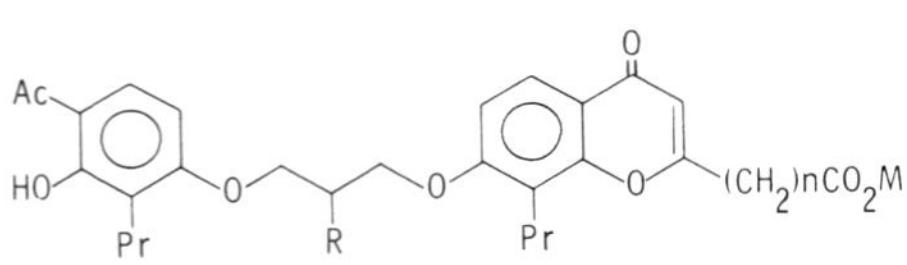

16 R=OH n=0 M=Na

17 R=H n=2 M=(Lysine)H

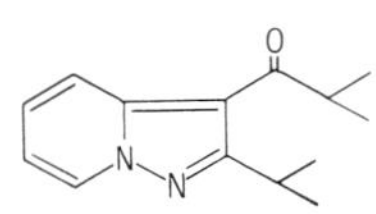

18

The pyridoquinazoline (19) inhibited both the SRS-A-induced contractions in the GP ileum (IC_{50} = 10^{-6}M) and the LTE_4-induced bronchoconstriction in the intact GP (10 mg/kg i.v.).[91] The LT analog (20) antagonized contractions induced by LTC_4, LTD_4 and LTE_4 in the GP trachea at 10^{-4}M and also blocked LTD_4-induced bronchoconstriction and microvascular permeability in the intact GP (5 mg/kg i.v.).[92] SKF 88046 (21) which was previously thought to be a selective antagonist of LTD_4 apparently acts indirectly as an end organ antagonist of Tx, $PGF_{2\alpha}$ and PGD_2.[93,94] Tranilast inhibited contractions induced by LTC_4 (IC_{50} = 2.2 x 10^{-4}M) and LTD_4 (IC_{50} = 2.0 x 10^{-4}M) in the isolated GP trachea.[95] Oxatomide antagonized contractile responses evoked by SRS-A in the sensitized GP lung at > 0.1 μM.[96] Verapamil, nicardipine and DSCG antagonized the effects of LTD_4 in isolated GP trachea (0.3, 1.0 and 3.0 mg/kg, respectively).[97]

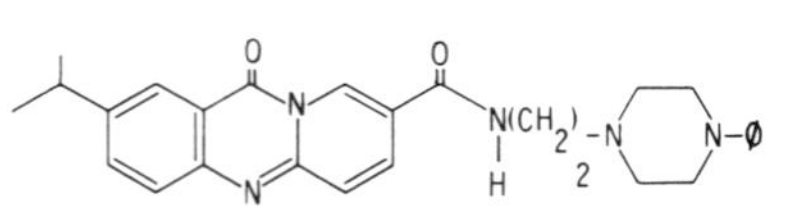

19

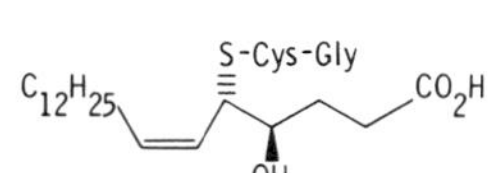

20

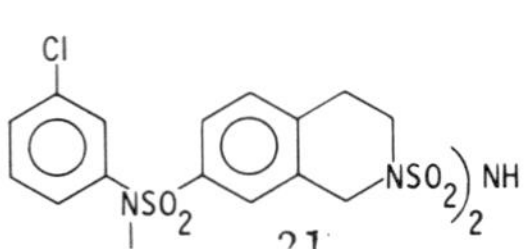

21

Mediator Release Inhibitors (MRI) - A comprehensive review of this class of compounds has appeared recently.[98] Non-linear regression analysis of a series of 51 drugs active in the rat PCA assay revealed that both the conformation of a drug and its capacity to act as an electron acceptor in charge-transfer interactions were critical for high activity.[99] Studies continue to elucidate the mode of action of the prototype MRI, disodium cromoglycate (DSCG) and a comprehensive review of the pharmacology of DSCG have recently appeared.[100] *In vitro* studies on GP trachea revealed that DSCG produced functional antagonism of antigen-induced mediators in addition to mediator-release inhibition.[101] Studies in non-atopic humans revealed that DSCG inhibited immediate and delayed skin wheal-flare responses to PAF.[102] The bronchoconstriction caused by LTD_4 was partially or completely inhibited in GPs[97] and sheep[103] by pretreatment with DSCG. Studies on rat basophils which were deficient in DSCG-binding protein revealed that implantation of this protein restored Ca^{2+} influx and degranulation capacity.[104] In rat peritoneal mast cells DSCG appears to activate an endogenous control mechanism which is mediated by cGMP.[105] Evidence exists that DSCG may inhibit neutrophil chemotactic-factor induced activation of leukocytes directly.[106] Contrary to previous reports, it is claimed that DSCG has significant bronchodilator activity in EIA.[107-109]

A number of new orally active MRI's have been described. The oxamic acid (22) (PRH-836-EA) has an ED_{50} of 0.6 mg/kg p.o. in the rat PCA assay.[110] The quinazoline carboxylate (23) was orally active (ED_{50} = 0.05 mg/kg) against anaphylactic bronchospasm in passively sensitized rats and inhibited the antigen-induced histamine release from passively sensitized rat peritoneal cells (IC_{50} = 0.01 μM).[111] The naphthinotriazole (24) (BRL 22321) possessed oral activity in the rat PCA (ED_{50} = 0.2 mg/kg) and also relaxed smooth muscle.[112] Both the pyridinecarboxamide (25) (ID_{50} = 0.8 mg/kg) and the benzopyranopyridine (26) (ID_{50} = 2.5 mg/kg) showed good oral activity in the rat PCA model.[113,114] RHC 3288 (27), a structurally novel MRI that contains an oxadiazolone as the acidic moeity, inhibited IgE-mediated release of histamine from rat mast cells (I_{50} = 0.9 μM) and was orally active as an inhibitor of IgE-mediated rat PCA (ED_{50} = 1.0 mg/kg).[115,116]

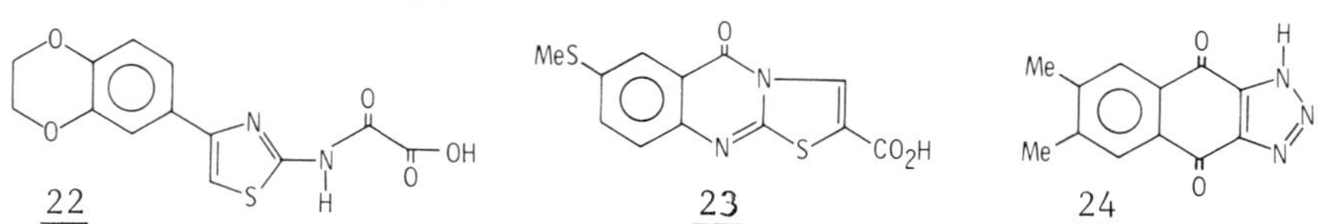

22 23 24

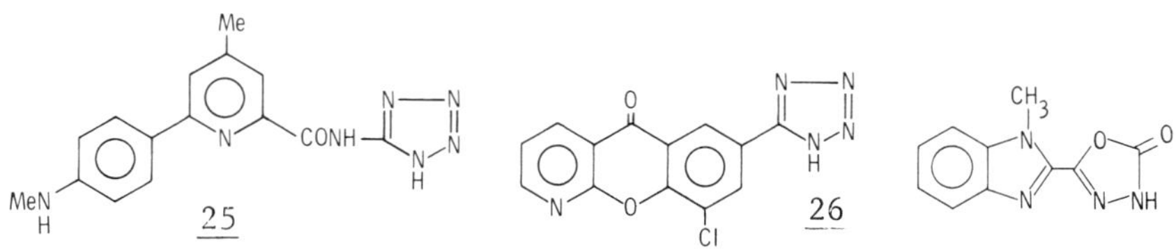

25 26 27

Lodoxamide (0.1 mg via aerosol) protected against grass pollen antigen in asthmatics.[117] The chromone derivative (28) (FPL 58668) given via aerosol inhibited *Ascaris*-induced bronchoconstriction and the concomitant increase in plasma histamine levels in laboratory-bred monkeys.[118] The pyrimidine (29) (BL 5255) inhibited mediator release from passively sensitized rat peritoneal mast cells, chopped monkey lung and human lung.[119] The azapurinone (30) (MB22,948) significantly reduced EIA but was ineffective against histamine-induced asthma.[120] Development of tiaramide has been stopped following reports of possible epileptiform seizures.[121]

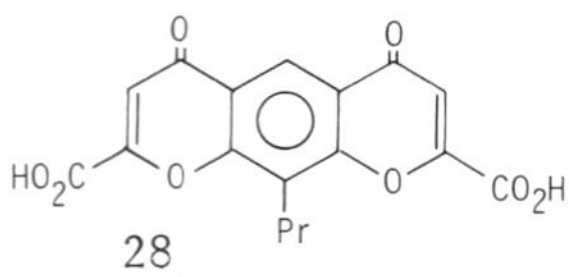

28

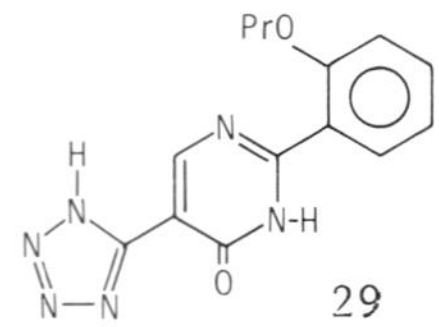

29

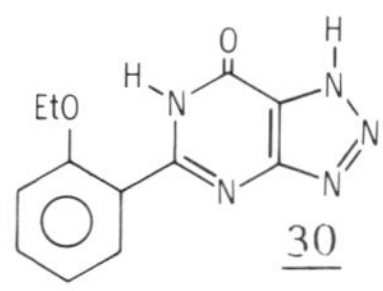

30

Calcium Channel Blockers - The therapeutic value of calcium channel blockers for the treatment of asthma remains unclear; however, they are important investigative tools for the study of airway smooth muscle and cell secretion.[122-125] Nifedipine inhibited calcium ionophore-induced release of PAF and SRSA from human PMNs.[126] Verapamil and nifedipine inhibited PGE_2, SRSA and 12-HETE synthesis in a dose dependent manner in RBL cells.[127] Nifedipine, nimodipine and to a lesser extent verapamil inhibited histamine release from human basophils stimulated by either antigen, anti-IgE or the calcium ionophore, A 23187.[128] Theophylline-induced airway smooth muscle relaxation is enhanced in the presence of nifedipine in the GP.[129] Although nifedipine blocked the contraction of the rat tracheal muscle to KCl and $CaCl_2$, it did not significantly inhibit histamine, methacholine or 5-HT-induced muscle contractions in the GP and the methacholine contraction in rats.[130] Diltiazem, verapamil, nicardipine and bepridil had a greater antagonistic effect on non-physiological agonists ($BaCl_2$, KCl, TEA) than on LTD_4, ACh, 5-HT and Hist in GP trachea.[131] Pretreatment with either nifedipine or verapamil provided significant inhibition in lung resistance (R_L) while neither drug affected dynamic compliance (C_{dyn}) in the dog. However, aerosolized verapamil significantly inhibited both the R_L and C_{dyn} response to antigen whereas a similar concentration of nifedipine was without effect.[132] In Ascaris suum sensitized sheep, verapamil (i.v.) pretreatment inhibited antigen-induced bronchoconstriction, whereas, verapamil did not modify bronchoconstriction produced by aerosols of histamine and carbachol.[133] A clinical study showed that verapamil, administered either orally or by aerosol, can prevent allergen-induced bronchoconstriction.[134] Although a bronchodilating effect in large airways and a diminution of the diurnal variation of peak expiratory flow rate could be shown with nifedipine in patients with chronic asthma, no significant long-term bronchodilating effect was observed.[135] In asthmatics sensitive to bronchoconstriction induced by cold air, nifedipine offered significant protection.[136] In a double-blind, placebo-controlled, cross-over study, a single dose of 20 mg nicardipine did not change the sensitivity or reactivity to carbachol.[137] The combination of verapamil and DSCG was superior to DSCG alone in inhibiting exercise induced asthma in the patients studied.

Antihistamines and anticholinergics - The role of histamine and its receptors (H_1 and H_2) in the pathogenesis of asthma was reviewed.[138,139] The effects of antihistamines in rhinitis has attracted attention and a combination of H_1 and H_2 receptor antagonists were additive in preventing either histamine or allergen provocation in the nose of humans.[140,141] Astemizole is a potent and long lasting H_1 antagonist and is without significant 5-HT antagonist, anticholinergic or mast cell protective effects.[142] In contrast to other antihistamines it did not change the EEG pattern of dogs. It was orally effective in the treatment of seasonal allergic rhinitis.[143] Aerosolized clemastine failed to alter bronchial parameters in asthmatic children.[144] The tripolidine derivative, BW825c, inhibits histamine induced skin flare in man without CNS impairment.[145] SKF 93319 (32) is a specific H_1 and H_2 receptor antagonist that inhibits histamine-induced bronchoconstriction in GPs.[146] The pharmacology of the anticholinergic bronchodilator oxitropium was reviewed.[147] Administered by aerosol, it was less effective than either fenoterol or ipratropium in EIA bromide (IB).[148] Thiazinamium chloride, an aerosol bronchodilator with anticholinergic and antihistaminic properties, was shown to antagonize bronchoconstriction induced by PAF and LTC_4 in guinea pigs, and it also inhibited the synthesis of TxB_2 from alveolar macrophages.[149-151]

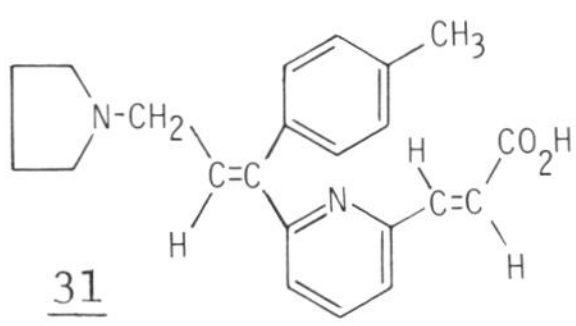

31

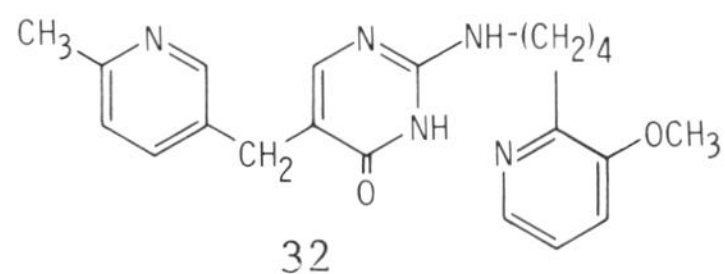

32

β-adrenoceptor agonists - Radioligand binding technology to study adrenoceptors has also heightened interest in β-adrenergic subsensitivity in atopic asthma.[152] Down regulation of β-adrenoceptors in lymphocytes and PMNs was shown by terbutaline.[152-154] Impairment of small airway bronchodilation was reported after 4 week therapy with inhaled salbutamol and s.c. terbutaline.[155] A selective β_2-adrenoceptor defect was observed in all allergic subjects not just asthmatics.[156] Ketotifen was shown to reverse β-receptor tachyphylaxis induced in rats.[157] The pharmacology of β-adrenergic pulmonary selective agents has been reviewed; furthermore, clinical reports on several of the new β_2-agonists have appeared.[158,159] Inhaled bitolterol mesylate, an inactive prodrug, produced a longer lasting protection of EIA than isoproterenol.[160] Procaterol was effective and well-tolerated orally with a long duration of action in asthmatics.[161] It also produced mild and transient tremor and nervousness. Fenoterol inhalation powder and spray were compared in a number of asthma trials.[162] Aerosolized metaproterenol was superior to isoetharine, isoproterenol and atropine in terms of onset of action, peak reversal of airway obstruction and duration, in a double blind trial in chronic asthma.[163] Bambuterol (KWD-2183), a lipophilic terbutaline ester prodrug, was shown to be a potent and highly selective inhibitor of pseudocholinesterase and acetylcholinesterase.[164]

Combination bronchodilator therapy - This topic has been recently reviewed.[165] In chronic asthma, the combination of IB and fenoterol was more effective administered by a metered dose inhaler than salbutamol.[166] Aerosol IB when administered concurrently with inhaled fenoterol and oral oxtriphylline increased bronchodilation with no detectable additional side effects.[167] Inhalation of IB followed by metaproterenol resulted in additive bronchodilation that was greater and longer lasting than IB alone, metaproterenol alone or metaproterenol followed by IB.[168] Neither inhaled terbutaline nor DSCG were very effective in cold air-induced bronchoconstriction whereas both drugs in combination were much more effective.[169] The interaction of methylxanthines and β-agonists continues to be a subject of interest. An increased incidence of cardiotoxicity in rats was observed with a combination of aminophylline and isoproterenol or terbutaline.[170] Salbutamol and theophylline administered orally to childhood asthmatics interacted to produce tachycardia and reduced therapeutic activity but did not result in elevated blood theophylline levels.[171]

Xanthines - The improved efficacy and safety of theophylline formulations in asthma was reviewed.[172] However, the mechanism of action of xanthines in asthma remains controversial. Since at therapeutic concentrations, theophylline is a poor PDE inhibitor but a potent adenosine (Ado) antagonist, Ado antagonism is thought to contribute significantly to its bronchodilator as well as diuretic and CNS stimulatory effects.[173] The role of Ado as a mediator of asthma has been the focus of attention. In asthmatics, but not normals, aerosolized Ado is a potent bronchoconstrictor.[174] *In vitro*, it can inhibit and potentiate IgE-mediated histamine release from human lung, mast cells and basophils.[175-178] Methylxanthines inhibit both forms of the extracellular Ado R-site receptors namely A_1 which inhibits adenylate cyclase and A_2 which is found in human lung mast cells and basophils and is associated with an increase in adenylate cyclase.[176,178,179] Of a series of alkylxanthines evaluated for Ado receptor binding, 1,3-dipropyl substituents enhanced binding potency compared with the 1,3-dimethyl substitution in theophylline.[180] Compound (33) possessed a K_1 for Ado A_1 receptors of 22 pM being 70,000 times more potent than theophylline. The relaxant effects of a series of xanthines in carbachol-constricted guinea pig trachea showed that substitution in the 1- and 3- positions improved relaxant potency.[181] Unsubstituted xanthine and all 9- methyl xanthines were weakly active. However, the bronchodilating xanthines lacked universal Ado antagonism. Enprofylline (34) is an example of a bronchodilator xanthine (5 x more active than theophylline) that is a weak cAMP-PDE inhibitor with negligible ability to antagonize Ado.[182,183] It is well absorbed, has a high renal clearance and a human plasma half life of 113 min.[184] After oral administration, it produced significant bronchodilation at plasma concentrations of 3 μg/ml.[185] It lacks theophylline-like extrapulmonary side effects but produces dose-related headache and nausea.[184] Thus, Ado antagonism is neither necessary nor a desirable property for xanthine antiasthmatics. Inhaled methylxanthines (including theophylline) possessed mild bronchodilator activity in asthmatics[186] against both Ado and histamine but were extremely bitter in taste.[187]

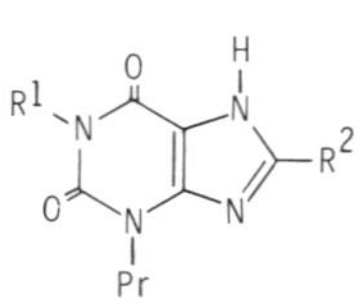

	R^1	R^2
33	Pr	$2NH_2.4ClØ$
34	H	H

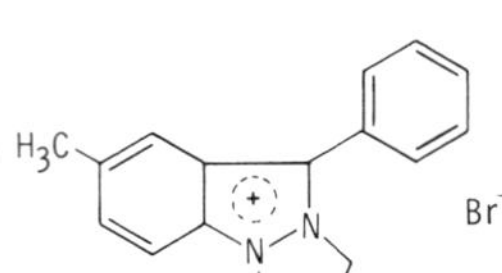

35

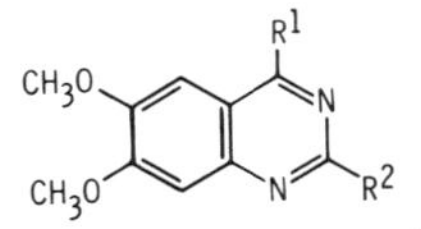

36

R^1 = $NHCH_2CH(OH)Aryl$ R^2 = $N(Et)_2$

R^1 = $N(Et)_2$ R^2 = $NHCH_2CH(OH)Aryl$

Miscellaneous - Treatment of severe steroid-dependent asthmatics with high daily doses of inhaled steroids in order to reduce oral steroid requirements continues to attract attention.[188,189] Budesonide was much more effective in chronic asthma by inhalation than by oral administration.[190] Its low oral bioavailability (11%), a short half life and extensive liver metabolism may explain why it produces few systemic side effects.[191]

FKK (35) is one of a series of indazole derivatives found to possess oral bronchodilator activity without effect on blood pressure or heart rate and is virtually devoid of CNS effects.[192] Its mechanism of action is unknown but it is devoid of anticholinergic, α-adrenergic blocking or β-adrenoceptor agonist properties. Trapidil, a coronary vasodilator, possesses bronchodilator activity and also inhibits IgE-mediated histamine release from human PMNs.[193] The diterpene forskolin, which stimulates adenylate cyclase by directly activating the catalytic subunit, inhibited acetylcholine, histamine and antigen-induced tracheal smooth muscle contractions in vitro and antigen-induced bronchospasm in vivo in guinea pigs.[194,195] Two series of aminoquinazolines (36) were claimed to be PDE inhibitors and tracheal smooth muscle relaxants.[196] Dazoxiben, a thromboxane synthetase inhibitor protective against bronchoconstriction in guinea pigs, depressed late phase cutaneous allergic reactions in man.[197] However, severe side effects accompanied this response.

The role of α-adrenergic receptor stimulation in asthma remains controversial.[198] Inhalation of phentolamine, the α-adrenergic antagonist, produced both bronchodilatation and bronchoconstriction in extrinsic asthmatics.[199] However, oral phentolamine partially blocked exercise and cold-induced bronchospasm in asthma without affecting reactivity to histamine, suggesting α-adrenergic mechanisms may be involved in airway cooling.[200]

References

1. J.M. Smith Thorax, 38, 244 (1983).
2. J. Morley and K.D. Rainsford, Eds., "Pharmacology of Asthma", Agents Actions, 13, Birkhauser Verlag, Basel, 1983.
3. P.J. Barnes, J.R. Soc. Med., 76, 580 (1983).
4. R. Pauwels and M. Van Der Straeten, Eds. Eur. J. Resp. Dis., 64, suppl. 129 (1983).
5. Eur. J. Resp. Dis., 64, Suppl 127 (1983).
6. B.R. Creese and M.K. Bach, Prostaglandins, Leukotrienes Med., 11, 161 (1983).
7. P. Nath, A.P. Joshi and K.P. Agrawal, J. Allergy Clin. Immunol., 72, 351 (1983).
8. R.M. Naclerio, H.L. Meier, A. Kagey-Sobotka, N.F. Adkinson, D.A. Meyers, P.S. Norman and L.M. Lichtenstein, Am. Rev. Respir. Dis., 128, 597 (1983).
9. V. St. Georgiev in "Survey of Drug Research in Immunologic Disease", Vol. 1, S. Karger AG, Basel, 1983, p. 103.
10. J. Ackroyd and F. Scheinmann, Chem. Soc. Rev. 11, 321 (1982).
11. R.H. Green and P.F.Lambeth, Tetrahedron, 39, 1687 (1983).
12. B. Samuelsson, Science, 220, 568 (1983).
13. S. Hammarstrom, Ann. Rev. Biochem., 52, 355 (1983).
14. S.E. Dahlen, Acta Physiol. Scand., Suppl. 512, 1, (1983).
15. E. Granstrom, Acta Obstet. Gynecol. Scand., Suppl. 113, 9 (1983).
16. P.J. Piper, Br. Med. Bull., 39, 255 (1983).
17. N. Cohen, B.L. Banner, R.J. Lopresti, F. Wong, M. Rosenberger, Y. Liu, E. Thom and A.A. Liebman, J. Am. Chem. Soc., 105, 3661 (1983).
18. D. Denis, S. Charleson, A. Rackham, T.R. Jones, A.W. Ford-Hutchinson, A.Lord, M. Cirino, Y. Girard, M. Larue and J. Rokach, Prostaglandins, 24, 801 (1982).
19. C. Fuganti and S. Servi, C. Zirotti, Tetrahedron Lett., 24, 5285 (1983).
20. E.J. Corey, S.G. Pyne and W. Su, Tetrahedron. Lett., 24, 4883 (1983).
21. R. Zamboni, S. Milette and J. Rokach, Tetrahedron Lett., 24, 4899 (1983).
22. E.J. Corey and P.T. Lansbury, J. Am. Chem. Soc., 105, 4093 (1983).
23. A. Panossina, M. Hamberg and B. Samuelsson, FEBS Lett., 150, 511 (1983).
24. R.N. Krell, B.S. Tsai, P. Berdoulay, M. Barone and R.e. Giles, Prostaglandins, 25, 171 (1983).
25. P. Sirois, S. Roy and P. Borgeat, Prostaglandins, 26, 91 (1983).
26. G.K. Hogaboom, S. Mong, H. Wu and S.T. Crooke, Biochem. Biophys. Res. Commun., 116, 1136 (1983).
27. A. Kreisle and C.W. Parker, J. Exp. Med., 157, 628 (1983).
28. C.M. Yang, S.S.M. and S.R. Wang, J. Allergy Clin. Immunol. 71, suppl. 153 (1983).
29. P. Godard, M. Damon, C. Chavis, C. Kervan and F.B. Michel, Am. Rev. Respir. Dis., 127, 73 (1983).
30. W.R. Henderson, J.B. Harley, A.S. Fauci and S.J. Klebanoff, J. Allergy Clin. Immunol. 71, 135 (1983).

31. M. Griffin, J.W. Weiss, A.G. Leitch, E.R. McFadden, E.J. Corey, K.F. Austen and J.M. Drazen, N. Eng. J. Med., 308, 436 (1983).
32. S.E. Dahlen, G. Hansson, P. Hedqvist, T. Bjorck, E. Granstrom and B. Dahlen, Proc. Nat. Acad. Sci., 80, 1712 (1983).
33. S.J. Coles, K.H. Neill, L.M. Reid, K.F. Austen, Y. Nii, E.J. Corey and R.A. Lewis, Prostaglandins, 25, 155 (1983).
34. L. Juhlin and S. Hammarstrom, Prostaglandins, Leukotrienes and Med., 11, 381 (1983).
35. N.A. Soter, R.A. Lewis, E.J. Corey and K.F. Austin, J. Invest. Dermatol., 80, 115 (1983).
36. M.A. Bray, Brit. Med. Bull., 39, 244 (1983).
37. T.E. Rollins, B. Zanolari, M.S. Springen, Y. Guindon, R. Zomboni, C.K. Lau and J. Rokach, Prostaglandins, 25, 281 (1983).
38. W. Jubiz, Biochem. Biophys. Res. Comm., 110, 842 (1983).
39. T.R. Martin and L.C. Altman, Clin. Res., 31, 73A (1983).
40. F.J. Zijlstra, J.E. Vincent and I.L. Bonta, Prostaglandins, Leukotrienes Med., 11, 385 (1983).
41. R.M. Muccitelli, R.R. Osborn and B.M. Weichman, Prostaglandins, 26, 197 (1983).
42. E.B. Cramer, L. Pologe, N.A. Pawlowski, Z.A. Cohn and W.P. Scott, Proc. Nat. Acad. Sci., 80, 4109 (1983).
43. E.A. Ham, D.D. Soderman, M.E. Zanetti, H.W. Dougherty, E. McCauley and F.A. Kuehl, Proc. Nat. Acad. Sci., 80, 4349 (1983).
44. A.W. Ford-Hutchinson, Int. J. Immunopharmacol., 5, 17 (1983).
45. B.B. Vargaftig and J. Benveniste, Trends In Pharmacol. Sci., 342 (1983).
46. G. Camussi, F. Bussolino, C. Tetta, W. Piacibello and M. Aglietta, Int. Archs. Allergy Appl. Immunol., 70, 245 (1983).
47. D.H. Albert and F. Snyder, J. Biol. Chem., 258, 97 (1983).
48. S.B. Hwang, C.S.C. Lee, M.J. Cheah and T.Y. Shan, Biochem., 22, 4756 (1983).
49. N.P. Stimler and J.T. O'Flaherty, Am. J. Pathol., 113, 75 (1983).
50. G.S. Worthen, A.J. Goins, B.C. Mitchel, G.L. Larsen, J.R. Reeves and P.M. Henson, Chest, 83, 13S (1983).
51. G. Camussi, I. Pawlowski, C. Tetta, C. Roffinello, M. Alberton, J. Brentgens and G. Andres, Am. J. Patholog., 112, 78 (1983).
52. J.E. Heffner, S.A. Shoemaker, M. Canham, M. Patel, I.F. McMurtry, H.G. Morris and J.E. Repline, J. Clin. Invest., 71, 351 (1983).
53. M. Mojarad, Y. Hamasaki and S.I. Said, Bull. Eur. Physiopathol. Respir. Clin. Res. Phys., 19, 253 (1983).
54. A. Denjean, B. Arnoux, R. Masse, A. Lockhart and J. Benviste, J. Appl. Physiol., 55, 799 (1983).
55. J. Bjork and G. Smedegard, J. Allergy Clin. Immunol., 71, 145 (1983).
56. H.P. Hartung, M.J. Parnham, J. Winkelmann and U. Hadding, Agents Actions, 13, 504 (1983).
57. H.P. Hartung, FEBS Lett., 160, 209 (1983).
58. L.M. McManus, F.A. Fitzpatrick, D.J. Hanahan and R.N. Pinckard, Immunopharmacol., 5, 197 (1983).
59. G.S. Worthen, S. Gumbay, G.L. Larsen and P.M. Hensen, Am. Rev. Respir. Dis., 127, Pt. 2, 57 (1983).
60. G.S. Basran and D. McGivern, Am. Rev. Respir. Dis., 127, Pt. 2, 78 (1983).
61. J.T. Flaherty, M.J. Thomas and M.J. Hammett, C. Carroll, C.E. McCall and R.L. Wykle, Biochem. Biophys. Res. Comm., 111, 1 (1983).
62. G.J. Blackwell and R.J. Flower, Br. Med. Bull., 39, 260 (1983).
63. P. Borgeat, B.F. Laclos and J. Maclout, Biochem. Pharmacol., 32, 381 (1983).
64. G.A. Higgs and J.R. Vane, Br. Med. Bull., 39, 265 (1983).
65. L.R. Ballou and W.Y. Cheung, Proc. Nat. Acad. Sci., 80, 5203 (1983).
66. P.V. Wichert and W. Meyer, Agents Actions, 3, 233 (1983).
67. H. Kunze, E. Bohn and B. Dameram, Pharmacol. Res. Comm., 15, 869 (1983).
68. F.B. Casey, B.J. Appleby and D.C. Buck, Prostaglandins, 25, 1 (1983).
69. J.R. Pfister and D.V. Krishna Murthy, J. Med. Chem., 26, 1099 (1983).
70. C.D. Perchonock, J.A. Finkelstein, I. Uzinskas, J.G. Gleason, H.M. Sarau and L.B. Cleslinski, Tetrahedron Lett., 24, 2457 (1983).
71. E.J. Corey, S.S. Kantner and P.T. Lawsbury, Tetrahedron Lett., 24, 265 (1983).
72. J.W. Patterson, J.R. Pfister, P.J. Wagner and D.V. Krishna Murthy, J. Org. Chem., 48, 2572 (1983).
73. Y. Arai, K. Shimoji, M. Konno, Y. Konishi, S. Okuyama, S. Iguchi, M. Hayashi, T.Miyamoto and M. Toda, J. Med. Chem., 26, 72 (1983).
74. Y. Koshihara, S. Murota and K.C. Nicolau, Adv. Prostaglandin, Thromboxane Leukotriene Res., 11, 163 (1983).
75. J.W. Patterson and D.V. Krishna Murthy, J. Org. Chem., 48, 4413 (1983).
76. M.O. Funk and A.W. Altenden, Biochem. Biophys. Res. Comm., 114, 937 (1983).
77. F.F. Sun and J.C. McGuire, Prostaglandins, 26, 211 (1983).

78. W.C. Hope, A.F. Welton, C. Fieldler-Nagy, C. Batula-Bernado and J.W. Coffey, Biochem. Pharmacol., 32, 367 (1983).
79. Y. Koshihara, T. Neichi, S. Murota, A. Lao, Y. Fujimoto and T. Tatsuno, FEBS Lett., 158, 41 (1983).
80. K. Sekiya, H. Okuda and S. Arichi, Biochim. Biophys. Acta., 713, 68 (1982).
81. T. Neichi, Y. Koshihara and S. Murota, Biochim. Biophys. Acta., 753, 130 (1983).
82. M. Shiraishi and S. Terro, J. Chem. Soc., Perkin Trans. 1, 1591 (1983).
83. M. Mardin, W.D. Busse, R. Grutzmann and R. Rochels, Arch. Pharmacol., 322, suppl. R114 (1983).
84. A.M. Magro and I. Hurtado, J. Immunopharmacol., 5, 191 (1983).
85. J.C. Sircar and C.F. Schwender, Prostaglandins, Leukotrienes Med., 11, 373 (1983).
86. C. Fiedler-Nagy, A.F. Welton, C. Batula-Bernardo, K. Sachelli and J.W. Coffey, Fed. Proc., 42, 2091 (1983).
87. C. Fiedler-Nagy, J.G. Hamilton, C. Batula-Bernardo and J.W. Coffey, Fed. Proc., 42, 919 (1983).
88. M.C. Holroyde and A.M. Ghelani, Eur. J. Pharmacol., 90, 251 (1983).
89. M. O'Donnell and A.F. Welton, Agents Actions, 14, 43 (1984).
90. K. Nishino, H. Ohkubo, M. Ohashi, S. Hara, J. Kito and T. Irikura, Jpn. J. Pharmacol., 33, 267 (1983).
91. J.W. Tilley, P. Levitan, A.F. Welton and H.J. Crowley, J. Med. Chem., 26, 1638 (1983).
92. B.M. Weichman, D.A. Holden, R.R. Osborn, D.F. Woodward, T.W. Ku, J.G. Gleason and M.A. Wasserman, Pharmacologist, 25, 122 (1983).
93. J.G. Gleason, R.D. Krell, B.M. Weichman, F.E. Ali and B. Berkowitz in "Advances in Prostaglandin, Thromboxane and Leukotriene Research", Vol. 9, B. Samuelsson and R. Paoletti, Eds., Raven Press, New York, 1982, p. 243.
94. M.A. Wasserman, B.M. Weichman and J.G. Gleason, Pharmacologist, 25, 122 (1983).
95. H. Komatsu, A. Ujiie, J. Naito, Nippon Yakurigaku Zasshi, 82, 47 (1983).
96. K. Ohmari, H. Ishii, M. Nito, K. Shuto and N. Nakamizo, Nipon Yakurigaku Zasshi, 81, 399 (1983).
97. C. Advenier, J. Cerrina, P. Duroux, A. Floch, J. Pradel and A. Reiner, Br. J. Phrmacol., 78, 301 (1983).
98. C.F. Schwender, Drugs Future, 8, 699 (1983).
99. B.V. Cheney and R.E. Christoffersen, J. Med. Chem., 26, 726 (1983).
100. B.A. Berman and R.N. Ross, Clin. Rev. Allergy, 1, 105 (1983).
101. A.G. Stewart and M.R. Fennessy, Clin. Exp. Pharmacol. Physiol., 10, 595 (1983).
102. G.S. Basran, C.P. Page, W. Paul and J. Morley, Eur. J. Pharmacol., 86, 143 (1983).
103. W.M. Abraham, J.C. Delehunt, E. Russi, A. Wanner, L. Yerger and G.A. Chapman, Am. Rev. Respir. Dis., 127, (4, Pt. 2), 65 (1983).
104. N. Mazurek, P. Bashkin, A. Loyter and I. Pecht, Proc. Nat. Acad. Sci. USA, 80, 6014 (1983).
105. E. Wells and J. Mann, Biochem. Pharmacol., 32, 837 (1983).
106. A.B. Kay in "Proceedings of the XI International Congress of Allergology and Clinical Immunology", J.W. Kerr and M.A. Ganderton, Eds. Macmillan, London, 1983, p. 245.
107. J.S.G. Cox in "Disodium Cromoglycate in Allergic Airways Disease", J. Pepys and A.W. Frankland, Eds. Butterworths, London, 1970.
108. S. Godfrey in "Asthma", T.J. H. Clark and S. Godfrey, Eds. Chapman and Hall, London, 1977, p. 273.
109. R.M. Jones, C.R. Horn, D.V. Lee and S.R. Brennan, Br. J. Dis. Chest, 77, 362 (1983).
110. K.D. Hargrave, F.K. Hess and J.T. Oliver, J. Med. Chem., 26, 1158 (1983).
111. R.A. LeMahieu, M. Carson, A.F. Welton, H.W. Baruth and B. Yaremko, Jr. Med. Chem., 26, 107 (1983).
112. B.A. Spicer, J.W. Ross, G.D. Clarke, E.J. Harling, P.A. Hassall, H. Smith and J.F. Taylor, Agents Actions, 13, 301 (1983).
113. Y. Honma, K. Oda, T. Hashiyama, K. Hanamoto, H. Nakai, H. Inoue, A. Ishida, M. Takeda, Y. Ono and K. Tsuzurahara, J. Med. Chem., 26, 1499 (1983).
114. T. Oe, M. Tsuruda, H. Matsuo, M. Terasawa and K. Goto, Yakugaku Zasshi, 103, 300 (1983).
115. A. Khandwala, S. Coutts, V. Dally-Meade, N. Jariwala, J. Musser, R. Brown, H. Jones, B. Loev and I. Weinryb, Biochem. Pharmacol., 32, 3325 (1983).
116. J.H. Musser, R.E. Brown, B. Loev, K. Bailey, H. Jones, R. Kahen, F. Huang, A. Kandwala, M. Leibowitz, P. Sonnino-Goldman and D. Ponigi-Ruzza, J. Med. Chem., 27, 121 (1984).
117. R.W. Parrish and B.H. Davies, Thorax, 38, 458 (1983).
118. I.M. Richards, R.P. Eady, D.M. Jackson, T.S.C. Orr, D.I. Pritchard, K. Vendy and E. Wells, Clin. Exp. Immunol., 54, 461 (1983).
119. F.C. Reed, H.H. Newball and P. Siminoff, Inter. Archs. Allergy Appl. Immunol., 71, 59 (1983).
120. R.M. Rudd, A.R. Gellert, P.R. Studdy and D.M. Geddes, Br. J. Dis. Chest, 77, 78 (1983).
121. E. Berglund, J. Orehek and P. Vermeire, Eur. J. Respir. Dis., 64, 321 (1983).
122. C.H. Fanta and J.M. Drazen, Am. Rev. Respir. Dis., 127, 673 (1983).
123. P.J. Barnes, Thorax, 38, 481 (1983).

124. E. Middleton, Eur. J. Respir. Dis., 64, suppl. 128, 123 (1983).
125. D.J. Triggle, Allergy, 38, 1 (1983).
126. E. Jouvin-Marche, J. Cerrina, E. Coeffier, P. Duroux and J. Benveniste, Eur. J. Pharmacol., 89, 19 (1983).
127. L. Levine, Biochem. Pharmacol., 32, 3023 (1983).
128. C. Jensen, P.S. Skov and S. Norn, Allergy, 38, 233 (1983).
129. A.S. Elguindi and A.S. Choudhry, Am. Rev. Respir. Dis., 127, 92 (1983).
130. J.B. Cheng and R.G. Townley, Arch. Int. Pharmacodyn., 263, 228 (1983).
131. J. Cerrina, C. Advenier, A. Renier, A. Floch and P. Duroux, Eur. J. Pharmacol., 94, 241 (1983).
132. P.E. Malo, M.A. Wasserman and R.L. Griffin, Eur. J. Pharmacol., 92, 69 (1983).
133. E.W. Russi, B. Marchette, L. Yerger, W.M. Abraham, T. Ahmed, Am. Rev. Respir. Dis., 127, 675 (1983).
134. A. Miadonna, A. Tedeschi, E. Leggieri, M. Cottini, M. Restuccia and C. Bianchini, Ann. of Allergy, 51, 201 (1983).
135. D. Patakas, E. Vlachoianni, U. Tsara, G. Louridas, P. Arginopoulou, J. Allergy, Clin. Immunol., 72, 269 (1983).
136. A.F. Henderson, R.W. Heaton, J.F. Costello, Thorax, 38, 512 (1983).
137. A. Baronti, A. Grieco, I.M. Lelli and C. Vibelli, Curr. Ther. Res., 34, 142 (1983).
138. P. Eyre and N. Chand, in "Pharmacology of Histamine Receptors", C.R. Ganellin and M.E. Parsons, Eds. Wright PSG, Bristol, 1982, pp. 298-322.
139. J. White and N.M. Eiser, Br. J. Dis. Chest, 77, 215 (1983).
140. C. Secher, J. Kirkegaard, P. Borum, A. Maansson, P. Osterhammel and N. Mygind, J. Allergy Clin. Immunol., 70, 211 (1982).
141. G.B. Carpenter, A.L. Bunker-Soler and H.S. Nelson, J. Allergy Clin. Immunol., 71, 412 (1983).
142. F.H.L. Awouters, C.J.E. Niemegeers and P.A.J. Janssen, Arz. Forsch., 33, 381 (1983).
143. J.D. Wilson and J.H. Hillas, Clin. Allergy, 13, 131 (1983).
144. R.L. Henry, I.G.C. Hodges, A.D. Milner and G.M. Stokes, Arch. Dis. Child., 58, 304 (1983).
145. A.F. Cohen, M. Hamilton, C. Burke, J. Findlay and A.W. Peck, Br. J. Pharmacol., (in press).
146. R.C. Blakemore, T.H. Brown, D.G. Cooper, G.J. Durant, C.R. Ganellin and R.J. Ife, Br. J. Pharmacol., 80, suppl., 437P (1983).
147. H. Koch, Med. Actual/Drugs Today, 19, 444 (1983).
148. K. Larsson, Respiration, 43, 57 (1982).
149. J. Chang, A. Dervinis and A.J. Lewis, Br.J. Pharmacol., 80, 439P (1983).
150. A. Dervinis, A. Blumenthal, S. Schwalm, J. Chang and A.J. Lewis, Pharmacologist, 25, 121 (1983).
151. J. Chang, S. Piperata, M. Skowronek and A.J. Lewis, Biochem. Pharmacol., 32, 2671 (1983).
152. J.J. Krzanowski and A. Szentivanyi, J. Allergy Clin. Immunol., 72, 433 (1983).
153. R.D. Aarons, A.S. Nies, J.G. Gerber and P.M. Molinoff, J. Pharmacol. Exp. Ther., 224, 1 (1983).
154. Y. Sano, G. Watt and R.G. Townley, J. Allergy Clin. Immunol., 72, 495 (1983).
155. M.E. Conolly, D.P. Tashkin, K.K.P. Hui, M.R. Littner and R.N. Wolfe, J. Allergy Clin. Immunol., 70, 423 (1982).
156. J.H. Shelhamer, Z. Morom and M. Kaliner, J. Allergy Clin. Immunol., 71, 57 (1983).
157. U. Bretz, U. Martin, L. Mazzoni and U.M. Ney, Eur. J. Pharmacol., 86, 321 (1983).
158. C.W. Bierman, Clin. Rev. Allergy, 1, 87 (1983).
159. J.W. Patterson, K.M. Lulich and R.G. Goldie, Trends Pharmacol. Sci., 4 (1983).
160. C.W. Bierman, T. Mingo, C.T. Furukawa, E.G. Shapiro and W.E. Pierson, J. Allergy Clin. Immunol., 71, 124 (1983).
161. S. Siegel, R. Katz, G. Rachelefsky, M. Brandon and L. Borgen, J. Allergy Clin. Immunol., 71, 125 (1983).
162. J.R. Vale and F. Molgaard, Eur. J. Resp., 64, Suppl. 130 (1983).
163. A.G. Peerless, G.S. Rachelefsky, M.R. Mickey, R.M. Katz and S.C. Siegel, J. Allergy Clin. Immunol., 72, 702 (1983).
164. L.A. Svensson, T. Olsson, K. Tegner and A. Tunek, 186th ACS Natl Mtg. Toronto, Abs. 18 (1983).
165. G.M. Shenfield, Drugs, 24, 414 (1982).
166. K. Flint, B. Hockley and N. McIJohnson, Eur. J. Respir. Dis., 64 (suppl. 128), 548 (1983).
167. A.S. Rebuck, M. Gent and K.R. Chapman, J. Allergy Clin. Immunol., 71, 317 (1983).
168. I. Bruderman, R. Cohen-Aronovski and J. Smorzik, Chest, 83, 208 (1983).
169. K.M. Latimer, P.M. O'Byrne, M.M. Morris, R. Roberts and F.E. Hargeave, Am. Rev. Respir. Dis., 128, 440 (1983).
170. V.E. Whitehurst, X. Joseph, R.A. Nicklas and T. Balazs, J. Allergy Clin. Immunol., 71, 123 (1983).

171. K.P. Dawson and D.M. Fergusson, Arch. Dis. Child, 57, 674 (1982).
172. L. Hendeles and M. Weinberger, Pharmac. Ther., 18, 91 (1983).
173. N. Svedmyr, in "Pharmacology of Asthma", p. 83 (in ref. 2).
174. M.J. Cushley, A.E. Tattersfield and S.T. Holgate, Br. J. Clin. Pharmacol., 15, 161 (1983).
175. S.P. Peters, E.S. Schulman, D.W. MacGlashan, R.P. Schleimer, H.H. Newball and L.M. Lichtenstein, J. Allery Clin. Immunol., 69, 150 (1982).
176. C.J. Vardey and I.F. Skidmore, Eur. J. Resp. Dis., 63, 157 (1982).
177. E.S. Schulman, D.W. MacGlashan, S.P. Peters, R.P. Schleimar, H.H. Newball and L.M. Lichtenstein, J. Immunol., 129, 2662 (1983).
178. M.K. Church, S.T. Holgate and P.J. Hughes, Br. J. Pharmacol., 80, 719 (1983).
179. J.W. Daly, J. Med. Chem., 25, 197 (1982).
180. R.F. Bruns, J.W. Daly and S.H. Snyder, Proc. Natl. Acad. Sci. USA, 80, 2077 (1983).
181. J.A. Karlsson, G. Kjellin and C.G.A. Persson, J. Pharm. Pharmacol., 34, 788 (1982).
182. C.G.A. Persson, J.A. Karlsson and I. Erjefalt, Life Sci., 30, 2181 (1982).
183. E. Lunnell, N. Svedmyr, K.E. Andersson and C.G.A. Persson, Eur. J. Respir. Dis., 64, 333 (1983).
184. L.C. Laursen, N. Johannesson, A. Dirksen, R. Djurup, E.P. Munch, E. Tandorf and B. Weeke, Allergy, 38, 75 (1983).
185. E. Lunnell, N. Svedmyr, K.E. Andersson and C.G.A Persson, Eur. J. Clin. Pharmacol., 22, 395 (1982).
186. M.J. Cushley, A.E. Tattersfield and S.T. Holgate, Am. Rev. Respir. Dis., 125, 62 (1982).
187. M.J. Cushley and S.T. Holtage, Thorax, 38, 223 (1983).
188. L. Rosenhall, G. Lundquist, E. Adelroth and C. Glennov, Eur. J. Resp. Dis., 122, 154 (1982).
189. M.J. Smith and M.E. Hodson, Lancet, 1, 265 (1983).
190. R. Ellul-Micallef and S.A. Johansson, Br. J. Clin. Pharmacol., 15, 419 (1983).
191. A. Ryrfeldt, P. Andersson, S. Edsbacker, M. Tonnesson, D. Davies and R. Pauwels, Eur. J. Resp. Dis., Suppl. 63, 86 (1982).
192. Y. Shiraki and K. Sakai, Arch. Int. Pharmacodyn, 262, 150 (1983).
193. Y. Morita, S. Inokima and T. Miyamoto, Int. Archs Allergy Appl., Immunol., 71, 289 (1983).
194. J. Chang, A. Dervinis, A. Blumenthal, S. Schwalm and A.J. Lewis, Pharmacologist, 25, 121 (1983).
195. J.F. Burka, J. Pharmacol., Exp. Ther., 225, 427 (1983).
196. T.N. Riley and J. Millen, In 186th ACS Natl Meet. Washington, DC, USA, Aug. 28, 1983, Medi, 21.
197. W. Dorsch, J. Ring and H. Melzer, J. Allergy Clin. Immunol., 72, 168 (1983).
198. S.R. Smith, J. Clin. Hosp. Pharm., 8, 201 (1983).
199. R.J. Shiner and M.I. Molho, Chest, 83, 602 (1983).
200. E.R. Bleeker, K.S. Chahal, P. Mason and S. Permutt, Eur. J. Respir. Dis., 64 (suppl.), 258 (1983).

SECTION III-Chemotherapeutic Agents

Editor: Frank C. Sciavolino, Pfizer Central Research
Groton, Connecticut 06340

Chapter 11. Antibacterial Agents

E. S. Hamanaka and M. S. Kellogg
Pfizer Central Research, Groton, CT 06340

General. Highlights of the antibacterial research literature during 1983 featured impressive developments in the cephalosporin and quinolone carboxylic acid classes of antibacterial agents. Among these were two new oral cephalosporin prodrugs and potent, parenteral cephalosporins possessing the broadest antibacterial activity yet observed in this series. An entire issue of *Tetrahedron* was devoted to recent aspects of the chemistry of β-lactam antibiotics.[1] Fervent interest in the quinolone carboxylic acids was evident in the sixty-two papers, featuring seven new agents, which were presented at the 23rd ICAAC. Reviews covering the SAR, pharmacokinetics and clinical studies of 3rd generation and investigational cephalosporins[2] and the antibacterial activity and therapeutic use of moxalactam[3] and cefotaxime[4] were published. Other areas of β-lactam antibiotics which were reviewed include prodrugs,[5] discovery in nature,[6] penicillin-binding proteins (PBP) and mechanism of action,[7-9] antimicrobial therapy,[10-12] effect of protein binding on antibacterial activity and pharmacology,[13,14] β-lactamase inhibitors[15] and mechanistic aspects of β-lactamase inhibitors.[16] Reviews on new antibacterial agents[17] and antimicrobial agents and various aspects of antimicrobial chemotherapy[18-20] were published, as were books devoted to antibiotics,[21-23] β-lactam antibiotics,[24] cephalosporins,[25] antimicrobial therapy,[26] antibiotic pharmacokinetics[27] and antimicrobial agents and clinical pharmacology.[28]

Monobactams. Like aztreonam, AMA-1080 (Ro 17-2301, **1**) and its congener **2** exhibited broad, potent activity against Gram-negative bacteria, good β-lactamase stability and weak activity against Gram-positive and anaerobic bacteria.[29,30] AMA-1080 is stable to β-lactamases produced by aztreonam-resistant *Klebsiella oxytoca* and *Proteus vulgaris* and is active against these pathogens *in vitro.*[31,32] Incorporation of fluorine in the 4-methyl group of aztreonam resulted in slightly reduced potency.[33] The syntheses and SAR of 3-acylamino,[34] 3-methoxylated[35] and 4-alkylated monobactams[36] were reported. Replacement of the $-SO_3^{(-)}$ residue of monobactams by $-OP(O)RO^{(-)}$ gave increased β-lactamase stability, but reduced antibacterial potency.[37] A mode of action study of monobactams with 3-acylamino side chains corresponding to penicillin G, piperacillin, azlocillin and cefotaxime showed that for antibacterial activity, binding to PBP 3 of *Escherichia coli* and to PBPs 1, 2 and 3 of *Staphylococcus aureus* were required.[38] SQ 26,917, the 4β-methyl congener of aztreonam, was comparable in activity to aztreonam *in vitro* with the exception of lower activity against *Enterobacter* spp. and *Morganella morganii* and greater activity against *Pseudomonas*.[39] SQ 28,332, SQ 28,502 and SQ 28,503, new nonmethoxylated monobactam antibiotics with oligopeptide side chains at C-3, were isolated from fermentations of a *Flexibacter* sp.[40,41] Only the structure of SQ 28,332 (**3**) has been completely determined.[40]

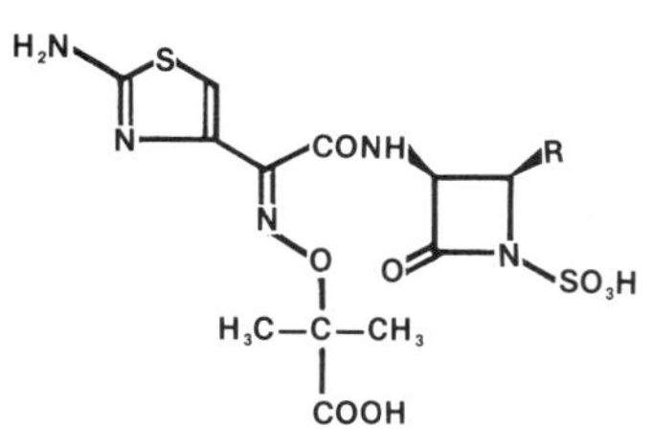

1 R = CH_2OCONH_2
2 R = $CONH_2$

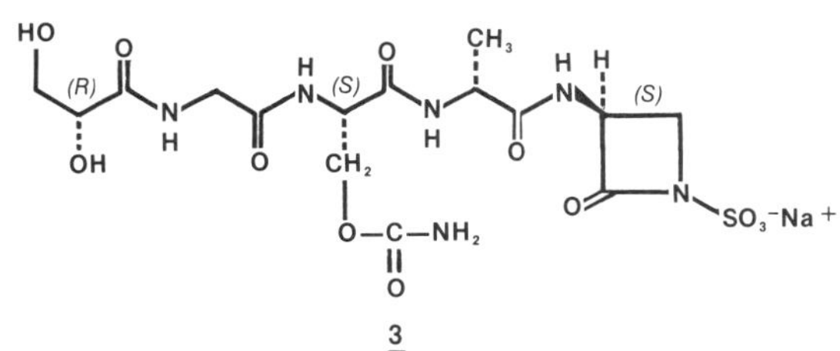

3

ISBN 0-12-040519-9

Cephalosporins. The orally active 3rd generation cephalosporin FK 027 (FR 17027) continued to be extensively studied *in vitro* and *in vivo*.[42-44] KY-109 (**4**), a double prodrug of KY-087 (**5**), gave high blood levels of **5** when administered orally to various animal species.[45] Oral administration of cefuroxime axetil, the 1-acetoxyethyl ester prodrug of cefuroxime, to human volunteers produced higher peak serum levels than an equivalent dose of ampicillin and urinary recovery of 35%.[46,47] CM 40874 (**6**), a new, parenteral cephalosporin 1-S-oxide derivative, had potent activity against Gram-negative bacteria, including *Pseudomonas*, but poor activity against *Staphylococcus* and anaerobes.[48,49] Another cephalosporin 42980 RP (**9**) possessed a structure and antibacterial profile similar to cefotaxime and ceftriaxone.[50] A novel oxacephem derivative (6315-S, **7**), exhibited broad, potent antibacterial spectrum with the exception of activity against *Pseudomonas*.[51] The effect on antibacterial activity of α-substituents in the 2-(2-amino-4-thiazolyl)acetyl side chain of ceftizoxime was studied.[52] Cefotaxime analogs in which the 2-aminothiazole moiety was replaced by 5-amino-1,2,4-thiadiazole possessed broad, potent antibacterial activity *in vitro*.[53] Among a variety of cefotaxime analogs bearing a tetrazolo[1,5-b]pyridazin-6-ylthiomethyl group at C-3′, FCE 20485 (**10**) was of interest because of its good *in vitro* and *in vivo* activity and its very long half-life in mice.[54] Modification of the C-3′ substituent of cefotaxime resulted in cephalosporins possessing the broadest antibacterial spectra encountered in this class of antibiotics.[55-57]

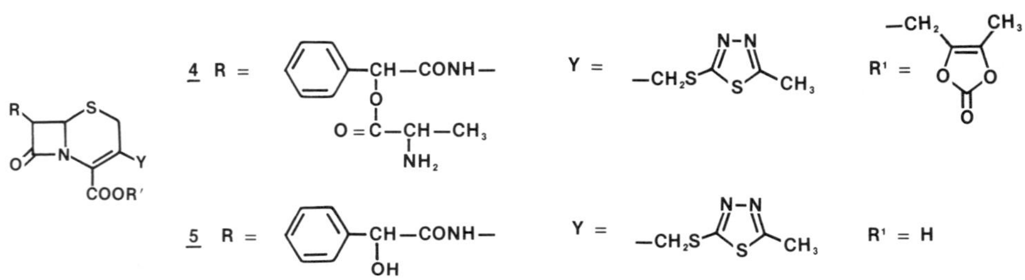

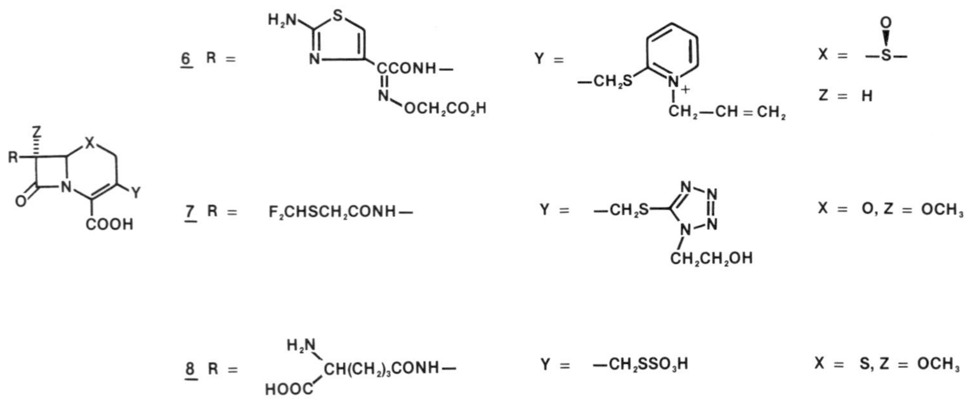

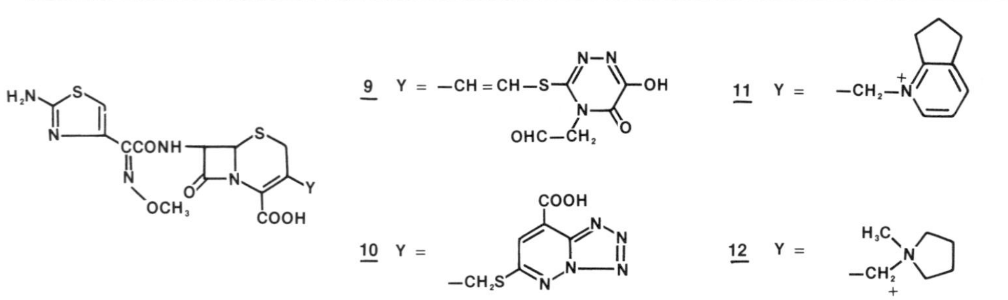

HR 810 (**11**) exhibited more potent activity than cefotaxime or ceftazidime against *S. aureus*, enterococci and *Enterobacter* spp. and was comparable in activity to ceftazidime against *Pseudomonas aeruginosa*.[58,59] Also HR 810 possessed good pharmacokinetic properties in several animal species and good activity against experimental bacterial infections in mice.[60] BMY-28142 (**12**), another cefotaxime analog, appeared to be similar in activity to HR 810.[61] Full papers describing the antibacterial properties of L-640,876 *in vitro* and *in vivo* and its mode of action were published.[62-64] The relationship between PBP affinities and antibacterial activity for cefotaxime and several closely related analogs was investigated.[65,66] The effect of various 7α substituents (H, OCH_3, OC_2H_5, SCH_3, CN, CH_3) on the β-lactamase stability of cephem derivatives and on their affinity for PBPs in *Morganella morganii* was described.[67] The syntheses and substituent effects on antibacterial activity, alkaline hydrolysis rates and infrared absorption frequencies of cephem analogs related to moxalactam (latamoxef) were reported.[68] SF-1623 (**8**), a new cephamycin, was isolated from fermentation broths of *Streptomyces chartreusis*.[69] Papers were published on the laboratory and clinical studies of cefotetan,[70] ceftazidime,[71,72] and cefoperazone[73] which were presented at their respective symposia.

Penicillins. BRL 20330 (**13**), an orally effective prodrug, delivered good blood levels of temocillin after oral administration to human volunteers.[74] The preparation and pharmacological properties of new prodrugs of ampicillin were described,[75-77] and among these KB-1585 (**14**) was well absorbed after oral administration to rats, dogs and humans.[76,77] The proceedings of symposia dealing with laboratory and clinical studies of amdinocillin (mecillinam),[78] mezlocillin[79] and azlocillin[80] were published. Another symposium was concerned with recent developments in oral antibiotic therapy and included an update on bacampicillin.[81]

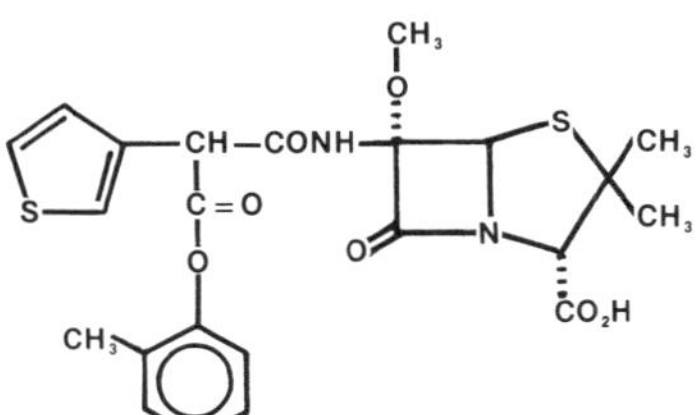

13

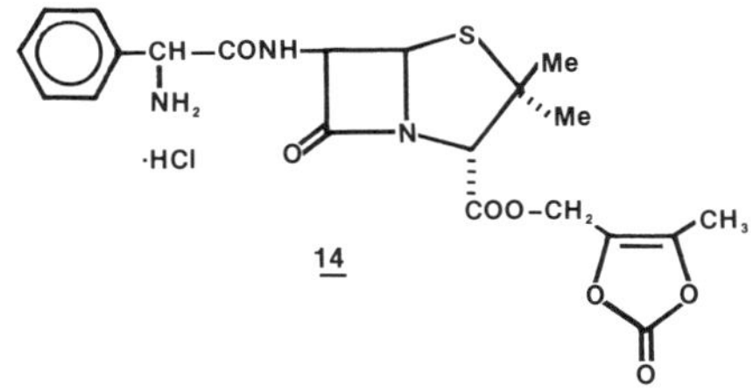

14

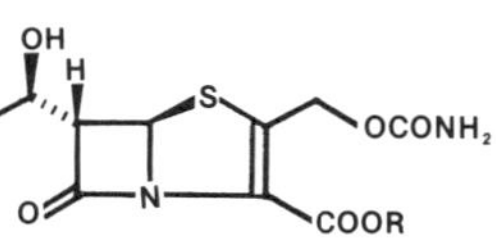

15 R = $CH_2OCOC(CH_3)_3$
16 R = CH_2OCOCH_3

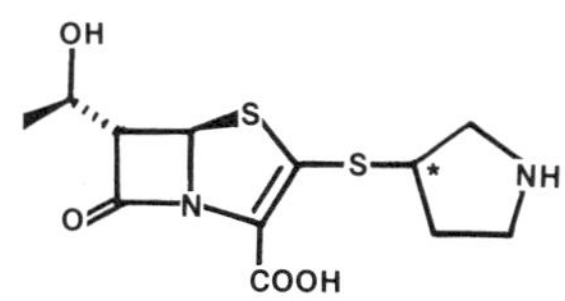

R-17 (*R*)
S-17 (*S*)

Penems. After oral administration to mice, FCE 22553 (**15**) and FCE 22891 (**16**), prodrug esters of FCE 22101, were well absorbed and effective against experimental bacterial infections.[82,83] Diastereomeric penems *R*-**17** and *S*-**17** exhibited antibacterial activity comparable to thienamycin.[84]

Carbapenems. L644,440 (**18**) and L646,591 (**19**) had the same antibacterial profile as imipenem (N-formidoyl thienamycin, MK0787) but exhibited much greater resistance to renal dehydropeptidase-I (DHP) and better urinary recovery than thienamycin.[85,86] Carbapenem **20** was 4–5 times more active than thienamycin against *E. coli*, *Klebsiella pneumoniae* and *Pseudomonas aeruginosa*. Although RS-533 was slightly less potent than **20** *in vitro*, it was more effective against experimental bacterial infections in mice.[84] Total syntheses of 2-aryl and 2-heteroaryl carbapen-2-em-3-carboxylic acids, with and without a 6-hydroxyethyl side chain, were described along with their antibacterial activity and DHP stability.[87] In a study of the effect of side chain substituents on antibacterial activity and metabolic stability of a series of thienamycin analogs, increased stability to DHP was reflected in improved *in vivo* activity and pharmacokinetic properties in mice.[88] The antibacterial activity and pharmacological properties of MM 13902 and its congeners, as well as pharmacokinetic studies of MM 13902 in various animal species and human subjects were reported.[89] Syntheses and SAR for a series of carbapenems related to C-19393 H_2 (carpetimycin A) were published.[90] Naturally occurring 5,6-*cis*-carbapenem antibiotics having the E-2-acetamidoethenylsulfinyl or E-2-acetamidoethenylthio side chain at C-2 were converted to the Z-isomers when refluxed in chloroform containing quaternary ammonium halides.[91] The Z-isomers were slightly less potent against Gram-negative bacteria but were 3-fold more stable to DHP enzyme and some were more effective *in vivo*. CD spectral studies of the natural 5,6-*cis*-carbapenem antibiotics bearing the E-2-acetamidoethenylsulfinyl side chain at C-2 showed that they all possess the R-configuration of the sulfoxide.[92] When DHP inhibitor MK0791 (cilastatin **21**) was intravenously coadministered with imipenem to human subjects, plasma levels and AUC values for imipenem were increased about 20%, half-life was not altered significantly and urinary recovery was markedly enhanced to about 72%.[93] The novel carbapenam No. 17927D (**22**) was isolated from culture broths of three Streptomycetes strains.[94] Two new carbapenem antibiotics, carpetimycins C (**23**) and D (**24**), possessing broad, potent antibacterial activity were isolated from fermentation broths of *Streptomyces* sp. KC-6643.[95] Details of the isolation, structure elucidation and antibacterial activity of northienamycin and 8-epi-thienamycin were published. Their N-formimidoyl derivatives exhibited broad antibacterial activity but lower potency than imipenem.[96] Syntheses of (-)-carpetimycin A (C-19393 H_2)[97] and (±)-C-19393 H_2 and several congeners[98] were reported, as were studies related to the structures of the olivanic acids MM 13902, MM 4550 and MM 17880.[99] Papers presented at a symposium on laboratory and clinical studies of imipenem were published.[100]

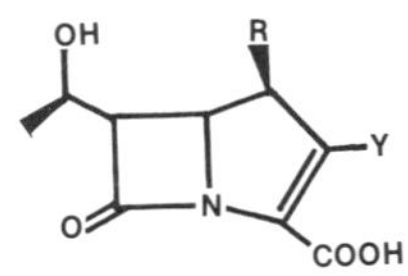

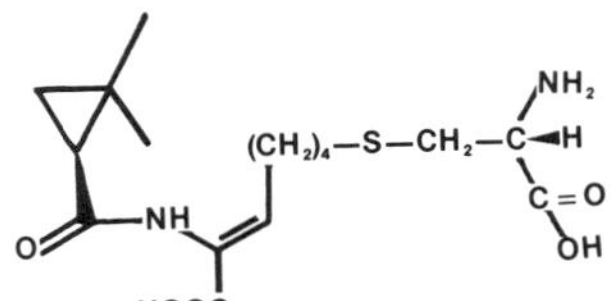

21

18 R = H, Y = —$SCH_2C(=NH)N(CH_3)_2$

19 R = CH_3, Y = —$SCH_2C(=NH)N(CH_3)_2$

20 R = H, Y = —S— 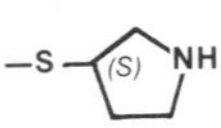

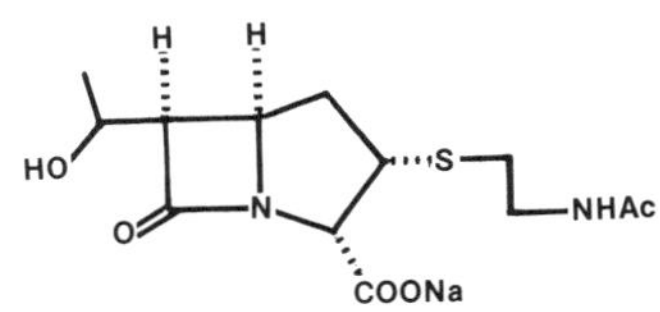

22

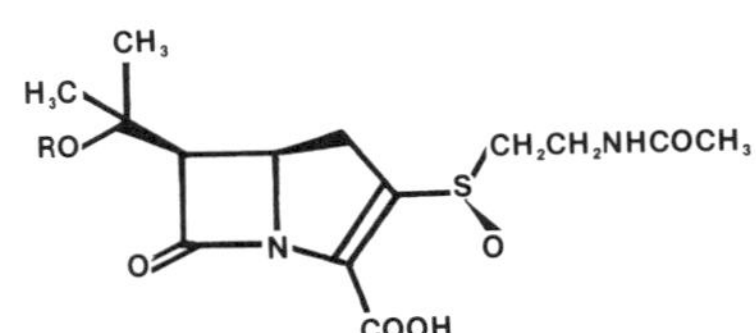

23 R = —H
24 R = —SO_3H

β-Lactamase Inhibitors. The mechanism of action of Ro 15-1903 and the relationship between β-lactamase inhibitory activity and C-6 side-chain structure of a series of penams related to Ro 15-1903 were studied.[101,102] Failure of the enzyme level β-lactamase inhibitory activity of Ro 15-1903 to translate into correspondingly potent activity at the whole cell level and *in vivo* was attributed to its instability in testing medium, mouse plasma and human sera.[103] The (2,3)-β-methylenepenam **25** was comparable to sulbactam and clavulanic acid in its ability to protect ampicillin from hydrolysis by β-lactamases from various bacteria, whereas 6β-bromo analog **26** and 6α-chlorosulfone **27** were somewhat less active than their penam counterparts.[104] Varying degrees of β-lactamase-inhibitory activity were reported for: a) clavulanic acid analogs lacking the C-3 carboxyl group;[105] b) 6β-sulfonyloxypenicillanic acid derivatives;[106] c) N-alkylaminopenicillanic acids;[107] and d) 3-ureidopenam sulfones.[108] FR 900318 (**28**), a penicillanic acid derivative isolated from fermentation broths of *Aspergillus candidus*, exhibited β-lactamase inhibitory activity.[109]

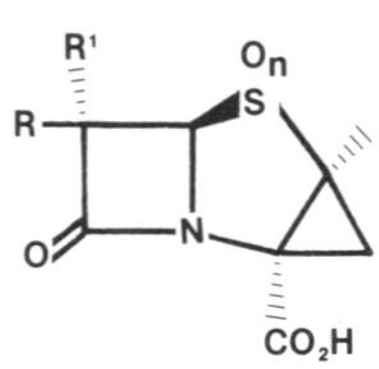

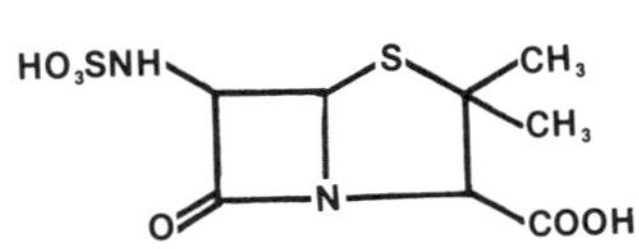

28

25 R = R' = H, n = 2
26 R = Br, R' = H, n = 0
27 R = H, R' = Cl, n = 2

Aminoglycosides. Biotransformation of kanamycin (KM) A to amikacin by a mutant strain of the butirosin producing organism,[110] suggested an alternative to chemical synthesis of amikacin. The relative importance of the amino groups to antibacterial activity in the aminoglycosides (AGs) was the subject of several studies. Heptadeoxy KM possessing no hydroxy functionality retained moderate antibacterial activity. With the 2″-hydroxyl group and introduction of the (S)-4-amino-2-hydroxybutyryl (HABA) moiety, activity improved markedly.[111] Chemical modifications of istamycin B also indicated a predominant role for the amino functionality.[112] Removal of the 1-amino moiety of gentamicin (GM) C[113] and selective deamination of neomycin B[114] resulted in decreased activity. 1-*epi* KM A and its HABA derivative were much less active than the natural epimers.[115] Substitution of KM and related AGs with 1-N-(2-aminoethanesulfonyl) gave products with improved activity against resistant strains.[116] The synthesis and SAR of guanidino derivatives of GM and KM A[117] were reported with the major interest related to reducing nephrotoxicity.[118] In the fortimicin (FM) series, 3-0-demethyl-2,3-di-*epi* FM and the 3-*epi* isomer were much less active than FM A.[119] Similarly, while 3-amino-3-demethoxy FM A and 2-amino-3-0-demethyl-2-deoxy FM A had activity comparable to FM A, the corresponding 2-amino- 3-0-demethyl-2-deoxy-2-*epi* FM A was inactive.[120] Halogenated analogs of lividomycin B were found to possess diminished activity.[121] Saccharocin (also KA-5685), a new member of the apramycin group, was isolated independently by two groups from *Saccharopolyspora* sp.[122,123] Synthetic modification at C-6′ of spectinomycin gave aliphatic side chain analogs with improved Gram positive potency and activity against *P. aeruginosa*.[124]

Macrolides. In the 16 membered ring macrolide series the relationship between antimicrobial and ribosome binding activities of tylosin and its derivatives was studied.[125] Affinity for ribosomes did not directly correlate with antibacterial activity and the introduction of amino substituents did not markedly improve potency for the strains studied.[126] Hybrid biosynthesis using protylonolide (PT) with daunomycin producer *Streptomyces* sp. KA-464 did not yield daunosaminylated derivatives but gave instead 19-hydroxy PT and 23-hydroxy PT.[127] Conversion of PT to chimeramycins A and B through biotransformation with a *S. ambofaciens* KA-448 culture was reported.[128] Further work on isolation and identification of macrolides from blocked mutants of *S. fradiae* was also described.[129,130] Two minor mycinamicin components were discovered and their structures determined.[131] A procedure was published for the synthesis of 20-dihydro-20-deoxy derivatives of 16-membered ring macrolides.[132] Bioconversion of a rosaranolide to incorporate the oral absorption characteristics of mycinamicins into rosaramicins was studied.[133] The products were less active than the parent antibiotics. Acetal analogs of neospiramycin at the 3 and/or 4 position were prepared and found to provide *in vivo* activity.[134] Microbial transformation of maridomycin III by *Serratia marcescens* gave chemically interesting products that possessed poor antibacterial activity.[135] Several new members of the milbemycin family were discovered and studied,[136-139] as were new 20-membered ring macrolides, the irumanolides I and II.[140] The biological conversion of erythromycin (EM) B produced new macrolides which were less active than EM A, but more stable to acid.[141] Chemical conversion of 3-0-mycarosyl EM B into its (8S)-8-fluoro analog was reported.[142] Application of the mutational biosynthesis approach gave four new (8S)-8- fluoro EM antibiotics with antibacterial activity similar to EM A and B but with high acid stability.[143,144] In a comparison of three EM A C-9 oxime derivatives with other orally administered drugs, **29** (RU 28965), **30** and **31** exhibited *in vitro* activity similar to EM A, but improved *in vivo* performance.[145,147] The first report of C-9 functionalization of EM A through carbon-carbon bond formation was published,[148] and interest in the total synthesis of macrolides[149,150] and new chemical methods for macrocyclic ring formation continued.[151,152]

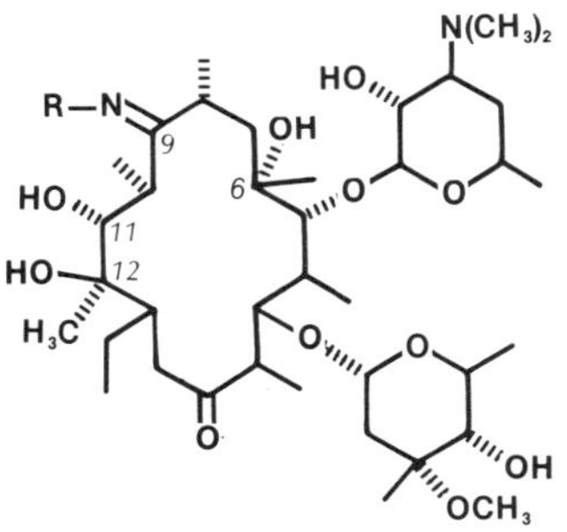

R	
$H_3CO(CH_2)_2OCH_2O$	29
$Me_2N(CH_2)_2O$	30
$Et_2N(CH_2)_2O$	31

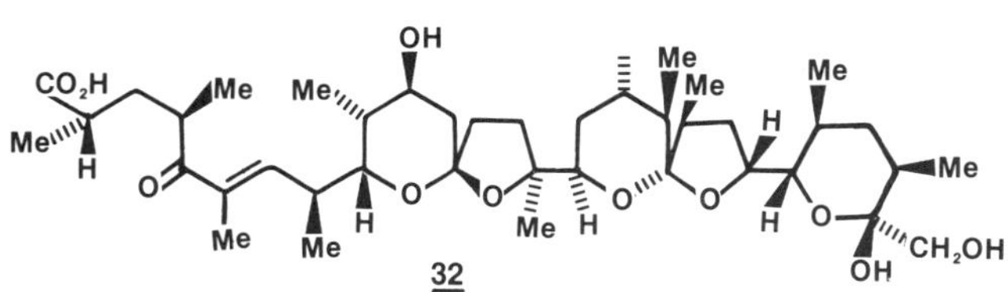

32

Polyether Ionophores. Isolation and identification of antibiotic X-14885A was reported.[153,154] Its structure is closely related to A-23187 and cezomycin, members of the pyrrolether class of natural ionophores. Aspects of the discovery and ionophoric properties of the X-14868 complex were described.[155] Another new naturally derived agent, CP-53607 (**32**), was reported to be effective against coccidia and as a rumen propionic acid stimulant.[156] The C-26 urethane analogs of monensin transported both Rb^+ and Ca^{2+}.[157] Their antibacterial and anticoccidial activities were also reported as greater than monensin. Microbial conversion of grisorixin by *Streptomyces rimosus* was found to produce an inactive and detoxified product.[158] A unified stereochemical model of polyether antibiotic structure and biogenesis was proposed.[159] Total synthesis of ionophores continued to be of interest[160,161] and the unnatural enantiomer of lasalocid A was found to have similar biological properties to the natural product.[162]

Quinolones. Interest remained high in the synthesis of nalidixic acid analogs.[163-165] Rosoxacin (**33**) derivative Win 35,349 (**34**) showed good activity *in vitro* and *in vivo* against strains of *Staphylococcus aureus* and *epidermidus*.[166,167] Because of improved potency and spectrum, most efforts centered on 6-fluoro analogs of pipemidic acid. The broad spectrum and potency of norfloxacin (**35**), which inhibited conjugal plasmid transfer[168] and appeared to augment antifungal activity of amphotericin B,[169] was mirrored in the discovery of several new agents. Ciprofloxacin (**36**, Bay 0 9867) has broad, potent, antibacterial activity[170-172] and appears to exhibit the most potent anti-pseudomonas activity yet found.[173] Enoxacin (AT 2266, CI 919, **37**) had good oral pharmacokinetics which translated to good *in vivo* protection in animal models.[173] Ofloxacin (DL 8280, HOE 280, **38**) was reported more active than norfloxacin against staphylococci.[174,175] Pefloxacin (1589 RB, **39**) penetrated inflammatory cerebrospinal fluid at effective antibacterial levels after oral or parenteral administration in patients with bacterial meningitis.[176] AM-833 (**40**), a difluoro quinolone, was similar to norfloxacin *in vitro*[177] but was reported to have superior animal pharmacokinetics and *in vivo* protection.[178] Win 49,375 (**41**) was described as a broad-spectrum, potent, antibacterial[179-181] for parenteral[182] and oral use.[183] A brief review of the ICAAC reports on these agents was published.[184]

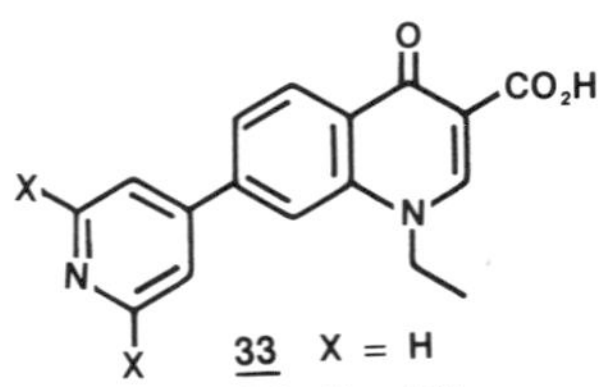

33 X = H
34 X = CH_3

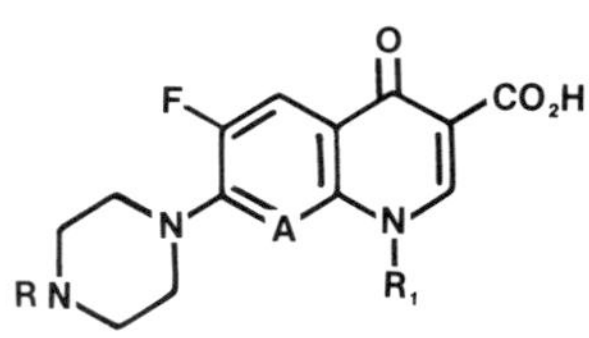

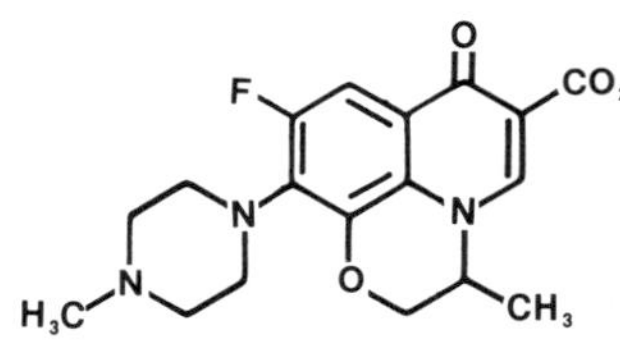

38

	A	R	R_1
35	CH	H	C_2H_5
36	CH	H	—◁
37	N	H	C_2H_5
39	CH	CH_3	C_2H_5
40	CF	CH_3	CH_2CH_2F
41	CH	CH_3	$NHCH_3$

Glycopeptides. The binding sites of vancomycin and ristocetin A were studied using time-dependent nuclear Overhauser difference spectroscopy.[185] The structure of vancomycin was reassigned and its rearrangement to CDP-I investigated.[186] Glycopeptides were featured at an ICAAC symposium and new agents, actaplanin (A4696),[187-189] complex A41030,[190] aglycones A41030 and A47934,[191-193] were described. Synthesis of aglycoristocetin derivatives with antibacterial activity cast doubt on the importance of a basic amino group at the N-terminus of ristocetin.[194] Isolation and purification of OA-7653, active against Gram positive bacteria, was reported.[195] The antibacterial activity of teichomycin relative to other antibiotics was described.[196-198]

Ansamycins. Synthesis of 25-deacetoxy-25-*epi*-hydroxy rifamycin S (RF S) was reported.[199] The additional hydroxyl function introduced by the transformation did not have the expected effect on antibacterial activity, although the proper conformation for interaction with the target enzyme was proved. RF derivatives containing 3-amidino and 4-aminoimidazolo [4,5-c] moieties were prepared and some showed good antimycobacterial activity in mice.[200] LM 427, a spiropiperidyl RF, was particularly effective in mouse tubercular infection models.[201] Microbial transformation of RF B to RF O and RF S, a key intermediate in the synthesis of rifampicin, was reported to proceed in high yield.[202]

New Antibiotics. Some novel compounds recently discovered in antibacterial screening are tabulated below, along with their source, producing organism, and general activity.

TABLE I

Antibiotic	Producing Organism	Activity	Reference
LL-B∅2964	*Streptomyces coeruleorubidus*	G^+, G^-	203, 204
Paulomycins	*Streptomyces paulus*	G^+, AA	205
Y-TO678H	*Chromobacterium violaceum*	G^-	206, 207
Thiotetromycin	*Streptomyces* sp. OM-674	AA	208–210
Myxopyronins	*Myxococcus fulvus*	G^+, G^-	211
Isohematinic Acid	*Actinoplanes phillippinensis*	AA	212, 213
FR-900109	*Streptomyces prunicolor*	G^+	214, 215
U-56,407	*Streptomyces hagronensis*	G^+	216

G^+ (Gram-positive), G^- (Gram-negative), AA (anaerobic)

REFERENCES

1. J. E. Baldwin, guest ed., *Tetrahedron*, **39**, 2445-2599 (1983).
2. P. Garzone, J. Lyon and V. L. Yu, *Drug Intell. Clin. Pharmac.*, **17**, 507, 615 (1983).
3. A. A. Carmine, R. N. Brogden, R. C. Heel, J. A. Romankiewicz, T. M. Speight and G. S. Avery, *Drugs*, **26**, 279 (1983).
4. A. A. Carmine, R. N. Brogden, R. C. Heel, T. M. Speight and G. S. Avery, *Drugs*, **25**, 223 (1983).
5. H. Ferres, *Drugs of Today*, **19**, 499 (1983).
6. K. Kitano in "Progress in Industrial Microbiology", M. E. Bushell, ed., Elsevier, New York, N.Y., 1983, p. 37.
7. D. J. Waxman and J. L. Strominger, *Ann. Rev. Biochem.*, **52**, 825 (1983).
8. B. G. Spratt, *J. Gen. Microbiol.*, **129**, 1247 (1983).
9. N. H. Georgopapadakou, *Annual Reports Med. Chem.*, **18**, 119 (1983).
10. T. R. Beam, Jr., *J. Med.*, **14**, 307 (1983).
11. R. L. Thompson and A. J. Wright, *Mayo Clin. Proc.*, **58**, 79 (1983).
12. A. J. Wright and C. J. Wilkowske, *Mayo Clin. Proc.*, 21 (1983).
13. R. Wise, *J. Antimicrob. Chemother.*, **12**, 1, 105 (1983).
14. D. J. Merrikin, J. Briant and G. N. Rolinson, *J. Antimicrob. Chemother.*, **11**, 233 (1983).
15. S. J. Cartwright and S. G. Waley, *Med. Res. Rev.*, **3**, 341 (1983).
16. K. Bush and R. B. Sykes, *J. Antimicrob. Chemother.*, **11**, 97 (1983).
17. J. R. Prous and N. E. Mealy, *Drugs of Today*, **19**, 327 (1983).
18. R. S. Edson and T. F. Keys, *Mayo Clin. Proc.*, **58**, 99 (1983).
19. W. R. Wilson and F. R. Cockerill III, *Mayo Clin. Proc.*, **58**, 92 (1983).
20. F. R. Cockerill III and R. S. Edson, *Mayo Clin. Proc.*, **58**, 147 (1983).
21. R. Reiner, "Antibiotics: An Introduction", Thieme-Stratton, New York, N.Y., 1982.
22. G. Lancini and F. Parenti, "Antibiotics: An Integrated View", Springer-Verlag, New York, 1982.
23. F. E. Hahn, ed., "Antibiotics VI: Modes and Mechanisms of Microbial Growth Inhibitors", Springer-Verlag, New York, N.Y., 1983.
24. D. Greenwood, "Antibiotics of the Beta-Lactam Group", John Wiley & Sons, New York, N.Y., 1982.
25. H. C. Neu, ed., "New Beta-Lactam Antibiotics: A Review from Chemistry to Clinical Efficacy of the New Cephalosporins", College of Physicians of Philadelphia, Philadelphia, Pa., 1982.
26. B. A. Cunha, guest ed., *Med Clin. North Am.*, **66** (1), (1982).
27. H. Schönfeld, ed., "Antibiotics and Chemotherapy, Vol. 31: Pharmacokinetics II", S. Karger, New York, N.Y., 1982.
28. H. P. Kuemmerle, ed., "Clinical Chemotherapy, Vol. 1: Fundamentals", Thieme-Stratton, New York, N.Y., 1983.
29. S. Kishimoto, M. Sendai, S. Hashiguchi, M. Tomimoto, Y. Satoh, T. Matsuo, M. Kondo and M. Ochiai, *J. Antibiot.*, **36**, 1421 (1983).
30. M. Arisawa, J. Ohshima, H. B. Maruyama and P. Angehrn, 23rd ICAAC, 581 (1983).
31. J. Siegel, P. Ryan, D. April and J. Christenson, 23rd ICAAC, 580 (1983).
32. K. Okonogi and M. Kuno, 23rd ICAAC, 325 (1983).
33. J. S. Skotnicki, T. J. Commons, R. W. Rees and J. L. Speth, *J. Antibiot.*, **36**, 1201 (1983).
34. T. Matsuo, T. Sugawara, H. Masuya, Y. Kawano, N. Noguchi and M. Ochiai, *Chem. Pharm. Bull.*, **31**, 1874 (1983).
35. T. Matsuo, H. Masuya, T. Sugawara, Y. Kawano, N. Noguchi and M. Ochiai, *Chem. Pharm. Bull.*, **31** 2200 (1983).
36. C. M. Cimarusti, D. P. Bonner, H. Breuer, H. W. Chang, A. W. Fritz, D. M. Floyd, T. P. Kissick, W. H. Koster, D. Kronenthal, F. Massa, R. H. Mueller, J. Pluscec, W. A. Slusarchyk, R. B. Sykes, M. Taylor and E. R. Weaver, *Tetrahedron*, **39**, 2577 (1983).
37. W. A. Slusarchyk, D. P. Bonner, T. Dejneka, W. H. Koster and E. R. Weaver, 23rd ICAAC, 328 (1983).
38. N. H. Georgopapadakou, S. A. Smith, C. M. Cimarusti and R. B. Sykes, *Antimicrob. Ag. Chemother.*, **23**, 98 (1983).
39. H. C. Neu and P. Labthavikul, *Antimicrob. Ag. Chemother.*, **24**, 227 (1983).
40. P. D. Singh, J. H. Johnson, P. C. Ward, J. S. Wells, W. H. Trejo and R. B. Sykes, *J. Antibiot.*, **36**, 1245 (1983).
41. R. Cooper, K. Bush, P. A. Principe, W. H. Trejo, J. S. Wells and R. B. Sykes, *J. Antibiot.*, **36** 1252 (1983).
42. M. Inoue and S. Mitsuhashi, 23rd ICAAC, 254 (1983).
43. T. Yokota, R. Sekiguch and E. Suzuki, 23rd ICAAC, 255 (1983).

44. H. C. Neu and P. Labthavikul, 23rd ICAAC, 256 (1983).
45. N. Kakeya, K. Nishimura, A. Yoshimi, S. Nakamura, S. Nishizawa, S. Tamaki, H. Matsui, T. Kawamura, M. Kasai, T. Tachi and K. Kitao, 23rd ICAAC, 257 (1983).
46. D. K. Sommers, M. Van Wyk, P. E. O. Williams and S. M. Harding, 23rd ICAAC, 720 (1983).
47. R. Wise, J. M. Andrews and S. Bennett, 23rd ICAAC, 721 (1983).
48. A. Salhi, J. F. Acar, T. Combes, E. Hadibi and F. Blanc, 23rd ICAAC, 259 (1983).
49. P. Courvalin, E. Derlot and Y. A. Chabbert, 23rd ICAAC, 261 (1983).
50. R. Williamson, L. Gutmann, M. D. Kitzis and J. F. Acar, 23rd ICAAC, 994 (1983).
51. T. Yoshida, T. Tsuji, S. Matsuura, T. Matsubara and Y. Harada, 23rd ICAAC, 265 (1983).
52. H. Takasugi, Z. Tozuka and T. Takaya, *J. Antibiot.*, **36**, 36 (1983).
53. I. Csendes, B. W. Müller and W. Tosch, *J. Antibiot.*., **36**, 1020 (1983).
54. M. Alpegiani, F. Casabuona, R. Giorgi, G. Nannini, E. Perrone, G. Meinardi, A. Bianchi and G. Monti, *J. Antibiot.*, **36**, 1013 (1983).
55. R. Lattrell, J. Blumbach, W. Dürckheimer, R. Kirrstetter, N. Klesel, W. Schwab, K. Seeger, G. Seibert and M. Wieduwilt, 23rd ICAAC, 571 (1983).
56. J. Okumura, S. Aburaki, H. Kamachi, Y. Narita, T. Naito and H. Kawaguchi, 23rd ICAAC, 576 (1983).
57. F. T. Counter, S. J. Bogard, R. F. Brown, P. W. Ensminger, A. S. Katner, M. D. Kinnick, R. E. Koehler, W. H. W. Lunn, J. L. Ott, D. A. Preston, J. K. Shadle, J. K. Swartzendruber, J. R. Turner, R. T. Vasileff and J. A. Webber, 23rd ICAAC, 995 (1983).
58. G. Seibert, N. Klesel, M. Limbert, E. Schrinner, K. Seeger, I. Winkler, R. Lattrell, J. Blumbach, W. Dürckheimer, K. Fleischmann, R. Kirrstetter, B. Mencke, B. C. Ross, K. -H. Scheunemann, W. Schwab and M. Wieduwilt, *Arzneim.-Forsch.*, **33**, 1084 (1983).
59. G. Seibert, M. Limbert, I. Winkler and T. Dick, *Infection*, **11**, 275 (1983).
60. N. Klesel and K. Seeger, *Infection*, **11**, 318 (1983).
61. M. Bies, R. E. Buck, T. A. Pursiano, D. R. Chisholm, Y. H. Tsai, M. Misiek, K. E. Price and F. Leitner, 23rd ICAAC, 577 (1983).
62. L. R. Koupal, B. Weissberger, D. L. Shungu, E. Weinberg and H. H. Gadebusch, *J. Antibiot.*, **36**, 47 (1983).
63. L. R. Koupal, B. A. Pelak, P. J. Cassidy and H. H. Gadebusch, *J. Antibiot.*, **36**, 54 (1983).
64. E. Celozzi, B. A. Pelak, E. O. Stapley and H. H. Gadebusch, *J. Antibiot.*, **36**, 64 (1983).
65. K. Seeger, W. Dürckheimer, F. Wengenmayer and H. Strecker, *Drugs Exptl. Clin. Res.*, **9**, 427 (1983).
66. A. Costa and G. A. Botta, *J. Antibiot.*, **36**, 1007 (1983).
67. S. Ohya, M. Yamazaki and S. Sugawara, *Antimicrob. Ag. Chemother.*, **23**, 522 (1983).
68. M. Narisada, T. Yoshida, M. Ohtani, K. Ezumi and M. Takasuka, *J. Med. Chem.*, **26**, 1577 (1983).
69. S. Inouye, M. Kojima, T. Shomura, K. Iwamatsu, T. Niwa, Y. Kondo, T. Niida, Y. Ogawa and K. Kusama, *J. Antibiot.*, **36**, 115 (1983).
70. I. Phillips, R. Wise and D. A. Leigh, eds., *J. Antimicrob. Chemother.*, **11** (Suppl. A), 1-233 (1983).
71. H. P. Lambert, I. Phillips and R. Wise, eds., *J. Antimicrob. Chemother.*, **12** (Suppl. A), 1–405 (1983).
72. F. O'Grady, W. Marget and P. B. Harper, eds., *Infection*, **11** (Suppl. 1), S3-S57 (1983).
73. J. O. Klein and H. C. Neu, guest eds., *Rev. Infect. Dis.*, **5** (Suppl. 1), S108-S200 (1983).
74. R. J. Ponsford, D. J. Merrikin, B. Slocombe, M. J. Basker and T. C. G. Tasker, 23rd ICAAC, 415 (1983).
75. F. Sakamoto, S. Ikeda and G. Tsukamoto, *Chem. Pharm. Bull.*, **31**, 2698 (1983).
76. G. Tsukamoto, T. Nose, A. Suzuki and I. Utsumi, in "Current Chemotherapy and Immunotherapy", Vol. 1, P. Periti and G. G. Grassi, eds., American Society for Microbiology, Washington, D.C. (1982), p. 142.
77. C. L. Affoo, F. Moosdeen, J. D. Williams and S. Yamabe, "Current Chemotherapy and Immunotherapy", p. 336.
78. R. C. Moellering, Jr., sr. guest ed., *Am. J. Med.*, **75** (2A), 1-130 (1983).
79. H. C. Neu, D. S. Reeves and D. A. Leigh, eds., *J. Antimicrob. Chemother.*, **11** (Suppl. C), 1-103 (1983).
80. H. C. Neu, D. S. Reeves and D. A. Leigh, eds., *J. Antimicrob. Chemother.*, (Suppl. B), 1-233 (1983).
81. W. D. Sharpe, ed., *Bull. N.Y. Acad. Med.*, **59**, 420–524 (1983).
82. G. Franceschi, M. Foglio, M. Alpegiani, C. Battistini, A. Bedeschi, E. Perrone, F. Zarini, F. Arcamone, C. Della Bruna and A. Sanfilippo, *J. Antibiot.*, **36**, 938 (1983).
83. M. Foglio, C. Battistini, F. Zarini, C. Scarafile and G. Franceschi, *Heterocycles*, **20**, 1491 (1983).
84. T. Miyadera, Y. Sugimura, T. Hashimoto, T. Tanaka, K. Iino, T. Shibata and S. Sugawara, *J. Antibiot.*, **36**, 1034 (1983).
85. H. Kropp, J. G. Sundelof, J. S. Kahan, J. Huber, D. Bohn, L. Gerckens, F. M. Kahan and J. Birnbaum, 23rd ICAAC, 331 (1983).
86. D. H. Shih, J. A. Fayter, F. Baker, L. Cama and B. G. Christenson, 23rd ICAAC, 333 (1983).
87. L. D. Cama, K. J. Wildonger, R. Guthikonda, R. W. Ratcliffe and B. G. Christensen, *Tetrahedron*, **39** 2531 (1983).
88. M. J. Basker, R. J. Boon, S. J. Box, A. G. Brown, P. Davis, R. J. Ponsford, R. Southgate and S. R. Spear, *J. Antibiot.*, **36**, 1357 (1983).
89. M. J. Basker, R. J. Boon, S. J. Box, E. A. Prestige, G. M. Smith and S. R. Spear, *J. Antibiot.*, **36**, 416 (1983).
90. H. Natsugari, Y. Matsushita, N. Tamura, K. Yoshioka, M. Kondo, K. Okonogi, M. Kuno and M. Ochiai, *J. Antibiot.*, **36**, 855 (1983).
91. S. Harada, S. Tsubotani, M. Asai, K. Okonogi and M. Kondo, *J. Med. Chem.*, **26**, 753 (1983).
92. S. Harada, S. Tsubotani, S. Shinagawa and M. Asai, *Tetrahedron*, **39**, 75 (1983).
93. S. R. Norrby, K. Alestig, B. Björnegård, L. Å. Burman, F. Ferber, J. L. Huber, K. H. Jones, F. M. Kahan, J. S. Kahan, H. Kropp, M.A.P. Meisinger and J. G. Sundelof, *Antimicrob. Ag. Chemother.*, **23**, 300 (1983).
94. T. Haneishi, M. Nakajima, N. Serizawa, M. Inukai, Y. Takiguchi, M. Arai, S. Satoh, H. Kuwano and C. Tamura, *J. Antibiot.*, **36**, 1581 (1983).
95. M. Nakayama, S. Kimura, T. Mizoguchi, S. Tanabe, A. Iwasaki, A. Murakami, M. Okuchi, H. Itoh and T. Mori, *J. Antibiot.*, **36**, 943 (1983).
96. K. E. Wilson, A. J. Kempf, J. M. Liesch and B. H. Arison, *J. Antibiot.*, **36**, 1109 (1983).
97. T. Iimori, Y. Takahashi, T. Izawa, S. Kobayashi and M. Ohno, *J. Am. Chem. Soc.*, **105**, 1659 (1983).
98. H. Natsugari, Y. Matsushita, N. Tamura, K. Yoshioka and M. Ochiai, *J. Chem. Soc. Perkin Trans. I*, 403 (1983).
99. A. G. Brown, D. F. Corbett, A. J. Eglington and T. T. Howarth, *Tetrahedron*, **39**, 2551 (1983).
100. R. Lüthy, H. C. Neu and I. Phillips, eds., *J. Antimicrob. Chemother.*, **12** (Suppl. D), 1-149 (1983).
101. M. Arisawa and S. Adam, *Biochem. J.*, **211**, 447 (1983).
102. M. Arisawa and R. Then, *Biochem. J.*, **209**, 609 (1983).
103. M. Arisawa and R. L. Then, *J. Antibiot.*, **36**, 1372 (1983).
104. D. D. Keith, J. Tengi, P. Rossman, L. Todaro and M. Weigele, *Tetrahedron*, **39**, 2445 (1983).
105. C. -P. Mak, K. Prasad and F. Turnowsky, *J. Antibiot.*, **36**, 398 (1983).
106. L. B. Crast, R. G. Graham, W. J. Gottstein, U. J. Haynes, T. A. Pursiano, K. L. Denbleyker, D. J. Phelps, D. N. McGregor and J. C. Sheehan, 23rd ICAAC, 991 (1983).
107. M. J. Calverley and M. Begtrup, *J. Antibiot.*, **36**, 1507 (1983).
108. D. Bouzard, M. Essiz, P. Remuzon and A. Weber, *Eur. J. Med. Chem.*, **18**, 405 (1983).
109. M. Yamashita, S. Hashimoto, M. Ezaki, M. Iwami, T. Komori, M. Kohsaka and H. Imanaka, *J. Antibiot.*, **36**, 1774 (1983).
110. L. M. Cappelletti and R. Spagnoli, *J. Antibiot.*, **36**, 328 (1983).
111. H. Umezawa, H. Iwasawa, D. Ikeda and S. Kondo, *J. Antibiot.*, **36**, 1087 (1983).
112. D. Ikeda, S. Gomi, S. Kondo and H. Umezawa, *J. Antibiot.*, **36**, 331 (1983).
113. M. Philippe, B. Quiclet-Sire, A. M. Sepulchre, S. D. Gero, H. Loibner, W. Streicher, P. Stütz and N. Moreau, *J. Antibiot.*, **36**, 250 (1983).

114. S. Toda, S. Nakagawa, T. Naito and H. Kawaguchi, *J. Antibiot.*, **36**, 87 (1983).
115. Y. Takahashi, T. Tsuchiya, Y. Suzuki, S. Umezawa, H. Umezawa and S. Fukatsu, *Bull. Chem. Soc. Jpn.*, **56**, 1807 (1983).
116. E. Akita, Y. Horiuchi, T. Miyazawa and H. Umezawa, *J. Antibiot.*, **36**, 745 (1983).
117. W. Streicher, H. Loibner, J. Hildebrandt and F. Turnowsky, *Drugs Exptl. Clin. Res.*, **9**, 591 (1983).
118. E. Wilmotte, P. Maldague, P. Tulkens, R. Baumgartner, F. Schmook, H. Walzl and H. Obenaus, *Drugs Exptl. Clin. Res.*, **9**, 467 (1983).
119. J. Tadanier and R. Hallas, *J. Antibiot.*, **36**, 256 (1983).
120. J. Tadanier, R. Hallas, J. Holms, L. Freiberg and D. Bacino, *J. Antibiot.*, **36**, 267 (1983).
121. T. Torii, T. Tsuchiya, S. Umezawa and H. Umezawa, *Bull. Chem. Soc. Jpn.*, **56**, 1522 (1983).
122. M. Awata, S. Satoi, N. Muto, M. Hayashi, H. Sagai and H. Sakakibara, *J. Antibiot.*, **36**, 651 (1983).
123. K. Kamiya, T. Deushi, A. Iwasaki, I. Watanabe, H. Itoh and T. Mori, *J. Antibiot.*, **36**, 738 (1983).
124. D. R. White, C. J. Maring and G. A. Cain, *J. Antibiot.*, **36**, 339 (1983).
125. S. Ōmura, J. Inokoshi, H. Matsubara and H. Tanaka, *J. Antibiot.*, **36**, 1709 (1983).
126. H. Matsubara, J. Inokoshi, A. Nakagawa, H. Tanaka and S. Ōmura, *J. Antibiot.*, **36**, 1713 (1983).
127. N. Sadakane, Y. Tanaka and S. Ōmura, *J. Antibiot.*, **36**, 921 (1983).
128. S. Ōmura, N. Sadakane, Y. Tanaka and H. Matsubara, *J. Antibiot.*, **36**, 927 (1983).
129. R. Okamoto, K. Kiyoshima, M. Yamamoto, K. Takada, T. Ohnuki, T. Ishikura, H. Naganawa, K. Tatsuta, T. Takeuchi and H. Umezawa, 23rd ICAAC, 448 (1983).
130. H. A. Kirst, G. M. Wild, R. H. Baltz, E. T. Seno, R. L. Hamill, J. W. Paschal and D. E. Dorman, *J. Antibiot.*, **36**, 376 (1983).
131. M. Hayashi, K. Kinoshita, Y. Sudate, S. Satoi, H. Sakakibara, K. Harada and M. Suzuki, *J. Antibiot.*, **36**, 175 (1983).
132. A. Ganguly, Y. Lui and O. Sarre, *J. Chem. Soc., Chem. Commun.*, 1166 (1983).
133. B. K. Lee, M. S. Puar, M. Patel, B. Bartner, J. Lotvin, H. Munayyer and J. A. Waitz, *J. Antibiot.*, **36**, 742 (1983).
134. H. Sano, M. Inoue, K. Yamashita, R. Okachi and S. Ōmura, *J. Antibiot.*, **36**, 1336 (1983).
135. M. Uyeda, K. Kubota, S. Miyamura and M. Shibata, *J. Antibiot.*, **36**, 1722 (1983).
136. Y. Takiguchi, M. Ono, S. Muramatsu, J. Ide, H. Mishima and M. Terao, *J. Antibiot.*, **36**, 502 (1983).
137. M. Ono, H. Mishima, Y. Takiguchi and M. Terao, *J. Antibiot.*, **36**, 509 (1983).
138. H. Mishima, J. Ide, S. Muramatsu and M. Ono, *J. Antibiot.*, **36**, 980 (1983).
139. M. Ono, H. Mishima, Y. Takiguchi, M. Terao, H. Kobayashi, S. Iwasaki and S. Okuda, *J. Antibiot.*, **36**, 991 (1983).
140. N. Sadakane, Y. Tanaka and S. Ōmura, *J. Antibiot.*, **36**, 931 (1983).
141. R. Spagnoli, L. Cappelletti and L. Toscano, *J. Antibiot.*, **36**, 365 (1983).
142. L. Toscano and E. Seghetti, *Tetrahedron Lett.*, **24**, 5527 (1983).
143. L. Toscano, G. Fioriello, R. Spagnoli, L. Cappelletti and G. Zanuso, *J. Antibiot.*, **36**, 1439 (1983).
144. L. Toscano, G. Fioriello, R. Spagnoli and L. Cappelletti, *J. Antibiot.*, **36**, 1585 (1983).
145. R. N. Jones, A. L. Barry and C. Thornsberry, *Antimicrob. Agents Chemother.*, **24**, 209 (1983).
146. T. Barlam and H. C. Neu, 23rd ICAAC, 336 (1983).
147. J. F. Chantot, J. C. Gasc, S. Gouin D'Ambrieres and A. Lutz, 23rd ICAAC, 447 (1983).
148. J. R. Hauske, M. Guadliana and G. Kostek, *J. Org. Chem.*, **48**, 5138 (1983).
149. R. E. Ireland, J. P. Daub, G. S. Mandel and N. S. Mandel, *J. Org. Chem.*, **48**, 1312, (1983).
150. R. E. Ireland and J. P. Daub, *J. Org. Chem.*, **48**, 1303 (1983).
151. K. Steliou and M. Poupart, *J. Am. Chem. Soc.*, **105**, 7130 (1983).
152. B. M. Trost and R. W. Warner, *J. Am. Chem. Soc.*, **105**, 5940 (1983).
153. J. W. Westley, C. Liu, J. F. Blount, L. H. Sello, N. Troupe and P. A. Miller, *J. Antibiot.*, **36**, 1275 (1983).
154. C. Liu, M. Chin, B. La T. Prosser, N. J. Palleroni, J. W. Westley and P. A. Miller, *J. Antibiot.*, **36**, 1118 (1983).
155. C. Liu, T. E. Hermann, A. Downey, B. La T. Prosser, E. Schildknecht, N. J. Palleroni, J. W. Westley and P. A. Miller, *J. Antibiot.*, **36**, 343 (1983).
156. J. Tone, R. Shibakawa, M. Ishiguro, H. Maeda, S. Nishiyama, K. Tsukuda, M. Yamada, W. P. Cullen, J. R. Oscarson, J. E. Bright, L. R. Chappel, L. H. Huang, C. J. La Plante and W. D. Celmer, 23rd ICAAC, 215 (1983).
157. J. H. Westley, C. Liu, R. H. Evans, Jr., L. H. Sello, N. Troupe and T. Hermann, *J. Antibiot.*, **36**, 1195 (1983).
158. A. Cuer, G. Dauphin and J. C. Beloëil, *J. Antibiot.*, **36**, 20 (1983).
159. D. E. Cane, W. D. Celmer and J. W. Westley, *J. Am. Chem. Soc.*, **105**, 3594 (1983).
160. M. P. Edwards, S. V. Ley, S. G. Lister, and B. D. Palmer, *J. Chem. Soc., Chem. Commun.*, 630 (1983).
161. R. E. Ireland, R. C. Anderson, R. Badoud, B. J. Fitzsimmons, G. J. McGarvey, S. Thaisrivongs and C. S. Wilcox, *J. Am. Chem. Soc.*, **105**, 1988 (1983).
162. R. E. Ireland, L. Courtney and B. J. Fitzsimmons, *J. Org. Chem.*, **48**, 5186 (1983).
163. B. Singh and G. Y. Lesher, *Heterocyclic Chem.*, **20**, 491 (1983).
164. K. Grohe, 23rd ICAAC, 371 (1983).
165. M. P. Wentland, D. M. Bailey, J. B. Cornett, R. A. Dobson, R. G. Powles and R. B. Wagner, 23rd ICAAC, 375 (1983).
166. N. Markowitz, D. Pohlod and L. Saravolatz, 23rd ICAAC, 656 (1983).
167. J. R. O'Connor, R. A. Dobson, R. B. Wagner and G. Y. Lesher, 23rd ICAAC, 657 (1983).
168. D. C. Hooper, J. S. Wolfson, C. Tung and M. N. Swarz, 23rd ICAAC, 1021 (1983).
169. T. J. Walsh, R. C. Wharton, S. Sprecher, M. R. Moody and S. C. Schimpff, 23rd ICAAC, 971 (1983).
170. M. E. Gombert and T. M. Aulicino, 23rd ICAAC, 649 (1983).
171. L. J. Chalkley, L. Schrire, J. Saunders and H. Koornhof, 23rd ICAAC, 650 (1983).
172. B. G. Painter, S. Demovick, F. Hodin and M. Loper, 23rd ICAAC, 651 (1983).
173. R. Wise, J. M. Andrews and L. J. V. Edwards, *Antimicrob. Ag. Chemother.*, **23**, 559 (1983).
174. C. L. Heifetz, T. J. Griffin, J. C. Sesnie, M. A. Shapiro and S. A. Wold, 23rd ICAAC, 380 (1983).
175. S. W. B. Newsom, J. Matthews and A. R. Rampling, 23rd ICAAC, 707 (1983).
176. T. Kumada and H. C. Neu, 23rd ICAAC, 708 (1983).
177. M. Wolff, B. Regnier, M. Nkam, K. E. Rochan, C. Daldoss and F. Vachon, 23rd ICAAC, 853 (1983).
178. K. Hirai, M. Hosaka, Y. Oomori, S. Murayama, A. Ito, K. Takagi, T. Irikura and S. Mitsuhashi, 23rd ICAAC, 658 (1983).
179. K. Takagi, M. Hosaka, H. Kusajima, Y. Oomori, K. Hirai, H. Uchida and T. Irikura, 23rd ICAAC, 659 (1983).
180. K. D. Thompson, J. P. O'Keefe and W. A. Tatarowicz, 23rd ICAAC, 376 (1983).
181. A. Iravani, G. Welty and G. A. Richard, 23rd ICAAC, 377 (1983).
182. I. Garcia, V. Fainstein, D. Ho, B. Leblanc and G. P. Bodey, 23rd ICAAC, 708B (1983).
183. J. B. Cornett, R. A. Dobson, R. B. Wagner, B. A. Mayes, M. P. Wentland, D. M. Bailey, J. G. Taggart and C. H. Nash, 23 ICAAC, 704 (1983).
184. R. B. Wagner, R. A. Dobson, M. P. Wentland, D. M. Bailey, C. H. Nash and J. B. Cornett, 23rd ICAAC, 378 (1983).
185. Anon., *Lancet*, 24 (1984).
186. D. H. Williams, M. P. Williamson, D. W. Butcher and S. J. Hammond, *J. Am. Chem. Soc.*, **105**, 1332 (1983).

187. C. M. Harris, H. Kopecka and T. M. Harris, *J. Am. Chem. Soc.*, **105**, 6915 (1983).
188. M. Debono, K. E. Merkel, R. M. Molloy, M. Barnhart, E. Presti, A. H. Hunt and R. L. Hamill, 23rd ICAAC, 437 (1983).
189. A. H. Hunt, M. Debono, K. E. Merkel and M. Barnhart, 23rd ICAAC, 438 (1983).
190. C. C. Scheifinger, D. B. Anderson, R. K. McGuffey, L. C. Pendlum, L. F. Richardson and R. H. Wellenreiter, 23rd ICAAC, 439 (1983).
191. J. H. Eggert, K. H. Michel, L. D. Boeck, W. M. Nakatsukasa and R. E. Kastner, 23rd ICAAC, 440 (1983).
192. A. H. Hunt, J. L. Occolowitz, M. Debono, R. M. Molloy and G. M. Maciak, 23rd ICAAC, 441 (1983).
193. J. N. Hobbs, Jr. and N. E. Allen, 23rd ICAAC, 442 (1983).
194. R. L. Hamill, L. D. Boeck, R. E. Kastner and R. M. Gale, 23rd ICAAC, 443 (1983).
195. T. R. Herrin, A. M. Thomas and S. W. Fesik, 23rd ICAAC, 445 (1983).
196. T. Kamogashira, T. Nishida and M. Sugawara, *Agr. Biol. Chem.*, **47**, 499 (1983).
197. G. Peters and G. Pulverer, *J. Antimicrob. Chem.*, **11**, 94 (1983).
198. R. Pallanza, M. Berti, B. P. Goldstein, E. Mapelli, E. Randisi, R. Scotti and V. Arioli, *J. Antimicrob. Chem.*, **11**, 419 (1983).
199. P. E. Varaldo and G. C. Schito, *Drugs Exptl. Clin. Res.*, **9**, 437 (1983).
200. V. Brizzi, M. Brufani, L. Cellai and A. L. Segre, *J. Antibiot.*, **36**, 516 (1983).
201. L. Marsili, G. Franceschi, M. Ballabio, G. Oronzo, A. Vigevani, D. Ungheri, C. Della Bruna and A. Sanfilippo, *J. Antibiot.*, **36**, 1495 (1983).
202. C. Della Bruna, G. Schioppacassi, D. Ungheri, D. Jabès, E. Morvillo and A. Sanfilippo, *J. Antibiot.*, **36**, 1502 (1983).
203. B. L. Seong, H. J. Son, T. Mheen and M. H. Han, *J. Antibiot.*, **36**, 1402 (1983).
204. M. D. Lee, D. B. Cosulich, J. H. Martin, M. Dann and D. B. Borders, 23rd ICAAC, 209 (1983).
205. A. A. Fantini, D. P. Labeda, N. A. Kuck, J. D. Korshalla, F. Pinho, J. P. Thomas and R. T. Testa, 23rd ICAAC, 210 (1983).
206. G. E. Zurenko, C. W. Ford, J. C. Hamel, B. R. Hannon, G. P. Li, K. F. Stern and R. J. Yancey, Jr., 23rd ICAAC, 216 (1983).
207. H. Imai, K. Suzuki, S. Miyazaki, K. Tanaka, S. Watanabe and M. Iwanami, *J. Antibiot.*, **36**, 911 (1983).
208. M. Hamada, S. Kondo, H. Nakamura, T. Ikeda, D. Ikeda, K. Iinuma, S. Gomi, Y. Ikeda, T. Takeuchi, H. Umezawa and Y. Iitaka, *J. Antibiot.*, **36**, 445 (1983).
209. S. Ōmura, Y. Iwai, A. Nakagawa, R. Iwata, Y. Takahashi, H. Shimizu and H. Tanaka, *J. Antibiot.*, **36**, 109 (1983).
210. K. Tsuzuki and S. Ōmura, *J. Antibiot.*, **36**, 1584 (1983).
211. S. Ōmura, A. Nakagawa, R. Iwata and A. Hatano, *J. Antibiot.*, **36**, 1781 (1983).
212. H. Irschik, K. Gerth, G. Höfle, W. Kohl and H. Reichenbach, *J. Antibiot.*, **36**, 1651 (1983).
213. M. Takeuchi, Y. Itoh, R. Enokita, A. Torikata, S. Iwado and T. Haneishi, *J. Antibiot.*, **36**, 493 (1983).
214. Y. Itoh, M. Takeuchi, K. Shimizu, S. Takahashi, A. Terahara and T. Haneishi, *J. Antibiot.*, **36**, 497 (1983).
215. M. Yamashita, M. Iwami, K. Ikushima, T. Komori, H. Aoki and H. Imanaka, *J. Antibiot.*, **36**, 1123 (1983).
216. S. Koda, Y. Morimoto, M. Yamashita, T. Komori and H. Imanaka, *J. Antibiot.*, **36**, 1237 (1983).

Chapter 12. Antiviral Agents

James L. Kelley
Wellcome Research Laboratories, Burroughs Wellcome Co.
Research Triangle Park, NC 27709

The previous review of antiviral agents in Annual Reports in Medicinal Chemistry emphasized compounds with activity against DNA viruses.[1] The focus of this year's chapter is on agents with activity primarily against RNA viruses. A brief update of this year's advances in anti-DNA virus agents is provided. Antiviral agents with activity against RNA viruses were previously reviewed in the 1981 volume.[2] Neither interferons nor interferon inducers are included in this review as they were covered in the 1981 and 1982 volumes, respectively. Several reviews have been published that give an overview of the most promising clinical and experimental antiviral agents[3-5] or provide background on viral diseases.[6-8] More comprehensive reviews dealing with specific drugs are cited in the sections where these compounds are discussed.

VIRAL RESPIRATORY DISEASE

RNA viruses are the major causative factors of the various forms of acute respiratory disease.[8] The communicable diseases of the respiratory tract are probably the most common cause of symptomatic human infections. Among both children and adults, acute respiratory tract illness results in significant morbidity, lost time from work, physician visits, and death. It has been calculated that throughout the world over two million deaths occur annually from acute respiratory disease.[9] The viruses that are causative agents for human respiratory disease comprise the five taxonomically distinct families Orthomyxoviridae, Paramyxoviridae, Picornaviridae, Coronaviridae and Adenoviridae.[9] The influenza viruses, which are comprised of types A, B and C, belong to the family Orthomyxoviridae. Types A and B have been associated with significant increases in mortality during epidemics. The disease may be asymptomatic or cause symptoms ranging from the common cold to fatal pneumonia. Immunization against influenza has been recommended for high-risk groups, and antiviral chemotherapy (amantadine) is available for the treatment and prophylaxis of all influenza A infections. The family Paramyxoviridae includes respiratory syncytial virus (RSV) and parainfluenza virus which are a major cause of lower respiratory tract infections. RSV is a factor in severe bronchiolitis and pneumonia in infants and young children. Efforts to develop a vaccine for RSV have been ineffectual, but recent clinical trials with ribavirin as a small particle aerosol have been promising. The parainfluenza viruses, of which there are four human serotypes, are second only to RSV as a cause of lower respiratory tract illness. There is both a great need for and interest in developing a chemotherapeutic agent for treatment of these two viral, respiratory tract pathogens. The family Picornaviridae contains the genus Rhinovirus, which is composed of over a hundred distinct serotypes. The rhinoviruses are recognized as the most important causative agents of the

ISBN 0-12-040519-9

common cold. The development of a vaccine has been precluded due to the many rhinovirus serotypes. Although a number of different chemical agents have been tested in man, none has shown much clinical efficacy. Coronaviruse (family Coronaviridae) also has an appreciable role in upper respiratory tract illness although technical difficulties appear to have hindered studies on vaccines and antiviral agents. The adenoviruses (family Adenoviridae) are a ubiquitous group of doublestranded DNA viruses which are responsible for a wide variety of respiratory illnesses. These infections are most common among children, although acute respiratory disease and pneumonia are also common among military recruits. Two comprehensive reviews of viral respiratory diseases and measures for their control and treatment have been recently published.[8,9]

AGENTS ACTIVE PRIMARILY AGAINST RNA VIRUSES

Amantadine (1-adamantanamine) and rimantadine (α-methyl-1-adamantanemethylamine) - Amantadine (1) and rimantadine (2) are specifically active against influenza A virus infections. In 1966 amantadine was licensed for general use against Asian (H2N2) influenza in the United States and since 1976 has had FDA approval for the prophylaxis and therapy of all influenza A infections.[10,11] Rimantadine is a closely related analog that has been widely used in the U.S.S.R. A review of the results of experimental and clinical studies performed in the U.S.S.R. with rimantadine has been published.[12] Several placebo-controlled, double-blind clinical trials of amantadine and rimantadine in the treatment of influenza A infections have been recently completed.[13-15] In a large-scale trial, both drugs were highly effective with no significant differences between the rates of illness or infection in the two drug-treated groups.[13] The amantadine recipients reported a higher incidence of side effects largely attributed to central nervous system (CNS) symptoms.[13,14] This difference in side effects may be a pharmacokinetic phenomenon which results in higher plasma concentrations of amantadine.[16] A controlled study of healthy, adult volunteers found 1 and 2 to have minor side effects comparable to those of a common antihistamine.[17] However, an open study involving hospital employees receiving amantadine showed a high incidence of CNS symptoms.[18] Guidelines for the use of amantadine in patients with impaired renal function have been formulated from the results of pharmacokinetic studies on amantadine in patients with normal and impaired renal function.[19,20] Since approximately 90% of an oral dose of amantadine is excreted unchanged in the urine, patients with renal insufficiency can accumulate the drug, resulting in toxic manifestations. The mechanism by which 1 and 2 inhibit influenza A virus replication had previously been narrowed to a virus-specific event after virus penetration but prior to primary transcription. A recent study with virus labeled with radioactive precursors seems to show that uncoating is a multistep process.[21] Rimantadine prevents the second step of uncoating, the release of matrix (M) protein from ribonucleoproteins (RNP).[22] This blocks the penetration of RNP into the nucleus and prevents the nuclear phase of virus reproduction. Amantadine produces the same effect on uncoating as rimantadine.[21] These adamantyl amines may also have utility in the treatment of other types of viral infections. Rimantadine has been shown to be an effective inhibitor of dengue virus replication *in vitro*.[23] The amelioration of post-herpetic neuralgia due to recurrent herpes simplex sciatica[24] and from acute herpes zoster[25] has been reported with early administration of amantadine.

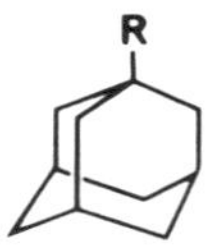

1, R = NH_2

2, R = $CHCH_3NH_2$

Ribavirin (1-β-D-ribofuranosyl-1,2,4-triazole-3-carboxamide) - This nucleoside has activity against a broad range of DNA and RNA viruses in tissue culture and in animal model systems.[2] An analysis of the status of ribavirin (3) as an antiviral agent has been published.[26] Its mechanism is still unresolved but may involve guanosine nucleotides and inhibition of inosine monophosphate dehydrogenase.[26] In a clinical trial against influenza A, oral ribavirin failed to alter clinical signs and symptoms of the disease.[2] However, it has recently been reported to have a therapeutic effect against both influenza A and influenza B virus infections when administered to patients by inhalation of small-particle aerosol through a face mask.[27,28] This method of treatment may find utility in those patients at high risk such as the elderly and the chronically ill. Ribavirin has also been shown to inhibit respiratory syncytial virus (RSV) infection in an animal model when administered i.p. or by aerosol treatment.[29] This in vivo activity has now been substantiated in two controlled, double-blind clinical trials.[30,31] Administration of a continuous aerosol of ribavirin to infants hospitalized with lower respiratory tract disease from RSV resulted in significant clinical improvement.[31] In a study of the effects of several antiviral agents against Colorado tick fever virus (CTFV), ribavirin inhibited CTFV in vitro.[32] Ribavirin did not protect mice inoculated intracerebrally with CTFV. However, the 2',3',5'-triacetate derivative 4 significantly increased the number of survivors when administered i.p., which suggests that this lipophilic prodrug of ribavirin is able to cross the blood-brain barrier.[32] Intraperitoneal administration of 4 was also found to protect mice inoculated intracerebrally with dengue virus, under conditions where ribavirin had no significant protective effect.[33] In another study, ribavirin reduced the growth of four types of dengue virus in vitro, but it had no effect on virus replication in human peripheral blood leukocytes (PBL) which have been implicated in the pathogenesis of dengue virus in vivo.[34] However, a combination of ribavirin with 6-mercapto-9-(tetrahydro-2-furyl)purine, an inhibitor of hypoxanthine-guanine phosphoribosyl transferase, resulted in a marked suppression of dengue virus replication in human PBL.[34]

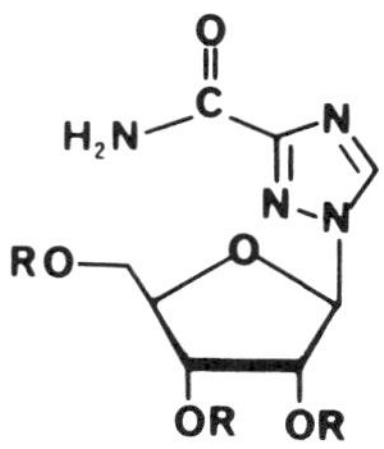

3, R=H

4, R=$COCH_3$

Selenazole (2-β-D-ribofuranosylselenazole-4-carboxamide) - The novel selenazole carboxamide nucleoside 5 was recently synthesized[35], and it has been reported to have in vitro antiviral activity superior to that of ribavirin.[36] Selenazole (5) has broad-spectrum antiviral activity against both RNA and DNA viruses, but it was most active against representative RNA viruses.[36] The virus families Paramyxoviridae, Reoviridae, Togaviridae, Bunyaviridae and Arenaviridae were particularly sensitive with ED_{50} values typically ranging from 1 to 8 μg/ml, although yellow fever virus was especially sensitive with an ED_{50} = 0.005 μg/ml. Although selenazole was relatively nontoxic for Vero and HeLa cells,[36] it was found to be cytotoxic towards P388 and L1210 cells in culture and exhibited marked activity against Lewis lung carcinoma in mice.[35]

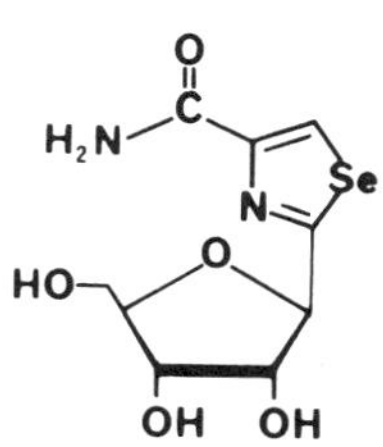

5

Carbocyclic 3-deazaadenosine ($C\text{-}c^3Ado$) - The carbocyclic nucleoside 6 has been reported to inhibit the *in vitro* replication of DNA and RNA viruses[1,37]. $C\text{-}c^3Ado$ was active against vaccinia, reo, measles, parainfluenza and vesicular stomatitis viruses (VSV) at concentrations from 0.2 to 1 μg/ml but exhibited no toxicity for the host cells at 400 μg/ml. *In vivo*, 6 protected newborn mice against a lethal infection of VSV when administered i.p. Its spectrum of antiviral activity was suggested to be compatible with an impairment of viral mRNA transcription and/or processing.[37]

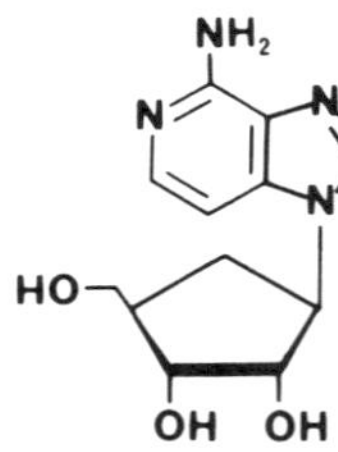

6

Enviroxime (LY 122772) - Enviroxime (7) is a potent inhibitor of rhinovirus replication in cell culture with a uniform level of activity against over sixty rhinovirus serotypes.[2] In clinical trials in England, intranasal and oral enviroxime had some beneficial effect on rhinovirus infections although the oral form was poorly tolerated.[38,39] In two U.S.A. based clinical trials, prophylactic administration of intranasal enviroxime failed to prevent infection or to reduce the frequency of colds after experimental rhinovirus challenge.[40,41] Enviroxime, as its syn-anti mixture, and the propenyl analog 8 (LY 127123) have been reported to be highly active against thirteen enteroviruses *in vitro*.[42] When injected s.c. into infant mice, both compounds significantly reduced the mortality caused by Echo and coxsackie viruses.[42] A number of analogs of the olefin 8 in which X was equal to groups as diverse as $CHCH_2CH_3$, $CHC{\equiv}N$, $CHCONH_2$, $CHCOOCH_3$ and CHCl were reported to have potent *in vitro* activity against poliovirus with IC_{50}'s ranging from 0.01 to 0.04 μg/ml.[43] Activity is highly specific for the *E*-isomer, a trend that parallels earlier findings in the enviroxime series.[44]

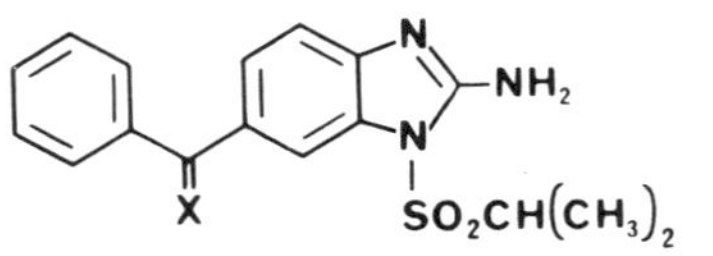

7, X = NOH—anti

8, X = $CHCH_3$—trans

The recent literature on 2-(α-hydroxybenzyl)benzimidazole (HBB) (9) and several analogs has been reviewed.[45] The *in vitro* antiviral activity of 9 against thirteen enteroviruses in comparison with enviroxime and in combination with guanidine has been published.[42] The mutual synergistic activity of the combination of 9 and guanidine was confirmed against Echo and coxsackie viruses in infant mice.[42,46]

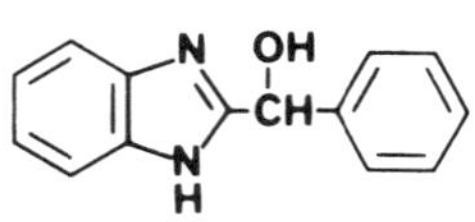

9

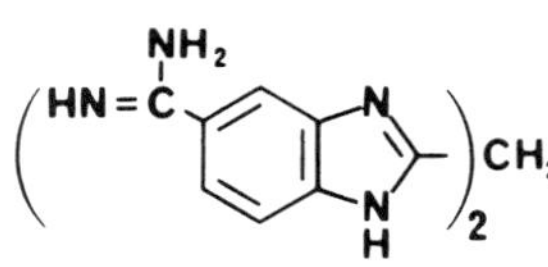

10

Mono- and diamidinobenzimidazoles such as bis(5-amidino-2-benzimidazolyl)methane (BABIM) (10) possess significant suppressive effects on the cytopathology and yield of respiratory syncytial virus (RSV).[47] The mechanism of action of BABIM may be related to its inhibition of certain trypsin-like proteases,[47] but a study of twenty-four amidine derivatives showed no correlation between RSV-induced cell fusion and inhibition of four protease enzymes.[48]

4',6-Dichloroflavan (BW683C) - BW683C (11) is a potent inhibitor of the in vitro replication of several serotypes of rhinovirus.[49] The IC_{50} value for 11 against serotype 1B is 7 nM. However, the sensitivity of the nineteen serotypes against which 11 was tested was highly variable. BW683C was found to be most active when added to the cell culture simultaneously with the virus.[50] The compound does not interfere with virus adsorption to cells nor with uncoating of the viral RNA. It appears to bind directly to rhinovirus 1B and may prevent virus replication immediately following uncoating of the viral RNA.[50] The compound is relatively nontoxic as no untoward effects were evident with HeLa cells at concentrations up to 3.6 μM. In rats, no ill effects were observed after oral administration of up to 1000 mg/kg or i.p. administration of up to 700 mg/kg.[49] In a double-blind, placebo-controlled volunteer trial, oral BW683C failed to produce a consistent reduction in any of the quantitative indices of rhinovirus infection.[51] Administration of this antiviral agent by inhalation may prove to be more efficacious.[52]

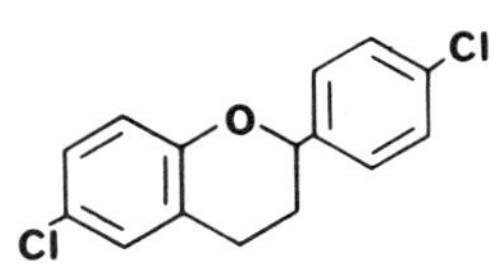

11

4',5-Dihydroxy-3,3',7-trimethoxyflavone (Ro 09-0179) - Ro 09-0179 (12), which was originally isolated from a Chinese medicinal herb, is a selective inhibitor of the replication of human picornaviruses.[53] Its IC_{50} against twenty serotypes of rhinovirus was 0.1 μg/ml in HeLa or WI-38 cells, and its 50% cytotoxic dose was about 15 μg/ml. The compound does not directly inactivate the virus, but it was most effective when present in the early stages of virus replication. Although 12 was not active in vivo, its diacetyl derivative 13 (Ro 09-0298) was orally effective in protecting mice from lethal infections with coxsackievirus B1.[53] The closely related natural product, quercetin, has antiviral activity both in vitro and in vivo.[54] Additional studies with quercetin against Mengo M virus infected mice have confirmed previous reports of its oral efficacy.[54,55]

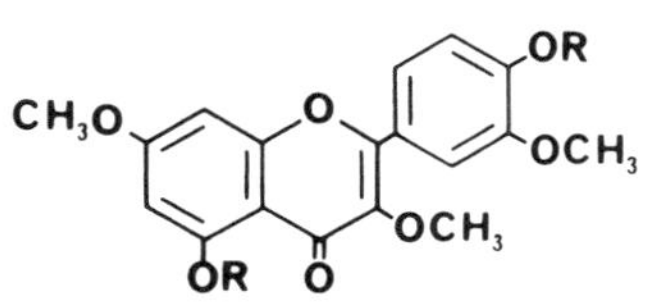

12, R=H

13, R=$COCH_3$

4'-Ethoxy-2'-hydroxy-4,6'-dimethoxychalcone (Ro 09-0410) - An agent active exclusively against rhinoviruses has emerged from a synthesis program on analogs of the flavone Ro 09-0179 (12). Ro 09-0410 (14), a chalcone, has activity against 46 of the 53 serotypes tested with IC_{50}s ranging from 3 μg/ml to less than 0.003 μg/ml.[56] In contrast, 14 was inactive against nine other DNA and RNA viruses at concentrations up to 3 μg/ml. Toxicological studies in rodents produced no adverse effects at a daily oral dose of 1000 mg/kg for two weeks. A phosphate ester (Ro 09-0415) of 14 has been evaluated in humans at the MRC Common Cold Unit in Salsbury, England.[57]

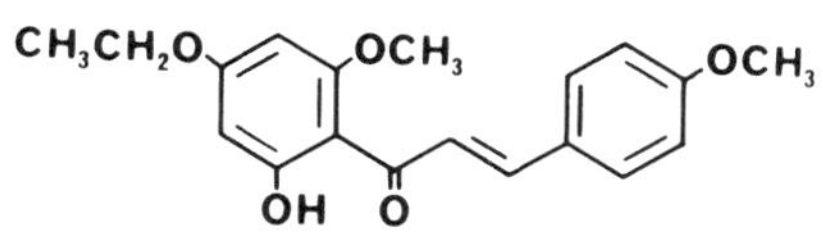

14

2-(3,4-Dichlorophenoxy)-5-nitrobenzonitrile (MDL-860) - MDL-860 (15) is a selective inhibitor of picornavirus replication in vitro.[58] At a concentration of 1.0 μg/ml, 15 caused a reduction in virus yield of at

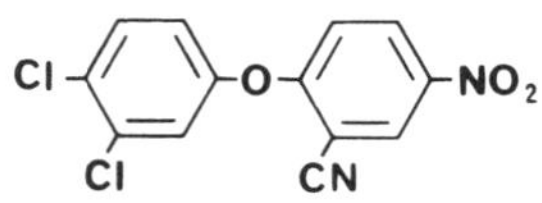

15

least ten fold in eight of ten enteroviruses and seventy-two of ninty rhinovirus serotypes, but it had no effect on six other types of viruses.[58] MDL-860 does not directly inactivate the virus nor does it inhibit virus uncoating.[59] Although 15 had no adverse effects on RNA or protein synthesis of uninfected HeLa cells, it was found to inhibit viral RNA synthesis when added shortly after virus absorption.

Sodium 5-aminosulfonyl-2,4-dichlorobenzoate (M12325) - A broad spectrum of activity against RNA viruses *in vitro* and *in vivo* has been reported for M12325 (16).[60] The MICs against several strains of influenza virus were comparable to the activity of amantadine, but 16 was less cytotoxic and had a higher LD_{50}. Oral doses of 16 were reported to be effective in reducing influenza induced mortality in mice.[60]

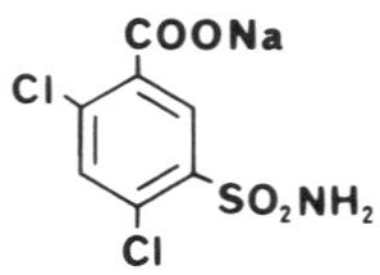

16

Arildone (WIN 38020) - Arildone (17) is a β-diketone which has broad-spectrum activity against several important DNA and RNA viral pathogens.[1,2] An isoxazole analog, WIN 49321 (18), has recently been reported to have antipicornavirus activity.[61] The isoxazole 18 inhibited plaque formation by 24 of 27 human rhinovirus serotypes, poliovirus and Echovirus by 50% at concentrations ranging from 0.03 to 15 μM. It was orally effective in protecting mice against poliovirus-induced death in a dose-dependent manner. Mechanism studies show that 18 prevents viral replication by inhibiting the virion uncoating process.[61]

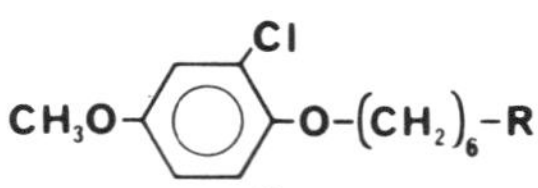

17, R = $CH(\overset{O}{\overset{\|}{C}}CH_2CH_3)_2$

18, R =

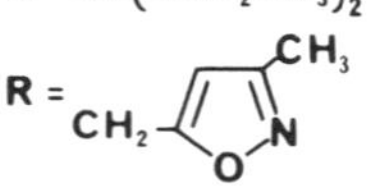

1-(5-Tetradecyloxy-2-furanyl)ethanone (RMI 15,731) - RMI 15,731 (19) has been reported to be a rhinovirus specific antiviral that directly inactivates virus *in vitro*.[2] In a recent clinical trial, intranasal 19 failed to show significant prophylactic activity in experimental human rhinovirus type 39 infections.[62]

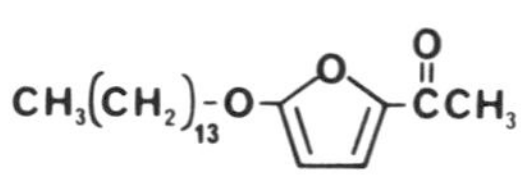

19

DNA VIRUSES

Significant progress continues to be made in the clinical use and development of agents active against DNA viruses. Acyclovir (9-(2-hydroxyethoxymethyl)guanine) has been the subject of several reviews[63-68] and of a symposium[69] during the past year. Clinical trial results have been published that attest to the efficacy of oral acyclovir in the treatment of primary[70] and recurrent[71] genital herpes simplex virus (HSV) infections and in the protection of bone marrow transplant patients from herpes infections.[72] Chronic, oral acyclovir treatment was highly effective in suppressing recurrent herpes simplex genitalis in a group of patients with unusually frequent episodes.[73] Topical acyclovir cream was effective for treatment of recurrent herpes labialis.[74] Considerable progress has been made in evaluating the clinical promise of acyclovir; however, there remains much to be learned concerning the best use of this drug in clinical practice.[75] Some studies on candidate prodrug forms of acyclovir have recently been published. Several esters of acyclovir were reported to have improved water solubility.[76] A diaminopurine analog A134U (2,6-diamino-9-(2-hydroxyethoxymethyl)purine), which is metabolically deaminated to acyclovir by adenosine deaminase,[77] has been reported to be better absorbed from the gut, resulting in higher plasma levels of acyclovir.[78] Several new reports have further documented the potent antiherpetic activity of 9-(2-hydroxy-1-(hydroxymethyl)ethoxymethyl)guanine. This compound has been reported to be highly effective in reducing the severity of both primary and recurrent lesions of HSV-2 in animal models.[79,80] It also appears to hold substantial promise in the treatment of human cytomegalovirus (HCMV) based on its specific anti-HCMV activity in vitro.[80,81] In an animal model, 9-(2-hydroxy-1-(hydroxymethyl)-ethoxymethyl)guanine was shown to be a potent, orally active, chemotherapeutic agent against Equid herpesvirus.[82] FIAC (2'-fluoro-5-iodo-1-β-D-arabinofuranosylcytosine) has been reported to stabilize cutaneous lesions in immunosuppressed patients suffering from acute herpesvirus infection.[84] A minor metabolite of FIAC, 2'-fluoro-5-methyl-1-β-D-arabinofuranosyluracil (FMAU), is also antiherpetic, but, in addition, it has i.p. and p.o. activity against murine leukemias resistant to 1-β-D- arabinofuranosylcytosine. Pharmacokinetic studies on FIAC and FMAU have been published,[85] and the synthesis and in vitro anti-HSV activities of a series of analogs have been reported.[86] A review of the antiviral activity of BVDU ((E)-5-(2-bromovinyl)-2'-deoxyuridine) and other 5-substituted pyrimidine nucleoside analogs has been published.[87] Several 2',3'-diester and 5'-monoester prodrug forms of ara-A (9-(1-β-D-arabinofuranosyl)adenine) have been evaluated with favorable results in HSV-2 induced genital infections in female guinea pigs.[88] The 5'-monophosphate of ara-A was found to be as effective as ara-A in the treatment of immunosuppressed patients suffering from varicella-zoster.[89] A carbocyclic analog of ara-A, cyclaradine, and its 5'-methoxyacetate ester have been reported to have activity in an HSV-1 encephalitis animal model that was comparable to the activity demonstrated by ara-A.[90] These two compounds may possess some clinical advantage over ara-A due to their low systemic toxicity. The antiviral effects of phosphonoformic acid (PFA, foscarnet sodium), which is presently in Phase III clinical trials in Europe,[83] have been discussed in a recent review.[91] The synthesis and structural requirements for antiherpes activity of a series of PFA esters have also been reported.[92]

APPROACHES TO THE DESIGN OF ANTIVIRAL AGENTS

Significant strides have been made in the development of clinically useful antiviral agents, especially against the DNA viruses of the herpes family. Most of these agents are directed against viral nucleic acid synthesis and require activation by a virus-induced thymidine kinase. Researchers are beginning to focus on other strategies which may produce broader spectrum antiviral agents with different mechanisms of action. Inhibition of polyamine biosynthesis may serve as a suitable target for antiviral drug design. The polyamine antimetabolite methylglyoxal bis(guanylhydrazone) and α-difluoromethylornithine, an inhibitor of polyamine biosynthesis, have recently been reported to inhibit replication of human CMV in vitro.[93] Another target is the inhibition of essential methylation reactions. For some viruses, efficient replication is dependent on viral mRNA that has been methylated at its 5'-terminal guanosine residue. Ribavirin 5'-triphosphate has been reported to be an effective inhibitor of this capping reaction which is catalyzed by a mRNA(guanine-7-)methyltransferase.[94] Another permutation of this theme is preservation of cellular S-adenosylhomocysteine (SAH) by inhibition of SAH hydrolase. SAH is an endogenous inhibitor of transmethylation reactions in which S-adenosylmethionine is the methyl donor.[95] The 2-5A system is one of the mechanisms by which interferon exerts its antiviral effect.[96,97] There has been substantial interest in exploiting this system as an approach to antiviral drug development. These efforts have primarily been aimed at the synthesis of 2-5A core analogs that are permeable to cells and stable to degradative enzymes. During the process of infection some viruses change the permeability of infected cells. This allows cellular penetration by compounds that are normally excluded. The selective antiviral activity of hygromycin B supports the suggestion that screening for agents which are selectively permeable to virus infected cells could result in broad-spectrum antiviral agents.[98] A review of the concept of selective delivery of antiviral agents by conjugation with protein carriers has been published.[99] These experiments suggest that the chemotherapeutic index of ara-A for the treatment of chronic hepatitis B could be improved by coupling the agent to lactosaminated albumin. A number of lipophilic amines of diverse structure have _in vitro_ antiviral activity against enveloped viruses including Semliki Forest virus, Sendai virus, vesicular stomatitis virus, and influenza A and B viruses.[100-103] This activity appears to be mediated by an obligatory lysosomal step in the uncoating of enveloped viruses. Amines such as chloroquine may prevent uncoating by increasing the lysosomal pH above a value required for release of the virus nucleocapsid into the cytoplasm. A better understanding of this process could lead to the development of broad-spectrum antiviral agents.

References

1. J.L. Kelley and L. Beauchamp, Annu. Rep. Med. Chem., 18, 139 (1983).
2. J.C. Drach and R.W. Sidwell, Annu. Rep. Med. Chem., 16, 149 (1981).
3. R.F. Schinazi and W.H. Prusoff, Pediatr. Clin. North Am., 30, 77 (1983).
4. P.E. Hermans and F.R. Cockerill, III, Mayo Clin. Proc., 58, 217 (1983).
5. R.G. Douglas, Jr., Medical Clin. North Am., 67, 1163 (1983).
6. A.G. Bukrinskaya, Advances in Virus Research, 27, 141 (1982).
7. F.J. Tyeryar, J. Infect. Dis., 148, 588 (1983).
8. L.J. Anderson, P.A. Patriarca, J.C. Hierholzer and G.R. Noble, Medical Clin. North Am., 67, 1009 (1983).
9. E.H. Lennette, Bull. W.H.O., 59, 305 (1981).
10. G.J. Galasso, Antiviral Res., 1, 73 (1981).
11. Y.J. Bryson, Pediatr. Infect. Dis., 1, 44 (1982).
12. D.M. Zlydnikov, O.I. Kubar, T.P. Kovaleva and L.E. Kamforin, Rev. of Infect. Dis., 3, 408 (1981).
13. R. Dolin, R.C. Reichman, H.P. Madore, R. Maynard, P.N. Linton and J. Webber-Jones, N. Engl. J. Med., 307, 580 (1982).
14. L.P. Van Voris, R.F. Betts, F.G. Hayden, W.A. Christmas and R.G. Douglas, J. Am. Med. Assoc., 245, 1128 (1981).
15. J.M. Quarles, R.B. Couch, J.R. Cate and C.B. Goswick, Antiviral Res., 1, 149 (1981).
16. F.G. Hayden, H.E. Hoffman and D.A. Spyker, Antimicrob. Agents Chemother., 23, 458 (1983).
17. V.M. Millet, M. Dreisbach and Y.J. Bryson, Antimicrob. Agents Chemother., 21, 1 (1982).
18. J.A. Flaherty and S.N. Bellur, J. Clin. Psychiatry, 42, 344 (1981).
19. V.W. Horadam, J.G. Sharp, J.D. Smilack, B.H. McAnalley, J.C. Garriott, M.K. Stephens, R.C. Prati and D.C. Brater, Ann. Intern. Med., 94, 454 (1981).
20. M.J. Wu, T.S. Ing, L.S. Soung, J.T. Daugirdas, J.E. Hano and V.C. Gandhi, Clin. Nephrol., 17, 19 (1982).
21. A.G. Bukrinskaya, N.K. Vorkunova, G.V. Kornilayeva, R.A. Narmanbetova and G.K. Vorkunova, J. Gen. Virol., 60, 49 (1982).
22. A.G. Bukrinskaya, N.K. Vorkunova and N.L. Pushkarskaya, J. Gen. Virol., 60, 61 (1982).
23. W.C. Koff, J.L. Elm, Jr. and S.B. Halstead, Am. J. Trop. Med. Hyg., 30, 184 (1981).
24. D.A. Fisher, Cutis, 29, 467 (1982).
25. A.W. Galbraith, Brit. J. Clinical Practice, September, 304 (1983).
26. T.W. Chang and R.C. Heel, Drugs, 22, 111 (1981).

27. V. Knight, H.W. McClung, S.Z. Wilson, B.K. Waters, J.M. Quarles, R.W. Cameron, S.E. Greggs, J.M. Zerwas and R.B. Couch, Lancet, 2, 945 (1981).
28. H.W. McClung, V. Knight, B.E. Gilbert, S.Z. Wilson, J.M. Quarles and G.W. Divine, J. Am. Med. Assoc., 249, 2671 (1983).
29. J.F. Hruska, P.E. Morrow, S.C. Suffin and R.G. Douglas, Jr., Antimicrob. Agents Chemother., 21, 125 (1982).
30. C.B. Hall, E.E. Walsh, J.F. Hruska, R.F. Betts and W.J. Hall, J. Am. Med. Assoc., 249, 2666 (1983).
31. C.B. Hall, J.T. McBride, E.E. Walsh, D.M. Bell, C.L. Gala, S. Hildreth, L.G. Ten Eyck and W.J. Wall, N. Engl. J. Med., 308, 1443 (1983).
32. D.F Smee, R.W. Sidwell, S.M. Clark, B.B. Barnett and R.S. Spendlove, Antimicrob. Agents Chemother., 20, 533 (1981).
33. W.C. Koff, R.D. Pratt, J.L. Elm, Jr., C.N. Venkateshan and S.B. Halstead, Antimicrob. Agents Chemother., 24, 134 (1983).
34. W.C. Koff, J.L. Elm and S.B. Halstead, Antiviral Res., 2, 69 (1982).
35. P.C. Srivastava and R.K. Robins, J. Med. Chem., 26, 445 (1983).
36. J.J. Kirsi, J.A. North, P.A. McKernan, B.K. Murray, P.G. Canonico, J.W. Huggins, P.C. Srivastava and R.K. Robins, Antimicrob. Agents Chemother., 24, 353 (1983).
37. E. De Clercq and J.A. Montgomery, Antiviral Res., 3, 17 (1983).
38. R.J. Phillpotts, R.W. Jones, D.C. DeLong, S.E. Reed, J. Wallace and D.A.J. Tyrrel, Lancet, 1, 1342 (1981).
39. R.J. Phillpotts, J. Wallace, D.A.J. Tyrrell and V.B. Tagart, Antimicrob. Agents Chemother., 23, 671 (1983).
40. F.G. Hayden and J.M. Gwaltney, Jr., Antimicrob. Agents Chemother., 21, 892 (1982).
41. R.A. Levandowski, C.T. Pachucki, M. Rubenis and G.G. Jackson, Antimicrob. Agents Chemother., 22, 1004 (1982).
42. E.C. Herrmann, Jr., J.A. Herrmann and D.C. DeLong, Antiviral Res., 1, 301 (1981).
43. C.J. Paget, D.C. DeLong, J. Nelson, and J.H. Wikel, Abstracts, 185th American Chemical Society National Meeting, Seattle, Washington, March 20-25, 1983; No. CARB 31.
44. C. Paget, J. Wikel, T. Morwick, T. Suarez, D. DeLong, J. Nelson, J. Quay and R. Bopp, Abstracts, North American Medicinal Chemistry Symposium, University of Toronto, Toronto, Canada, June 20-24, 1982; Poster 94.
45. H.J. Eggers, Handb. Exp. Pharmacol., 61, 377 (1982).
46. E.C. Herrmann, Jr., J.A. Herrmann and D.C. DeLong, Antiviral Res., 2, 339 (1982).
47. E.J. Dubovi, J.D. Geratz, S.R. Shaver and R.R. Tidwell, Antimicrob. Agents Chemother., 19, 649 (1981).
48. R.R. Tidwell, J.D. Geratz and E.J. Dubovi, J. Med. Chem., 26, 294 (1983).
49. D.J. Bauer, J.W.T. Selway, J.F. Batchelor, M. Tisdale, I.C. Caldwell and D.A.B. Young, Nature, 292, 369 (1981).
50. M. Tisdale and J.W.T. Selway, J. Gen. Virol. 64, 795 (1983).
51. R.J. Phillpotts, J. Wallace, D.A.J. Tyrrell, D.S. Freestone and W.M. Shepherd, Arch. Virol., 75, 115 (1983).
52. Drugs of the Future, 7, 542 (1982).
53. H. Ishitsuka, C. Ohsawa, T. Ohiwa, I. Umeda and Y. Suhara, Antimicrob. Agents Chemother., 22, 611 (1982).
54. J. Güttner, A. Veckenstedt, H. Heinecke and R. Pusztai, Acta Virol., 26, 148 (1982).
55. A. Veckenstedt and R. Pusztai, Antiviral Res., 1, 249 (1981).
56. H. Ishitsuka, Y.T. Ninomiya, C. Ohsawa, M. Fujiu and Y. Suhara, Antimicrob. Agents Chemother., 22, 617 (1982).
57. Scrip: World Pharmaceutical News, No. 689, p. 13 (1982).
58. R.D. Powers, J.M. Gwaltney, Jr. and F.G. Hayden, Antimicrob. Agents Chemother., 22, 639 (1982).
59. H.L. Torney, J.K. Dulworth and D.L. Steward, Antimicrob. Agents Chemother., 22, 635 (1982).
60. H. Ohnishi, K. Yamaguchi, S. Shimada, S. Himuro and Y. Suzuki, Antimicrob. Agents Chemother., 22, 250 (1982).
61. M.A. McKinlay, C.J. Brisson, J.V. Miralles and G.D. Diana in "Program and Abstracts, 23rd ICAAC", American Society of Microbiology, Las Vegas, Nevada, 1983, Abstract 160.
62. J.M. Gwaltney, Jr. and F.G. Hayden in "Program and Abstracts, 23rd ICAAC", American Society for Microbiology, Las Vegas, Nevada, 1983, Abstract 931.
63. D. Barry and R. Blum, in "Recent Advances in Clinical Pharmacology", Vol. 3, D.G. Shand and P. Turner, Ed., Churchill Livingstone Inc., New York, 1983, p. 57.
64. B. Bean, Postgrad. Med., 73(3), 297 (1983).
65. D.M. Richards, A.A. Carmine, R.N. Brogden, R.C. Heel, T.M. Speight and G.S. Avery, Drugs, 26, 378 (1983).
66. O.L. Laskin, Clin. Pharmacokinet., 8, 187 (1983).
67. C.C. Rudd, Drug. Ther. Hosp., 8(9), 41 (1983)

68. H.J. Rogers and A.S.E. Fowle, J. Clin. Hosp. Pharm., 8, 89 (1983).
69. J. Antimicrob. Chemother., 12, Suppl. B, (1983).
70. Y.J. Bryson, M. Dillon, M. Lovett, G. Acuna, S. Taylor, J.D. Cherry, B.L. Johnson, E. Wiesmeier, T. Creagh-Kirk and R. Keeney, N. Engl. J. Med., 308, 916 (1983).
71. R.C. Reichman, G.J. Badger, G. J. Mertz, L. Corey, D. Richman, M. Oxman, J. Connor, Y. Bryson, D. Tyrrell and J. Partnoy, Clin. Res., 31, 373A (1983).
72. E. Gluckman, J. Lotsberg, A. Devergie, X.M. Zhao, R. Melo, M. Gomez-Morales, T. Nebout, M.C. Mazeron and Y. Perol, Lancet, 2, 706 (1983).
73. S. Straus, M. Seidlin, H. Takiff, S. Bachrach, J. DiGiovanna, K. Western, T. Creagh-Kirk, L. Lininger and D. Alling, Clin. Res., 31, 543A (1983).
74. A.P. Fiddian, J.M. Yeo, R. Stubbings and D. Dean, Br. Med. J., 286, 1699 (1983).
75. D. Brigden, R.E. Keeney and D.H. King, J. Antimicrob. Chemother, 12, Suppl. B, 195 (1983).
76. L. Colla, E. De Clercq, R. Busson and H. Vanderhaeghe, J. Med. Chem., 26, 602 (1983).
77. T. Spector, T.E. Jones and L.M. Beacham, III, Biochem. Pharmac., 32, 2505 (1983).
78. H.C. Krasny, S.H.T. Liao and S.S. Good, Clin. Pharmac. Ther., 33, 256 (1983).
79. E.B. Fraser-Smith, D.F. Smee and T.R. Matthews, Antimicrob. Agents Chemother., 24, 883 (1983).
80. A.K. Field, M.E. Davies, C. DeWitt, H.C. Perry, R. Liou, J. Germershausen, J.D. Karkas, W.T. Ashton, D.B.R. Johnston and R.L. Tolman, Proc. Natl. Acad. Sci. USA, 80, 4139 (1983).
81. E.-C. Mar, Y.-C. Cheng and E.-S. Huang, Antimicrob. Agents Chemother., 24, 518 (1983).
82. E.A. Rollinson and G. White, Antimicrob. Agents Chemother., 24, 221 (1983).
83. Scrip: World Pharmaceutical News, No. 834, p. 14 (1983).
84. C.W. Young, R. Schneider, B. Leyland-Jones, D. Armstrong, C.T.C. Tan, C. Lopez, K.A. Watanabe, J.J. Fox and F.S. Philips, Cancer Res., 43, 5006 (1983).
85. F.S. Philips, A. Feinberg, T.-C. Chou, P.M. Vidal, T.-L. Su, K.A. Watanabe and J.J. Fox, Cancer Res., 43, 3619 (1983).
86. K.A. Watanabe, T.-L. Su, R.S. Klein, C.K. Chu, A. Matsuda, M.W. Chun, C. Lopez and J.J. Fox, J. Med. Chem., 26, 152 (1983).
87. E. De Clercq, Pure and Appl. Chem., 55, 623 (1983).
88. W.M. Shannon, G. Arnett, D.C. Baker, S.D. Kumar and W.I. Higuchi, Antimicrob. Agents Chemother., 24, 706 (1983).
89. J.L. Vilde, F. Bricaire, A. Huchon and F. Brun-Vezinet, J. Med. Virology, 12, 149 (1983).
90. W.M. Shannon, L. Westbrook, G. Arnett, S. Daluge, H. Lee and R. Vince, Antimicrob. Agents Chemother., 24, 538 (1983).
91. B. Oberg, Pharmac. Ther., 19, 387 (1983).
92. J.O. Norén, E. Helgstrand, N.G. Johansson, A. Misiorny and G. Stening, J. Med. Chem., 26, 264 (1983).
93. A.S. Tyms and J.D. Williams, Nature, 297, 690 (1982).
94. B.B. Goswami, O.K. Sharma and E. Borek, Recent Results in Cancer Research, 84, 275 (1983).
95. E. De Clercq, Nucl. Acid Res., Series No. 11, 203 (1982).
96. P.F. Torrence, K. Lesiak, J. Imai, M.I. Johnston and H. Sawai, in "Nucleosides, Nucleotides and Their Biological Applications", J.L. Rideout, D.W. Henry and L.M. Beacham, III, Ed., Academic Press, New York, 1983, p. 67.
97. R.J. Suhadolnik, Y. Devash, P. Doetsch, E.E. Henderson, J.M. Wu, W. Pfleiderer and R. Charubala, in "Nucleosides, Nucleotides and Their Biological Applications", J.L. Rideout, D.W. Henry and L.M. Beacham, III, Ed., Academic Press, New York, 1983, p. 147.
98. J.-C. Lacal and L. Carrasco, Antimicrob. Agents Chemother., 24, 273 (1983).
99. L. Fiume, C. Busi and A. Mattioli, FEBS Lett., 153, 6 (1983).
100. D.K. Miller and J. Lenard, Proc. Natl. Acad. Sci. USA, 78, 3605 (1981).
101. J. Lenard and D.K. Miller, Cell, 28, 5 (1982).
102. A. Helenius, M. Marsh and J. White, J. Gen. Virol., 58, 47 (1982).
103. M. Shibata, H. Aoki, T. Tsurumi, Y. Sugiura, Y. Nishiyama, S. Suzuki and K. Maeno, J. Gen. Virol., 64, 1149 (1983).

Chapter 13. Antifungal Chemotherapy

Michael B. Gravestock and John F. Ryley

ICI Pharmaceuticals Division, Mereside, Alderley Park, Macclesfield, Cheshire SK10 4TG, England

Introduction - The last two years have seen a continued growth in this field, principally due to the research interest in azole antifungal drugs. General reviews have been published on the place of imidazole antifungals in chemotherapy,[1,2] their place in the treatment of systemic fungal infections in comparison with amphotericin B (AMB)[3] and a comparison of their physicochemical and pharmacokinetic properties, potency and pharmacological activity.[4] The substantial effort which has gone into the synthesis of imidazole- and triazole-containing potential drugs and plant fungicides is illustrated by the length of two substantial reviews on the synthesis[5] and patent literature[6] of 1-substituted (1H)-azoles.

Ketoconazole (KC) as the first systemically-active azole antifungal has been the subject of a number of reviews including the general management of fungal disease,[7,8,9] and its efficacy in superficial and systemic infections;[10] it was the subject of a number of papers at a recent symposium.[11] The use of KC in the treatment of coccidioidomycosis has been recently reviewed in detail,[12] and parallel reviews of the indications for coccidioidomycosis treatment in general,[13] and the use of AMB[14] and miconazole (MC)[15] in particular have been presented.

The clinical pharmacokinetics of the systemic antifungals AMB, 5-fluorocytosine (5-FC), MC and KC were reviewed[16] with their individual limitations and preferred protocols for administration. Methods for measurement of the activity of antifungal drugs have been assessed[17] and the in vitro activity of nystatin, AMB, 5-FC and five azoles measured against 151 yeast strains and two strains of Geotrichum candidum.[18] A discussion of experimental approaches to the discovery of antifungal drugs with a survey of post-1970 synthetic and natural product chemicals with antifungal activity has been made.[19]

A monograph on fungal infection in the compromised patient[20] provides timely information on this area of increasing clinical concern. The antifungal action of the intraleukocytic antibacterial rifampin was reviewed[21] and the antimycotic activity and therapeutic effectiveness of 5-bromosalicyl-4'-chloranilide and its derivatives assessed in laboratory animals and man.[22]

New antifungal agents - Bifonazole (Bay-h-4502; 1) showed a broad spectrum of activity in vitro with particularly marked fungicidal effects on dermatophytes.[23] Topical applications of the drug were extremely effective against trichophytosis in guinea pigs. Comparison of the drug in vitro with MC and clotrimazole (CL) showed that potency relative to both varied from strain to strain and species to species of yeasts and dermatophytes.[24] In another in vitro comparison[25] with CL the latter was 2-4 fold more active against both groups of fungi in various media. Surveys of the compound's efficacy in the topical treatment of

ISBN 0-12-040519-9

dermatomycoses, particularly pityriasis versicolor and tinea cruris, have appeared.[23,24,26,27]

A new 1,2,4-triazole- rather than imidazole-based antifungal Bay-n-7133 (2a) which is orally active has been studied in vitro in comparison with KC,[28,29] MC[29] and a chemically related imidazole

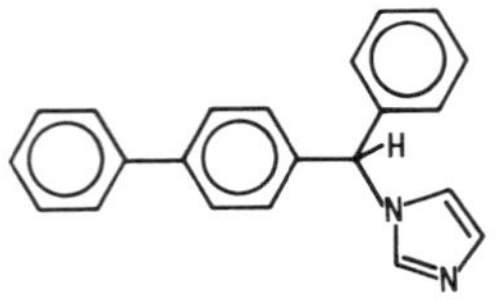

1

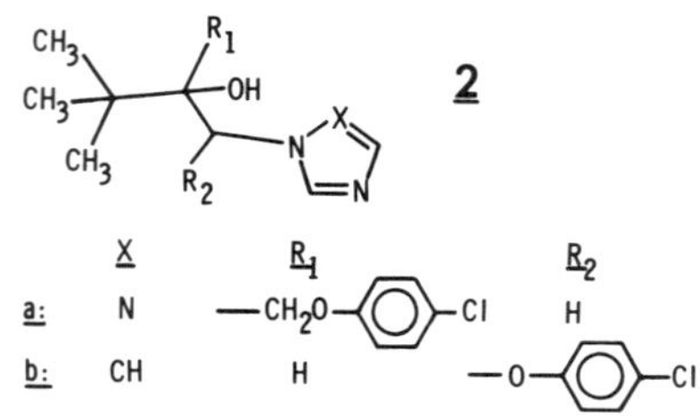

2

Bay-l-9139 (2b).[29] Both new compounds are comparable to the standards in spectrum, but 2a was more potent. Antimycotic results in vivo for 2a showed activity equal to or less than KC.[30] The compound was active against murine aspergillosis[31] but less active against murine candidosis than 2b.[32]

Another triazole-based drug itraconazole (R51211; 3a) shows considerable promise as a safe orally absorbed antifungal. R51211 was highly active in vitro against dermatophytes, yeasts, aspergilli, dematiaceous, dimorphic and various other fungi.[33] Oral treatment of aspergillosis and cryptococcosis in mice and sporotrichosis in guinea pigs was successful as was treatment of disseminated trichophytosis, systemic candidosis and superficial mycoses in guinea pigs; solution in PEG was necessary for good cure rates with this rather insoluble compound. Initial clinical results in the treatment of vaginal candidosis and pityriasis versicolor showed good response at one quarter the dose of KC for 5 days;[34] a two day dosing schedule was also promising. Morphological studies on the yeast to mycelial transformation of Candida albicans showed similar dose/inhibition properties to KC. The superior in vivo properties may therefore be due to pharmacokinetic factors, distribution etc.[35]

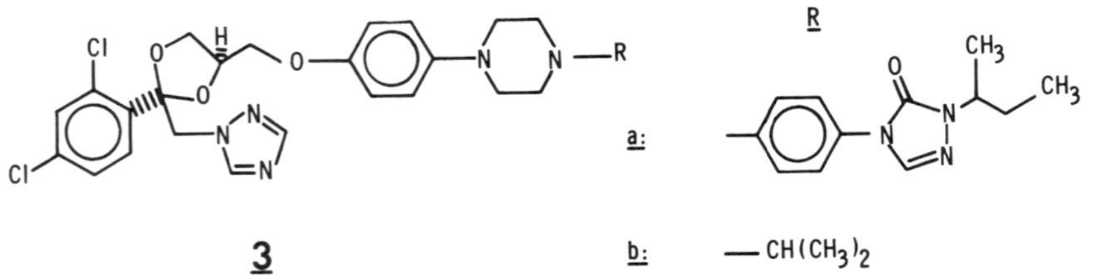

3

A triazole-based tertiary alcohol ICI 153,066 (4) was shown to have therapeutic activity against vaginal candidosis in mice and rats and against ringworm in mice and guinea pigs following oral administration.[36] In these animal models the compound was 10-100 times more potent than KC. A topical formulation in PEG was 10 times more potent than a corresponding KC or MC preparation. Gross toxicity of the compound was twice that of KC, and it showed considerably greater persistence in the blood of mice and rats. The spectrum of activity against a range of clinical isolates of yeasts and dermatophytes was good although certain KC-resistant C. albicans strains were also resistant to 4.[37] The efficacy of 4 in experimental coccidioidomycosis in mice was compared with KC.[38] Although the compound showed potency advantages over KC, improved survival times only, rather than biocure, were observed with both agents.

Fungicidal activity has been claimed for a novel imidazole-based drug zinoconazole (SC 38390; 5).[39] In vitro experiments with dermatophytes and C. albicans at levels no greater than the MIC showed marked reductions in viability within a few hours of treatment.[39,40]

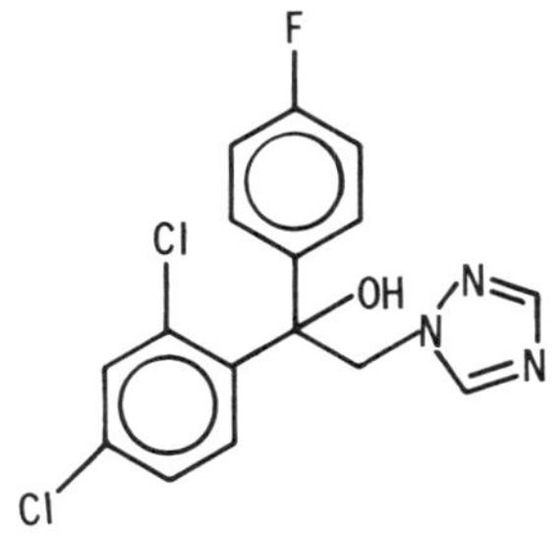

4

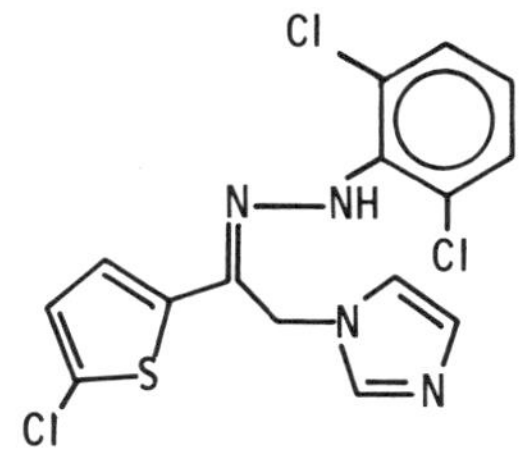

5

The results closely resembled earlier observations of fungicidal effects with high concentrations of CL and MC. Oral administration in rat vaginal candidosis and mouse systemic candidosis models produced results approaching KC.[40] Topical administration was comparable to CL. The compound was Ames test negative.

Terconazole (R42470; 3b) a triazole analogue of KC with a more lipophilic substituent on the terminal piperidine nitrogen is highly active in vitro against a wide range of yeasts and mycelium-forming fungi.[41,42] This in vitro activity depends largely on the medium used. In vaginal candidosis in rats and dermatophytosis in guinea pigs 3b was more potent than CL.[43] A double blind clinical trial for the topical treatment of vaginitis showed that the two treatments were equally effective and acceptable.[44]

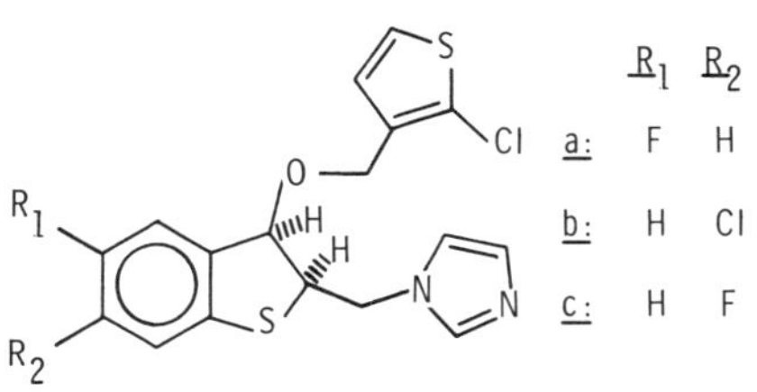

6

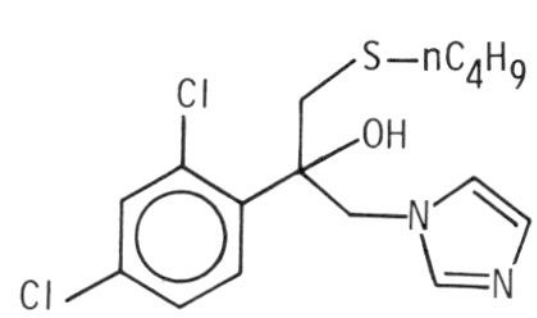

7

Two other imidazole-based antifungals have recently been highlighted, both containing a sulfur atom in the side chain. Sch 31153 (6a) was the best of three dihydrobenzthiophen derivatives (6a; Sch 30566, 6b; Sch 31498, 6c) which showed generally excellent broad spectrum activity in vitro and topical activity against C. albicans and Trichophyton mentagrophytes.[45,46] Comparison with CL, MC and tioconazole in a hamster vaginal candidosis model showed activity greater than the former two compounds but similar to tioconazole.[46] Synthetic procedures and structure-activity relationships were also presented.[45] SM-4470 (7) was active against Cryptococcus and Aspergillus species[47] and was active orally against C. albicans in mice (potency twice that of KC), against C. albicans in the rat (similar to KC), against cryptococcosis in mice (similar to KC) and was much better than KC versus aspergillosis in mice and T. mentagrophytes in the guinea pig.

Departing from the azole-based antifungals, an interesting new agent has emerged from the naftifine series; SF 86-327 (8) is an allylamine with significant advantages over the former compound, showing oral activity in the guinea pig dermatophytosis model.[48] It has high activity in vitro against dermatophytes, but much less activity against yeasts such as C. albicans. An essential part of the structure is the conjugated enyne group with trans stereochemistry. Preliminary clinical data on treatment of dermatophytosis in volunteers and patients showed successful control with oral doses of 250 mg/kg b.i.d.[48] A comprehensive series of papers on structure-activity relationships, synthesis, in vitro data, animal data, mode of action (see below), ultrastructural studies on T. mentagrophytes and C. albicans, metabolism, pharmacokinetics and toxicological evaluation were presented recently.[49]

The antifungal nitropyrrole antibiotic pyrrolomycin shows weak activity and high toxicity. A program of synthesis/structure optimisation has led to a related tetrazole, CN-146 (9) which has been selected for development as a topical antimycotic.[50] The compound is active topically but not systemically in animal models and also has gram-positive/negative antibacterial properties.

8

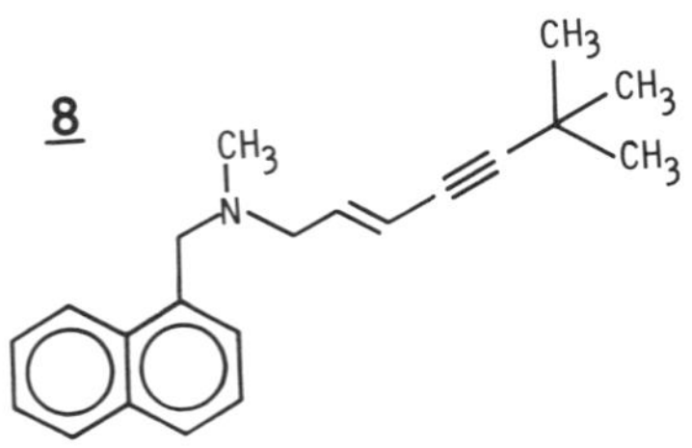

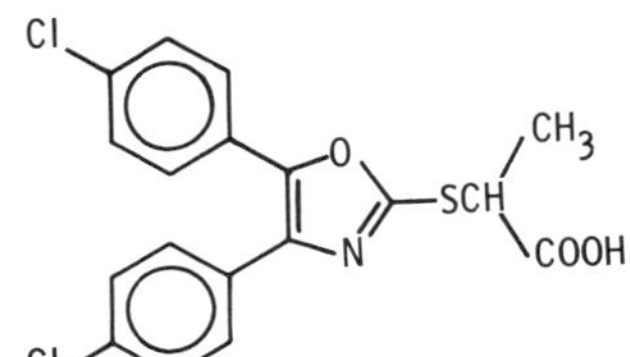

10

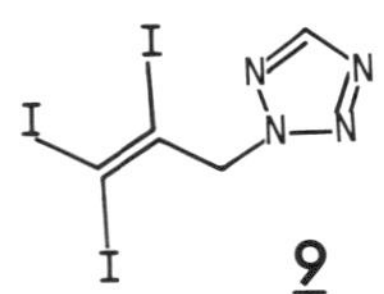

9

Tioxaprofen (10) was synthesised as a non-steroidal anti-inflammatory but has been found to have antimycotic activity against T. mentagrophytes, T. rubrum, Microsporum canis and C. albicans.[51] The compound appears to act by uncoupling mitochondrial respiration.

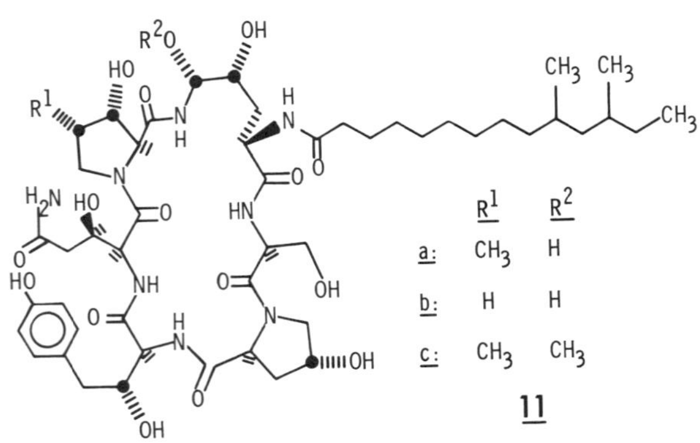

11

A group of new cyclic peptides containing a fatty acid side chain has been isolated.[52] The complex has been separated into three components, sporiofungins A, B and C (11a,b,c) which are closely related to the echinocandin glucan synthesis inhibitors. The sporiofungins are active against a broad range of yeasts and filamentous fungi with MIC values for Candida spp. in the μg range. Acute toxicity in mice is low and in vivo activity in experimental Candida infections has been demonstrated; chronic liver toxicity however precludes development.

Mechanisms of Action - Some of the factors influencing the C-14 demethylase activity of azoles in Candida were investigated;[53] interspecies variability in enzyme activity, ability to penetrate the fungal cell wall as well as structural features were found to be important. A number of papers have reviewed the potency of azoles and other agents as inhibitors of sterol C-14 demethylation, although many of the compounds considered are plant fungicides.[54,55,56] QSAR studies on phenoxytriazolyl methanes as sterol inhibitors have been made based on MIC values against Saccharomycopsis lipolytica.[57] The resistance of a mutant strain of the plant pathogen Ustilago maydis to fenarimol, etaconazole, MC and the aza-steroid A25822B was shown to be due to a deficiency in sterol C-14 demethylase.[58] However, two C. albicans clinical isolates resistant to KC and cross-resistant to 4 were apparently resistant due to cell membrane changes rather than internal enzymology.[59] The overall significance of C-14 demethylase blockade in the growth inhibition of Saccharomyces cerevisiae by CL and MC has been challenged following comparative experiments with the 15-azasteroid sterol Δ^{14}-reductase inhibitors, where a demethylase-deficient mutant strain was inhibited by the azoles.[60] However, this result may well reflect the additional mode(s) of action often cited[58,61] for the above compounds relative to agents such as KC. Thus at higher than MIC concentrations, CL and MC are fungicidal to S. cerevisiae, apparently as a result of direct membrane damage.[61,62]

Possible clues to the factors responsible for this membrane damage have emerged from studies on gram-positive bacteria such as Staphylococcus aureus which is inhibited by some azoles but contains no membrane sterols. Thus MC is bactericidal at low concentrations and causes release of cellular K^+. KC has little or no such effects, and slows growth only at high concentrations.[63] The hypothesis that K^+ release is due to a direct membrane damaging effect of MC not shared by KC is supported by other studies on Candida species.[64,65] S. aureus was protected against growth inhibitory effects of MC and KC by unsaturated fatty acids, and partly by oleic acid.[66] This acid however had no effect on KC's action on C. albicans. Cytochrome oxidase and ATPase activities were more sensitive to MC than to KC. It was postulated that MC may change lipid organisation whereas KC is localized in the multilayer without such effect.[66]

A study of nystatin, AMB and five imidazole drugs on C. albicans[67] by scanning electron microscopy showed differential effects on the cytoplasmic membrane and on the form and release of spheroplasts from young yeast cells. Ultrastructural changes produced by KC in Cryptococcus neoformans and Sporothrix schenckii were examined.[68] Surface changes, fatty degeneration of the cytoplasm and lysis of subcellular organelles were observed in some but not all the cells. At a concentration of 1 μg/ml S. schenckii was more sensitive than C. neoformans judged by the numbers of abnormal cells.

The effects of isoconazole on cell membranes in several yeasts resulted in leakage of K^+, adenosine, adenine nucleotides and α-glucosidase.[69] At fungicidal concentrations respiration of both intact yeast cells and isolated mammalian mitochondria was inhibited.

In another study of isoconazole, fungistatic or fungicidal concentrations inhibited protein, RNA, DNA, mannan and glucan biosynthesis in C. albicans.[70] However, no correlation was observed between suppression of respiration and the sensitivity of yeast cells. Econazole was the subject of two detailed studies on its cell membrane damaging effects. Effects on rat liver mitochondria occurred at or above 10μg/ml, the degree of damage being assessed by liberation of acid phosphatase and β-glucuronidase.[71] In S. cerevisiae the drug attaches first to particulate and later to soluble fractions after penetration into the cell.[72] The highest concentration was found in the mitochondria, but no inhibition of mitochondrial enzymes such as cytochrome oxidase, succinate dehydrogenase or phenylalanyl-tRNA synthetase was observed. Biosynthesis of the mitochondrial membrane enzymes was affected without that of the synthetase, a matrix enzyme originating in the cytoplasm. In a study of KC on intact human fibroblasts and C. albicans cells the differences in sensitivity observed in vivo and in vitro correlated well with the effects of KC on sterol synthesis.[73] In human fibroblasts, 50% inhibition of cholesterol synthesis was achieved at 0.7μM while the same effect on ergosterol synthesis in C. albicans was found at a 140 times lower concentration.

An evaluation of oxiconazole showed a broad fungistatic spectrum and fungicidal activity in selected species - Aspergillus fumigatus, C. neoformans, C. albicans and T. mentagrophytes; DNA synthesis was more sensitive at sub-inhibitory concentrations than RNA, protein or carbohydrate synthesis.[74] Studies on a MC derivative, R24571[75] and KC[76] as inhibitors of Ca^{2+} binding to calmodulin failed to show effects at physiologically significant levels in skeletal muscle sarcoplasmic reticulum. KC and MC were studied for their ability to inhibit dexamethasone binding in cell cytosol.[77] Basal tyrosine aminotransferase activity (TAT) was not affected, but partial antagonism of dexamethasone-induced TAT activity was noted, which correlated well with azole drug-binding potency. Glucocorticoid receptor antagonism in target tissues therefore seems possible.[77]

The potential for antifungal drugs to stimulate host defenses as a possible component of their fungicidal action has been considered in earlier work, particularly for the polyenes. The relationship between adjuvant and mitogenic effects for amphotericin B methyl ester (AME) was considered.[78] AMB and AME showed potent murine immunostimulant and B-lymphocyte activating activity. The latter was a more potent polyclonal B-cell activator, but only at high doses. The evidence suggests that B-cell activating properties of AME are not involved in the cellular mechanism of adjuvant activity in vivo, but there was a strong correlation between the in vitro polyclonal B-cell activating effects and in vivo adjuvant effects of AME in various mouse strains. KC has been examined for its ability to prevent further development of the invasive form of C. albicans in deep and superficial infections.[79] Synergism with host defense cells was demonstrated in culture and the compound's unexpected potency in vivo was ascribed to facilitation of the host defense cells in view of its relatively poor in vitro potency. This synergism between polymorphonuclear granulocytes and KC against C. albicans in vitro has been examined in another study.[80] Treatment of granulocytes did not influence their activity, but treatment of C. albicans with KC led to significantly higher killer activity but unchanged phagocytosis. This result appears to show that antimycotic effects per se rather than immunomodulating properties are responsible for the observed synergism.[80]

Naftifine interferes with sterol biosynthesis in C. parapsilosis and T. mentagrophytes. It produced an increase in cellular squalene content and resulted in decreased amounts of sterols including ergosterol. Squalene is the only lipid intermediate of ergosterol biosynthesis which accumulates in treated cells and results from the inhibition of squalene epoxidase.[81] Dose-dependent ultrastructural changes in C. parapsilosis include increased accumulation of lipid particles in the cytoplasm, thickening of the cell wall and alterations of the plasma membrane.[82] Studies on the mode of action of 8 on C. albicans, C. parapsilosis and T. mentagrophytes indicated that it too acts by inhibition of squalene epoxidase, and that accumulation of free squalene results in the extremely low MIC values and fungicidal effects seen against dermatophytes in general.[49]

Therapeutic, Pharmacological and Toxicological Studies - The disposition of KC in patients was studied after daily doses of 50-200mg;[83] peak serum concentrations of up to 50μg/ml and potential therapeutic concentrations up to 26 h after the higher doses were detected. Possible interactions with other drugs, particularly antacids, were investigated; they did not significantly affect KC pharmacokinetics. A study of KC in recurrent vaginal candidosis showed 97% cure at follow-up (7 days) and a relapse rate of 10% at 35 days; the oral treatment was preferred by the majority of patients over previously used intravaginal preparations.[84] Other recent clinical experience with KC has been reviewed including use of a topical preparation.[85] A study of the drug in control of systemic C. albicans in mice dosed with cytarabine showed good limitation of the spread of infection, suggesting prophylactic use in patients on cytarabine chemotherapy.[86] The drug has good prophylactic potential and therapeutic efficacy in keratomycosis in rabbits.[87]

The use of tioconazole in gynecology[88] and in the treatment of superficial mycoses[89] has been reviewed. The compound gave good control of dermatophyte infections and was well tolerated in animals indicating minimal risk of side effects in man.[90] The pharmacokinetics of CL following intra-vaginal administration were studied; a new formulation gave fungicidal levels in the vagina 48 h after administration with no detectable blood levels at 72 h.[91] Topical administration of isoconazole inhibited T. mentagrophytes infections on guinea pig skin.[92] Oral or ip administration decreased mortality in mice having systemic candidosis but the kidney infection was not eradicated. The extent of inter-species variation in percutaneous absorption of sulconazole has been investigated in rabbits, rats, dogs and guinea pigs.[93] A double-blind study in 114 patients with 2% butoconazole cream showed it to be more effective than MC cream against vaginal candidosis.[94]

Several studies on 5-FC combined with other agents have been made. 5-FC serum levels were not affected by the presence of KC.[95] The MIC of 5-FC against 18 pathogenic fungal isolates was not altered by 100μM allopurinol or oxypurinol; the potential for allopurinol reduction of 5-FU (5-FC metabolite) myelotoxicity in the clinical setting was discussed.[96] KC, AMB, 5-FC and combinations of these drugs were compared in chronic cryptococcal meningitis therapy.[97] The addition of KC to AMB therapy was at least as effective as the combination AMB/5-FC over a 2 week period. Other workers have found mutual antagonism between AMB and KC against C. albicans.[98]

The bioavailability of topically applied econazole, oxiconazole and dimethylmorpholine was studied in human skin and nails.[99] Tinctures were better for nails, ointments for skin to aid penetration. The importance of the right base in topical ciclopiroxolamine formulations has been emphasised[100] and a general review made of its topical effectiveness versus bifonazole and CL.[101] Naftifine was found to be an effective topical agent as a 1% gel in the treatment of dermatophytosis.[102]

Isolated reports of liver toxicity with azoles have continued. Severe liver toxicity developed in one patient on valconazole[103] and the UK-CSM have issued warnings on the risk of hepatitis with KC following 15 yellow card reports.[104] Similar warnings have been issued in the USA,[105] but it should be emphasised that although transient increases in liver enzymes may be detected, symptomatic hepatitis occurs in only 1 in 12,000 patients.[106] KC also blocks testosterone[107] and adrenal steroid synthesis[108] and as a result may cause gynecomastia, and with high doses decreased libido and effects on sperm production. An interesting consequence of these antihormonal effects is its utility in the treatment of prostatic cancer.[109] A potentially dangerous interaction between KC and cyclosporin A, an immunosuppressive drug used in kidney and marrow transplant patients has been reported; cyclosporin blood levels increase and renal impairment develops.[110,111]

References

1. J. R. Prous and N. E. Mealy, Med.Actual., 19, 327 (1983).
2. T. Arai and J. Uno, Nippon Saikingaku Zasshi, 38, 525 (1983).
3. G. Medoff, J. Brajtburg, G. S. Kobayashi and J. Bolard, Annu.Rev.Pharmacol.Toxicol., 23, 303 (1983).
4. M. Plempel, Shinkin to Shinkinsho, 23 17 (1982). (C.A. 98 22120q).
5. G. J. Ellames in "Topics in Antibiotic Chemistry", vol. 6, P. Sammes, ed., Ellis Horwood, Chichester, 1982, p.14.
6. L. Zirngibl, Prog.Drug Res., 27, 253 (1983).
7. J. Symoens and G. Cauwenbergh, Prog.Drug Res., 27, 63 (1983).
8. H. B. Levine, "Ketoconazole in the Management of Fungal Disease", ADIS Press, Australia, 1982 (Revised Edition).
9. H. Koch, Pharm.Int., 4, 151 (1983).
10. R. C. Heel, R. N. Brogden, A. Carmine, P. A. Morley, T. M. Speight and G. S. Avery, Drugs, 23, 1 (1982).
11. 13th ICC (1983)*, SS 4.8/1, Part 40, Abstracts 1-10.
12. J. N. Galgiani, Drugs, 26, 355 (1983).
13. D. A. Stevens, Drugs, 26, 334 (1983).
14. D. J. Drutz, Drugs, 26, 337 (1983).
15. D. A. Stevens, Drugs, 26, 347 (1983).
16. T. K. Daneshmend and D. W. Warnock, Clin.Pharmacokinet., 8, 17 (1983).
17. G. S. Kobayashi and G. Medoff, Mycol.Ser., 3, 357 (1983).
18. T. Bergan and M. Vangdal, Chemotherapy, 29, 104 (1983).
19. J. F. Ryley, R. G. Wilson, M. B. Gravestock and J. P. Poyser, Adv.Pharmacol. Chemother., 18, 49 (1981).
20. D. W. Warnock and M. D. Richardson, "Fungal Infection in the Compromised Patient", Wiley, Chichester, 1982.
21. G. Medoff, Rev.Infect.Dis., 5, S614 (1983).
22. W. Wohlrab, K. D. Wozniak, K. Schwabe, B. Tschiersch, R. P. Zaumseil and K. M. Taube, Abh.Akad.Wiss.DDR, Abt.Math.,Naturwiss.,Tech., 1N, 193 (1983). (C.A. 99 63694u).
23. M. Plempel, E. Regel and K. H. Büchel, Arzneim.-Forsch., 33, 517 (1983).
24. H. Yamaguchi, T. Hiratani and M. Plempel, Arzneim.-Forsch., 33, 546 (1983).
25. S. Shadomy, D. M. Dixon and R. May, Sabouraudia, 20, 313 (1982).
26. S. Stettendorf, 13th ICC (1983)*, SY 49/2, Part 20, Abstract No. 6.
27. C. L. Meneghini, G. Angelini and G. A. Vena, 13th ICC (1983)*, SY 49/2, Part 20, Abstract No. 5.
28. H. Yamaguchi, T. Hiratani and M. Plempel, J.Antimicrob.Chemother., 11, 135 (1983).
29. R. A. Fromtling, H. P. Yu and S. Shadomy, Sabouraudia, 21, 179 (1983).
30. M. Plempel, G. Holmwood and K. H. Büchel, 13th ICC (1983)*, SY 49/1, Part 20, Abstract No. 6.
31. W. Ritter, M. Plempel, G. Gruenwaldt and J. Sommer, 13th ICC (1983)*, SY 49/1, Part 20, Abstract No. 7.
32. P. D. Hoeprich and J. Merry, Clin.Res., 31, 91A (1983).

33. J. Van Cutsem, F. Van Gerven, R. Zaman, J. Heeres and P. A. J. Janssen, 13th ICC (1983)*, SS 4.8/1, Part 40, Abstract No. 11.
34. G. Cauwenbergh, 13th ICC (1983)*, SS 4.8/1, Part 40, Abstract No. 12.
35. M. Borgers, M. Van de Ven, G. Willemsens and H. Van den Bossche, 13th ICC (1983)*, SE 4.8/2, Part 61, Abstract No. 13.
36. J. F. Ryley and R. G. Wilson, Proc.22nd ICAAC, Miami Beach, 4-6 Oct.1982, Abstract No. 477.
37. J. F. Ryley and R. G. Wilson, 13th ICC (1983)*, SY 49/1, Part 20, Abstract No. 9.
38. H. B. Levine and J. M. Cobb, 13th ICC (1983)*, SY 49/1, Part 20, Abstract No. 10.
39. G. J. Ellames and A. M. S. Pope, 13th ICC (1983)*, PS 4.8/3, Part 115, Abstract No. 4.
40. R. L. Dyer, G. J. Ellames, B. J. Hamill, P. W. Manley and A. M. S. Pope, J.Med.Chem., 26, 442 (1983).
41. J. Van Cutsem, F. Van Gerven, R. Zaman and P. A. J. Janssen, Chemotherapy, 29, 322 (1983).
42. J. Heeres, R. Hendrickx and J. Van Cutsem, J. Med.Chem., 26, 611 (1983).
43. J. Heeres and J. Van Cutsem, 13th ICC (1983)*, PS 4.8/3, Part 115, Abstract No. 5.
44. A. Rodriguez-Noriega, A. Palacio-Hernanz, F. Sanz-Sanz, S. Reyes, A. Bartlett-Coma and M. Bris-Orche, 13th ICC (1983)*, SE 4.8/2, Part 61, Abstract No. 8.
45. R. E. Pike, D. F. Rane, J. J. Wright, R. Parmegiani and D. Loebenberg, Proc.22nd ICAAC, Miami Beach 4-6 Oct.1982, Abstract No. 476.
46. D. Loebenberg, R. Parmegiani, R. Antonacci, T. Yarosh-Tomaine, D. Rane, J. J. Wright and G. H. Miller, ibid, Abstract No. 474.
47. T. Okuda, K. Ichise, T. Tanio and I. Saji, Proc.23rd ICAAC, Las Vegas 24-26 Oct.1983, Abstract No. 265B.
48. G. Petranyi, A. Stütz and U. Ganzinger, 13th ICC (1983)*, SY 49/1, Part 20, Abstract No. 5.
49. Poster Session, 13th ICC (1983)*, PS 4.8/4, Part 116, pp.1-60.
50. M. Koyama, H. Fukuyasu, N. Ezaki, K. Miyauchi, T. Tsuruoka, F. Kai, S. Inouye, T. Ito and T. Niida, Proc.23rd ICAAC, Las Vegas 24-26 Oct.1983, Abstract No. 217.
51. K. Kawai, R. Watanabe, Y. Nozawa, M. Nozaki, K. Tsurumi and H. Fujimura, Experientia, 39, 889 (1983).
52. W. Pache, M. Dreyfuss, R. Traber and H. Tscherter, 13th ICC (1983)*, PS 4.8/3, Part 115, Abstract No. 10.
53. G. W. Pye and M. S. Marriott, Sabouraudia, 20, 325 (1982).
54. P. Gadher, E. I. Mercer, B. C. Baldwin and T. E. Wiggins, Pestic.Biochem.Physiol., 19, 1 (1983).
55. H. D. Sisler, R. C. Walsh and B. N. Ziogas, Pestic.Chem.:Hum.Welfare Environ., 3, 129 (1983).
56. T. Kato, ibid, 3, 33 (1983).
57. D. Berg, K. H. Büchel and W. Kraemer, ibid, 3, 297 (1983).
58. R. C. Walsh and H. D. Sisler, Pestic.Biochem.Physiol., 18, 122 (1982).
59. J. F. Ryley, R. G. Wilson and K. J. Barrett-Bee, Sabouraudia, 22, 53 (1984).
60. F. R. Taylor, R. J. Rodrigue and L. W. Parks, Antimicrob.Ag.Chemother., 23, 515 (1983).
61. I. J. Sud and D. S. Feingold, ibid, 20, 71 (1981).
62. W. H. Beggs, IRCS Med.Sci.:Libr.Compend., 11, 262 (1983).
63. I. J. Sud and D. S. Feingold, Antimicrob.Ag.Chemother., 22, 470 (1982).
64. J. Cope, J.Gen.Microbiol., 119, 245 (1980).
65. W. H. Beggs, J.Antimicrob.Chemother., 11, 381 (1983).
66. H. Van den Bossche, J. M. Ruysschaert, F. Defrise-Quertain, G. Willemsens, F. Cornelissen, P. Marichal, W. Cools and J. Van Cutsem, Biochem.Pharmacol., 31, 2609 (1982).
67. M. Bastide, S. Jouvert and J. M. Bastide, Can.J.Microbiol., 28, 1119 (1982).
68. M. B. Negroni de Bonvehi, M. Van de Ven, M. Borgers and R. Negroni, Clin.Res.Rev., 3, 5 (1983).
69. K. Nishiki, D. Kudoh and N. Oshino, Shinkin to Shinkinsho, 23, 227 (1982). (C.A. 98, 212726q).
70. H. Yamaguchi, T. Hiratani, N. Suegara and K. Iwata, Chemotherapy (Tokyo), 30, 1439 (1982).
71. W. Raab and F. Högl, Mykosen, 25, 431 (1982).
72. K. Wilm and A. J. C. Stahl, Biochem.Pharmacol., 32, 1825 (1983).
73. J. Heeres, M. De Brabander, H. Van den Bossche in "Curr.Chemother.Immunother"., Vol. 2, P. Periti, G. G. Grassi, Eds., Am.Soc.Microbiol.Washington DC., 1982, p.1007.
74. A. Polak, Arzneim.-Forsch., 32, 17 (1982).
75. C. F. Louis, J. Turnquist and B. Jarvis, Cell Calcium, 4, 107 (1983).
76. A. M. Cheah, Experientia, 38, 445 (1982).
77. D. S. Loose, E. P. Stover and D. Feldman, J.Clin.Invest., 72, 404 (1983).
78. J. R. Little, A. Abegg, E. Plut, Cell.Immunol., 78, 224 (1983).
79. M. Borgers, H. Van den Bossche and M. De Brabander, Am.J.Med., 74(1B), 2 (1983).
80. B. Farkas and A. Dobozy, Mykosen, 26, 22 (1983).

81. F. Paltauf, G. Daum, G. Zuder, G. Högenauer, G. Schulz and G. Seidl, Biochim.Biophys.Acta, 712, 268 (1982).
82. J. G. Meingassner and U. B. Sleytr, Sabouraudia, 20, 199 (1982).
83. C. Brass, J. N. Galgiani, T. F. Blaschke, R. Defelice, R. A. O'Reilly and D. A. Stevens, Antimicrob.Ag.Chemother., 21, 151 (1982).
84. J. Scudamor and P. Tooley, Curr.Ther.Res.,Clin.Exp., 33, 917 (1983).
85. G. Cauwenbergh, V. Schuermans, M. Moens and A. De Dier, 13th ICC (1983)*, SY 49/2, Part 20, Abstract No. 3.
86. R. F. Hector and J. E. Domer, J.Infect.Dis., 147, 946 (1983).
87. E. O. Oji, Int.Ophthamol., 5, 163 (1982).
88. J. T. Henderson (Ed.), Gynäk.Rdsch., 23, Suppl.1, pp.1-61 (1983).
89. R. J. Hay, 13th ICC (1983)*, SY 49/2, Part 20, Abstract No. 4.
90. M. O'Neill (Ed.), Dermatologica, 166, Suppl.1, pp.1-33 (1983).
91. W. Ritter, K. Patzschke, U. Krause and S. Stettendorf, Chemotherapy (Basel), 28, Suppl.1, 37 (1982).
92. K. Iwata and K. Uchida, Chemotherapy (Tokyo), 30, 1216 (1982).
93. B. A. Zaro, K. M. Hama and M. D. Chaplin, Proc.West.Pharmacol.Soc., 25, 357 (1982).
94. M. L. De Natale, M. E. Page and G. D. Adamson, Clin.Res., 31, 36A (1983).
95. G. A. Roselle and P. Warth, J.Antimicrob.Chemother., 12, 189 (1983).
96. T. M. Kerkering, P. M. Schwartz, A. Espinel-Ingroff, P. J. Turek and R. B. Diasio, Antimicrob.Ag.Chemother., 24, 448 (1983).
97. J. R. Perfect and D. T. Durak, J.Infect.Dis., 146, 429 (1982).
98. I. J. Sud and D. S. Feingold, Antimicrob.Ag.Chemother., 23, 185 (1983).
99. G. Stuettgen and E. Bauer, Mykosen, 25, 74 (1982).
100. W. Petri, 13th ICC (1983)*, SY 49/2, Part 20, Abstract No. 1.
101. W. Dittmar, 13th ICC (1983)*, SY 49/2, Part 20, Abstract No. 2.
102. U. Ganzinger, A. Stephen and G. Gumhold, Clin.Trials J., 19, 342 (1982).
103. Anon, Scrip, 663, 1st.Feb., 4 (1982).
104. Anon, Scrip, 792, 9th.May, 7 (1983).
105. Anon, Scrip, 790, 2nd.May, 10 (1983).
106. P. A. J. Janssen and J. E. Symoens, Amer.J.Med., 74(1B), 80 (1983).
107. A. Pont, P. L. Williams, S. Azhar, R. E. Reitz, C. Bochra, E. R. Smith and D. A. Stevens, Arch.Intern.Med., 142, 2137 (1982).
108. A. Pont, P. L. Williams, D. S. Loose, D. Feldman, R. E. Reitz, C. Bochra and D. A. Stevens, Ann.Intern.Med., 97, 370 (1982).
109. J. M. Allen, D. J. Kerle, H. Ware, A. Doble, G. Williams and S. R. Bloom, Brit.Med.J., 287, 1766 (1983).
110. R. M. Ferguson, D. E. R. Sutherland, R. L. Simmons and J. S. Najarian, Lancet, ii, 882 (1982).
111. G. R. Morgenstern, R. Powles, B. Robinson and T. J. McElwain, Lancet, ii, 1342 (1982).

* Proc. 13th.Int.Cong.Chemother., Vienna 28th.Aug.-2nd.Sept., 1983, K. H. Spitz and K. Karrer, Eds.

Chapter 14. Antineoplastic Agents

Terrence W. Doyle

Bristol-Myers Pharmaceutical R&D Division,
P.O. Box 4755, Syracuse, N.Y. 13221

Introduction. The search for more selective antitumor agents through natural products screening, the synthesis of target-specific compounds and analogs of known agents, and by the use of prodrugs and novel delivery systems, continued at an increased pace in 1983. Reviews have appeared outlining the preclinical and early clinical data on 4'-epidoxorubicin[1] and mitoxantrone.[2] A comparison of the preclinical and early clinical results of four second generation analogs of cis-diamminedichloro platinum II (cis-DDP), has been published.[3] The solution chemistry of cis-DDP, its reactions with α-pyridone to form platinum "blues", and the implications of this chemistry with respect to the mechanism of action of platinum antitumor agents was reviewed.[4] A review of the mechanism of action of the quinone antibiotics (anthracyclines, mitomycins, streptonigrin, and saframycin) has been published.[5]

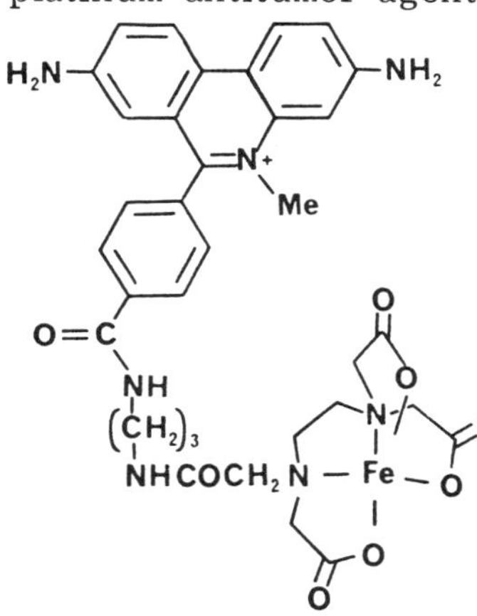

1

In 1983 further details of important new techniques for the determination of the binding sites of small molecules on naturally occurring DNA fragments were published. The two approaches to "footprinting" involve partial digestion of DNA-drug complexes using MPE Fe II, 1,[6,7] or DNAse I[8] followed by analysis of the sequence of those portions of the DNA protected by drug. The techniques have been compared and shown to give complementary information concerning drug-DNA interactions.[9] A recent review outlines the utility of the method.[10]

Antimetabolites - Two reports on the metabolism of methotrexate (MTX) to 7-hydroxy-methotrexate (7-OH MTX) have appeared. The major metabolite of MTX in rabbits was found to be 7-OH MTX. The clearance of 7-OH MTX from plasma was slower than that of MTX itself.[11] In a study of the uptake and metabolism of MTX in a MTX-sensitive human acute lymphoblastic leukemia cell line, MOLT4, it was shown that 7-OH MTX and MTX are converted to polyglutamates and that 7-OH MTX interferes with MTX accumulation and metabolism.[12] In Phase I clinical trials of 10-deazaaminopterin, marked leukemic cell kill was noted in CML.[13] Synthetic studies on the 1-deaza-7,8-dihydropteridines have helped to further define the SAR's in this system. Most modifications of the heterocycle or 4-amino function led to decreased activity; however, the substitution of a methyl group in position 7 led to increased activity.[14] A report on the mechanism of action of 5,8-dideazaisopteroylglutamate (IAHQ) has demonstrated that, unlike MTX, IAHQ polyglutamates inhibit thymidylate synthase.[15]

ISBN 0-12-040519-9

Further data concerning the mechanism of action of 2-β-D-ribofuranosylthiazole-4-carboxamide 2 (tiazofurin), in comparison with ribavirin and pyrazofurin have appeared. Marked inhibition of DNA and RNA synthesis by 2 was noted.[16] Synthesis of the selenium analog of 2 was also reported. This compound, 3, was ∿5 fold more potent than 2 in cell tissue culture (P388, L1210 cells) and showed good Lewis Lung Carcinoma (LLC) activity.[17] In order to overcome deactivation of ARA-C by cytidine deaminase, the preparation of 2'-O-nitro-1-β-D-arabinofuranosylcytosine (nitrara-C) was undertaken. Nitrara C was >15 times less susceptible to the enzyme and exhibited comparable antitumor activity in L1210 leukemia.[18] The synthesis of 3'-amino-2',3'-dideoxycytidine and its potent activity in L1210 leukemia was reported.[19] A comparison of the *in vivo* L1210 activity of sangivamycin with its 2'-deoxy, 3'-deoxy, and 5'-deoxy analogs revealed that only the 3'-deoxy analog retained the activity of the parent albeit at an enormous loss of potency.[20]

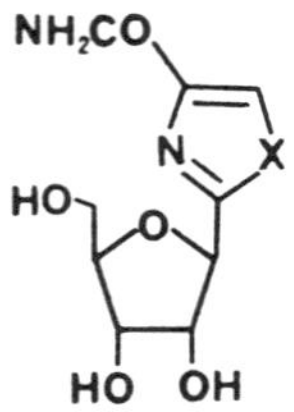

2 X=S
3 X=Se

Several papers have appeared detailing the mechanisms by which acivicin acts to reduce pyrimidine nucleotide content in affected cells. In rats inplanted with hepatoma 3924A (s.c.), acivicin acts to rapidly inactivate GMP synthetase, CTP synthetase, carbamoyl-phosphate synthetase II and amidophosphoribosyl transferase.[21] In independent studies it was shown that acivicin in combination with D-galactosamine produced a synergistic decrease in UTP content of AS-30D rat hepatoma cells.[22] It was suggested that the free carboxyl function of acivicin is essential to its ability to bind to γ-glutamyl transferase and by inference to other enzymes inactivated by the drug.[23] A report of the Phase II trial of acivicin in heavily pretreated metastatic breast cancer patients indicated no activity.[24]

Alkylating Agents - Phase I clinical trials of SOAz (1,3,3,5,5 pentakis(aziridino)-1λ^6,2,4,6,3λ^5,5λ^5 thiatriazadiphosphorine-1-oxide) have been completed. Cumulative myelosuppression was observed to be dose limiting-no responses were seen.[25] Metabolic studies on cyclophosphamide and ifosfamide have indicated the necessity for metabolic activation at C4. A number of putative metabolites of ifosfamide have been synthesized and demonstrated to be present in the urine of patients receiving drug.[26] The crystal structures of the isomeric 4-hydroperoxy ifosfamides 4 and 5 have been determined. In both isomers the 4-hydroperoxy group is axial.[27] The synthesis and *in vivo* evaluation of a 1,3-dipalmitoylglycerol ester of chlorambucil has been reported. The compound showed reduced toxicity and significantly greater activity against ip implanted P388 with oral dosing relative to chlorambucil. Evidence was presented for selective uptake into the lymphatic system.[28] Analysis of the activity of a series of nitrosoureas at the cellular level has shown that activity is not influenced by the structure of the N3 substituent, lipophilicity, or carbamoylating activity. The observed differences in *in vivo* bioactivity can most likely therefore be

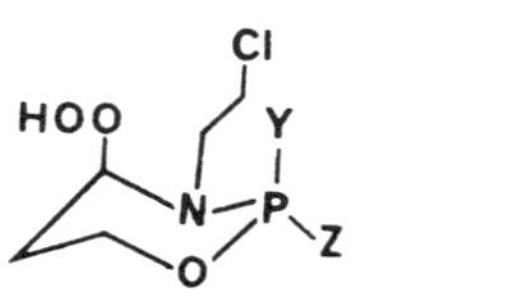

4 Y=O Z=NH⌄⌃Cl
5 Y=NH⌄⌃Cl Z=O

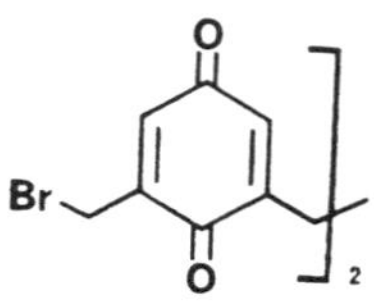

6

explained by differences in biodistribution.[29] The synthesis of a bis bioreductive alkylating agent 6 has been reported.[30] The compound is inactive in cell tissue culture but exhibits good *in vivo* activity in mice inoculated with D98/HRI cells. The authors propose that the compound is bioactivated by reduction in the hypoxic core of the tumor.

Metal Complexes - The Fourth International Symposium on Platinum Coordination Complexes in Cancer Chemotherapy covered various aspects of the synthesis, biochemistry, pharmacology, toxicology and clinical evaluation of platinum containing drugs.[31] A study correlating the structure of platinum complexes with activity, toxicity, and lability has appeared.[32] It was shown in cultured L1210 cells that growth inhibitory action is correlated with the nature of the inert ligands and the toxicity is correlated with the rate of hydrolysis of the labile ligands. A study reported the synthesis and activity of a series of complexes containing 2,3-diaminopropanol as the amine ligand. Both the R and S complexes were prepared and their activities compared.[33] The R complexes were generally more active and potent than the S complexes. It has been generally accepted that the ability of *cis*-DDP to interact with DNA and to form crosslinks is responsible for the observed antitumor effects. Recently the ability of both *cis*-DDP and *trans*-DDP to inhibit DNA synthesis in cultured L1210 cells has been studied.[34] It was found that the ability of the *trans* isomer to penetrate the cell is 8 fold greater than that of the *cis* isomer and that the ability of both *cis*- and *trans*-DDP to inhibit DNA replication is approximately equal suggesting that quantitative DNA synthesis inhibition is not the whole story. A study of *cis*- and *trans*- DDP on RNA transcription has also indicated that the *trans* isomer is more potent *in vitro* and more toxic *in vivo*[35] This was interpreted as implying that *cis*-DDP may react with a highly specific DNA configuration. The use of terbium fluorescence as a probe of the interaction of *cis*- and *trans*-DDP with naturally occurring DNA fragments has indicated that *cis*-DDP causes the DNA-helix to open up to permit terbium-guanine interaction to occur.[36] The *trans* isomer has no effect. Some light may have been shed on the remarkable synergism seen between *cis*-DDP and bleomycin. Treatment of two restriction fragments of pBR322 DNA with *cis*-DDP resulted in the activation of several new cutting sites in GC rich regions of the DNA for bleomycin.[37] One concern of *cis*-DDP therapy has been the nephrotoxicity seen in clinical use and the subsequent necessity to hydrate patients. A report of the Phase I clinical trial of *cis*-dichloro-*trans*-dihydroxy-*bis*-isopropylamine platinum IV (CHIP) has shown that this platinum IV analog of *cis*-DDP was devoid of significant nephrotoxicity as predicted by preclinical models.[38]

A number of reports of non-platinum antitumor metal complexes have appeared. A comparison of a number of rhodium I, iridium I and ruthenium II complexes with *cis*-DDP in an LLC metathesis model showed that rhodium and ruthenium complexes were as effective as *cis*-DDP in reducing lung metastases.[39] The localization of titanium and vanadium metallocenes by electron energy loss spectroscopy within Ehrlich ascites tumor cells has been reported.[40] The drugs accumulate in the nuclear heterochromatin. A new class of titanium containing compounds was reported. Compounds of general formula 7 (X=F, Cl, Br) were shown to be curative in L1210 leukemia.[41] The results of a Phase II trial of gallium nitrate in malignant lymphoma using

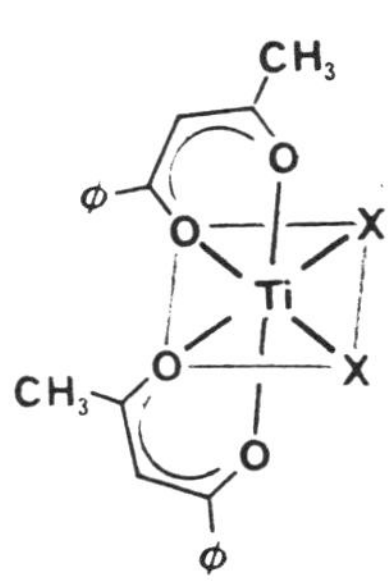

7

heavily pretreated patents showed an 18% response rate.[42] The use of sodium hexachloroiridate to treat mouse ovarian tumors has been reported to be curative.[43]

Natural Products - Research on natural products continued to supply interesting new lead structures in 1983. In addition, work on the mechanism of action of natural products has provided new insights useful in analog design. Two papers have appeared which will be quite useful to those responsible for the discovery and evaluation of new agents. The biochemical prophage induction assay has been developed as a useful prescreen for antitumor agents and has been adapted for bioautography.[44] A quite useful compilation of data on the properties of 59 antitumor agents relevant to in vitro human tumor stem cell assays has appeared.[45]

The interaction of bleomycin with metal ions and the chemistry associated with this interaction has been the focus of numerous studies aimed at elucidating the detailed mechanism of action of this drug. A recent paper notes that the interaction of bleomycin with iron gives a reactive intermediate which cleaves DNA.[46] Ferrous ion-bleomycin interaction results in chemiluminescence suggesting that the activated intermediate may be electronically excited. Mossbauer studies of the ferrous and ferric bleomycin complexes indicated that the latter complex may exist in two distinct conformations.[47] In a study of cobalt-bleomycin complexes, the existence of a cobalt-bleomycin-hydroperoxide complex capable of nicking DNA has been demonstrated.[48]

The isoquinolinequinones related to saframycin have generated a good deal of interest in the past few years. A review covering isoquinolinequinones from both marine and fermentation sources has appeared.[49] Several papers reporting saframycin-like antitumor agents having reduced quinone rings, or precursors to quinone rings, have been published. Details of the structure of saframycin R 8 and its interaction with

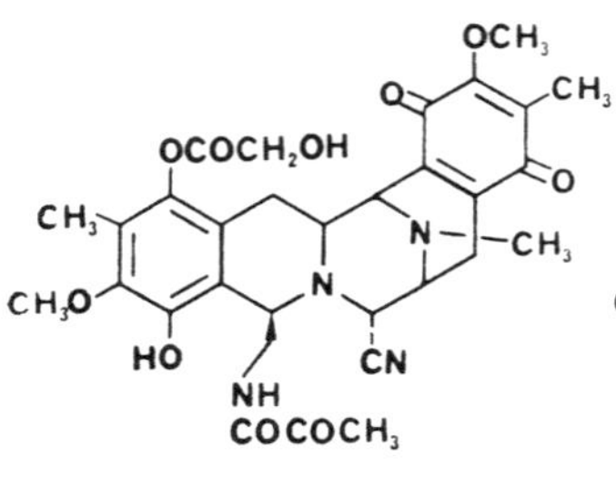

8

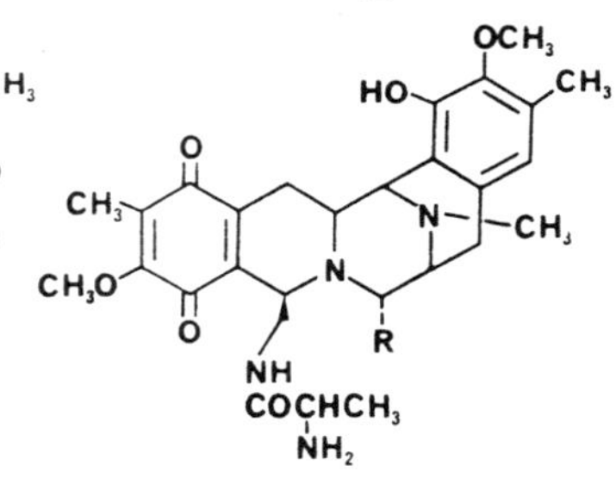

9

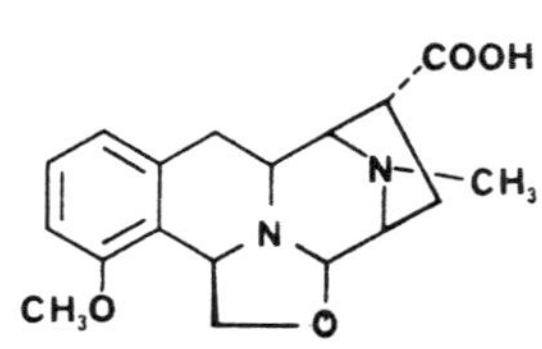

10

DNA suggest that cleavage of the glycolic acid function may occur to activate the drug.[50] The structures of two new compounds related to saframycin, i.e. safracin A (R=H) and B (R=OH) 9 have been reported.[51a] Safracin A and B are both active on P388 leukemia (T/C 150 at 50 mg/kg and T/C at 170 1.0 mg/kg respectively).[51b] The highly simplified saframycin-like DC-52 10 has also been isolated (T/C 147 at 12.5 mg/kg in P388).[52]

A comparison of the antitumor activities of mitomycin C and M-83 (7-N-(-p-hydroxyphenyl)mitomycin C) has appeared.[53] Compound M-83 is presently in Phase I clinical trials in Japan.The results of a structure-activity study of β-substituted 7-ethylamino mitomycins have been published. Two analogs, 7-N-β-fluoroethyl mitomycin C and 7-N-β-mercapto-ethyl mitomycin C, gave exceptionally good activity in the P388 and B16 murine models.[54] The products of reductively activated mitomycin C with d(GpC) have been isolated.[55] The major product was derived from reaction of the O-6 of guanosine at the 1 position of the mitosene derived from mitomycin C. The question of mitomycin C activation by the hypoxic component of tumors has received a good deal of attention. A recent paper concludes that while mitomycin C is active against hypoxic cells, it has only minor specificity for hypoxic as compared with aerobic cells *in vivo*.[56] The absolute configuration of mitomycin C has been revised based on x-ray determination of the structure of 1-N-(p-bromobenzoyl)-mitomycin C.[56]

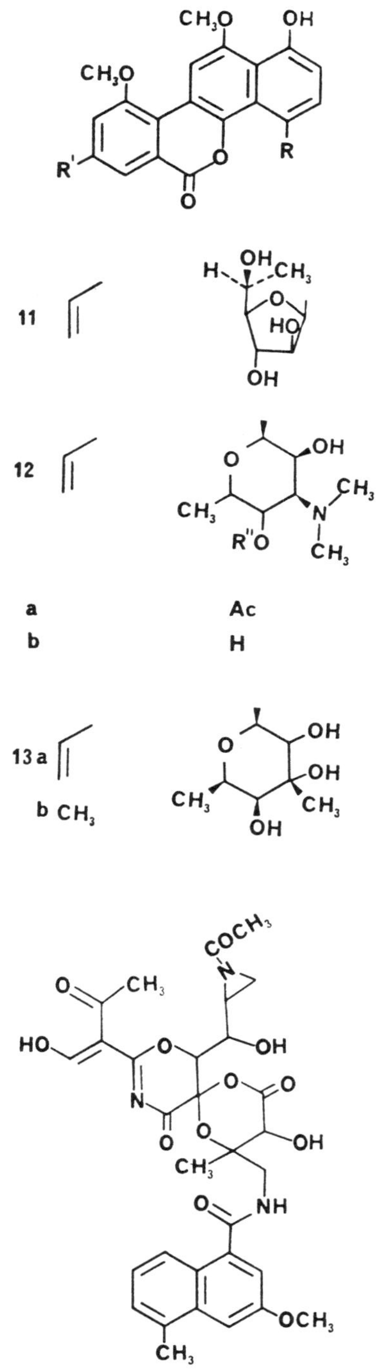

14

Research into the nature of the chromophore(s) associated with the chromopeptides auromomycin and neocarzinostatin has resulted in further insights into the mechanism of action of these molecules. It has been shown that the DNA cleaving action of auromomycin and its chromophore is antagonized by ethidium binding[57] and further that a free radical may be involved in the DNA breakage.[58] A partial structure has been proposed for the chromophore of neocarzinostatin which accounts for all the heteroatoms and proposes the presence of a strained oxygen ring (possibly an epoxide).[59] Mechanistic studies on CC-1065 (7-[[1,6-dihydro-4-hydroxy-5-methoxy-7-[(4,5,8,8atetrahydro-7-methyl-4-oxocyclopropa[c]benzo[1,2-b:4,3-b']dipyrrol-2(-1H)-yl)carbonyl]benzo[1,2-b:4,3-b']dipyrrol3(2H)-yl]carbonyl]-1,6-dihydro-4-hydroxy-5-methoxy-benzo[1,2-b:4,3-b']dipyrrole-3(2H)-carboxamide) focusing on cell cycle effects have indicated a slowing through S phase and blockage of cell progression from G_2 to M.[60] Two papers have appeared which demonstrate that the DNA strand scission is responsible for the cytotoxicity of VP-16 (4'-demethylepipodophyllotoxin-9-(4,6-O-ethylidene-β-D-glucopyranoside)).[61,62] Using alkaline-elution techniques both single and double strand breaks were detected.[61]

The biosynthesis of gilvocarcin V 11 has been shown to proceed through a polyketide chain with the vinyl group arising from propionate.[63] Details of the fermentation, isolation and initial biological evaluation of ravidomycin 12a have been published.[64,65] The isolation of virenomycin V and M 13a,b from Streptomyces albaduncus and their P388 activity were presented.[66] Interestingly, both the vinyl and methyl substituted compound showed good P388 activity.

A revised structure for carzinophilin (14) has been proposed based upon a revision of the molecular weight as determined by molecular secondary-ion mass spectrometry.[67] A screening program of extracellular polysaccharides of marine bacteria against sarcoma-180 in mice has led to the discovery of marinactin which is a heteroglycan consisting of glucose, mannose and fucose (7:2:1).[68] Marinactan gave 70-90% inhibition of growth in solid sarcoma 180 when dosed at 10-50 mg/kg on a QD 1→10 schedule. The structure determination of kijanimicin, a novel tetronic acid, was reported.[69] Recently it has been shown to have modest activity in P388 and B16 tumor models.[70]

Anthracyclines - Two reports of Phase II clinical trials with 4'-epidoxorubicin (4'-ED) in a variety of malignancies have appeared. The response rate in previously untreated breast cancer patients was 54%. Responses were also seen in endometrial carcinoma, cervical carcinoma, non-Hodgkins lymphoma, melanoma, and skin cancer.[71] Only a single partial responder was seen in 34 evaluable cases of non-small cell lung cancer.[72]

The earlier discovery that replacement of the 3'-amino group by a morpholine in daunomycin gave a compound significantly more potent than the parent drug has led to an investigation of the cellular pharmacology of 3'-desamino-3'-morpholinodaunorubicin (3'-MD) in human colon carcinoma cells. It has been found that 3'-MD preferentially inhibits RNA synthesis vs DNA synthesis[73] and the cellular uptake of 3'-MD is approximately 15 times faster than that of adriamycin.[74] Despite the greater bioavailability 3'-MD was less cytotoxic in vitro than adriamycin. Replacement of the 3'-amino function of adriamycin by a cyano-morpholine group led to a compound 15 which has 600 times the potency of adriamycin in P388 leukemia.[75] The antitumor efficacy of 15 and adriamycin are comparable in P388 with 15 being less active in L1210 and B16 melanoma. The quinomethide tautomer 16 has previously been proposed as a reactive intermediate capable of alkylating biological macromolecules. Evidence for the existence of 16 has been obtained via trapping experiments.[76] The cardiotoxicity of anthracycline antitumor agents has been postulated to be due to special biochemical conditions occurring in cardiac tissue which favor an anthraquinone catalyzed electron shuttle to H_2O_2 resulting in the generation of OH radicals.[77] A study of liposome-encapsulated adriamycin has indicated that less cardiotoxicity is associated with this formulation.[78]

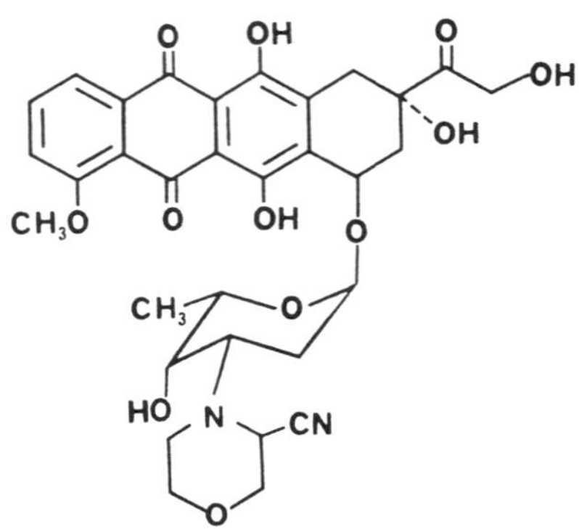

15

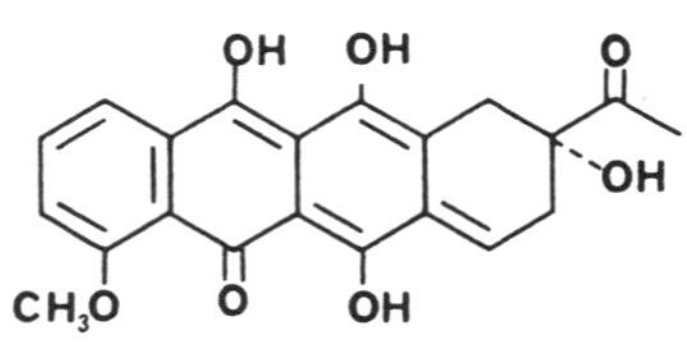

16

Two new anthracyclines related to nogalamycin have been reported.

Arugomycin has seven sugar residues glycosidically attached at the 7 and 4' positions on the aglycone.[79] Decilorubicin has four sugar residues and is missing the phenolic hydroxy at C4 and the C10-carbomethoxy function.[80] Both compounds contain an unusual nitrosugar in the sugar chain.

Analog studies based on the anthracyclines have resulted in the discovery of mitoxantrone (DHAQ) and bisantrene. A comparison of the cytotoxicities of several analogs related to DHAQ and bisanthrene appeared.[86] Experimental and computer graphics studies of DHAQ gave results consistent with DHAQ being an intercalative binder to DNA.[81] A comparative study of daunorubicin and DHAQ has shown that DHAQ does not cause lipid peroxidation in contrast to daunorubicin.[82] The authors postulate that this may explain the relatively low cardiotoxicity of DHAQ. It was shown that the phenol groups of DHAQ influence both drug uptake and interaction with DNA.[83] The 1,2-dihydroxy analog was equiactive with DHAQ although approximately 8-fold less potent at the optimum dose. Methylation of the phenol functions led to a substantial loss in potency.[84] Reports of a clinical trial of bisanthrene in metastatic breast cancer revealed a 21% partial response rate in heavily pretreated patients.[85]

Miscellaneous Synthetic Agents - Several papers have appeared on the mechanism of action and synthesis of new AMSA (4'-[(9-acridinyl)amino]-methanesulphon-m-anisidide) analogs. Evidence has been advanced that mAMSA intercalation into DNA induces a type II topoisomerase to break the DNA.[86] The use of known topoisomerase inhibitors, i.e. novobiocin and coumermycin, negated the DNA cleavage activity of mAMSA. A series of 3',5'-substituted analogs had lower binding constants with DNA and were inactive.[87] The synthesis of a series of 3-substituted-5-carboxamido analogs of mAMSA has helped to further define the SAR's of this interesting class of AMSA derivatives.[88]

The mechanism of action of a number of new analogs of ellipticine has been the subject of several papers. Activation of (9-OH-NME$^+$) 17 by a peroxidase-H_2O_2 system gives the quinone imine 18 (R=H) and eventually the orthoquinone 19.[89] In the presence of nucleophiles, substitution of H10 was observed leading to the formation of adducts 18 (R= methoxy, cysteinyl, pyridinium).[90] Compound 18 has been shown to be mutagenic in Salmonella.[91] A second ellipticine analog BD40 (20) has good antitumor activity but is not mutagenic in yeast.[92] Modifications of 20 in the sidechain or intercalating heterocycle failed to give improvements in the in vivo activity.[93]

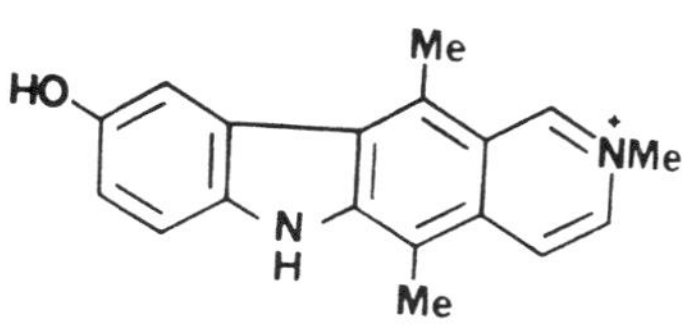

17

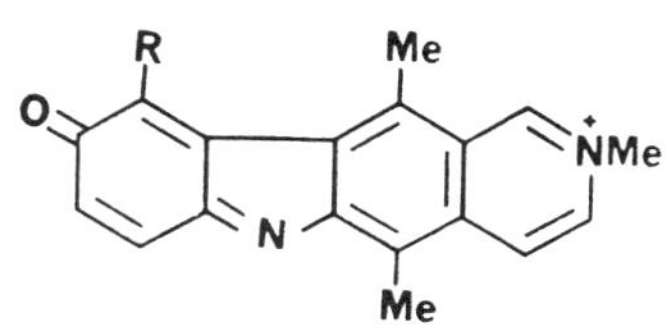

18

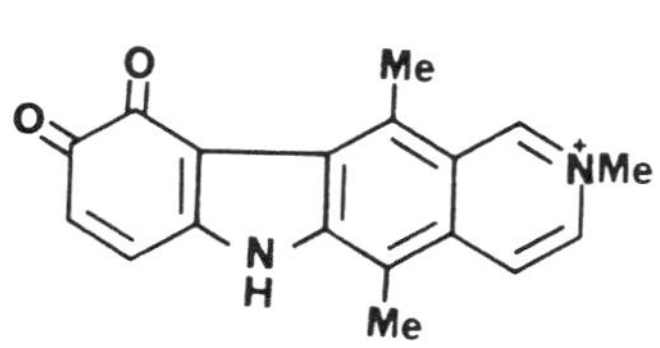

19

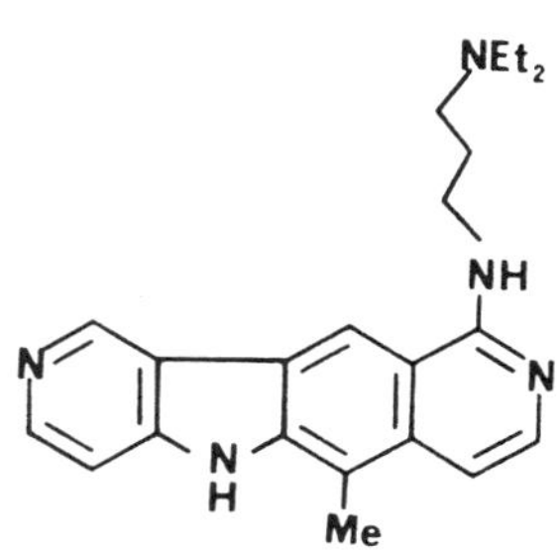

20

The role of retinoids in differentiation and carcinogenesis was the subject of a review article.[94] Results of a clinical trial of the oral retinoid Ro13-6298 (21a) in the treatment of solar keratoses and squamous cell epithelioma have been published.[95] Ro13-6298 produced a reduction in the number of lesions. The results of an SAR study of compounds related to 21 was published.[96] When tested *in vitro* in the differentiation assays using murine F9 teratocarcinoma cells and human HL-60 promyelocytic leukemia cells, compound 21 was more active than its 3-methyl analog in the F9 assay while the order of activity was reversed in the HL60 assay.

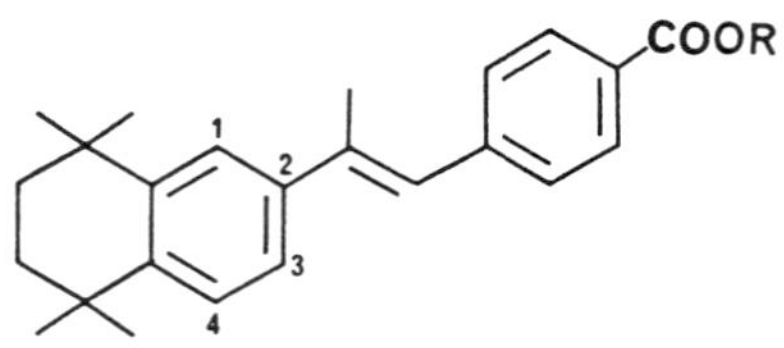

21 a R=Et
b R=H

References

1. F. Ganzina, Cancer Treat. Reviews, 10, 1-22 (1983).
2. I.E. Smith, Cancer Treat. Reviews, 10, 103-115 (1983).
3. F.H. Lee, R. Canetta, B.F. Issell, and L. Lenaz, Cancer Treat. Reviews, 10, 39-51 (1983).
4. S.J. Lippard, Science, 218, 1075-1082 (1982).
5. J.W. Lown, Mol. and Cellular Biochem., 55, 17-40 (1983).
6. M.W. Van Dyke and P.B. Dervan, Cold Spring Harbor Symp. on Quant. Biol., 347-353 (1982).
7. M.W. Van Dyke and P.B. Dervan, Biochemistry, 22, 2373-2377 (1983).
8. M.J. Lane, J.C. Dabrowiak, and J.N. Vournakis, Proc. Natl. Acad. Sci., USA, 80, 3260-3264 (1983).
9. M.W. Van Dyke and P.B. Dervan, Nucleic Acids Res., 11, 5555-5567 (1983).
10. J.C. Dabrowiak, Life Sciences, 32, 2915-2931 (1983).
11. K. Sasaki, R. Hosoya, Y-M. Wang, and G.L. Raulston, Biochem. Pharm., 32, 503-507 (1983).
12. G. Fabre, L.H. Matherly, R. Favre, J. Catalin, and J. Cano, Cancer Res., 43, 4648-4652 (1983).
13. V.E. Currie, R.P. Warrell, Jr., Z. Arlin, C. Tan, F.M. Sirotnak, G. Greene, and C.W. Young, Cancer Treat. Rep., 67, 149-154 (1983).
14. C. Temple, Jr., G.P. Wheeler, R.D. Elliott, J.D. Rose, R.N. Comber, and J.A. Montgomery, J. Med. Chem., 26, 91-95 (1983).
15. D.J. Fernandes, J.R. Bertino, and J.B. Hynes, Cancer Res., 43, 1117-1123 (1983).
16. P.P. Saunders, R. Kuttan, M.M. Lai, and R.K. Robins, Mol. Pharm., 23, 534-539 (1983).
17. P.C. Srivastara and R.K. Robins, J. Med. Chem., 26, 445-448 (1983).
18. T.L. Chwang, A. Fridland, T.L. Avery, J. Med. Chem., 26, 280-283 (1983).
19. T-S. Lin and W.R. Mancini, J. Med. Chem., 26, 544-548 (1983).
20. T. Maruyama, L.L. Wotring and L.B. Townsend, J. Med. Chem., 26, 25-29 (1983).
21. G. Weber, 48.Konferenz der Gesellschaft für Biologische Chemie, 364, 1254 (1983).
22. K. Leube and D.O.R. Keppler, Biochem. Pharm., 32, 1865-1869 (1983).
23. C.S. Schasteen, N.P. Curthoys, and D.J. Reed, 112, 564-570 (1983).
24. G. Fleishman, H-Y. Yap, W.K. Murphy, and G. Bodey, Cancer Treat. Rep., 67, 843-844 (1983).
25. S. Rodenhuis, N.H. Mulder, D.T. Sleijfer, H.S. Koops, and J.C. van de Grampel, Cancer Chemother. Pharmacol., 10, 178-181 (1983).
26. K. Misiura, A. Okruszek, K. Pankiewicz, W.J. Stec, Z. Czownicki, and B. Utracka, J. Med. Chem., 26, 674-679 (1983).
27. A. Camerman, H.W. Smith, and N. Camerman, J. Med. Chem., 26, 679-683 (1983).
28. A. Garzon-Aburbeh, J.H. Poupaert, M. Claesen, P. Dumont, and G. Atassi, J. Med. Chem., 26, 1200-1203 (1983).
29. R.J. Weinkam and M.E. Dolan, J. Med. Chem., 26, 1656-1659 (1983).
30. D.T. Witiak, P.L. Kamat, D.L. Allison, S.M. Liebowitz, R. Glaser, J.E. Holliday, M.L. Moeschberger, and J.P. Schaller, J. Med. Chem., 26, 1679-1683 (1983).
31. Plat. Coord. Comp. in Cancer Chemother., M.P. Hacker, E.B. Douple, and I.H. Krakoff, Eds., Martinus Nijhoff Publishing Co. (1984).
32. J-P. Macquet and J.L. Butour, J. Natl. Can. Inst., 70, 899-905 (1983).
33. M. Noji, S. Motoyama, T. Tashiro, and Y. Kidani, Chem. Pharm. Bull., 31, 1469-1473 (1983).
34. B. Salles, J.L. Butour, C. Lesca, and J.P. Macquet, Biochem. Biophys. Res. Comm., 112, 555-563 (1983).

35. D. Miller, D.M.A. Minahan, M.E. Friedman, H.H. Kohl, and C.A. McAuliffe, Chem.-Biol. Interactions, 44, 311-316 (1983).
36. M. Arquilla, L.M. Thompson, L.F. Pearlman, and H. Simpkins, Cancer Res., 43, 1211-1216 (1983).
37. P.K. Mascharak, Y. Sugiura, J. Kuwahara, T. Suzuki, and S.J. Lippard, Proc. Natl. Acad. Sci. USA, 80, 6795-6798 (1983).
38. P.J. Creaven, S. Madajewicz, L. Pendyala, A. Mittelman, E. Pontes, M. Spaulding, S. Arbuck, and J. Solomon, Cancer Treat. Rep., 67, 795-800 (1983).
39. G. Sava, T. Giraldi, G. Mestroni, and G. Zassinovich, Chem.-Biol. Interactions, 45, 1-6 (1983).
40. P. Köpf-Maier and D. Krahl, Chem.-Biol. Interactions, 44, 317-328 (1983).
41. H.J. Keller, B. Keppler, and D. Schmahl, J. Cancer Res. Clin. Oncol., 105, 109-110 (1983).
42. J.K. Weick, R.L. Stephens, L.H. Baker, and S.E. Jones, Cancer Treat. Rep., 67, 823-825 (1983).
43. G.M. Kolodny and C. Rose, Cancer Res., 43, 3101-3103 (1983).
44. R.K. Elespuru and R.J. White, Cancer Res., 43, 2819-2830 (1983).
45. E.J. Pavlik, D.E. Kenady, J.R. van Nagell, K. Keaton, M.B. Hanson, E.S. Donaldson, W.O. Griffen, and R.C. Flanigan, Cancer Chemother. Pharmacol., 11, 8-15 (1983).
46. M.A. Trush, E.G. Mimnaugh, Z.H. Siddik, and T.E. Gram, Biochem. Biophys. Res. Comm., 112, 378-383 (1983).
47. B. Balko and G.W. Liesegang, Biochem. Biophys. Res. Comm., 110, 827-836 (1983).
48. C-H. Chang, J.L. Dallas and C.F. Meares, Biochem. Biophys. Res. Comm., 110, 959-966 (1983).
49. T. Arai and A. Kubo, "The Alkaloids," Vol. XXI, Academic Press, New York, N.Y., 1983, p. 56-100.
50. J.W. Lown, C.C. Hanstock, A.V. Joshua, T. Arai, and K. Takahashi, J. Antibiotics, 36, 1184-1194 (1983).
51a. Y. Ikeda, H. Matsuki, T. Ogawa, and T. Munakata, J. Antibiotics, 36, 1284-1289 (1983).
51b. Y. Ikeda, Y. Shimada, K. Honjo, T. Okumoto, and T. Munakata, J. Antibiotics, 36, 1290-1295 (1983).
52. K. Takahashi and F. Tomita, J. Antibiotics, 36, 468-470 (1983).
53. R. Imai and M. Morimoto, J. Antibiotics, 36, 559-565 (1983).
54. B.S. Iyengar, S.M. Sami, W.A. Remers, W.T. Bradner, and J.E. Schurig, J. Med. Chem., 26, 16-20 (1983).
55. M. Tomasz, R. Lipman, J.K. Snyder, and K. Nakanishi, J. Am. Chem. Soc., 105, 2059-2063 (1983).
56a. A.M. Rauth, J.K. Mohindra, and I.F. Tannock, Cancer Res., 43, 4154-4158 (1983).
56b. K. Shirahata and N. Hirayama, J. Am. Chem. Soc., 105, 7199-7200 (1983).
57. H. Suzuki and N. Tanaka, J. Antibiotics, 36, 575-582 (1983).
58. H. Suzuki, Y. Kirino, and N. Tanaka, J. Antibiotics, 36, 583-587 (1983).
59. O.D. Hensens, R.S. Dewey, J.M. Liesch, M.A. Napier, R.A. Reamer, J.L. Smith, G. Albers-Schönberg, and I.H. Goldberg, Biochem. Biophys. Res. Comm., 113, 538-547 (1983).
60. B.K. Bhuyan, S.L. Crampton, and E.G. Adams, Cancer Res., 43, 4227-4232 (1983).
61. A.J. Wozniak and W.E. Ross, Cancer Res., 43, 120-124 (1983).
62. D.K. Kalwinsky, A.T. Look, J. Ducore, and A. Fridland, Cancer Res., 43, 1592-1597 (1983).
63. K. Takahushi and F. Tomita, J. Antibiotics, 36, 1531-1535 (1983).
64. S.N. Sehgal, H. Czerkawski, A. Kudelski, K. Pandev, R. Saucier, and C. Vezina, J. Antibiotics, 36, 355-361 (1983).
65. S. Rakhit, C. Eng, H. Baker and K. Singh, J. Antibiotics, 36, 1490-1494 (1983).
66a. J.A. Matson, R.W. Myllymaki, T.W. Doyle, W.T. Bradner, and J.A. Bush, Proceedings of Symposium on Discovery and Development of Naturally Occurring Antitumor Agents, (1983), NCI, Frederick Cancer Res. Facility.
b. J.A. Bush, J.A. Matson, T.W. Doyle, and W.T. Bradner, Proc. XXIII ICAAC, Abst. 218 (1983).
67. M. Onda, Y. Konda, A. Hatano, T. Hata, and S. Omura, J. Am. Chem. Soc., 105, 6311-6312 (1983).
68. H. Umezawa, Y. Okami, S. Kurasawa, T. Ohnuki, M. Ishizuka, T. Takeuchi, T. Shiio, and Y. Yugari, J. Antibiotics, 36, 471-477 (1983).
69. A.K. Mallams, M.S. Puar, R.R. Rossman, A.T. McPhail, R.D. Macfarlane, and R.L. Stephens, J. Chem. Soc. Perkin Trans. I, 1497-1534 (1983).
70. W.T. Bradner, C.A. Claridge, J.B. Huftalen, J. Antibiotics, 36, 1078-1079 (1983).
71. P. Hurteloup, P. Cappelaere, J.P. Armand, and G. Mathé, Cancer Treat. Rep., 67, 337-341 (1983).
72. L.A. Kalmon, M.G. Kris, R.J. Gralla, D.P. Kelsen, E.S. Casper, R.T. Heelan, and R.E. Wittes, Cancer Treat. Rep., 67, 591-592 (1983).

73. J.B. Johnston and R.I. Glazer, Cancer Res., 43, 1606-1610 (1983).
74. J.B. Johnston and R.I. Glazer, Cancer Res., 43, 1044-1048 (1983).
75. E.M. Acton, G.L. Tong, and R.L. Wolgemuth, Proc. Am. Ass. Cancer Res., 24, 252 (1983).
76. D.L. Kleyer and T.H. Koch, J. Am. Chem. Soc., 105, 5154-5155 (1983).
77. H. Nohl and W. Jordan, Biochem. Biophys. Res. Comm., 114, 197-205 (1983).
78. E.H. Herman, A. Rahman, V.J. Ferrans, J.A. Vick, and P.S. Schein, Cancer Res., 43, 5427-5432 (1983).
79. H. Kawai, Y. Hayakawa, M. Nakagawa, K. Imamura, K. Tanabe, A. Shimazu, H. Seto, N. Ōtake, J. Antibiotics, 36, 1569-1571 (1983).
80. K. Ishii, S. Kondo, Y. Nishimura, M. Hamada, T. Takeuchi, H. Umezawa, J. Antibiotics, 36, 451-453 (1983).
81. S.A. Islam, S. Neidle, B.M. Gandecha, and J.R. Brown, Biochem. Pharmacol., 32, 2801-2808 (1983).
82. L.H. Patterson, B.M. Gandecha, and J.R. Brown, Biochem. Biophys. Res. Comm., 110, 399-405 (1983).
83. A. Nishio and E.M. Uyeki, Cancer Res., 43, 1951-1956 (1983).
84. K. Nakamura, T. Asai, T. Hara, N. Minorikawa, and T. Suami, Bull. Chem. Soc. Jpn., 56, 1812-1816 (1983).
85. H-Y. Yap, B-S. Yap, G.R. Blumenschein, B.C. Barnes, F.C. Schell, and G.P. Bodey, Cancer Res., 43, 1402-1404 (1983).
86. B. Marshall, S. Darkin, and R.K. Ralph, FEBS Letters, 161, 75-78 (1983).
87. W.A. Denny, G.J. Atwell, and B.C. Baguley, J. Med. Chem., 26, 1625-1630 (1983).
88. W.A. Denny, G.J. Atwell, and B.C. Baguley, J. Med. Chem., 26, 1619-1625 (1983).
89. J. Bernadou, G. Meunier, C. Paoletti, and B. Meunier, J. Med. Chem., 26, 574-579 (1983).
90. G. Meunier, B. Meunier, C. Auclair, J. Bernadou, and C. Paoletti, Tetrah. Letters., 24, 365-368 (1983).
91. D.M. DeMarini, S. Cros, C. Paoletti, P. Lecointe, and A.W. Hsie, Cancer Res., 43 3544-3552 (1983).
92. E. Moustacchi, V. Favaudon, and E. Bisagni, Cancer Res., 43, 3700-3706 (1983).
93. C. Rivalle, F. Wendling, P. Tambourin, J-M. Lhoste, E. Bisagni, and J-C. Chermann, J. Med. Chem., 26, 181-185 (1983).
94. M.B. Sporn and A.B. Roberts, Cancer Res., 43, 3034-3040 (1983).
95. T. Kingston, S. Gaskell, and R. Marks, Eur. J. Cancer Clin. Oncol., 19, 1201-1205 (1983).
96. S. Strickland, T.R. Breitman, F. Frickel, A. Nurrenbach, E. Hadicke, and M.B. Sporn, Cancer Res., 43, 5268-5272 (1983).

Chapter 15. Antiparasitic Agents

Bernard J. Banks,
Pfizer Central Research, Sandwich, Kent, England.

General: A useful introduction to parasite chemotherapy has been published.[1] The target sites, modes of action and mechanisms of the selective toxicities of drugs used in the treatment of protozoan, platyhelminth and nematode infections are clearly presented in tabular form. A fuller account of human parasitic diseases has also appeared.[2] Advances in the knowledge of human intestinal protozoan and helminth infections have been reviewed[3] and some current problems, rational approaches and recent progress in antiparasitic chemotherapy were concisely discussed.[4] Objectives and priorities for the control of internal parasites of ruminants in the U.S. have been reported.[5] Proceedings of the Fifth International Congress of Pesticide Chemistry have been published.[6]

HELMINTH AND ECTOPARASITE INFECTIONS

Avermectins: The screening strategy that led to discovery of the avermectins, their fermentation development and the utility of the major component of the complex, avermectin B_{1a} (1) has been described.[7] Summarized information on ivermectin (22,23-dihydroavermectin B_1) as an agent for the control of animal endo- and ecto-parasites has been updated.[8] Patents have appeared describing new avermectins obtained by fermentation,[9] microsomal oxidation of earlier compounds[10] and semi-synthesis.[11]

	R_5	R_{13}	R_{25}	R	X-Y
1	H	A	sBu	H	CH=CH
2	H	A	sBu	OH	CH=CH
3	H	A	sBu	H	Z
4	H	A	sBu	H	CH_2CO
5	Me	A	sBu	H	Z
7	H	B	sBu	H	CH=CH
8	Me	B	sBu	H	Z
9	H	A	iPr	H	CH_2CH_2
10	H	H	iPr	H	CH_2CH_2

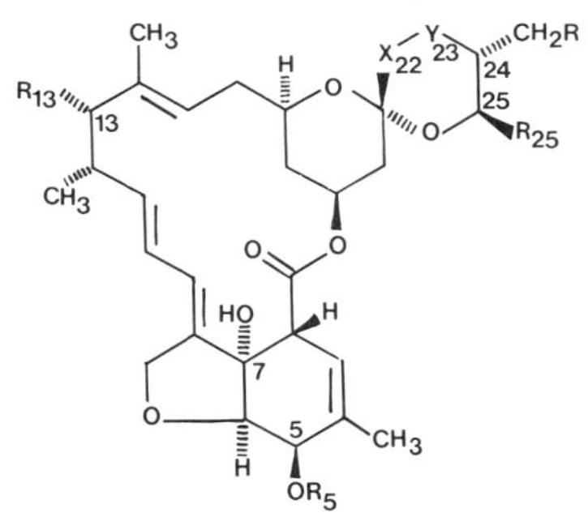

A = S, $R_{4''}$ = H
B = S, $R_{4''}$ = Ac

S = 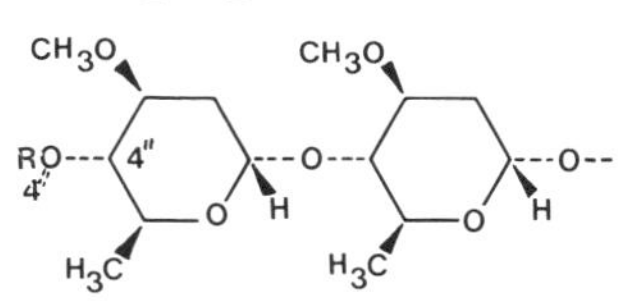

Z = CH_2-CH(OH)

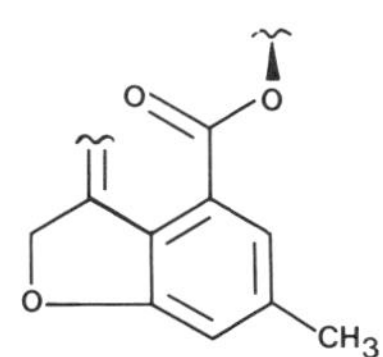

6

The metabolism of [^{3}H]avermectin B_{1a} by steer or rat liver microsomes has been studied.[12] The major polar metabolites accounting for 50-80% of the total radioactivity were identified as 2 and its monosaccharide. 22,23-Dihydroavermectin B_{1a} gave a similar result. 22,23-Dihydro-2 can also be recovered from livers of steers given ivermectin subcutaneously 14 days previously.[13] Soil microbial oxidation of avermectin B_{2a} (3), recently

ISBN 0-12-040519-9

reported[13] effective against the root-knot nematode, Meloidogyne incognita, yields the ketone 4. This metabolite, also nematicidal, was shown to be more persistent than 3 and is responsible for the prolonged activity of 3 when applied to soil.[14] The anthelmintic activities and potencies of acyl derivatives of avermectins have been reported.[15] Of the possible mono- and di-acetates of avermectins B_{1a} and A_{2a} (5) only the 4"-O-acetates 7 and 8 retained the high potency of 1 and 5 against a Trichostrongylus colubriformis infection in gerbils. Acylation of the 5- or 23-hydroxyl groups resulted in substantial reductions in potency whilst the aromatic derivative 6 was inactive. The potencies of further 4"-O-acylated materials decreased with increasing lipophilicity of the substituent. No advantage for 7 over 1 was found when given orally to sheep infected with Haemonchus contortus, Ostertagia circumcincta, T. axei, T. colubriformis, Cooperia oncophora and Oesophagostomum columbianum. Ivermectin, containing principally 22,23-dihydroavermectin B_{1a}, has received approval for use in pigs in the U.K. An injected dose of 0.3mg/kg is employed for the treatment and control of nematodes (Ascaris suum, Hyostrongylus rubidus, Oesophagostomum spp and Strongyloides ransomi), adult lungworms (Metastrongylus spp), lice (Haematopinus suis), and mange mites (Sarcoptes scabiei var suis).[16] Trichuris suis is refractory at doses up to 0.5mg/kg.[17] Efficacy of ivermectin in treating Onchocerca volvulus infection (river blindness) in man at single oral doses of up to 0.05mg/kg was assessed by determining the number of microfilariae recovered from skin snips. Great reduction in the numbers were seen in all subjects receiving 0.03 or 0.05mg/kg and complete elimination was observed in 6 of 8 patients receiving the upper dose.[18] The practical value of the microfilaricidal action of ivermectin in endemic areas remains to be established.[19,20]

The actions of avermectins on the GABA receptor-chloride ionophore complex have been reviewed.[21] Avermectin B_{1a} at concentrations as low as 50nM is reported to inhibit the binding of strychnine to the postsynaptic glycine receptor - chloride ionophore complex in membrane bound or solublized rat spinal chord preparations.[22]

Other Endectocides: The structures of the avermectins and the milbemycins were interrelated by the conversion of 22,23-dihydroavermectin B_{1b} (9) to milbemycin B-41D (10).[23] Compound 10, the first of the C-25 isopropyl-bearing milbemycins is more potent as an anthelmintic than mixtures of milbemycins α_2 and α_4.[24] Review articles describing the producing organism,[25] structure determination[26] and biosynthesis[27] of the milbemycins are available as is a brief account of all three aspects.[28]

	R_1	R_2	R_3	R_4	R_5	R_6
11	Me	H	MeO	H	A	iPr
12	Me	H	MeO	Me	A	iPr
13	Me	H	MeO	H	Me	iPr
14	H	Me	Me	H	B	Et

A=COCH=CHCO$_2$H;
B=COCH=CHCONHC=C(OH)CH$_2$CH$_2$CO

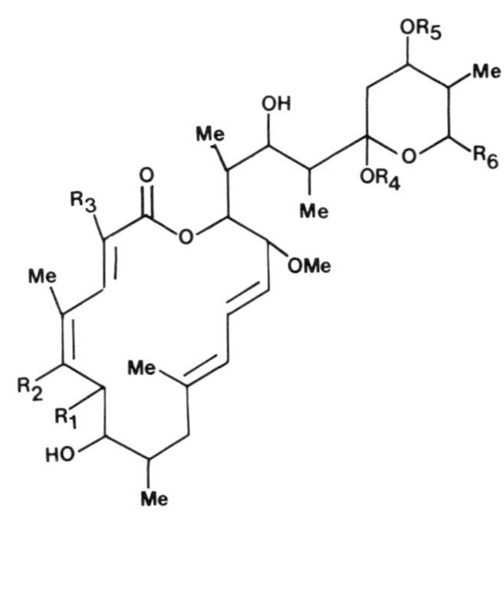

Fermentation-derived lactones with proposed structures 11, 12 and 13, have been patented as parasiticides effective against insects, arthropods, nematodes and cestodes.[29] They show activity against the free living, soil nematode Caenorhabditis elegans at less than 30mcg/ml. In common with the avermectins they cause the release of GABA from isolated rat brain synaptosomes though the tapeworm activity of these compounds is not shared by the avermectins. A fourth lactone of the series, 14 has both tapeworm and antiprotozoal activity with Giardia lamblia being employed to detect 14 during isolation.[30]

Endoparasitic activity for the known ectoparasiticide 15 was reported against artificially established Haemonchus contortus infections in sheep with doses in the range 0.5-1.0mg/kg providing 90% reduction in helminth egg excretion.[31] Compound 15 was equally effective against a strain of H. contortus resistant to thiabendazole, parbendazole and albendazole. Similar activities were also reported for the closely related 2-arylazo and 2-arylhydrazinoimidazoline[32,33] and thiazoline[34] derivatives 16-24 vs. H. contortus in sheep as were activities against Boophilus microplus Biarra strain.

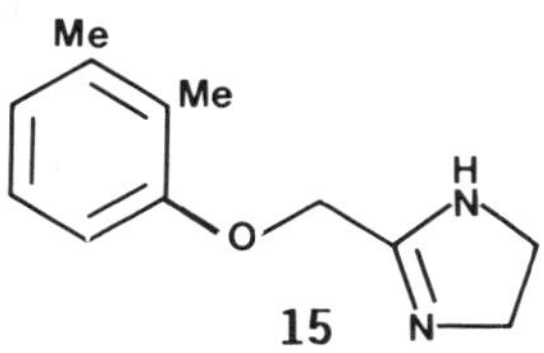

15

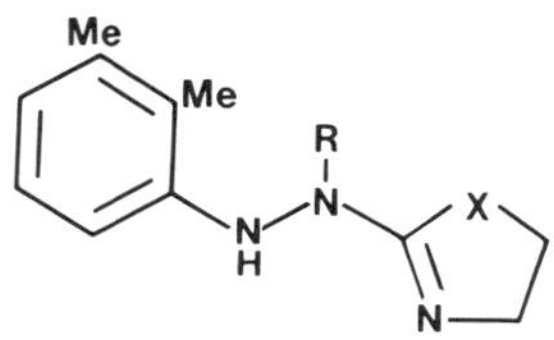

	R	X
16	H	NH
17	$C_6H_5SO_2$	NH
18	p$ClC_6H_4SO_2$	NH
19	H	S
20	$C_6H_5SO_2$	S

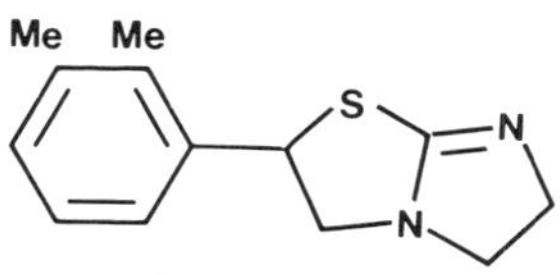

25

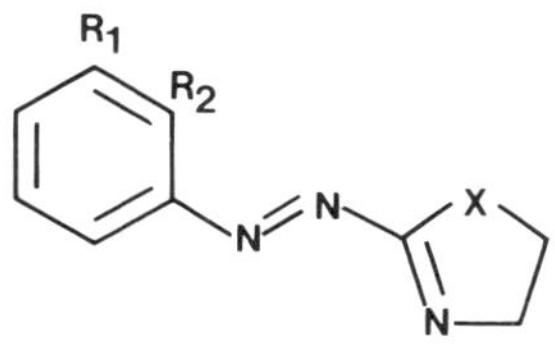

	R_1	R_2	X
21	Me	Me	NH
22	Me	Me	S
23	Cl	Cl	NH
24	Cl	Me	NH

The egg laying of B. microplus was completely inhibited by the topical application of 25 at a dose of 10mcg/tick. Contact insecticidal activity vs. Lucilia cuprina larvae was found at a dose of 100mg/m^2 of compound impregnated surface and anthelmintic effects on C. elegans were observed at a concentration of 10mcg/ml.[35]

Praziquantel: A major review describing synthesis, SARs, pharmacokinetics, metabolism, toxicology, pharmacology, toleration, mode of action and efficacy of praziquantel vs. trematodes and cestodes was published.[36] Following the report[37] of synergy for oxamniquine and praziquantel in the therapy of Schistosoma mansoni infections in mice, a clinical study of patients infected with S. mansoni and S. haematobium has been undertaken.[38] Mean egg count reductions of 95 and 97%, for S. mansoni and S. haematobium, respectively, were recorded six months post-treatment following the simultaneous administration of single oral doses of oxamniquine (7.5mg/kg) and praziquantel (15mg/kg). The authors conclude that the incidence and severity of side effects, observed following treatment with oxamniquine (30mg/kg) or praziquantel (40mg/kg) in areas of very heavy S. mansoni infections, are reduced by the combination therapy.

Benzimidazole Carbamates: An extensive survey of mebendazole and related 5-substituted benzimidazole anthelmintics is available.[39] The biochemical characterisation of a helminth (Ascaridia galli) tubulin, the proposed binding site of the benzimidazole carbamates, has been achieved following purification by column chromatography on DEAE-Sephadex.[40] A single peak was responsible for both colchicine and parbendazole binding activity.

Two dimensional polyacrylamide gel electrophoresis of helminth and mammalian tubulin revealed a difference in their α-subunits. This was confirmed by limited proteolytic peptide mapping where it was consistently observed that some peptides were novel to each tubulin.[41] It was suggested that the difference in the cytoplasmic α-tubulin subunit described compared with that of mammalian cells could account for the selective toxicity of the benzimidazole carbamates without the need to postulate differential pharmacokinetics between parasite and host.

Morantel: The conventional use of veterinary anthelmintics is for the therapy of existing infections and does not prevent rapid reinfection in circumstances of continuous environmental challenge such as pasture grazing. A new sustained release morantel tartrate delivery system for cattle has been described.[42] A single oral treatment administered at turnout provides protection against gastrointestinal nematode infection for the entire grazing season. Improved performance compared with untreated or conventionally medicated cattle was also observed.

Nematodes: Reviews have been published describing the problems of drug resistance and the search for more effective therapy against resistant nematode disease.[43,44] Although parasites are being continually added to the list of resistant field strains, e.g. *Nematodirus spathiger* in lambs,[45] reports of novel agents have been scarce. A series of 2-(perhydro-1,4-thiazepin-4-yl)-ethylbenzyl and benzhydryl ethers were reported to be active against *Rhabditis strongyloides*, with compound 26 causing paralysis at a potency approximately one third that of thiabendazole.[46] The SARs of a series of 3,6-diaryl-(7H)-s-triazolo[3,4-b][1,3,4]thiadiazines were elucidated using *Ascaris vitulorum in vitro*.[47] Some aryl clopidol derivatives (27) were active against *Nippostrongylus brasiliensis* and *Hymenolepis nana*.[48] Poor correlation between potency in the *Nematospiroides dubius* model and activity against a natural mixed trichostrongylid nematode population in sheep was reported for 2-hetero-aromatic-5 or 6-isocyanato-benzoxazoles and benzthiazoles.[49] Two recent series of nitro-2-naphthofurans[50] and nitroindazoles[51] have shown activity against *Syphacia obvelata* in mice.

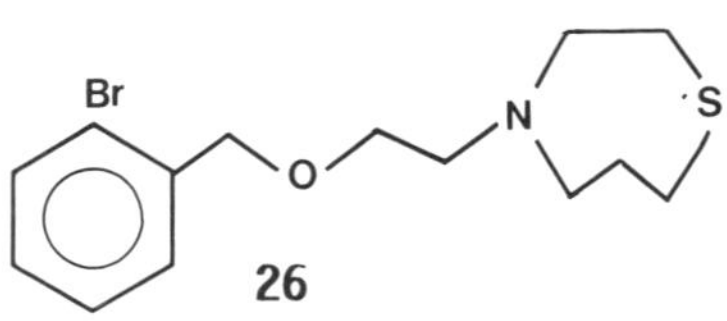

26

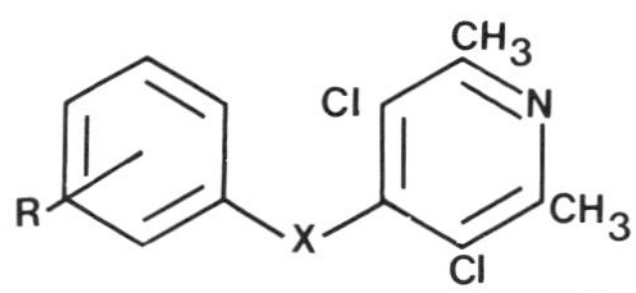

27

X=O; NH
R=Me; Cl;
MeO; etc.

Filaria: The weak microfilaricidal effects of mebendazole and its pronounced embryotoxic action on *Onchocerca volvulus* are synergized by levamisole.[52] Macrofilaricidal activity for the combination was not observed. While the mechanism of this synergy is unknown, it has been proposed that the immunomodulating properties of levamisole are responsible rather than a change in mebendazole pharmacokinetics. A flubendazole/levamisole combination was without significant effect on *O. volvulus*. Of a series of N-[4-[[4-alkoxy-3-[(dialkylamino)methyl] phenyl]amino]-2-pyrimidyl]-N'-phenylguanidines evaluated either orally or subcutaneously in jirds carrying infections of *Litomosoides carinii* and *Brugia pahangi*, compound 28 was the most promising against adult *L. carinii* giving 100% reduction in live worm numbers at 3mg/kg/day p.o. for 5 days.[53] Microfilaricidal activity was not observed nor was activity vs. *B. pahangi*, a model thought to be more relevant to *O. volvulus* infections. The corresponding benzimidazoles 29 were inactive in both test systems.[54]

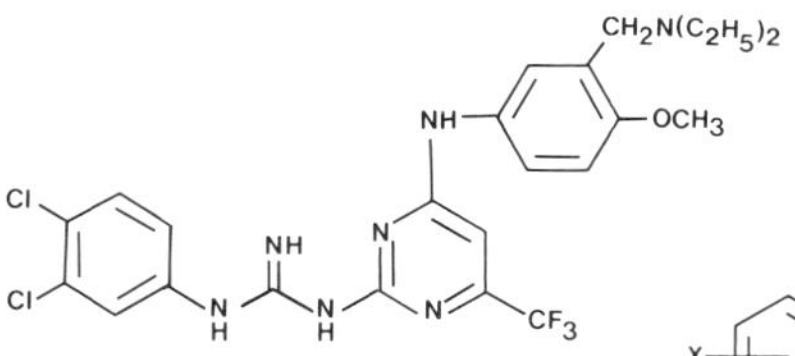

28

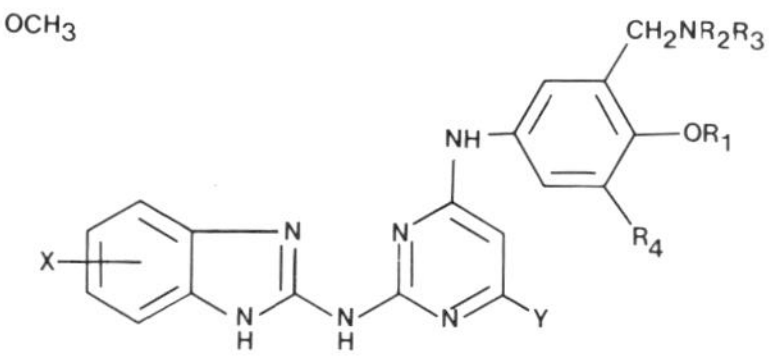

29

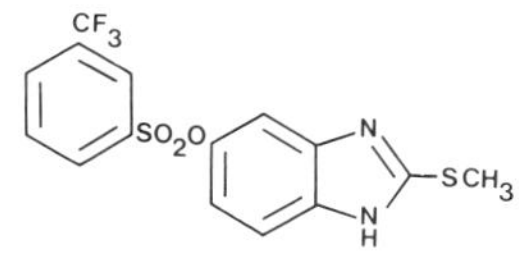

30

Fasciola: The chemotherapy of fascioliasis has been reviewed.[55] Of the agents currently available or under development, triclabendazole (CGA-89317) shows clear superiority in the control of immature *Fasciola hepatica*.[55] The obvious advantages of early control coupled with a wide safety index at the recommended dose of 10mg/kg could make triclabendazole an attractive agent for fascioliasis in sheep and calves. A reduced dose of 5mg/kg removes adult flukes from goats.[56] Triclabendazole is weakly active against *Dicrocaelium* and nematodes but can be applied with other nematocides. Synergy with benzimidazole carbamates has been claimed.[57] The closely related sulphonyl esters, e.g. 30, have been patented.[58]

MK-401 (31) and analogues are reported as competitive inhibitors of both 3-phosphoglycerate and ATP binding to phosphoglycerate kinase isolated from *F. hepatica*.[59] Good correlation of *in vitro* and *in vivo* potency, and the size of the 6-substituent has led to a proposed mechanism of action.

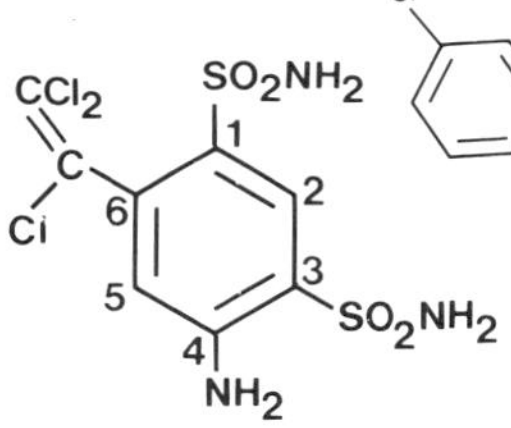

31

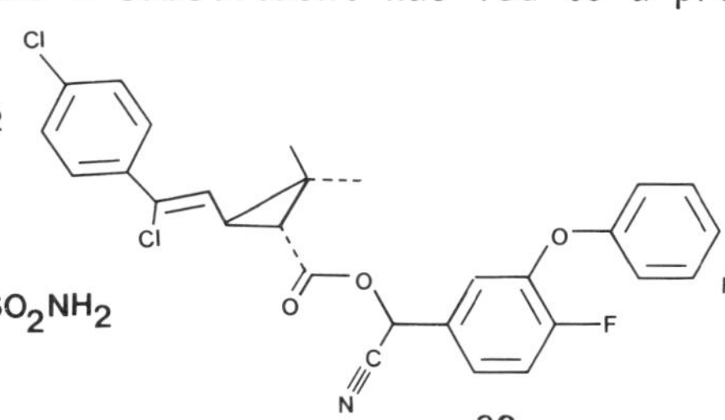

32

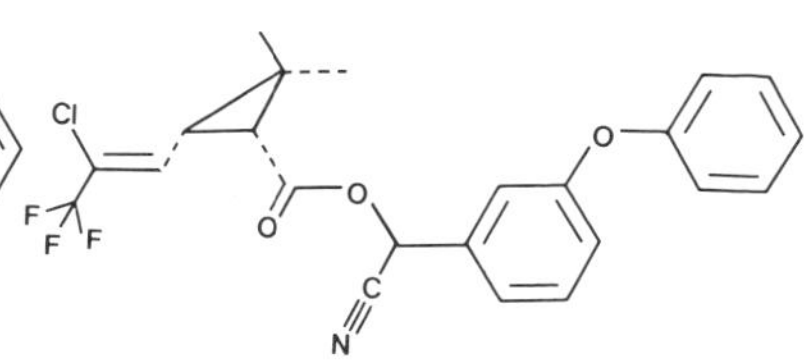

33

Pyrethroids: The spread of organophosphate resistance amongst tick populations and the recent emergence of an Australia tick strain resistant to the amidine acaricides provide the background to the introduction of pyrethroids into the acaricide market. Whilst some compounds first found use in crop protection, others, e.g. flumethrin (32) have been developed for acaricide use. They offer control of resistant strains and their high acaricidal potency affords extended protective periods (7-15 days) which permit less frequent dipping.[60,61] In contrast to the amidine class, protection from nuisance flies is achievable, with as little as 200mg/animal of cyhalothrin (33) giving greater than 28 days protection against reinfestation by *Haematobia irritans exigua*.[61]

The mechanisms of selective action of pyrethroids have been reviewed.[62] The specific binding of [^{35}S]-4-*t*-butyl-1-phospha-2,6,7-trioxabicyclo [2.2.2]octane-1-sulphide, a radioligand for the picrotoxinin binding site of the GABA receptor-chloride ionophore complex, is inhibited by pyrethroid esters of (S)-α-cyano-3-phenoxybenzyl alcohol that cause type II syndrome.[62] The *in vitro* potency of this group of pyrethroids in this assay is well correlated with their intracerebral toxicity in mice. This observation coupled with the protective effect of diazepam for type II poisoning[63] suggests the GABA receptor/chloride ionophore complex to be a primary target for type II pyrethroids, with their binding site closely associated with that of picrotoxinin.[64] The broad ectoparasiticidal spectrum of the pyrethroids continues to attract a great deal of synthetic attention with more than 100 patents published during 1983. A useful summation of the vast literature dealing with the well-established pesticides has been published which also discusses compound classes that affect insect development (chitin synthesis inhibitors, juvenile hormones, precocenes and xanthene dyes) and speculates as to their potential insecticidal usage.[65]

PROTOZOAL DISEASE

Coccidia: A textbook on the biology of coccidia has been published with key chapters describing the chemotherapy[66] and drug resistance[67] phenomena associated with coccidiosis. The use of polyether ionophores as coccidiostats is increasing as newer agents like salinomycin and narasin reach the market. Further agents in this class have been reported[68] and the pharmacology and toxicology of the monovalent carboxylic acid ionophores have been reviewed.[69]

Sporozoites of *Eimeria tenella* survive exposure to 100ppm of monensin, salinomycin or lasalocid *in vitro* for a period of 30 minutes, a time sufficient to penetrate host cells.[70] Merozoites were rapidly destroyed by pellicular rupture at these concentrations. Prophylactic-only use of drugs in this class might thus be rationalized since continuous administration is required to destroy the periodically released infectious merozoites.

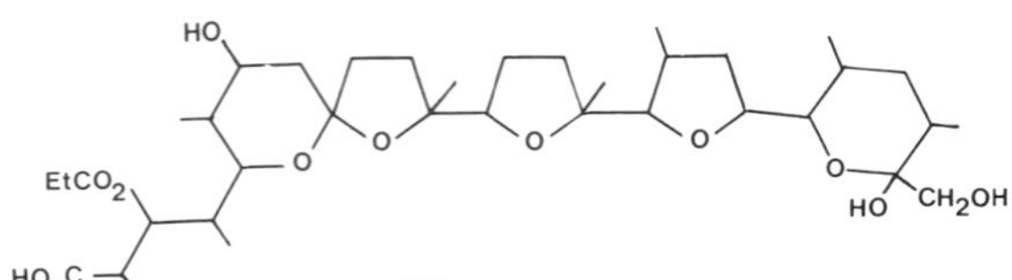

34

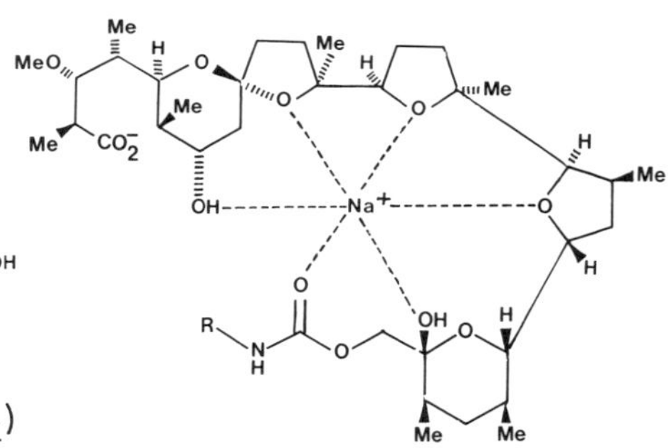

35

The *in vivo* activity of laidlomycin (34) against *E. tenella* is enhanced by acylation, particularly butyration, of the primary hydroxyl group.[71] Urethane derivatives of monensin (35) retain the anticoccidial activity of the parent compound and the p-chlorophenyl analogue had an ED_{50} of 2mg/kg when evaluated as an antimalarial in the *Plasmodium berghei* mouse model.[72,73] Recent novel structural types from fermentation sources with activity against *E. tenella* include lactones 36[74] and 37,[75] and althiomycin (38).[76] Interest in synthetic anticoccidial agents continues to be strong and a review covering the period to 1981 is available.[77] Ribonucleosides of 1H-pyrazolo[3,4-d]pyrimidines are reported to inhibit the development of *E. tenella in vivo*.[78-80] Of particular interest was the cinnamylthio analogue 39 which controlled *E. tenella*, *E. necatrix*, *E. brunetti* and *E. maxima* at doses of 25-100ppm. A dose of 800ppm was however needed for *E. acervulina* activity. N-1-p-Benzophenone derivatives of 6-azauracil are highly effective *vs. E. tenella in vivo* with the most potent exhibiting broad spectrum activity.[81] In common with previously reported series, adverse toxicology precluded development. The reduction in potency against *E. tenella in vivo* found with the uracil analogue of tiazuril was thought to be related to the reduced acidity of the imide proton.[82]

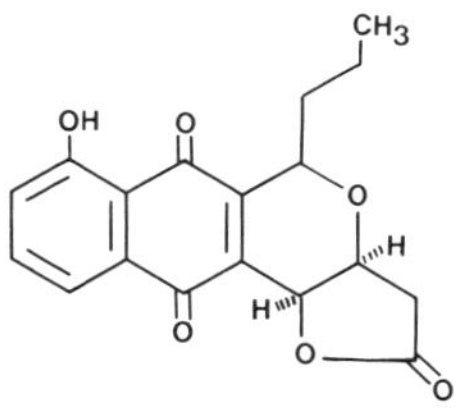

36

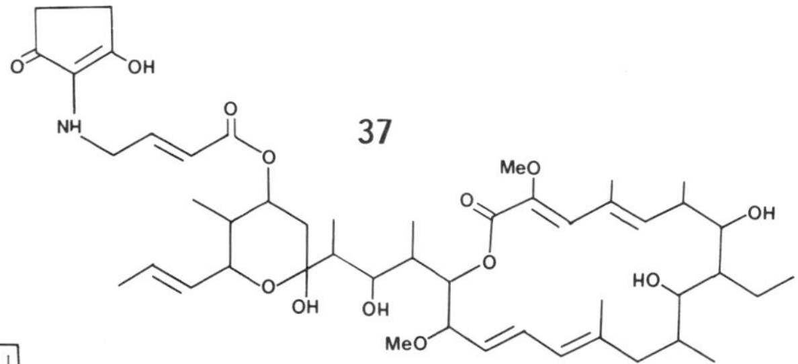

37

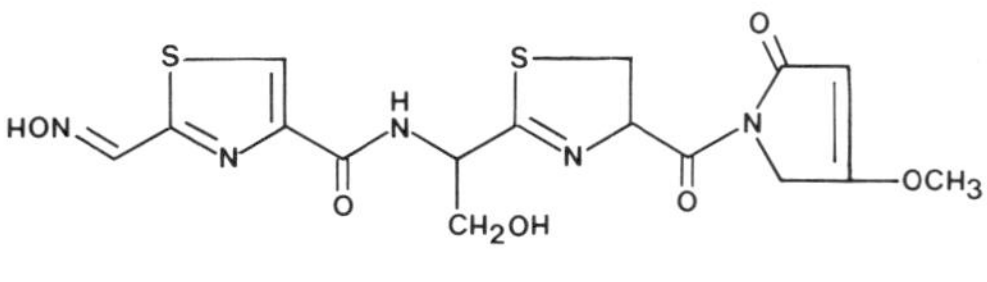

38

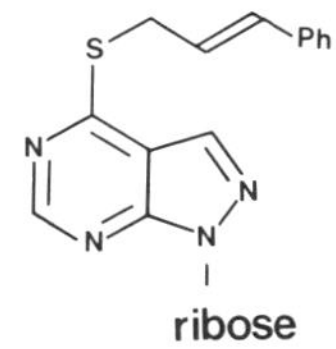

39

Malaria: A succinct account of the problems in malaria treatment provides a useful introduction to the topic.[83] More detailed reviews covering chemotherapeutic advances,[84] prospective drug developments,[85] chemoprophylaxis[86] and the Seventh Asian Malaria Conference[87] are available. Stringent precautions have been recommended[88,89] for the introduction of mefloquine in the treatment of multi-resistant falciparum malaria to protect this agent against resistance development. Greater binding of mefloquine to phospholipids has been proposed to account for its superiority in the treatment of chloroquine-resistant malaria.[90] Consistent with this hypothesis are the enriched levels of phosphatidylinositol in malaria parasites.

In a major study, numerous agents from the 8-aminoquinoline class were found inactive against Plasmodium berghei yoelli/P. y. nigeriensis in rodents although they displayed moderate to high activity against P. cynomolgi infections in the rhesus monkey. An abbreviated simian model has been proposed as scientifically essential and economically feasible.[91] Modification of existing continuous culture methods for P. falciparum allows rapid and accurate determination of drug action in vitro against the human pathogen by monitoring the incorporation of ^{3}H-hypoxanthine into parasite nucleic acids.[92] Using this procedure, the anticancer drug spirogermanium (40) was found active in the range 2.5-40nM/ml, with chloroquine-resistant strains tending to be one-to-two fold more susceptible.[93]

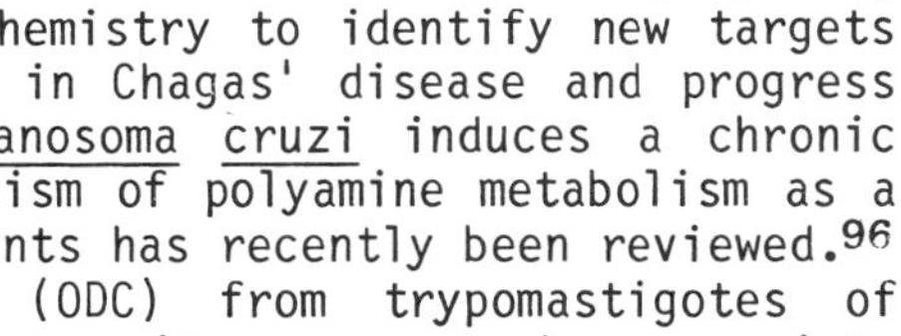

40

Trypanosomiasis: With no drugs introduced for over twenty years, attention has focussed on parasite biochemistry to identify new targets for drug action.[94] Recent developments in Chagas' disease and progress towards an animal model in which Trypanosoma cruzi induces a chronic infection have been reported.[95] Antagonism of polyamine metabolism as a possible approach to antitrypanosomal agents has recently been reviewed.[96] Comparison of ornithine decarboxylase (ODC) from trypomastigotes of T. brucei brucei and rat liver show the parasite enzyme to be over sixty times more sensitive to the suicide inhibitor DL-α-difluoromethyornithine (DFMO).[97] ODC blockade by DFMO on T. b. brucei in rats has been shown to block cytokinesis and induce changes in morphology resembling transformation of the bloodstream trypomastigotes.[98] Calcium has been identified as the serum synergistic factor responsible for the increased activity of salicylhydroxamic acid/glycerol combination.[99] The amine 353C (41) was approximately 10-20 times more effective than the generally used nifurtimox or benznidazole in producing a radical cure of T. cruzi in mice.[100] Well tolerated neuroleptic phenothiazines could provide a lead for new trypanocidal drugs, rapid disintegration of the pellicular layer of microtubules of T. b. brucei in vitro having been reported.[101]

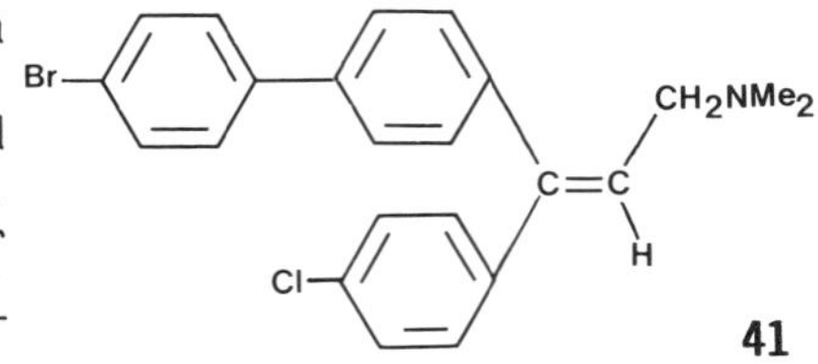

41

Leishmaniasis: A literature review on the search and development of antileishmanial drugs is available.[102] Pyrazolopyrimidines continue to be a major area of interest. Compounds with activities against several Leishmania parasites have been described[103-107] and their metabolism reviewed.[108] Neuroleptic phenothiazines were reported effective against L. donovani in vitro.[109] Importantly chlorpromazine was found to kill amastigotes within human monocyte-derived macrophages.

Other infections: Chemotherapy of the tick-transmitted infections, Anaplasma, Babesia, Cowdria and Theileria, has been reviewed.[110] Halofuginone at 1.2mg/kg p.o. in two doses, given on the first and fourth days of fever, was curative in corridor disease (T. parva lawrencei infections), though recovered animals became carriers.[111] CG 10213-GO (42) was superior to currently used 5-nitroimidazoles against hepatic and caecal infections of Entamoeba histolytica in hamsters[112] and Trichomonas vaginalis or T. foetus in mice.[113]

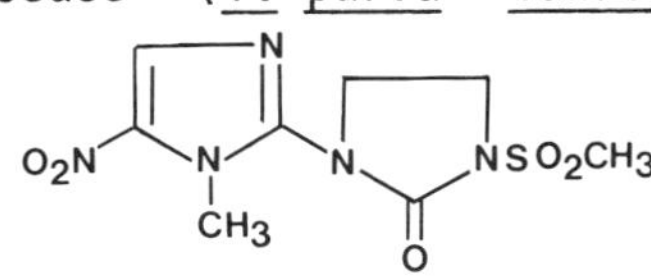

42

References

1. W.E. Gutteridge in "Modern Parasitology. A Textbook of Parasitology", F.E.G. Cox, Ed., Blackwell Scientific Publications, Oxford, 1982, p.287.
2. "Handbook Series in Zoonoses", Section C, Vols. 1-3, J.H. Steele, Ed.-in-chief, CRC Press, Boca Raton, Florida, 1982.
3. WHO Tech. Rep. Ser., no. 666 (1981).
4. C.C. Wang, Trends in Biochem. Sci., 7, 354 (1982).
5. J.B. Malone, J.C. Williams, R.P. Herd, R.B. Wescott, J.F. Williams, P.R. Fitzgerald, R. Fayer, D.A. Armstrong, N.F. Baker, R.R. Bell, G.W. Benz, R.E. Bradley, R.C. Bergstrom, H. Ciordia, R.M. Corwin, T.M. Craig, J.P. Dubey, W.J. Foreyt, N.L. Gates, H.C. Gibbs, R.A. Knight, R.S. Rew, B.E. Stromberg and H.H. Vegors, Am. J. Vet. Res., 44, 1836 (1983).
6. "Pesticide Chemistry: Human Welfare and the Environment", Vols. 1-4, J. Miyamoto and P.C. Kearney, Eds.-in-chief, Pergamon Press, Oxford, 1983.
7. E.O. Stapley and H.B. Woodruff in "Trends In Antibiotic Research", H. Umezawa, A.L. Demain, T. Hata and C.R. Hutchinson, Eds., Japan Antibiotics Research Association, Tokyo, 1982, p. 154.
8. W.C. Campbell, M.H. Fisher, E.O. Stapley, G. Albers-Schönberg and T.A. Jacob, Science, 221, 823 (1983).
9. US Patent, US 4285963 to Merck and Co. Inc., 25.08.1981; European Patent, EP 58518 to Merck and Co. Inc., 25.08.1982; European Patent, EP 73660 to Merck and Co. Inc., 9.03.1983.
10. US Patent, US 4333925 to Merck and Co. Inc., 8.06.1982.
11. European Patent, EP 74758 to Merck and Co. Inc., 23.03.1983; European Patent, EP 89202 to Merck and Co. Inc., 21.09.1983.
12. G.T. Miwa, J.S. Walsh, W.J.A. Vandenheuvel, B. Arison, E. Sestokas, R. Buhs, A. Rosegay, S. Avermitilis (sic), A.Y.H. Lu, M.A.R. Walsh, R.W. Walker, R. Taub and T.A. Jacob, Drug Metab. Dispos., 10, 268 (1982).
13. I. Putter, J.G. MacConnell, F.A. Preiser, A.A. Haidri, S.S. Ristich and R.A. Dybas, Experientia, 37, 963 (1981).
14. V.P. Gullo, A.J. Kempf, J.G. MacConnell, H. Mrozik, B. Arison and I. Putter, Pestic. Sci., 14, 153 (1983).
15. H. Mrozik, P. Eskola, M.H. Fisher, J.R. Egerton, S. Cifelli and D.A. Ostlind, J. Med. Chem., 25, 658 (1982).
16. Anonymous, Animal Pharm, No. 41, 10 (1983).
17. T.B. Stewart, O.G. Marti and O.M. Hale, Am. J. Vet. Res., 42, 1425 (1981).
18. M.A. Aziz, S. Diallo, I.M. Diop, M. Lariviere and M. Porta, Lancet, II, 171 (1982).

19. A. Rougemont, Lancet, II, 1158 (1982).
20. M.A. Aziz, S. Diallo, M. Lariviere, I.M. Diop, M. Porta and P. Gaxotte, Lancet, II, 1456 (1982).
21. C.C. Wang and S.S. Pong in "Membranes and Genetic Disease", J.R. Sheppard, V.E. Anderson, J.W. Eaton, Eds., Alan R. Liss, New York, 1982, p. 373.
22. D. Graham, F. Pfeiffer and H. Betz, Neurosci. Lett., 29, 173 (1982).
23. H. Mrozik, J.C. Chabala, P. Eskola, A. Matzuk, F. Waksmunski, M. Woods and M.H. Fisher, Tetrahedron Lett., 24, 5333 (1983).
24. UK Patent, GB 2056986 to Sankyo KK., 25.03.1981.
25. T. Okazaki, M. Ono, A. Aoki and R. Fukuda, J. Antibiot., 36, 438 (1983).
26. H. Mishima, J. Ide, S. Muramatsu and M. Ono, J. Antibiot., 36, 980 (1983).
27. M. Ono, H. Mishima, Y. Takiguchi and M. Terao, J. Antibiot., 36, 991 (1983).
28. H. Mishima in Ref. 6, Vol. 2, 129 (1983).
29. European Patent, EP 50964 to Merck and Co. Inc., 5.05.1982.
30. US Patent, US 4390546 to Merck and Co. Inc., 28.06.1983.
31. European Patent, EP 49797 to Bayer AG., 21.04.1982.
32. European Patent, EP 73394 to Bayer AG., 9.03.1983.
33. German Patent, DE 3133887 to Bayer AG., 10.03.1983.
34. European Patent, EP 73393 to Bayer AG., 9.03.1983.
35. European Patent, EP 80359 to Pfizer Inc., 1.06.1983.
36. P. Andrews, H. Thomas, R. Pohlke and J. Seubert, Med. Res. Rev., 3, 147 (1983).
37. J.R. Shaw and K.W. Brammer, Trans. R. Soc. Trop. Med. Hyg., 77, 39 (1983).
38. R.N.H. Pugh and C.H. Teesdale, Br. Med. J., 287, 877 (1983).
39. H. Van den Bossche, F. Rochette and C. Hörig, Adv. Pharmacol. Chemother., 19, 67 (1982).
40. C.M. Ireland, L. Clayton, W.E. Gutteridge, C.I. Pogson and K. Gull, Mol. Biochem. Parasitol., 6, 45 (1982).
41. P.J. Dawson, W.E. Gutteridge and K. Gull, Mol. Biochem. Parasitol., 7, 267 (1983).
42. D.H. Bliss and R.M. Jones, Vet. Parasitol., 12, 219 (1983) and following papers.
43. H.J. Boersema, Tijdschr. Diergeneeskd., 108, 426 (1983).
44. D.G. Bennett, Compend. Cont. Ed. Pract. Vet., 5,6 (Suppl.), 343 (1983).
45. A. Middleberg and P.B. McKenna, N.Z. Vet. J., 31, 65 (1983).
46. M. Strzelczyk, Pol. J. Pharmacol. Pharm., 34, 265 (1982).
47. M.A. El-Dawy, A-M.M.E. Omar, A.M. Ismail and A.A.B. Hazzaa, J. Pharm. Sci., 72, 45 (1983).
48. A. Svab, V. Zikan, D. Nemcova and J. Danek, Ceskoslov. Farm., 32, 49 (1983).
49. R.D. Haugwitz, R.G. Angel, G.A. Jacobs, B.V. Maurer, V.L. Narayanan, L.R. Cruthers and J. Szanto, J. Med. Chem., 25, 969 (1982).
50. R. Cavier, J-P. Buisson and R. Royer, Eur. J. Med. Chem., 17, 91 (1982).
51. J-M. Dumont, T. Yanes, J. Paris and A-F. Petavy, Eur. J. Med. Chem., 18, 469 (1983).
52. K. Awadzi, H. Schulz-Key, R.E. Howells, D.R.W. Haddock and H.M. Gilles, Ann. Trop. Med. Parasitol., 76, 459 (1982).
53. M. Angelo, D. Ortwine, D. Worth, L.M. Werbel and J.W. McCall, J. Med. Chem., 26, 1258 (1983).
54. M.M. Angelo, D. Ortwine, D.F. Worth and L.M. Werbel, J. Med. Chem., 26, 1311 (1983).
55. J.C. Boray, N.S. Wales Vet. Proc., 18, 42 (1982).
56. K. Wolff, J. Eckert, G. Schneiter and H. Lutz, Vet. Parasitol., 13, 145 (1983).
57. European patent, EP 84516 to Ciba Geigy AG., 27.07.1983.
58. European patent, EP 72532 to Hoechst AG., 23.02.1983.
59. M.D. Schulman, D.A. Ostlind and D. Valentino, Mol.Biochem.Parasitol., 5, 133 (1982).
60. T.J. Hopkins and I.R. Woodley, Vet. Med. Nachr., 130 (1982).
61. V.K. Stubbs, C. Wilshire and L.G. Webber, Aust. Vet. J., 59, 152 (1982).
62. J.E. Casida, D.W. Gammon, A.H. Glickman and L.J. Lawrence, Ann. Rev. Pharmacol. Toxicol., 23, 413 (1983).
63. D.W. Gammon, L.J. Lawrence and J.E. Casida, Toxicol.Appl.Pharmacol., 66, 290 (1982).
64. L.J. Lawrence and J.E. Casida, Science, 221, 1399 (1983).
65. "Insecticide Mode of Action", J.R. Coats, Ed., Academic Press, New York, 1982.
66. L.R. McDougald in "The Biology of the Coccidia", P.L. Long, Ed., University Park Press, Baltimore, 1982, p. 373.
67. H.D. Chapman in "The Biology of the Coccidia", P.L. Long, Ed., University Park Press, Baltimore, 1982, p. 429.
68. Japanese Patent, J 57085390 to Kaken Chem. KK., 28.05.1982; European Patent, EP 63238 to Kaken Chem. KK., 27.10.1982, European Patent, EP 88446 to Hoffman-La Roche AG., 14.09.1983.
69. B.C. Pressman and M. Fahim, Ann. Rev. Pharmacol. Toxicol., 22, 465 (1982).
70. H. Mehlhorn, H. Pooch and W. Raether, Z. Parasitenkd., 69, 457 (1983).
71. R.D. Clark, G.L. Hedden, A.F. Kluge, M.L. Maddox, H.R. Spires and P.F. Long, J. Antibiot., 35, 1527 (1982).
72. J.W. Westley, C-M. Liu, R.H. Evans, Jr., L.H. Sello, N. Troupe and T. Hermann, J. Antibiot., 36, 1195 (1983).

73. E.G. Schildknecht, D. Siegel and R.W. Richle, Chemotherapy, 29, 145 (1983).
74. Japanese Patent, J 58088313 to Asahi Chemical Ind. KK., 26.05.1983.
75. European Patent, EP 85859 to Kitasato Inst., 17.08.1983.
76. Japanese Patent, J 58043922 to Asahi Chemical Ind. KK., 14.03.1983.
77. Y. Morisawa, Med. Res. Rev., 2, 63 (1982).
78. T.A. Krenitsky, J.L. Rideout, G.W. Koszalka, R.B. Inmon, E.Y. Chao, G.B. Elion, V.S. Latter and R.B. Williams, J. Med. Chem., 25, 32 (1982).
79. J.L. Rideout, T.A. Krenitsky, G.W. Koszalka, N.K. Cohn, E.Y. Chao, G.B. Elion, V.S. Latter and R.B. Williams, J. Med. Chem., 25, 1040 (1982).
80. J.L. Rideout, T.A. Krenitsky, E.Y. Chao, G.B. Elion, R.B. Williams and V.S. Latter, J. Med. Chem., 26, 1489 (1983).
81. R.D. Carroll, M.W. Miller, B.L. Mylari, L.R. Chappel, H.L. Howes, Jr., M.J. Lynch, J.E. Lynch, S.K. Gupta, J.J. Rash and R.C. Koch, J. Med. Chem., 26, 96 (1983).
82. M.W. Miller and L.R. Chappel, J. Med. Chem., 26, 1075 (1983).
83. P. Nickel, Pharm. Int., 4, 37 (1983).
84. R.E. Howells, Br. Med. Bull., 38, 193 (1982).
85. R.O. Pick, Scand. J. Infect. Dis., Suppl., 36, 31 (1982).
86. L.J. Bruce-Chwatt, Br. Med. J., 285, 674 (1982).
87. WHO Tech. Rep. Ser., no. 680 (1982).
88. T. Harinasuta, D. Bunnag and W.H. Wernsdorfer, Bull. WHO, 61, 299 (1983).
89. UNDP/WORLDBANK/WHO UPDATE, Bull. WHO, 61, 169 (1983).
90. R. Chevli and C.D. Fitch, Antimicrob. Agents Chemother., 21, 581 (1982).
91. L.H. Schmidt, Am. J. Trop. Med. Hyg., 32, 231 (1983).
92. T.G. Geary, A.A. Divo and J.B. Jensen, J. Parasitol., 69, 577 (1983).
93. J.E.K. Mrema, M. Slavik and J. Davis, Int. J. Clin. Pharmacol., Ther. Toxicol., 21, 167 (1983).
94. A. Fairlamb, Trends Biochem. Sci., 7, 249 (1982).
95. Z. Brener, Bull. WHO, 60, 463 (1982).
96. C.J. Bacchi, P.P. McCann, H.C. Nathan, S.H. Hutner and A. Sjoerdsma, Adv. Polyamine Res., 4, 221 (1983).
97. J. Garofalo, C.J. Bacchi, S.D. McLaughlin, D. Mockenhaupt, G. Trueba and S.H. Hunter, J. Protozool., 29, 389 (1982).
98. C.J. Bacchi, J. Garofalo, D. Mockenhaupt, P.P. McCann, K.A. Diekema, A.E. Pegg, H.C. Nathan, E.A. Mullaney, L. Chunosoff, A. Sjoerdsma and S.H. Hunter, Mol. Biochem. Parasitol., 7, 209 (1983).
99. A.B. Clarkson, Jr. and B.O. Amole, Science, 216, 1321 (1982).
100. P.A. Barrett, E. Beveridge, D. Bull, I.C. Caldwell, P.J. Islip, R.A. Neal and N.C. Woods, Experientia, 38, 338 (1982).
101. T. Seebeck and P. Gehr, Mol. Biochem. Parasitol., 9, 197 (1983).
102. D.A. Chistyakov, A.A. Drabkina and L.V. Rusak, Med. Parazitol. Parazit. Bolezni, 51, 57 (1982).
103. J.D. Berman and H.K. Webster, Antimicrob. Agents Chemother., 21, 887 (1982).
104. J.L. Avila and M.A. Casanova, Antimicrob. Agents Chemother., 22, 380 (1982).
105. J.J. Marr, R.L. Berens, D.J. Nelson, T.A. Krenitsky, T. Spector, S.W. LaFon and G.B. Elion, Biochem. Pharmacol., 31, 143 (1982).
106. B.C. Walton, J. Harper III and R.A. Neal, Am. J. Trop. Med. Hyg., 32, 46 (1983).
107. J.D. Berman, L.S. Lee, R.K. Robins and G.R. Revankar, Antimicrob. Agents Chemother., 24, 233 (1983).
108. J.J. Marr and R.L. Berens, Mol. Biochem. Parasitol., 7, 339 (1983).
109. R.D. Pearson, A.A. Manian, J.L. Harcus, D. Hall and E.L. Hewlett, Science, 217, 369 (1982).
110. N. McHardy in "Parasites - Their World and Ours", D.F. Mettrick and S.S. Desser, Eds., Elsevier Biomedical Press, Amsterdam, 1982, p 330.
111. A.J. De Vos and J.A. Roos, Onderstepoort J. Vet. Res., 50, 33 (1983).
112. D.K. Ray, V.B. Shrivastava, J.S. Tendulkar, A.K. Dutta, K.K. Bhopale, D.K. Chatterjee and K. Nagarajan, Ann. Trop. Med. Parasitol., 77, 287 (1983).
113. D.K. Ray, D.K. Chatterjee and J.S. Tendulkar, Ann. Trop. Med. Parasitol., 76, 175 (1982).

Section IV - Metabolic Diseases and Endocrine Function

Editor: Beverly A. Pawson, Roche Research Center, Hoffmann-La Roche, Inc., Nutley, New Jersey 07110

Chapter 16. Progress in the Development of Antiobesity Drugs

Ann C. Sullivan, Susan Hogan and Joseph Triscari
Roche Research Center
Hoffmann-La Roche Inc., Nutley, New Jersey 07110

Introduction - Obesity results when energy intake exceeds energy expenditure. Since the therapeutic developments in this field were last reviewed here,[1] efforts directed at the pharmacological manipulation of food intake have continued with an increasing emphasis on peripherally acting appetite suppressants. Selective modulation of the intestinal absorption of dietary carbohydrate and/or lipid is now possible with several new agents. Inhibition of lipid synthesis concomitant with a stimulation of lipid oxidation can now be achieved pharmacologically. Significant progress to design agents which enhance energy expenditure has occurred. Drugs in each of these mechanistic classes are now undergoing early clinical evaluation and will be described in this review. The future availability of safe and effective pharmacologic agents with different mechanisms of action for the treatment of obesity is encouraging.

More extensive reviews describing antiobesity drugs which function by modulating energy intake, intestinal absorption of dietary carbohydrate or lipid, fatty acid synthesis, lipid utilization and energy expenditure have appeared recently.[2-5]

ANORECTIC AGENTS

The classical approach to the pharmacological treatment of obesity has relied predominantly on the use of appetite suppressants. These agents, primarily phenethylamine derivatives, lower food intake by potentiating central dopaminergic, serotoninergic or β-adrenergic mechanisms or antagonizing α-adrenergic pathways. Since the last review in this series[1] several monographs on obesity and appetite regulation have appeared.[6-10] However, few studies have been reported on the clinical efficacy of new antiobesity agents.

Anorectics Acting Through Serotoninergic Mechanisms - Fenfluramine produces anorexia by mediating the release and inhibiting the reuptake of serotonin, although it also affects lipid and carbohydrate metabolism. The fenfluramine-induced inhibition of gastric emptying is not antagonized by midbrain lesions of the raphe nuclei in free feeding rats, although similar lesions antagonize its anorectic effect in food deprived rats.[11] These data and the observed absence of a correlation between fenfluramine levels and weight loss in patients[12] suggest a complex mechanism of action which may involve peripheral activity.

The antidepressant serotonin uptake inhibitor femoxitine (1) decreased body weight in obese patients treated for 12 weeks. Femoxitine

ISBN 0-12-040519-9

is more effective in patients less than 45 years old.[13] It remains to be determined whether norfemoxitine, a metabolite found in serum of treated patients,[14] also has antiobesity activity. Zimelidine (2) decreased

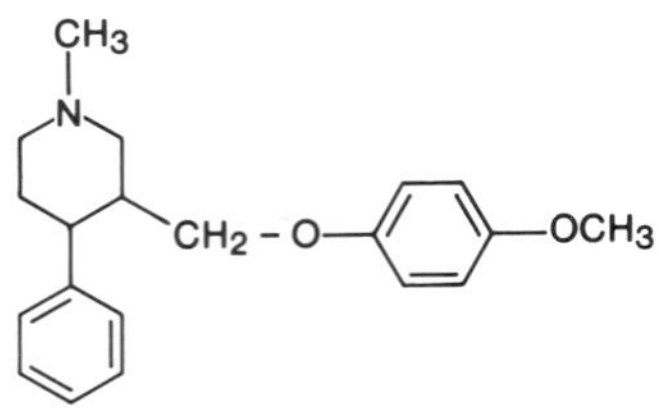

1

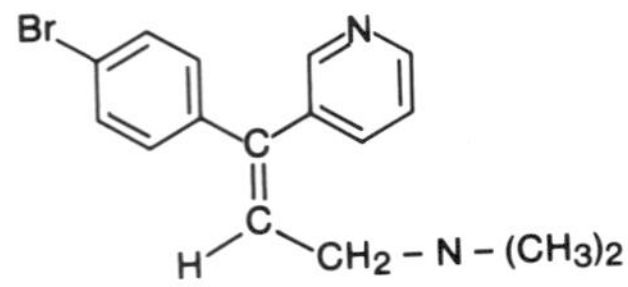

2

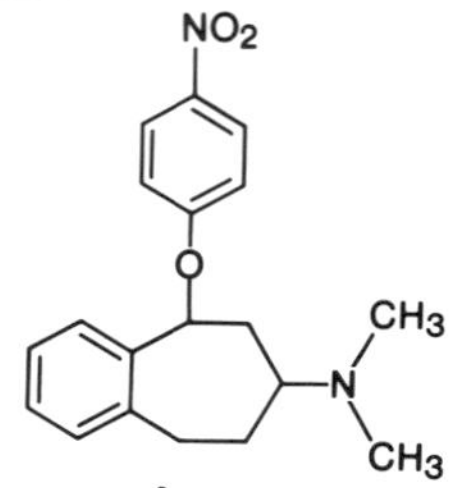

3

weight gain and subjective hunger ratings in obese patients.[15] Some patients treated for depression with bupropion report decreased appetite.[16]

RU 25591 (3), also a potent serotonin uptake inhibitor, suppresses food intake in dogs, rats and pigs.[17] The trans isomer has no effect on food intake in dogs. Tolerance to the anorexigenic activity of RU 25591 developed during a two-week course of treatment. The duration of action for the appetite suppression is shorter than that observed for the inhibition of serotonin uptake, suggesting the involvement of other pathways in its mechanism of action.

Two recently described anorectics, CM 57373 (4) and CM 57493 (5) reduce 5-hydroxyindoleacetic acid in rats, suggesting a serotoninergic mechanism.[18] The effect of CM 57373 is incompletely antagonized by two centrally acting serotoninergic antagonists, metergoline and methysergide, whereas the peripheral serotoninergic antagonist xylamidine has no effect. The anorectic potency of CM 57493 is decreased in cross-tolerance studies with fenfluramine and is antagonized by propranolol, but not by phentolamine or penfluridol.

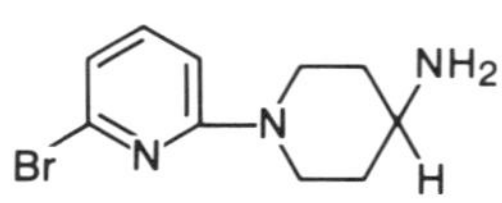

4

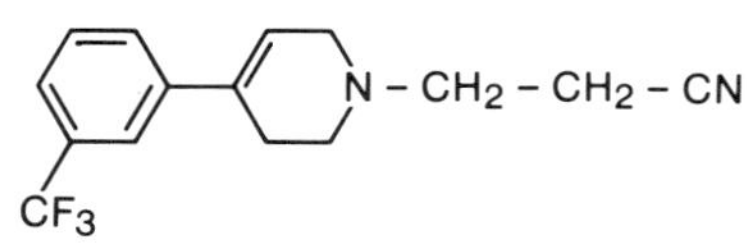

5

Anorectics Acting Through Dopaminergic Mechanisms - The dopaminergic pathway is thought to play a role in the anorectic activities of mazindol and amphetamine, which stimulate the release of dopamine and may also interfere with its reuptake.[19] The involvement of dopaminergic pathways in feeding behavior is supported by the observation that gamma-hydroxybutyric acid (γ-HBA), which decreased the firing rate of dopaminergic neurons by increasing the dopamine concentration in nerve terminals, increased food intake in rats.[20] Haloperidol antagonizes the increased food intake and fixed ratio response rate induced by γ-HBA.[20] RDS-127 (6) produces a dose-dependent suppression of food intake in rats trained to a daily 4-hour meal and is more active than amphetamine, at doses which do not increase motor activity.[21] Its effect on food intake is antagonized by pimozide but not by propranolol or phentolamine, consistent with a dopaminergic mechanism. Pimozide but not α-methyl-p-tyrosine antagonizes the increased motor activity produced by high doses of RDS-127. Two other 2-aminoindanes, JPC-60-36 (7) and JPC-211 (8), have little or no anorectic effect.[21] Another dopamine agonist, SKF 38393 (9) decreases food intake in rats without producing any central side effects.[22]

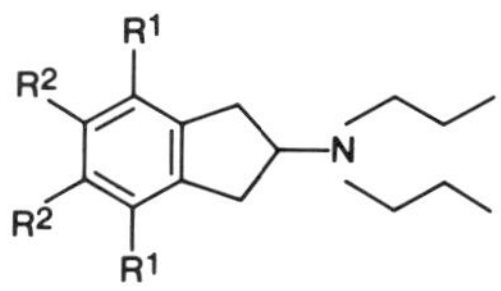

R1	R2	
OCH_3	H	6
H	H	7
H	OCH_3	8

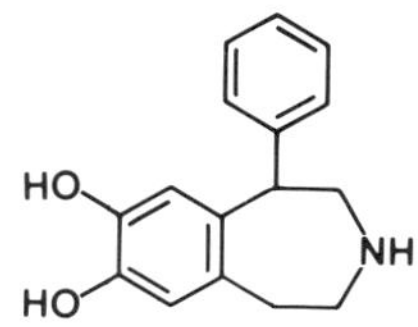

9

Piribedil (10), a dopamine agonist, suppresses food intake in rats and this effect is not antagonized by α-methyl-p-tyrosine, metergoline, propranolol, phentolamine or domperidone but is blocked by pimozide.[23] Motor activity is increased at 2 to 3 times the doses needed for food intake suppression.

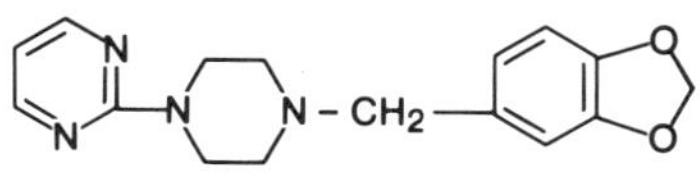

10

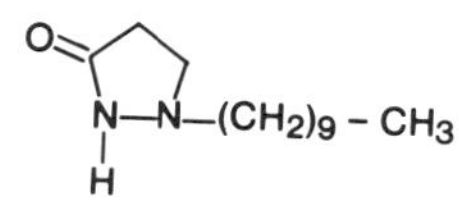

11

Anorectics Acting Through GABA-ergic Mechanisms - There is some evidence that brain GABA (γ-amino butyric acid) is elevated in satiated rats;[24] therefore, inhibitors of GABA metabolism would be expected to affect appetitive behavior. Ethanolamine-O-sulfate (EOS) is an inhibitor of GABA-transaminase which increases brain GABA levels and inhibits food intake at doses which do not alter locomotor activity in rats.[25] The intraventricular injection of EOS decreases 2-deoxyglucose uptake in the brain at doses which decrease food intake but do not alter sensorimotor function or startle response.[26] The GABA antagonists muscimol[25] and THIP[27] decrease food intake in rats. The anorectic effect of THIP is antagonized by bicuculline but not by the peripheral antagonist bicuculline-methobromide.[27] Bicuculline also antagonizes the diazepam-induced increase in food intake suggesting the involvement of benzodiazepine receptors in the regulation of feeding behavior by GABA.[28] Another inhibitor of GABA-transaminase, BW 357U (11) increases brain GABA levels and decreases food intake and body weight without producing any other behavioral changes.[29] γ-Vinyl GABA (GVG), also an inhibitor of GABA-transminase, decreases food intake in rats in a dose-dependent manner.[30] Depletion of brain dopamine with 6-OH-dopamine and serotoninergic blockade with metergoline do not attenuate the anorexigenic effect of GVG.[30]

Anorectics Acting Through Unspecified Mechanisms - A group of phenylpiperazinyl analogs are potent anorectic agents.[31] The most active of these compounds, 12 also decreases water intake and produces hypercholesterolemia in addition to decreasing food intake in rats. MB 11008 (13) decreases food intake and body weight in hens.[32]

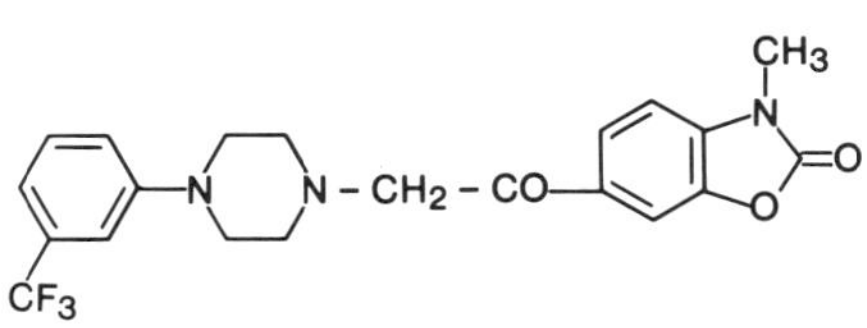

12

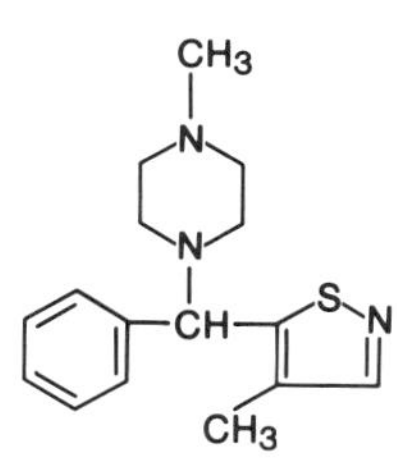

13

A number of 1,3-diaryltriazines show varying degrees of anorexigenic activity in rats, dogs and monkeys.[33] Chronic treatment with 14 decreases food intake and weight gain in several species without producing tolerance. A slight depression of motor activity, decreased hematocrit values and liver lesions are observed in rats following 30 days of treatment.[33]

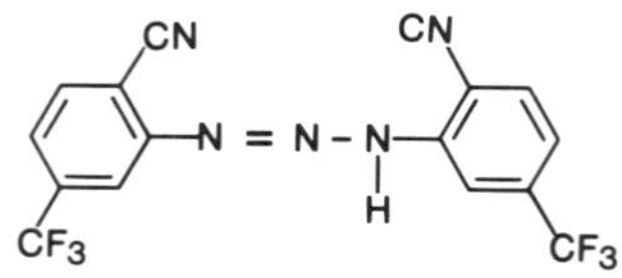

14

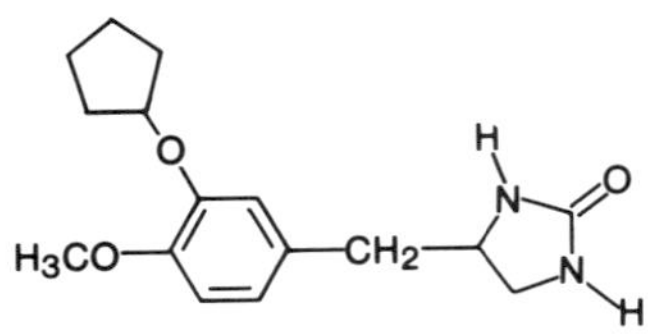

15

An imidazolidinone, GYK1 13 380 (15), a putative inhibitor of phosphodiesterase activity, produces anorexia in rats without causing CNS stimulation or depression. Tolerance to the anorexia does not develop during a one-month period of treatment.[34]

Peptide Anorectics - The role of cholecystokinin-octapeptide (CCK-8) in the control of appetitive behavior has been reviewed recently[35,36] and its clinical efficacy has been demonstrated in lean and obese subjects.[37-39] Ceruletide, which is structurally related to CCK-8, decreases solid food intake and subjective hunger ratings in patients.[40] A smaller fragment of cholecystokinin, CCK-4 inhibits food intake in rats at doses approximately 10-fold greater than those needed to produce the same effect with CCK-8.[41]

Bombesin (BBS) decreases food intake in normal[42] and obese[43] rats. In vagotomized animals, it retains its satiating effect[44] but it no longer inhibits gastric emptying.[45] Litorin, a BBS-like nonapeptide, also decreases food intake in rats in a dose-dependent manner at doses which do not decrease water intake.[46] Gastric Releasing Peptide (GRP), believed to be the mammalian counterpart of BBS, decreases food intake without decreasing water intake in rats.[47] Somatostatin[48] neurotensin,[49] satietin,[50] calcitonin,[51] TRH,[52] pancreatic polypeptide,[53] glucagon[54] and centrally administered insulin[55] also reduce food intake and/or decrease body weight in a variety of animal models.

Opioid Anorectics - Comprehensive reviews have been published about the importance of opioids in the regulation of appetite and drinking behavior.[56,57] It is known that opioid agonists such as morphine[58,59] and ethylketocyclazocine[60] as well as endogenous opioids such as beta-endorphin[61] can stimulate food and water intake. However, the response to opioid agonists is not always uniform or consistent and may depend on the class of opioid receptors involved. For instance, bremazocine, a kappa agonist increases food and water intake in mice while it decreases food intake in rats.[62] In addition, the ob/ob strain of genetically obese mice is moderately resistant to the kappa agonists butorphanol, trifluadom and ketocyclazocine[63] and does not demonstrate greatly elevated food intakes in response to subcutaneous injections of these opioid agonists.

The opioid antagonist naloxone is known to suppress food intake in animals[58] and more recently it has been shown to be effective in normal and obese humans.[64] Subjective hunger ratings and satiety were not affected. Obese subjects consume 30% fewer calories during the infusion of naloxone than do saline-infused controls.[65]

A study of the effect of several opioid antagonists on food intake and their ability to block opioid agonist-induced behaviors, demonstrates a degree of specificity which suggests that the mechanisms for these two activities may differ.[66] Naloxone, naltrexone, diprenorphine, Mr 1452 (16) and Mr 2266 (17) decrease food intake whereas WIN 44,441 (18) does not. Diprenorphine is the most potent of these agents in blocking the antinociceptive effect of morphine, whereas (16), (17) and (18) are weaker antagonists in this test.[66]

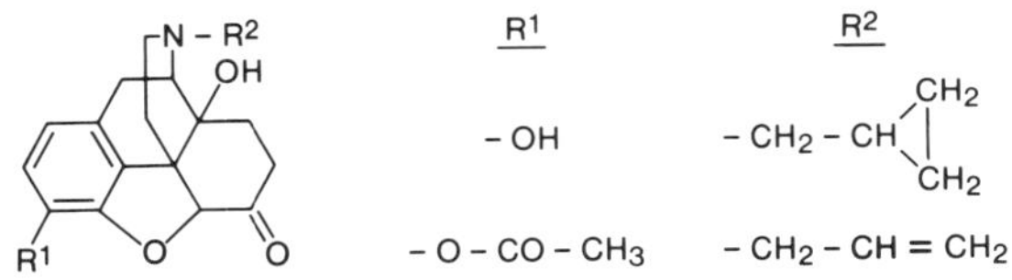

16

17

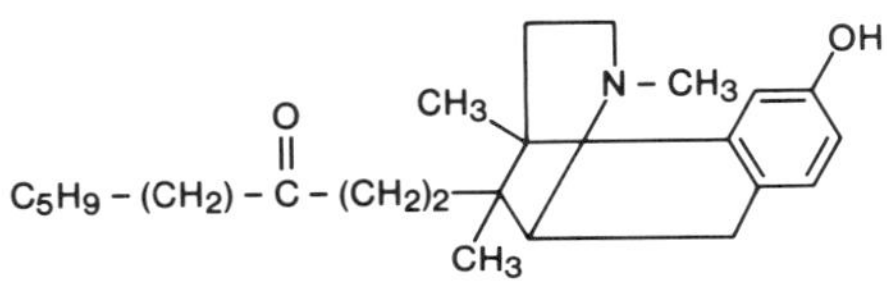

18

Peripherally Acting Anorectics - Fenfluramine[12] and CCK-8[36] are known to inhibit gastric emptying. (−)-threo-Chlorocitric acid[67] (19) is a new anorectic which has been shown to inhibit gastric emptying,[68] food intake and body weight gain through a selective reduction in body fat.[69] Tolerance to the anorexia does not develop during chronic treatment. Dogs are more sensitive than rats to its anorectic action, suggesting species-specific effects. It does not appear that the anorexia is mediated through CCK-8 since the anorectic effect of CCK-8,[36] but not that of 19 is abolished by vagotomy.

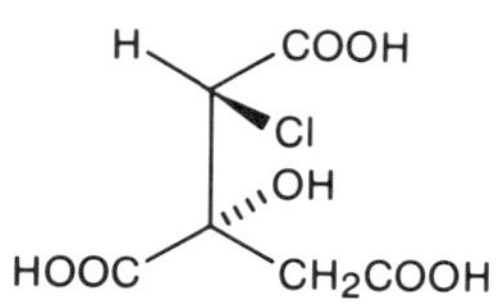

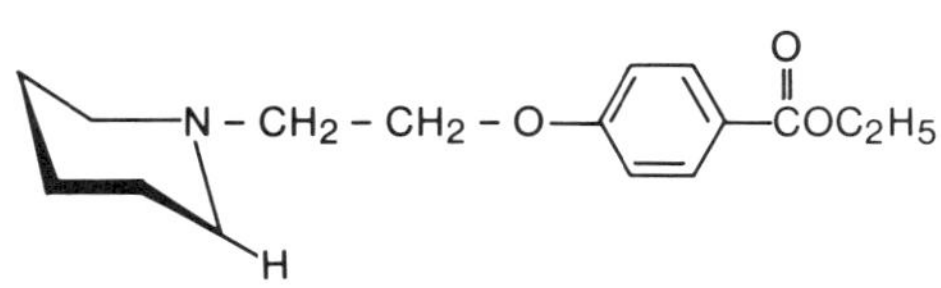

19 20

INHIBITION OF THE INTESTINAL ABSORPTION OF CARBOHYDRATES AND LIPIDS

Inhibition of Carbohydrate Absorption - The concept that retardation of the digestion of dietary carbohydrates would lead to a diminished post-prandial glycemia, insulinemia and triglyceridemia in patients with diabetes and/or obesity has stimulated the development of a number of inhibitors. Dietary di-, oligo- and polysaccharides must be degraded to monosaccharides by intestinal glucosidases before they are absorbed.

A number of amylase inhibitors have been reported; these are proteins, glycoproteins or pseudo-oligosaccharides isolated originally from grains or microorganisms.[70-74] Acarbose, Bay g 5421 is the most extensively studied of the glucosidase inhibitors. This pseudotetrasaccharide is a potent inhibitor of maltase, sucrase, glucoamylase and dextrinase.[75] In carbohydrate loading experiments in rats and humans, acarbose reduces the postprandial increases in circulating insulin and glucose levels.[76] Acarbose improves the metabolic status of both insulin-independent and insulin-dependent diabetics by lowering the hyperglycemia and glucosuria.[77,78] Studies in genetically obese Zucker rats[75] or diet-induced obese rats[79] treated with acarbose show reduced body weight gain and decreased sucrose-induced hypertriglyceridemia.[80] Symptoms of carbohydrate malabsorption are reported consistently with high doses of acarbose.[78]

Recently, a semisynthetic deoxynorjirimycin derivative Bay o 1248 (20), which is absorbed from the intestine and does not produce carbohydrate malabsorption, was reported as a potent glucosidase inhibitor exhibiting strong sucrase and maltase inhibitory activity and no amylase inhibition.[81,82] Administration of 20, which has a longer duration of action than acarbose, reduces food intake, body weight gain and epididymal fat pad weight in rats.[82]

Inhibition of Lipid Absorption - Agents which decrease the absorption of dietary lipids specifically by inhibiting pancreatic lipase may be suitable for the treatment of obesity. The reduced rate of lipid absorption produced by fenfluramine in rats is probably due to an inhibition of pancreatic lipase.[83,84] The structural features of a series of phenethylamines, including fenfluramine, required for lipase inhibition have been described using partially purified rat and human pancreatic lipase.[85] Pluronic L-101, a hydrophobic surfactant is a potent inhibitor of pancreatic lipase *in vitro* and reduces body weight gain and carcass lipid after chronic administration to rats.[86] Fecal excretion of dietary lipid is enhanced by Pluronic L-101, thus supporting its action as a lipase inhibitor.

A novel inhibitor of pancreatic lipase, Bay n 4605 (21) , was described recently.[82,87] This compound reduces post prandial hypertriglyceridemia and body weight in fat-fed rats.[87]

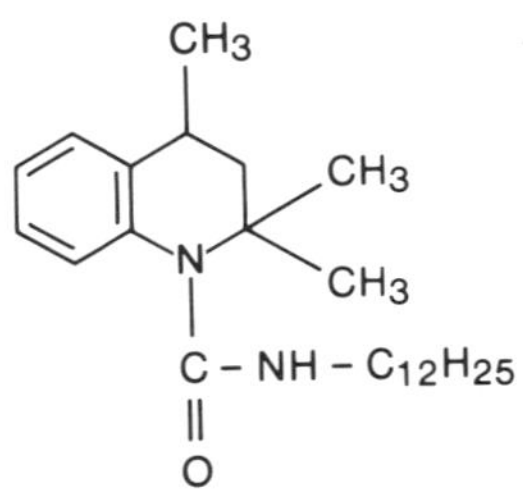

21

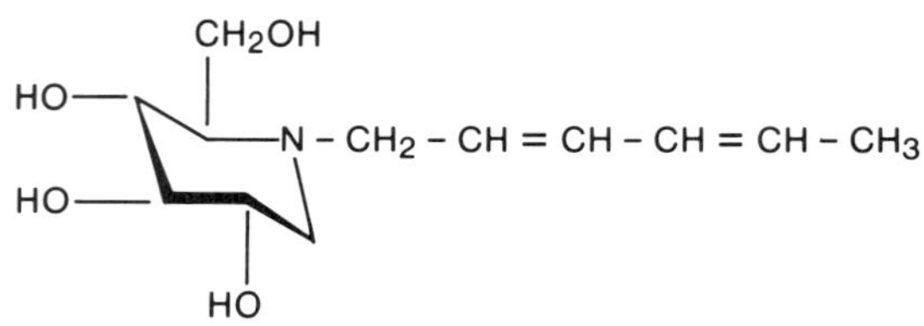

22

Another agent used for the modulation of intestinal absorption of lipids is sucrose polyester (SPE), a synthetic lipid which is not hydrolyzed by pancreatic lipase and hence is nonabsorbable. By providing a persistent lipophilic phase in the intestine, SPE reduces the absorption of lipophilic substances such as cholesterol. Mild decreases in low-density lipoprotein cholesterol levels were reported in hypercholesterolemic subjects.[88] Recently, the substitution of dietary fat with SPE in obese hypercholesterolemic patients produced a reduction in calories as well as low-density lipoprotein cholesterol levels.[89] These and other studies on SPE have been reviewed recently.[90]

Inhibition of Carbohydrate and Lipid Absorption - Another recently described deoxynorjirimycin derivative, Bay n 2920 (22), inhibits both dietary carbohydrate and lipid absorption.[82,91] As is the case with 20, this compound is absorbed almost completely from the intestine of rats and recovered in the urine.[82]

INHIBITORS OF LIPID BIOSYNTHESIS

Since one of the metabolic changes associated with some types of obesity in human beings and rodents is increased fatty acid synthesis, inhibitors of lipogenesis are sought as antiobesity agents. This area has been reviewed recently.[3]

Ro 22-0654 (23), an inhibitor of fatty acid synthesis, which possesses antiobesity activity in the rat has been described.[92,93] Chronic administration to both lean and genetically obese Zucker rats produces a significant reduction in body weight gain, which was accounted for by a selective reduction in body fat stores. Rats treated with 23 become less energy efficient as the amount of weight gained per quantity of food consumed is reduced significantly.

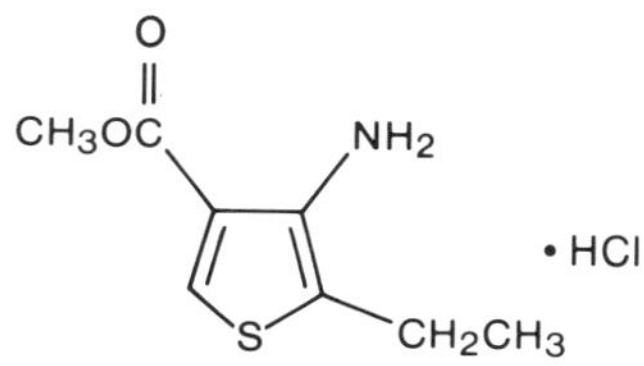

23

STIMULATION OF LIPID MOBILIZATION

Phenoxybenzamine, a non-selective α-antagonist, stimulated lipolysis in genetically obese rats,[94] and on chronic administration decreased body weight gain, food consumption and carcass lipid.[95] Fat cell proliferation is reduced during treatment, but returns to obese control levels following cessation of drug administration.

The new thermogenic agents which are described below also produce a marked stimulation of lipolysis in addition to their enhancement of energy expenditure.

REGULATION OF THERMOGENESIS

Excessive food intake does not always lead to obesity nor does moderate food intake guarantee leanness.[96-100] Indeed, the high energetic efficiency of several obese rodents[101-106] indicate that obesity can in part result from impaired energy expenditure. Recently, research has focused on the relationship between thermogenic activity in brown adipose tissue (BAT) and energy balance.

BAT is specialized for heat production and as such is fundamentally different from white adipose tissue which serves primarily as lipid stores. When activated by the sympathetic nervous system,[107] BAT mitochondrial substrate oxidation becomes uncoupled from ATP formation.[108,109] This is achieved by operation of a unique and regulated proton conductance pathway which permits protons to re-enter the mitochondrial matrix independent of ATP synthetase.[108,109] Respiratory control is restored *in vitro* when purine nucleotides block operation of the proton conductance pathway[110-112] by binding to a protein on the inner mitochondrial membrane[113] which is specific for BAT.[114,115] Measurements of GDP binding by BAT mitochondria and the relative amount of the uncoupling protein are commonly used as a qualitative index of the thermogenic capacity of BAT.

Recent studies show that BAT thermogenesis can be a significant component of overall energy expenditure for rats living in the cold[116,117] or during prolonged overfeeding of a palatable diet.[118,119] In contrast, many rodents with hypothalamic or genetic obesity exhibit defective BAT thermogenesis in response to diet and/or cold.[120,121]

While an extensive body of research demonstrates a causal relationship between BAT thermogenesis and energy balance in animals, it is not clear whether defective thermogenesis in BAT is relevant to human obesity.[121-123] There is histological evidence that BAT is present in adult man[124-126] as well as in neonates[127] and evidence that BAT exhibits biochemical properties similar to those of animals,[126,128] but currently there is no method available to directly assess the functional capacity of BAT in man.

THERMOGENIC AGENTS

Pharmaceutical research efforts are seeking thermogenic agents to treat obese individuals to counteract defective heat producing mechanisms as well as the inevitable reduction in metabolic rate which occurs following reducing dietary regimens. The major pharmacological approach has been to mimic sympathetic activation of thermogenic tissue by developing compounds that selectively stimulate adrenergic receptors or increase synaptic levels of noradrenaline by inhibiting its reuptake at nerve terminals.

Noradrenaline Uptake Inhibitors - Ciclazindol (24)[129-134] and its structural analog mazindol (25)[131,134-136] depress appetite and promote weight loss in lean and obese mammals, including man. Acute treatment with either ciclazindol[134,137,138] or mazindol[131,134,137,138] increases the resting metabolic rate (RMR) of lean rats. While mazindol is more potent and efficacious than ciclazindol at stimulating RMR in rats, ciclazindol has a far smaller effect on CNS arousal.[134,137,138] The peripheral action of mazindol may involve BAT since both mitochondrial GDP binding and protein content are increased in the tissue of cafeteria-fed rats.[134] Although energy balance studies of mazindol-treated rats fed a cafeteria diet fail to demonstrate an increase in energy expenditure, metabolic efficiency is reduced and the normal decline in RMR which occurs with reduced food intake is prevented.[134] It is not known whether either of these compounds exhibits thermogenic activity in human beings. However, ciclazindol induces weight loss in Type II diabetics, despite special efforts to maintain normal food intake.[132]

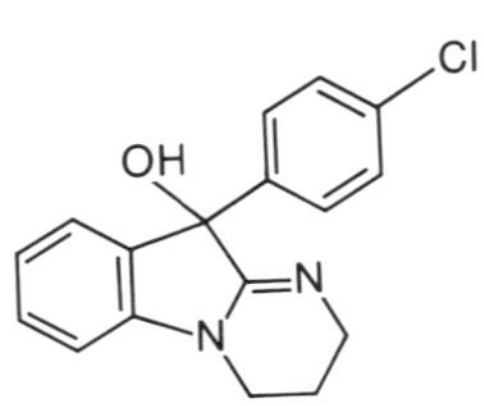

24

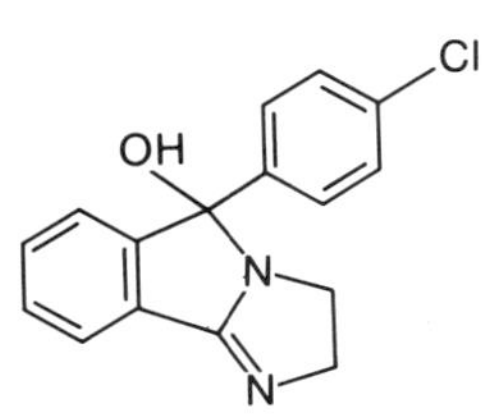

25

β-Adrenergic Agonists - BRL 26830A (26), Ro 16-8714 (27), LY104119 (28) and LY79771 (29) enhance thermogenesis by directly stimulating β-adrenoceptors. When chronically administered, these drugs cause weight loss or reduced weight gain in fa/fa rats, ob/ob mice and the A^{vy}/a strain of genetically obese mice, but not in their lean controls.[139-143]

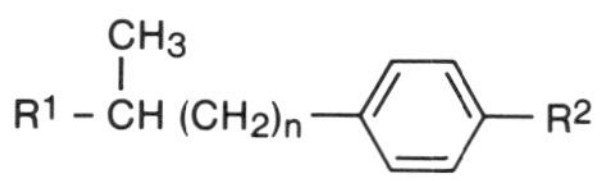

R^1	n	R^2	
$C_6H_4CH(OH)CH_2NH$ - .hemifumarate	1	$-COOCH_3$	26
$[C_6H_4CH(OH)CH_2]_2N$ -	2	$-CONH_2$	27
$C_6H_4CH(OH)CH_2NH$ - .HCl	2	$-CONH_2$	28
$C_6H_4CH(OH)CH_2NH$ -	2	-OH	29

Whole body calorimetry studies[139-141,143] and measurement of rectal temperature[142] show that obese rodents exhibit a rise in heat production following acute or chronic dosing with 26, 27, 28, or 29. In contrast to obese rodents which deplete lipid stores by increasing energy expenditure, lean rodents treated with 26, 28, 29 adjust both energy intake and energy expenditure to protect body fat content.[139,140] Prolonged dosing of lean mice with 28 or 29 promotes increases in both food intake and heat production.[142,143] The thermogenic effect of these compounds is due to direct stimulation of β-adrenoceptors, since propranolol inhibits the rise in metabolic rate of lean rats induced by a single dose of either 26 or 27 and blocks the 28-induced increase in CO_2 expiration of A^{vy}/a mice.[139,141,143] Propranolol also blocks the 29-stimulated increase of cyclic AMP and lipolysis in white adipose tissue of A^{vy}/a mice and their lean controls.[144]

Studies show that BAT is an effector tissue for the thermogenic action of 26, 27 and 29.[140,141,145] A single administration of 26 to lean rats markedly depletes BAT lipid content and preferentially increases BAT temperature.[140] Binding of GDP to BAT mitochondria increases in both lean and ob/ob mice following administration of 26. Treatment with 26 also causes an increase in mitochondrial protein and in the relative proportion of uncoupling protein, similar to that of cold-acclimated mice.[140] The effect of 27 on BAT metabolism is also well characterized. Respiration is stimulated by 27 in isolated brown adipocytes of lean rats and is completely blocked by propranolol.[141] As with 26, chronic treatment with 27 causes hypertrophy of BAT and increases mitochondrial GDP binding in fa/fa rats and ob/ob mice.[141] Compound 29 also exerts its thermogenic effects through increased BAT metabolism since it stimulates respiration in isolated brown adipocytes of hamsters.[145]

BAT has a substantial capacity for lipogenesis which increases during acclimation to cold[146,147] and sucrose overfeeding.[148] Administration of both ^3H_2O and U-^{14}C glucose to animals indicates that glucose is a major substrate for BAT lipogenesis in cold-acclimated rats.[149] Additionally, hexokinase and phosphofructokinase activities are unusually high in BAT of rats and increase during cold acclimation.[150] Activities of these key glycolytic enzymes are normal in liver, cardiac muscle and skeletal muscle of db/db mice, but are markedly lower in BAT than those of control mice.[151] However, prolonged administration of 26 to db/db mice increases the activities of hexokinase and phosphofructokinase in BAT thereby increasing the capacity of this tissue for utilization of blood glucose.[151]

REFERENCES

1. A.C. Sullivan, H.W. Baruth and L. Cheng, Ann. Rep. Med. Chem., 15, 172 (1980).
2. A.C. Sullivan and R. Gruen. Fed Proc. (in press).
3. A.C. Sullivan, J.G. Hamilton, J. Triscari, in "Biochemical Pharmacology of Obesity" P.B. Curtis-Prior, Ed. Elsevier Science Publishers, 1983, pp. 311-337.
4. A.C. Sullivan, J. Triscari, K. Comai. International Symposium on Novel Approaches and Drugs for Obesity. Int. J. Obesity (in press).
5. A.C. Sullivan, J. Triscari, L. Cheng, in "Curr. Concepts Nutr." Vol. 12, M. Winick, Ed. John Wiley and Sons, Inc., New York, 1983, pp. 139-167.
6. "Biochemical Pharmacology of Obesity", P.B. Curtis-Prior, Ed. Elsevier, Amsterdam, 1983.
7. "Drugs and Appetite", T. Silverstone, Ed. Academic Press, New York, 1982.
8. "Anorectic Agents: Mechanisms of Action and Tolerance", S. Garattini and R. Samanin, Eds. Raven Press, New York, 1981.
9. "Obesity: Pathogenesis and Treatment", G. Enzi, G. Crepaldi, G. Pozza, and A.E. Renold, Eds. Academic Press, New York, 1981.
10. J.G. Douglas, J.F. Munro. Pharmac. Ther. 18, 351 (1982).
11. R. Pietrusko, A. Stunkard, K. Brownell, D.B. Campbell. Int. J. Obes. 6, 567 (1982).
12. R.F. Davies, J. Rossi,3d,. J. Panksepp, N.J. Bean, A.J. Zolovick. Physiol. Behav. 30, 723 (1983).
13. J. Smedegaard, P. Christiansen, B. Skrumsager. Int. J. Obes. 5, 377 (1981).
14. H. Mengel, J. Lund, J. Lange. Arzneim. Forsch. 33, 462 (1983).
15. R.J. Simpson, D.J. Lawton, M.H. Watt, B. Tiplady. Brit. J. Clin. Pharmacol. 11, 96 (1981).
16. N. Harto-Truax, J.V.W. Fleet, W.C. Stern. Psychopharmacol. Bull. 18, 72 (1982).
17. C. Dumont, J. Laurent, A. Grandodam, J.R. Boissier. Life Sci. 28, 1939 (1981).
18. R. Roncucci, G.F. Miranda. International Symposium on Novel Approaches and Drugs for Obesity. Int. J. Obes. (in press).
19. R. Samanin, in "Medicinal Chemistry Advances" F.G. de las Heras, S. Vega, Eds. Pergamon Press, New York, 1981, p. 271.
20. P. Redgrave, E.B. Talia, L. White, P. Dean. Psychopharmacol. 76, 273 (1982).
21. S.P. Arneri, A. Roetker, J.P. Long. Neuropharmacol. 21, 885 (1982).
22. A. Girdhar, D.C. Sharma, K.S. Sachdev. Indian J. Physiol. Pharmacol. 26, 317 (1982).
23. M.O. Carruba, S. Ricciardi, P. Mantegazza. Life Sci. 27, 1131 (1980).
24. F. Cattabeni, A. Maggi, M. Monduzzi, L. Deangelis, C. Racagni. J. Neurochem. 31, 565 (1978).
25. B.R. Cooper, J.L. Howard, H.L. White, F. Soroko, R. Ingold, R.A. Maxwell. Life Sci. 26, 1997 (1980).
26. J.N. Nobrega, D.V. Coscina. Brain Res. 7, 243 (1983).
27. N. Blavet, F.V. DeFendis, F. Clostre. Pshycopharmacol. 76, 75 (1982).
28. J. Birk, R.G. Noble. Life Sci. 30, 321 (1982).
29. H.L. White, J.L. Howard, B.R. Cooper, F.E. Soroko, J.D. McDermed, K.J. Ingold, R.A. Maxwell. J. Neurochem. 39, 271 (1982).
30. S. Huot, M.G. Palfreyman. Pharmacol. Biochem. Behav. 17, 99 (1982).
31. M.C. Bermann, P. Berthelot, J.P. Bonte, M. Debaert, D. Lesieur, C. Brunet, M. Cazin, I. Lesieur, M. Lucyckx, J.C. Cazin. Arzneim. Forsch. 32, 604 (1982).
32. A.H. Sykus. Proc. Nutr. Soc. 42, A93 (1983).
33. D.T. Hill, K.G. Stanley, J.E.K. Williams, B. Loev, P.J. Fowler, J.P. McCafferty, E. Macko, C.E. Berkoff, C.B. Ladd. J. Med. Chem. 26, 865 (1983).
34. F. Andrasi. Neurosci. Letts. Suppl. 10, S44 (1982).
35. G.P. Smith, J. Gibbs, C. Jerome, F.X. Pi-Sunyer, H.R. Kissileff, J. Thornton. Peptides 2, 57 (1981).
36. G.P. Smith. Lancet 2(8341), 88 (1983).
37. H.R. Kisseleff, F.X. Pi-Sunyer, J. Thornton, G.P. Smith. Am. J. Clin. Nutr. 34, 154 (1981).
38. G. Stacher, H. Steinringer, G. Schmierer, C. Schneider, S. Winklehner. Peptides 3, 133 (1982).
39. F.X. Pi-Sunyer, H.R. Kissileff, J. Thornton, G.P. Smith. Physiol. Behav. 29, 627 (1982).
40. G. Stacher, H. Steinringer, G. Schmierer, C. Schneider, S. Winklehner. Peptides 3, 607 (1982).
41. L. Halmy, C. Nyakas, J. Walter. Experientia 38, 873 (1982).
42. J. Gibbs, P.J. Kulkosky, G.P. Smith. Peptides 2, 179 (1981).
43. C.L. McLaughlin, C.A. Baile. Physiol. Behav. 25, 341 (1980).
44. G.P. Smith, C. Jerome, J. Gibbs. Peptides 2, 409 (1981).
45. F. Porreca, T.F. Burks. Gastroenterol. 85, 313 (1983).
46. P.J. Kulkosky, J. Gibbs. Life Sci. 31, 685 (1982).
47. L.J. Stein, S.C. Woods. Peptides 3, 833 (1982).
48. A.S. Levine, J.E. Morley. Pharmacol. Biochem. Behav. 16, 897 (1982).
49. D. Luttinger, R.A. King, D. Sheppard, J. Strupp, C.B. Nemeroff, A.J. Prange Jr. Eur. J. Pharmacol. 81, 499 (1982).

50. J. Knoll. Pol. J. Pharmacol. Pharm. 34, 3 (1982).
51. A.S. Levine, J.E. Morley. Brain Res. 222, 187 (1981).
52. J.E. Morley, A.S. Levine, C. Prasad. Brain Res. 210, 475 (1981).
53. J.P. Mordes, G.L. Eastwood, S. Loo, A.A. Rossini. Peptides 3, 873 (1982).
54. N. Geary, G.P. Smith. Physiol. Behav. 283, 13 (1982).
55. S.C. Woods, E.C. Lotter, L.D. McKay, D. Porte Jr. Nature 282, 503 (1979).
56. D.J. Sanger. Appetite 2, 193 (1981).
57. J.E. Morley, A.S. Levine, G.T. Yim, M.T. Lowry. Neurosci. Behav. Rev. 7, 281 (1983).
58. J.E. Jalowiec, J. Panksepp, A.J. Zolovick, N. Najam, B.H. Herman. Pharmacol. Biochem. Behav. 15, 477 (1981).
59. F.S. Tepperman, M. Hirst. Pharmacol. Biochem. Behav. 17, 1141 (1982).
60. J.D. Sanger, P.S. McCarthy. Psychopharmacol. 74, 217 (1981).
61. L.D. McKay, N.J. Kenney, N.K. Edens, R.H. Williams, S.C. Woods. Life Sci. 29, 1429 (1981).
62. U. Hartig, K. Opitz. Arch. Int. Pharmacodyn. 262, 4 (1983).
63. J.E. Morley, A.S. Levine, B.A. Goswell, J. Kneip, M. Grace. Physiol. Behav. 31, 603 (1983).
64. E. Trenchard, T. Silverstone. Appetite 4, 43 (1983).
65. R.L. Atkinson. J. Clin. Endocrinol. Metab. 55, 196 (1982).
66. D.J. Sanger, P.S. McCarthy, J.A.H. Lord, C.F.C. Smith. Drug Develop. Res. 3, 137 (1983).
67. A.C. Sullivan, R.W. Guthrie, J. Triscari, in "Anorectic Agents: Mechanisms of Action and Tolerance", S. Garattini, R. Samanin, Eds. Raven Press, New York, 1981, p. 143.
68. J. Triscari, A.C. Sullivan. Pharmacol. Biochem. and Behav. 15, 311 (1981).
69. A.C. Sullivan, W. Dairman, J. Triscari. Pharmacol. Biochem. and Behav. 15, 303 (1981).
70. E. Truscheit, D.D. Schmidt, A. Arens, H. Lange, W. Wingender, in "Regulators of Intestinal Absorption in Obesity, Diabetes and Nutrition", Vol. 2, P. Berchtold, M. Cairella, A. Jacobelli, V. Silano, Eds. Societa editrice universo, Roma, 1981, pp. 157-179.
71. L. Vertesy, V. Oeding, R. Bender, G. Nessemann, D. Sukatsch, K.H. Zepf, in "Regulators of Intestinal Absorption in Obesity, Diabetes and Nutrition", Vol. 2, P. Berchtold, M. Cairella, A. Jacobelli, V. Silano, Eds. Societa editrice universo, Roma, 1981, pp. 269-274.
72. D.D. Schmidt, W. Frommer, B. Junge, L. Müller, E. Truscheit, W. Wingender, in "Regulators of Intestinal Absorption in Obesity, Diabetes and Nutrition", Vol. 2, P. Berchtold, M. Cairella, A. Jacobelli, V. Silano, Eds. Societa editrice universo, Roma, 1981, pp. 203-230.
73. K. Yokose, K. Ogawa, T. Sano, K. Watanabe, H.B. Maruyama, Y. Suhara. J. Antibiotics 36(9), 1157 (1983).
74. K. Yokose, K. Ogawa, Y. Suzuki, I. Umeda, Y. Suhara. J. Antibiotics 36, 1166 (1983).
75. W. Puls, U. Keup, H.P. Krause, L. Müller, D.D. Schmidt, G. Thomas, E. Truscheit. Front. Hormone Res. 7, 235 (1980).
76. W. Puls, U. Keup, H.P. Krause, G. Thomas, F. Hoffmeister. Naturwissenschaften 64, 536 (1977).
77. W. Creutzfeldt, in "First International Symposium on Acarbose", W. Creutzfeldt, Ed. Excerpta Medica. International Congress Series 594, Amsterdam-Oxford-Princeton, 1982.
78. R. Vargas, F.G. McMahon, J.R. Ryan, R. Michael, M.D. Jain. Clin. Pharmacol. Ther. 33, 216 (1983).
79. J.R. Vasselli, E. Haraczkiewicz, F.X. Pi-Sunyer. Nutr. Behav. 1, 21 (1982).
80. H.P. Krause, U. Keup, G. Thomas, W. Puls. Metab. 31, 710 (1982).
81. B. Junge, J. Stoltefub, L. Müller, H.P. Krause, R. Sitt. Ger. Offen. DE 3,007,078 (1981).
82. W. Puls, H.P. Krause, L. Müller, H. Schutt, R. Sitt, G. Thomas. International Symposiun on Novel Approaches and Drugs for Obesity. Int. J. Obesity (in press).
83. K. Comai, J. Triscari, A.C. Sullivan. Biochem. Pharmacol. 27, 1987 (1978).
84. P.B. Curtis-Prior, A.R. Oblin, S. Tan. Int. J. Obesity 4, 111 (1980).
85. K. Comai, A.C. Sullivan. J. Pharma. Sci. 71, 418 (1982).
86. K. Comai, A.C. Sullivan. Int. J. Obesity 4, 33 (1980).
87. H.J. Kabbe, H.P. Krause, R. Sitt. Ger. Offen. DE 2,945,238 (1981).
88. M.J. Mellies, R.J. Jandacek, J.D. Taulbee, M.B. Tewksbury, G. Lamkin, L. Baehler, P. King, D. Boggs, S. Goldman, A. Gouge, R. Tsang, C.J. Glueck. Am. J. Clin. Nutr. 37, 339 (1983).
89. C.J. Glueck, R. Jandacek, E. Hogg, C. Allen, L. Baehler, M. Tewksbury. Am. J. Clin. Nutr. 37, 347 (1983).
90. R.J. Jandacek. International Symposium on Novel Approaches and Drugs for Obesity. Int. J. Obesity (in press).
91. B. Junge, L. Müller, R. Sitt, G. Thomas, H.P. Krause, W. Puls. Eur. Pat. Appl. EP 22,192 (1981).
92. J. Triscari, A.C. Sullivan. Fed. Proc. 42, 664 (1983).

93. J. Triscari, A.C. Sullivan. Life Sci. (in press).
94. B.E. Levin, K. Comai, A.C. Sullivan. Pharm. Biochem. and Behav. 14, 517 (1981).
95. K. Comai, A.C. Sullivan. Am. J. Phys. 243, E398 (1982).
96. H. Keen, B.J. Thomas, R.J. Jarrett, J.H. Fuller. Brit. Med. J. 1, 655 (1979).
97. G.A. Rose, R.T. Williams. Brit. J. Nutr. 15, 1 (1961).
98. E.A.H. Sims, E. Danforth Jr., E.S. Horton, G.A. Bray, J.A. Glennon, L.B. Salans. Rec. Prog. Horm. Res. 29, 457 (1973).
99. P.A. Stefanik, F.P. Heald, J. Mayer. Am. J. Clin. Nutr. 7, 55 (1959).
100. E.M. Widdowson. Proc. Nutr. Soc. 21, 121 (1962).
101. P.L Thurlby, P. Trayhurn. Br. J. Nutr. 42, 377 (1979).
102. J.G. Vander Tuig, D.R. Romsos, G.A. Leveille. J. Nutr. 110, 35 (1980).
103. P. Trayhurn, L. Fuller. Diabetologia 19, 148 (1980).
104. S. Deb, R. Martin. J. Nutr. 105, 543 (1975).
105. A. Djazayery, D.S. Miller, M.J. Stock. Nutr. Metab. 23, 357 (1979).
106. J.E. Cox, T.L. Powley. Am. J. Physiol. 240, E566 (1981).
107. L. Landsberg, J.B. Young. Life Sci. 28, 1801 (1981).
108. D.G. Nicholls. Biochim. Biophys. Acta 549, 1 (1979).
109. D.G. Nicholls, R.M. Locke. Physiol. Rev. 64, 1 (1984).
110. D.G. Nicholls. Eur. J. Biochem. 49, 573 (1974).
111. R.M. Locke, E. Rial, I.D. Scott, D.G. Nicholls. Eur. J. Biochem. 129, 373 (1982).
112. R.M. Locke, E. Rial, D.G. Nicholls. Eur. J. Biochem. 129, 381 (1982).
113. M. Heaton, R.J. Wagenvoord, A. Kemp, D.G. Nicholls. Eur. J. Biochem. 82, 515 (1978).
114. B. Cannon, A. Hedin, J. Nedergaard. FEBS Letters 150, 129 (1982).
115. D. Ricquier, J.P. Barlet, J.M. Garel, M. Combes-George, M.P. Dubois. Biochem. J. 210, 859 (1983).
116. D.O. Foster, M.L. Frydman Can. J. Physiol. Pharmacol. 56, 110 (1978).
117. D.O. Foster, M.L. Frydman. Can. J. Physiol. Pharmacol. 57, 257 (1979).
118. N.J. Rothwell, M.J. Stock. Pflügers Arch. 389, 237 (1981).
119. N.J. Rothwell, M.J. Stock. Nature 281, 31 (1979).
120. J. Himms-Hagen. New Engl. J. Med. (in press).
121. J. Himms-Hagen. Nutr. Rev. 41, 261 (1983).
122. B.E. Levin, A.C. Sullivan. International Symposium on Novel Approaches and Drugs for Obesity. Int. J. Obesity (in press).
123. S. Blaza. J. Roy. Soc. Med. 76, 213 (1983).
124. J.M. Heaton. J. Anat. 112, 35 (1972).
125. P. Huttunen, J. Hirvonen, V. Kinsula. Eur. J. Appl. Physiol. 46, 339 (1981).
126. D. Ricquier, M. Nechad, G. Mory. J. Clin. Endocrinol. Metab. 54, 803 (1982).
127. B. Cannon, B.W. Johansson. Molec. Aspects Med. 3, 119 (1980).
128. M.E.J. Lean, W.P.T. James. FEBS. Letters 163, 235 (1983).
129. E.M. Chart, A. Fletcher, M.G. Wyllie. Brit. J. Pharm. 75, 111P (1982).
130. K. Ghose V.A.R. Rao, J. Bailey, A. Coppen. Psychopharm. 57, 109 (1978).
131. S.S. Lang, E. Danforth Jr., E.L. Lien. Life Sci. 33, 1269 (1983).
132. M.E.J. Lean, L.J. Borthwick. Eur. J. Clin. Pharmacol. 25, 41 (1983).
133. W.P. Leary, K. Van der Byl, M. Howlett. Current Therapeutic Res. 34, 728 (1983).
134. M.G. Wyllie, A. Fletcher, N.J. Rothwell, M.J. Stock. International Symposium on Novel Approaches and Drugs for Obesity. Int. J. Obesity (in press).
135. J.R.S. Arch. Am. J. Clin. Nutr. 34, 2763 (1981).
136. P.J. Miach, W. Thomson, A.E. Doyle, W.J. Louis. Med. J. Aust. 2, 278 (1976).
137. A. Fletcher, D. Green, N.J. Rothwell, R.J. Stephens, M.J. Stock, M.G. Wyllie. Br. J. Pharmacol. 74, 770P (1981).
138. N.J. Rothwell, M.J. Stock, M.G. Wyllie. Br. J. Pharmac. 74, 539 (1981).
139. J.R.S. Arch, A.T. Ainsworth. Am. J. Clin. Nutr. 38, 549 (1983).
140. J.R.S. Arch, A.T. Ainsworth, R.D.M. Ellis, V. Piercy, V.E. Thody, P.L. Thurlby, C. Wilson, S. Wilson, P. Young. International Symposium on Novel Approaches and Drugs for Obesity. Int. J. Obesity, (in press).
141. M.K. Meier, L. Alig, M.E. Bürgi-Saville, M. Müller. International Symposium on Novel Approaches and Drugs for Obesity. Int. J. Obesity (in press).
142. W.N. Shaw, K.K. Schmiegel, T.T. Yen, R.E. Toomey, D.B. Meyers, J. Mills. Life Sci. 29, 2091 (1981).
143. T.T. Yen, M.M. McKee, N.B. Stamm. International Symposium on Novel Approaches and Drugs for Obesity. Int. J. Obesity (in press).
144. T.T. Yen, M.M McKee, N.B. Stamm, K.G. Berris. Life Sci. 32, 1515 (1983).
145. J.S. Horng, R.E. Toomey. Fed. Proc. 42, 379 (1983).
146. L. Aguis, D.H. Williamson. Biochim. Biophys. Acta. 666, 127 (1981).
147. P. Trayhurn. Biochim. Biophys. Acta 664, 549 (1981).
148. J.G. Granneman, G.N. Wade. Metabolism 32, 202 (1983).
149. J.G. McCormack. Prog. Lipid Res. 21, 195 (1982).
150. G.J. Cooney, E.A. Newsholme. FEBS. Letters. 148, 198 (1982).
151. M.A. Cawthorne, M.J. Carroll, A.L. Levy, C.A. Lister, M.V. Sennitt, S.A. Smith, P. Young. International Symposium on Novel Approaches and Drugs for Obesity. Int. J. Obesity (in press).

CHAPTER 17. ALDOSE REDUCTASE INHIBITORS AS A NEW APPROACH TO THE TREATMENT OF DIABETIC COMPLICATIONS

Christopher A. Lipinski and Nancy J. Hutson
Pfizer Central Research, Groton, CT 06340

INTRODUCTION

Diabetic Complications and Current Therapy. Research on the potential importance of the polyol pathway and the effects of aldose reductase (AR) inhibitors as an approach to the treatment of diabetic complications has markedly escalated since diabetes mellitus was last reviewed in Vol. 16 of *Annual Reports.*[1] Accordingly, recent developments in this area constitute the major focus of this review. Several excellent reviews of the recent advances in the understanding of the biochemical events contributing to the pathophysiology of diabetic complications have recently been published.[2-7] A recent comprehensive review summarizes the literature on medical approaches claimed to be effective in treatment of cataract.[8] A review on the biochemistry of the polyol pathway is in preparation.[9].

Diabetes mellitus (DM) which consists of an absolute (Type I or insulin-dependent, IDDM) or relative (Type II or non-insulin-dependent, NIDDM) lack of insulin is a disease of increasing incidence which affects five to six million people in the United States[10] and 20 to 30 million people worldwide. Currently available treatments of the disease can correct acutely life-threatening symptoms but do not prevent diabetic complications which are largely responsible for the large degree of morbidity and mortality present in the diabetic population. Diabetic complications consist of several different types. Diabetes mellitus is responsible for loss of vision in 12% of the total blind U.S. population and 84% of diabetic loss of sight is due to proliferative retinopathy.[11] Diabetics are at greater risk of developing senile cataracts than non-diabetics[12] and these along with glaucoma and optic neuropathy, account for the remaining blindness in the diabetic population. Chronic renal failure and peripheral neuropathy, which leads to a high incidence of motor, sensory and autonomic dysfunction, are also important and debilitating problems that diabetics develop. Macroangiopathy in medium and large arteries make stroke and coronary artery disease two times as prevalent in diabetics. Diabetics also have a three or four times greater chance of developing peripheral arterial disease and a five times greater chance than the general population of developing gangrene which leads to amputation. The end result of all of these complications of the disease is that the average diabetic has a life expectancy only two-thirds that of non-diabetics.[10]

Much recent evidence points to the fact that the metabolic consequences of insulin lack and, more specifically, hyperglycemia lead to the pathophysiology of diabetic complications.[13,14] Currently available treatments of diabetes that seek to correct this insulin lack and resulting hyperglycemia consist of oral hypoglycemic drugs, diet and insulin injection.

A more recent treatment consisting of constant insulin delivery over 24 hours by a mechanized and computerized pump (CSII, continuous subcutaneous insulin infusion), has been developed in an attempt to achieve better glycemic control. The efficacy of this relatively new treatment on the prevention of diabetic complications is still unknown with both positive[15] and negative results[16] thus far reported. There is also some risk in achieving strict glycemic control as these patients are in danger of developing ketoacidosis which could lead to death.[17] Pancreatic and islet transplantation are also new methods of treatment which are meeting with limited degrees of success mostly due to the rejection of the transplanted tissue and the difficulty of obtaining tissue for transplant.[18] Whether successful or not these latter two newer treatments are targeted towards the minority of diabetic patients suffering from Type I (IDDM) as opposed to the 85-90% of the total diabetic patient population who suffer

ISBN 0-12-040519-9

from Type II (NIDDM) diabetes. In summary, none of the currently available treatments appear to achieve the necessary control of glycemia (or the appropriate delivery of insulin) to prevent the development of diabetic complications in diabetic patients.

ALDOSE REDUCTASE AND THE POLYOL PATHWAY

Polyol Pathway. Recently, theories and data have been emerging which implicate aldose reductase (AR), increased flux through the polyol pathway due to hyperglycemia and accumulation of sorbitol and fructose in the pathophysiology of diabetic complications. The biological rationale and chemical progress towards design of AR inhibitors constitute the major focus of this review.

The polyol pathway consists of two enzymes, AR and sorbitol dehydrogenase. AR has a rather broad substrate specificity for sugars and a fairly high Km for glucose.[19-24] These two enzymes catalyze the following reactions:

$$\text{Glucose} \xrightarrow{\text{NADPH} \quad \text{NADP}^+} \text{Sorbitol} \xrightarrow{\text{NAD}^+ \quad \text{NADH}} \text{Fructose}$$

Because of the high Km of AR for glucose, it is hypothesized that under normoglycemic conditions there is very little flux through the polyol pathway. On the other hand, under conditions of hyperglycemia, as in diabetes, the flux through this pathway can be considerable and the products of the pathway, sorbitol and fructose, are slowly metabolized and not freely diffusable through the cell plasma membrane, and thus tend to accumulate in tissues.[1,2] The metabolic and biochemical consequences of flux through this pathway and the contribution of these consequences to the pathophysiology of diabetic complications are currently under intense study. The very extensively studied AR inhibitor, sorbinil (**1**), has aided greatly in the elucidation of the role of the polyol pathway in the etiology of diabetic complications.

Tissue Location and Role of Aldose Reductase in Animal Models of Diabetic Complications. Aldose reductase (AR) has been located immunohistochemically in many tissues of the dog and rat, most notably, in corneal epithelium, retina, optic nerve, kidney papillae, aortic endothelium and smooth muscle cells as well as peripheral nerve and lens.[25-28] AR has also been measured in human and monkey retinal mural cells.[29-30] These cells are thought to provide the structural support for retinal capillaries and their loss is the first abnormality seen in clinical diabetic retinopathy. In addition, AR-like activity has been reported in a human retinoblastoma cell line and sorbinil inhibits this activity in these cells.[31] Finally, a recent report has demonstrated that AR is present in isolated capillaries from bovine retina and cerebral cortex.[32] Therefore, AR appears to be present in all tissues which are uniquely susceptible to deterioration during prolonged exposure to the hyperglycemia of diabetes. Accumulation of the products of the polyol pathway, sorbitol and fructose, has been demonstrated in these tissues[2] and, where tested, sorbinil and other AR inhibitors have been shown to inhibit this accumulation.

Data have recently been published which enhance the potential importance of the polyol pathway in different animal models of diabetic complications. Long-term diabetic dogs (3-5 years) are the only known animal model which seems relevant for clinical diabetic retinopathy. In this model dogs are fed a high galactose diet (galactose is an excellent substrate for aldose reductase) and have normal insulin and glucose levels but nevertheless develop lesions similar to those in clinical diabetic retinopathy.[33] Retinal capillary basement membrane thickening is seen in galactose-fed rats and prevention of this with sorbinil treatment has been reproted by two separate investigators.[34,35] While perhaps this animal model is not as relevant to retinopathy as the galactose-fed dogs, basement membrane thickening has been implicated in the pathogenesis of nephropathy and the generalized deterioration of both large and small arteries. In addition, poor wound healing is also a major problem in diabetics.

Sorbinil and alrestatin (AY-22,284, **7**) have been shown to normalize the impaired ability of diabetic and galactosemic rats to undergo corneal re-epithelialization after complete corneal denudation.[36,37]

Polyol Pathway and Neuropathy in Diabetic Animals. Peripheral neuropathy is a common and serious complication of diabetes, especially in patients with long-standing disease who are poorly controlled. Decreased motor and sensory nerve conduction velocity can be seen in diabetic patients of short duration before the appearance of any clinical neurological symptoms, but it is not known if the early electrophysiological abnormalities are directly linked to the later neuropathy. The relationship, however, between the early electrophysiological changes and abnormalities in nerve membrane phospholipids containing *myo*-inositol important to nerve transmission is becoming more clearly understood. Chemically-induced diabetes in animals leads to increased nerve sorbitol and fructose, loss of nerve *myo*-inositol and decreased motor nerve conduction velocity (MNCV) within two weeks of the onset of hyperglycemia.[38] It is as yet uncertain if and how these different biological effects of diabetes on nerve function and metabolism are interrelated, but recent studies on the biochemical effects of AR inhibitors have contributed to further understanding of these relationships. First, it has recently been shown that glucose directly inhibits *myo*-inositol transport in rabbit endoneural preparations[39] and therefore, hyperglycemia itself could directly contribute to nerve *myo*-inositol loss. Sorbinil *in vitro* in rat endoneurial preparations prevents glucose from having this effect.[40]

The importance of *myo*-inositol loss to the overall biochemical and functional changes in diabetic nerve is supported by the finding that feeding diabetic rats a relatively high *myo*-inositol diet corrects defects in MNCV,[38,41] Na^+-K^+ ATPase[42] and restores *myo*-inositol levels in peripheral nerves even though nerve sorbitol and fructose remain elevated under these conditions.[43,44] There may be a further link between *myo*-inositol loss and Na^+-K^+ ATPase activity since *myo*-inositol transport is Na^+-dependent and loss of Na^+-K^+ ATPase activity could further reduce the ability of the nerve cell to accumulate *myo*-inositol.[45]

Depending upon the experimental protocol, sorbinil treatment prevents or reverses all of these effects of diabetes on rat peripheral nerve, that is sorbinil restores nerve *myo*-inositol and Na^+-K^+ ATPase activity and corrects impaired MNCV.[43,46,47] ICI-105,552 (**8**) reduces nerve sorbitol, normalizes nerve *myo*-inositol and prevents changes in MNCV when animals are treated from the initiation of diabetes.[48] When given after diabetes was established **8** moderately inhibited sorbitol and fructose accumulation in nerve, but was unable to restore nerve sorbitol accumulation after 2 weeks of treatment in diabetic rats.[50] Whatever the mechanistic link between sorbitol accumulation, *myo*-inositol and Na^+-K^+ ATPase loss and nerve conduction velocity changes, it is clear from the data that at least with sorbinil all of these pathological changes in the diabetic rat can be prevented. These effects of AR inhibitors may also translate into prevention or reversal of longer term neuropathic changes.

Polyol Pathway and Clinical Diabetic Neuropathy. Recent clinical data strengthens the argument for a role of AR in diabetic neuropathy in man. Firstly, elevated fructose and sorbitol and lowered *myo*-inositol levels are found in post mortem sciatic nerve samples obtained from diabetic patients.[51] Secondly, current clinical results with two AR inhibitors support common mechanisms for the impact of diabetes on nerve function in humans and in animal models of diabetes. After nine weeks of treatment, sorbinil significantly improved nerve conduction velocity in the peroneal, median sensory and median motor nerves in insulin-dependent diabetics.[52] This effect disappeared when patients were taken off drug. Recently patients with painful diabetic neuropathy previously unresponsive to other drugs, were treated with sorbinil, and moderate to marked symptomatic relief was noted after 3 or 4 days of medication. On stopping medication, pain returned.[53] Similar results have also been reported by another investigator.[54] The relatively weak AR inhibitor, alrestatin, when given over 12 wks at high doses to diabetic patients with clinical signs of polyneuropathy, produced significant differences over placebo for many of the measured variables.[55] Improvements in sensory nerve conduction velocity were observed when alrestatin was given intravenously for 5 days [56] but in another study no beneficial effect was found.[57] Treating diabetic patients with *myo*-inositol has met with limited success in affecting symptoms of diabetic peripheral neuropathy.[58-60] This may be due to difficulty in providing the appropriate dose as too much *myo*-inositol has been shown to be harmful to nerve function in diabetic rats.[38] AR inhibition, therefore, may provide the more simple and direct approach to the treatment of diabetic polyneuropathy.

Polyol Pathway and Cataracts. Studies with the orally active AR inhibitor sorbinil at the National Eye Institute have demonstrated the importance of the polyol pathway to cataract formation. The proposal that the accumulation of sugar alcohol and fructose (via the polyol pathway) that occurs in lenses of diabetic or galactosemic animals was the crucial biochemical event precipitating the cataractogenic process was not convincingly shown in initial studies with less potent AR inhibitors (i.e. alrestatin, quercitrin (**21**)) which delayed but did not prevent sugar cataract formation.[61-62] More recently, using sorbinil as the AR inhibitor, the NEI group has been able to substantiate their hypothesis. Sorbinil treatment totally prevented cataract development for periods of up to 12 months in both diabetic and galactosemic animals.[63-65] A sorbinil analog M79175 (**3**) also inhibited the formation of cataract in galactosemic rats for a period of 30 days and essentially normalized lens sorbitol and fructose in diabetic rats.[66] The carboxylic acid ICI-105,552 also delayed the appearance of cataract in diabetic rats.[67]

Biochemical studies support the premise that sufficient AR inhibition can abolish the other pathological changes associated with cataractogenesis. For example, data from several laboratories shows that sorbinil not only prevents[68] but also reverses [69] lens ultrastructural changes and prevents the loss of Na^+-K^+ATPase[70] activity and tissue *myo*-inositol[71] that occurs during sugar cataract formation. The total prevention of galactosemic cataract by sorbinil is not associated with a reduction in the increase in non-enzymatic glycosylation of lens proteins seen under these conditions.[72]

In summary, it is yet unknown whether a common pathophysiological mechanism underlies all diabetic complications. However, it is clear from the literature that aldose reductase inhibition is a therapeutic approach to these chronic disorders.

ALDOSE REDUCTASE INHIBITOR SAR

Acidic Cyclic Amides Sorbinil (**1**) represents the prototype member of this class of AR inhibitors. The extensive biological data base on this agent as a probe for the role of AR inhibition in treatment of diabetic complications is discussed earlier in this chapter. A chiral synthesis of the S enantiomer has been described.[73] The relative potency of the S to R enantiomer is 45 using rat lens AR (RLAR) and 20 using human placental AR (HPAR).[74] The oxazolidinedione (**2**) gave *in vitro* activity (calf lens AR) similar to that of **1** but reduced *in vivo* activity as measured by prevention of diabetic rat sciatic nerve sorbitol accumulation. As with the hydantoin series activity in the oxazolidinedione class was improved by 6-halosubstitution and resided in the S enantiomer.[75] M79175 (**3**) inhibits bovine lens AR (BLAR) at low doses, inhibits lens sorbitol accumulation in streptozotocin diabetic rats and given orally delays rat galactose induced cataract formation.[66] The best of the series of isoxazolidine-3,5-diones **4** showed marginal *in vivo* activity, which was attributed to a pKa significantly different from that of sorbinil.[76] 3,4-Dialkoxy substitution as in **5** gave optimum activity in an extensive series of 5-aryl-thiazolidine-2,4-diones which inhibited HPAR and swelling in a rat lens culture assay.[77] Among a series of 1-phenylsulfonyl hydantoins the best activity in inhibition of RLAR and BLAR was found for **6.** Based on inhibition studies at pH 6 and 7.5, the conclusion was reached that 5-substituted hydantoins inhibit AR in their ionized forms while 1-substituted hydantoins inhibit AR in the non-ionized form.[78]

Carboxylic Acids Alrestatin (**7**) was much less active against AR from human brain than against AR from calf lens or human placenta.[79] The R enantiomer of a methyl alrestatin analog showed a slight activity decrease over alrestatin and was only slightly more active than the S enantiomer.[74]

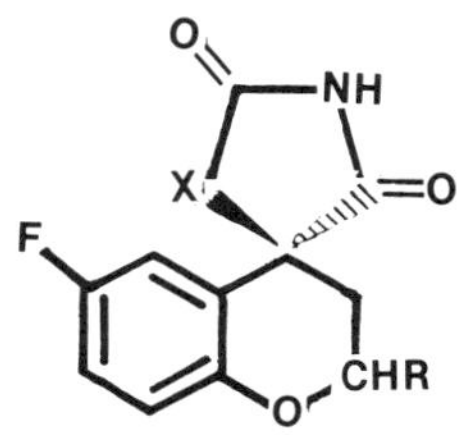

1, X = NH R = H
2 X = O R = H
3 X = NH , R = CH_3

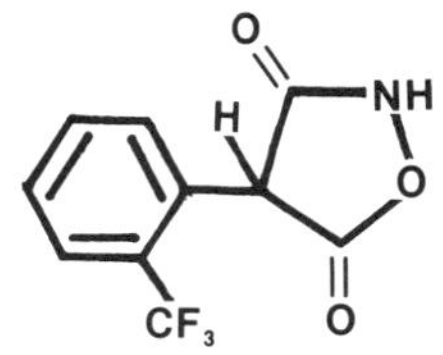

4

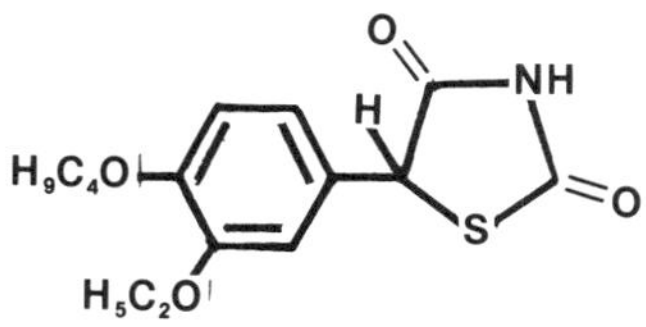

5

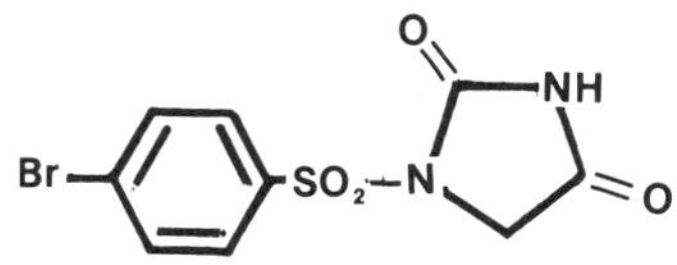

6

ICI-105,552 (**8**) reduced sorbitol levels in sciatic nerve and lens of streptozotocinized diabetic rats, but was without effect on reversing the reduction in *myo*-inositol;[80,49] however, in another study *myo*-inositol loss was prevented but only when drug was given from the onset of diabetes.[44] Elevated sorbitol levels were completely decreased in rat sciatic nerve and MNCV decreases were prevented.[44,48] ICI-105,552 markedly delayed the onset of cataract development in rats in an 11-month study and prevented occular haemorrhage.[67] As opposed to cataract prevention, retinal sorbitol levels were not significantly decreased with this compound in diabetic rats after 3 weeks[49,80] and consistent with this finding retinal lesions were not prevented even after 1 year.[81] Similarly, even though **8** prevented sorbitol accumulation in monkey kidney cells cultured in high glucose medium,[82] no inhibition of rat kidney basement membrane thickening was seen after 1 year of oral therapy.[81] Elevations in liver and kidney weight were seen both at 3 weeks and 1 year.[81]

ONO-2235 (**9**) is the predominant E,E geometric isomer resulting from condensation of α-methylcinnamaldehyde with rhodanine-N-acetic acid and not the E,Z isomer as incorrectly shown in reference 83. This agent is a potent inhibitor of RLAR and given orally for two weeks markedly reduced sorbitol levels in rat sciatic nerve and improved MNCV[50] and has completed three months clinical evaluation in neuropathy patients.[83]

ICI-128,436 (**10**) a phthalazine acetic acid inhibits RLAR or human lens AR(HLAR). With plasma t½ in rat of 10 hr, single doses reduced sorbitol and fructose values to the normal range in streptozotocin-diabetic rats and chronic doses prevented the development of lens sugar cataracts and deterioration of sciatic MNCV.[84]

AY-27,773-tolrestat (**11**) is a potent inhibitor of BLAR and given orally in a galactosemic rat model decreases sciatic nerve galactitol; this agent is currently in clinical trials.[85]

NSAI Agents Several carboxylic acids in clinical use as nonsteroidal antiinflammatory (NSAI) agents inhibit AR. Sulindac (**12**) and its human sulfide metabolite **13** inhibit HLAR.[86] Sorbitol accumulation in rat lens or nerve incubated in high glucose media is weakly inhibited by **12**. Topical application of **13** to rat eye also weakly inhibits sorbitol build.up.[87] Like HLAR, AR from cataractous human lens is inhibited by **12**, which is more active than indomethacin or the aspirin metabolite sodium salicylate.[88] Aspirin has been proposed for the treatment of human cataracts based on epidemiological considerations[89] and the fact that it penetrates into rabbit lens.[90] The lysine salt of the clinical NSAI bendazac (**14**) has been reported to improve visual acuity in cataractous patients in a small double blind study.[91,92] Drug levels in rat lens are far lower than in serum after chronic dosing. A possible mechanism based on preventing lens protein precipitation was proposed.[93]

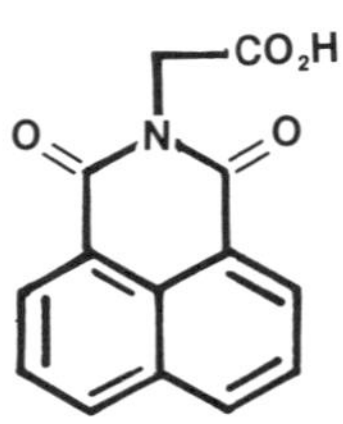

7

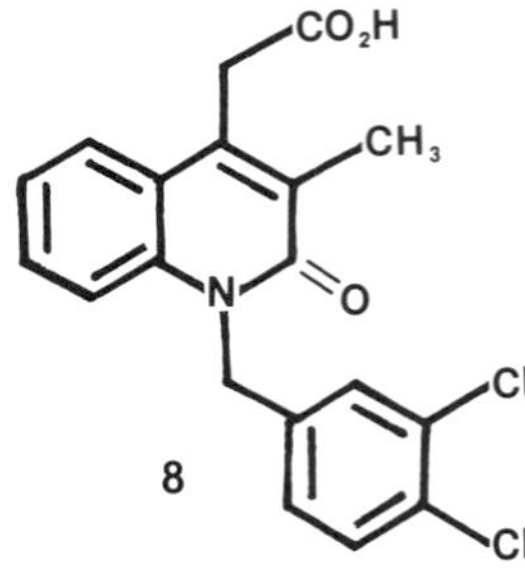

8

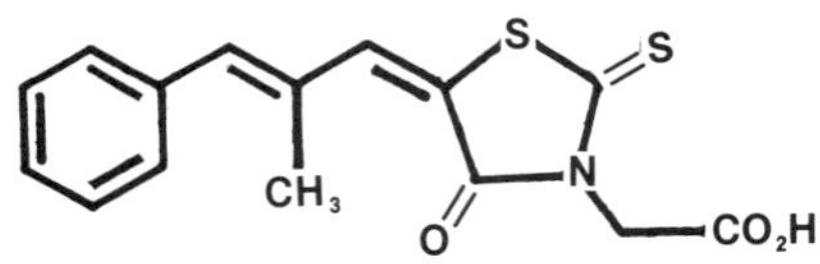

9

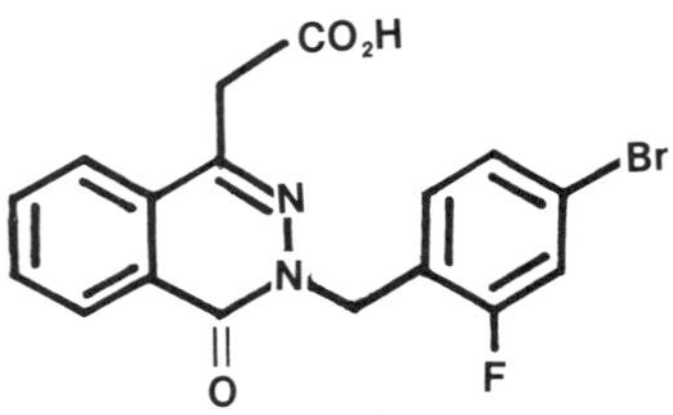

10

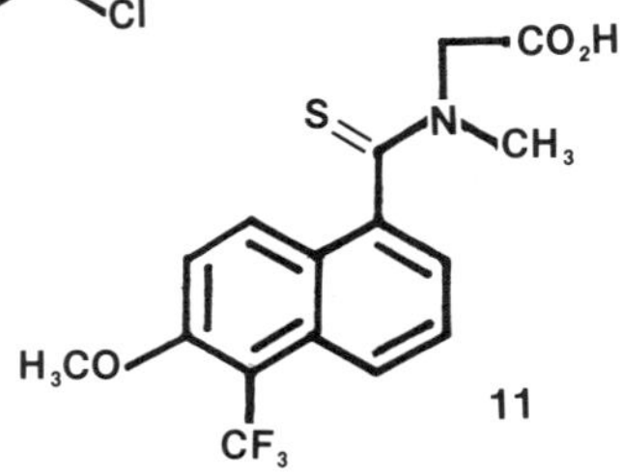

11

Antiallergy Agents. 7-Dimethylsulfamoyl-xanthone-2-carboxylic acid (**15**) an antiallergy agent inhibits rabbit lens AR[94] and when applied to the closed eye pouch of neonatal galactosemic rats, partially inhibits galactose induced adverse lens changes.[95] A correlation has been noted between antiallergy activity, xanthine oxidase inhibitory and AR inhibitory activity.[96] Since literature exists correlating antiallergy activity with electron accepting ability of an antiallergy inhibitor carbonyl group, a similar correlation was sought among a series of antiallergy agents which were examined as inhibitors of RLAR and HPAR. Although a quantitative relationship between carbonyl group electron accepting ability as measured by the carbonyl group lowest unoccupied molecular orbital (LUMO) and AR inhibition was not observed, the widespread AR activity among several antiallergy series was interpreted as supporting a charge transfer interaction between an enzyme nucleophile acting as an electron pair donor towards an AR inhibitor reactive carbonyl moiety.[97] All the inhibitors tested were lipophilic and planar carboxylic or enolic acids many of which possessed the molecular fragment **16**.

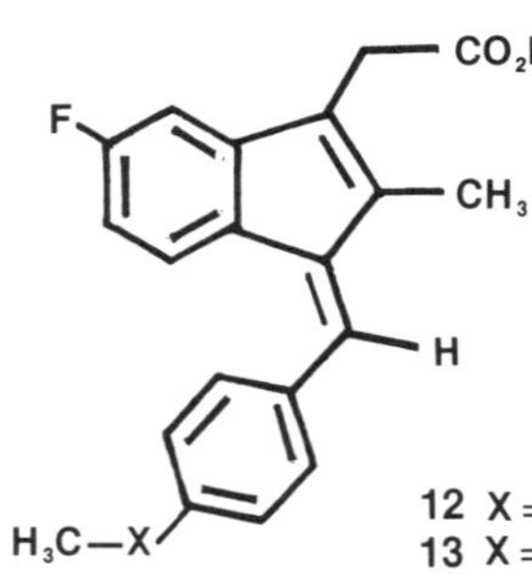

12 X = SO
13 X = S

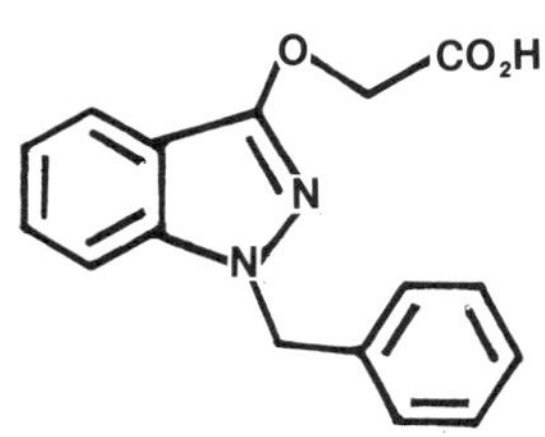

14

Among a smaller, more homogeneous series of 4-oxo-4H-chromens **17**, a significant correlation between LUMO energies and log % inhibition of RLAR was observed and a similar charge transfer mechanism was proposed.[98] Extension of LUMO calculations to a wide variety of structurally diverse AR inhibitors led to a proposed model of the AR inhibitor site having as key features two parallel lipophilic regions and a charge transfer pocket. Predictions were made as to structural features required for effective inhibition.[99]

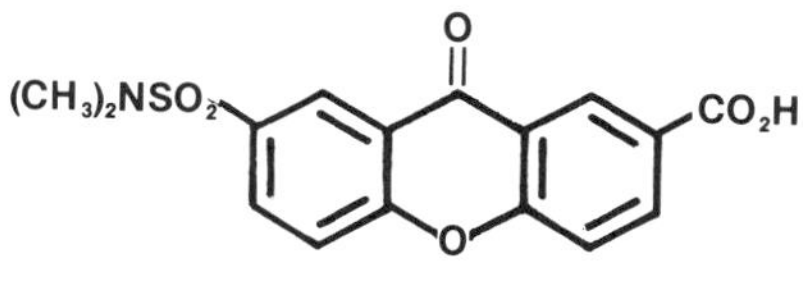

15

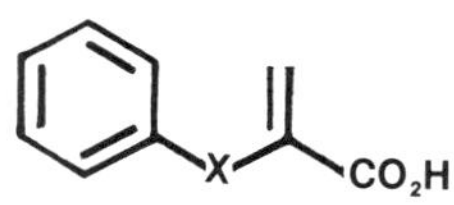

16 X = O, NH

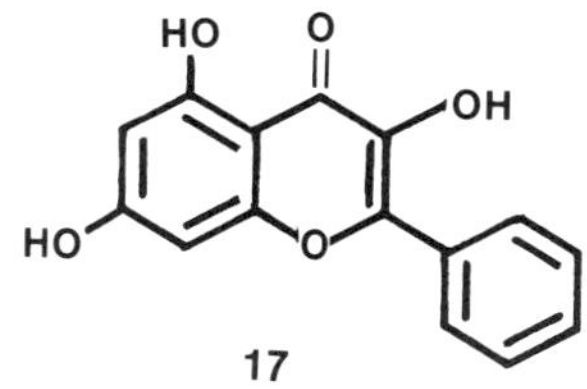

17

Phenols. The flavonoids (**18**) and axillarin (**19**) are potent *in vitro* inhibitors of RLAR.[100] 3-Substitution with groups as diverse as hydrogen (**18**), methoxy (**19**), hydroxy (**20**-quercetin), glycosidic (**21**-quercitrin), and acetate (**22**) is compatible with activity as long as rings A and C contain hydroxy or ethereal sulfate groups.[86,100] In contrast to this tenuous SAR, coumarins and isoflavones are inactive. In addition to RLAR, AR activity of flavonoids can be detected using BLAR and HLAR but not with flavonoid (**23**) when tested against AR derived from monkey kidney epithelial cells.[82] The tetrahydroxybenzophenone 70A196 (**24**) active against RLAR orally inhibits xylose induced cataracts in rats albeit at high dose.[101]

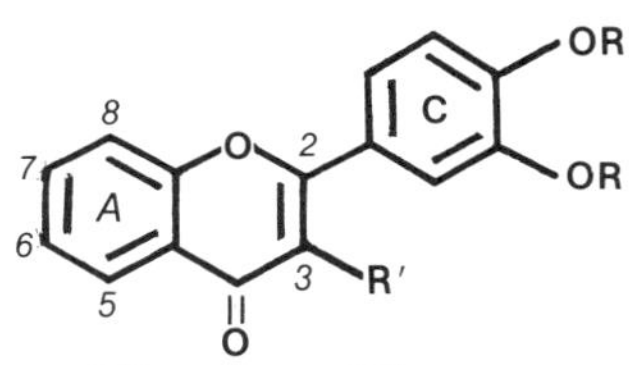

	R	R′	5	6	7	8
18	H	H	CH_3O	HO	CH_3O	CH_3O
19	H	CH_3O	HO	CH_3O	HO	HO
20	H	HO	HO	H	HO	H
21	Compd **20** 3-O-*L*-Rhamnoside					
22	HO_3S	H_3CCO_2	H	H	HO_3SO	H
23	H	HO	HO	H	$HO(CH_2)_2O$	H

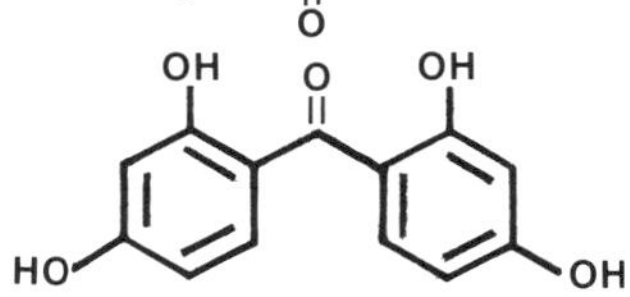

24

Significant differences in susceptibility of AR enzymes to inhibition have been observed. No trends can be predicted other than that in general **HPAR** is less susceptible to inhibition than other AR enzymes. From a drug discovery viewpoint fluctuation in inhibitor potency to aldose reductase from various sources complicates the discovery of a clinically effective AR and may require the use of human aldose reductase from the appropriate target tissue.[102]

REFERENCES

1. C.R. Rasmussen, B.E. Maryanoff and G.F. Tutwiler, *Annu. Reports Med. Chem.*, **16**, 173 (1981).
2. M. Brownlee, and A. Cerami, *Annu. Rev. Biochem.*, **50**, 385 (1981).
3. O.P. Ganda, *Diabetes*, **29**, 931 (1980).
4. S.M. Mauer, M.W. Steffes, F.C. Goetz, D.E.R. Sutherland, and D.M. Brown, *Diabetes*, **32**, Supl. 2, 52, (1983).
5. L.I. Rand, *Am. J. Med.*, **70**, 595 (1981).
6. J. Hilsted, *Diabetes*, **31**, 730 (1982).
7. M.J. Brown, and A.K. Asbury, *Ann. Neurol.*, **15**, 2 (1984).
8. P.F. Kador, *Ophthalmology (Rochester, Minn.)*, **90**, 352 (1983).
9. P.F. Kador, *Annu. Rev. Pharmacol*, 1984 in preparation.
10. O. Crofford in, *"Report of the National Comission on Diabetes"* DHEW Publ. No. (N.I.H.) **76-1018**, (1975).
11. J.C. Liang, and M.F. Goldberg, *Diabetes*, **29**, 841 (1980).
12. F. Ederer, R. Hiller, and H.R. Taylor, *Am. J. Opthalmol.*, **91**, 381 (1981).
13 K.M. West, L.J. Erdreich and J.A. Stober, *Diabetes*, **29**, 501 (1980).
14. K.M. West, M.M.S. Ahuja, P.H. Bennett, B. Grab, V. Grabauskas, O. Mateo-de-Acosta, J.H. Fuller, R.J. Jarrett, H. Kien, K. Kosaka, A.S. Krolewski, E. Mik, V. Schliak and A. Teuscher, *Diabetologia*, **22**, 42, (1982).
15. The Steno Study Group, *Lancet*, 16 Jan, 121 (1982).
16 T. Lauritzen, K. Frost-Larsen, H.-W. Larsen, T. Deckert, and The Steno Study Group, *Lancet*, Jan. 29, 200 (1983).
17. R.H. Unger, *Diabetes*, **31**, 479 (1982).
18. D.R. Sutherland, *Annu. Rev. Med.*, **33**, 271 (1982).

19. C.M. Sheaff, and C.C. Doughty, *J. Biol. Chem.*, **251**, 2696 (1976).
20. H.-H. Tinge, and M.J.C. Crabbe, *Biochem. J.*, **215**, 351 (1983).
21. H.-H. Tinge, and M.J.C. Crabbe, *Biochem. J.* **215**, 361 (1983).
22. R.K. Herman, P.F. Kador, and J.H. Kinoshita, *Exp. Eye Res.*, **37**, 467 (1983).
23. K.H. Gabbay and J.H. Kinoshita, *Methods Enzymol.* **XLI**, 159 (1975).
24. P.F. Kador, D. Carper and J.H. Kinoshita, *Anal. Biochem.* **114**, 53 (1981).
25. T.S. Kern, and R.L. Engerman, *Histochem. J.* **14**, 507 (1982).
26. M.A. Ludvigson, and R.L. Sorenson, *Diabetes*, **29**, 438 (1980).
27. M.A. Ludvigson, and R.L. Sorenson, *Diabetes*, **29**, 450 (1980).
28. C.C. Corder, J.G. Collins, T.S. Brannan and J. Sharma, *J. Histochem. Cytochem.*, **25**, 1 (1977).
29. Y. Akagi, P.F. Kador, T. Kuwabara, and J.H. Kinoshita, *Invest. Ophthalmol. Vis. Sci.*, **24**, 1516 (1983).
30. S.M. Buzney, R.N. Frank, S.D. Varma, T. Tanishima and K.H. Gabbay, *Invest. Opthalmol. Vis. Sci.*, **16**, 392 (1977).
31. P. Russell, L.O. Merola, Y. Yajima and J.H. Kinoshita, *Exp. Eye Res.*, **35**, 331 (1982).
32 A. Kennedy, R.N. Frank and S.D. Varma, *Invest. Opthalmol. Vis. Sci.*, **24** 1250 (1983).
33. R.L. Engerman and T.S. Kern, *Diabetes*, **33**, 97 (1984).
34. W.G. Robison, P.F. Kador and J.H. Kinoshita, *Science*, **221**, 1177 (1983).
35. R.N. Frank, R. Keirn, A. Kennedy and K.W. Frank, *Invest. Ophthalmol. Vis. Sci.*, **24**, 1519 (1983).
36. S. Fukushi, L.O. Merola, M. Tanaka, M. Datiles and J.H. Kinoshita, *Exp. Eye Res.*, **31**, 611 (1980).
37. M.B. Datiles, P.F. Kador, H.N. Fukui, T.S. Hu and J.H. Kinoshita, *Invest. Ophthalmol. Vis. Sci.*, **24**, 563 (1983).
38. D.A. Greene, P.V. DeJesus, and A.I. Winegrad, *J. Clin. Invest.*, **55**, 1326 (1975).
39. D.A. Greene and S.A. Lattimer, *J. Clin. Invest.*, **70**, 1009 (1982).
40. K.R.W. Gillon and J.N. Hawthorne, *Biochem. J.*, **210**, 775 (1983).
41. D.A. Greene, R.A. Lewis, S.A. Lattimer and M.J. Brown, *Diabetes*, **31**, 573 (1982).
42. D.A. Greene and S.A. Lattimer, *J. Clin. Invest.*, **72**, 1058 (1983).
43. K.R.W. Gillon and J.N. Hawthorne, *Life Sci.*, **32**, 1943 (1983).
44. J.H. Mayer and D.R. Tomlinson, *Diabetologia*, **25**, 433 (1983).
45. D.A. Greene, *Metabolism*, **7** Suppl. 1, 118 (1983).
46. D.K. Yue, A. Hanwell, P.M. Satchell, and J.R. Turtle, *Diabetes*, **31**, 789 (1982).
47. D. Finegold, S.A. Lattimer, S. Nolle, M. Bernstein and D.A. Greene, *Diabetes*, **32**, 988 (1983).
48. D.R. Tomlinson, P.R. Holmes, and J.H. Mayer, *Neurosci. Letters*, **31**, 189 (1982).
49. R. Poulsom, D.J. Mirrlees, D.C.N. Earl and H. Heath, *Exp. Eye Res.*, **36**, 751 (1983).
50. R. Kikkawa, I. Hatanaka, H. Yasuda, N. Kobayashi, Y. Shigeta, H. Terashima, I. Morimura and M. Tsuboshima, *Diabetologia*, **24**, 290 (1983).
51. J.A. Mayhew, K.R.W. Gillon and J.N. Hawthorne, *Diabetologia*, **24**, 13 (1983).
52. R. Judzewitsch, J.B. Jaspan, K.S. Polonsky, C.R. Weinberg, J.B. Halter, E. Halar, M.A. Pfeiffer, C. Vukadinovich, L. Bernstein, M. Schneider, K.Y. Liange, K.H. Gabbay, A.H. Rubenstein, D.Porte Jr., *N. Engl. J. Med.*, **308**, 119 (1983).
53. J. Jaspan, R. Maselli, K. Herold and C. Bartkus, *Lancet*, 758 (1983).
54. R.J. Young, D.J. Ewing and B.F. Clarke, *Diabetes*, **32**, 938 (1983).
55. J. Fagius and S. Jameson, *J. Neurol. Neurosurg. Psychiatry* **44**, 991 (1981).
56. A. Culebras, J. Alio, J.L. Herrara, *I. Lopez-Fraile, Arch. Neurol.*, **38**, 133 (1981).
57. K.H. Gabbay, N. Spack, S. Loo, H.J. Hirsch, A.A. Ackil, *Metabolism*, **28**, 471 (1979).
58. R.S. Clements, B. Vourganti, T. Kuba, S.J. Oh, and B. Darnell, *Metabolism*, **28**, 477 (1979).
59. G. Gregerson, B. Bertelsen, H. Harlio, E. Larsen, J.R. Andersen, A. Helles, M. Schmiegelow and J.E.J. Christensen, *Acta Neurol. Scand.*, **67**, 164 (1983).
60. J.G. Salway, J.A. Finnegan, D. Barnett, L. Whitehead, A. Karuanayaka and R.B. Payne, *Lancet*, 1282 (1978).
61. S.D. Varma, A. Mizuno and J.H. Kinoshita, *Science*, **195**, 205 (1977).
62. D. Dvornik, N. Simard-Duquesne, M. Krami, K. Sestanj, K.H. Gabbay, J.H. Kinoshita, S.D. Varma and L.O. Merola, *Science*, **182**, 1146 (1973).
63. M.J. Peterson, R. Sarges, C.E. Aldinger and D.P. MacDonald, *Metabolism*, **28**, 456 (1979).
64. S. Fukushi, L.O. Merola and J.H. Kinoshita, *Invest. Ophthalmol. Vis. Sci.* **19**, 313 (1980).
65. M.B. Datiles, H. Fukui, T. Kuwabara and J.H. Kinoshita, *Invest. Ophthalmol. Vis. Sci.* **22**, 174 (1982).
66. H. Ono, Y. Nozawa and S. Hayano, *Nippon Ganka Gakkai Zasshi*, **86**, 253 (1982).
67. R. Poulsom, R.P. Boot-Handford and H. Heath, *Curr. Eye Res.* **2**, 351 (1983).
68. A. Beyer-Mears, E. Cruz, J. Nicolas-Alexandre and E. Varagiannis, *Pharmacology*, **24**, 193 (1982).
69. T.-S. Hu, M. Datiles and J.H. Kinoshita, *Invest. Ophthalmol. Vis. Sci.*, **24**, 640 (1983).
70. N.J. Unaker and J.Y. Tsui, *Exp. Eye Res.*, **36**, 685 (1983).
71. N. Freinkel and L. Cook, *Clin. Res.* **28**, 389A (1980).
72. S.-H. Chiou, L.T. Chylack Jr., H.F. Bunn and J.H. Kinoshita, *Biochem. Biophys. Res. Commun.*, **95**, 894 (1980).
73. R. Sarges, H.R. Howard Jr. and P.R. Kelbaugh, *J. Org. Chem.*, **47**, 4081 (1982).
74. P.F. Kador, J.D. Goosey and N.E. Sharpless, *Eur. J. Med. Chem.*, **16**, 293 (1981).
75. R.C. Schnur, R. Sarges and M.J. Peterson, *J. Med. Chem.*, **25**, 1451 (1982).
76 A.B. Richon, M.E. Maragoudakis and J.S. Wasvary, *J. Med. Chem.*, **25**, 745 (1982).
77. T. Sohda, K. Mizuno, E. Imamiya, H. Tawada, K. Meguro, Y. Kawamatsu and Y. Yamamoto, *Chem. Pharm. Bull.*, **30**, 3601 (1982).
78. K. Inagaki, T. Miwa, T. Yashiro and J. Okuda, *ibid.*, **30**, 3244 (1982).
79. M. M. O'Brien, P.J. Schofield and M.R. Edwards, *J. Neurochem.*, **39**, 810 (1982).
80. R. Poulsom and H. Heath, *Biochem. Pharmacol.*, **32**, 1495 (1983).
81. R. Poulsom, R.P. Boot-handford and H. Heath, *Exp. Eye. Res.*, **37**, 507 (1983).
82. R. Boot-handford and H. Heath, *Biochem. Pharmacol.*, **30**, 3065 (1981).
83. N. Sakamoto and N. Hotta, *Farumashia*, **19**, 43 (1983).
84. D. Stribling, D.J. Mirrlees and D.C.N. Earl, *Diabetologia*, **25**, 196 (1983).
85. K. Sestanj, F. Bellini, S. Fung, N.A. Abraham, A. Treasurywala, L. Humber, N. Simard-Duquesne and D. Dvornik, *Abstract of Papers 186th ACS National Meeting, MEDI* **9** (1983).
86. P.S. Chaudhry, J. Cabrera, H.R. Juliani and S.D. Varma, *Biochem. Pharmacol.*, **32**, 1995 (1983).
87. M. Jacobson, Y.R. Sharma, E. Cotlier and J.D. Hollander, *Invest. Ophthalmol. Vis. Sci.*, **24**, 1426 (1983).
88. Y.R. Sharma and E. Cotlier, *Exp. Eye Res.*, **35**, 21 (1982).
89. E. Cotlier and Y.R. Sharma, *Lancet*, **1**, 358 (1981.
90. E. Cotlier, Y.R. Sharma, T. Niven and M. Brescia, *Am. J. Med.*, **83** (1983).

91. L. Bonomi, G. Marchini, I. DeFranco, S. Perfetti and M. DeGregorio, *Curr. Ther. Res.* **33**, 727 (1983).
92. M. Testa, G. Iuliano and B. Silvestrini, *Lancet*, **1**, 849 (1982).
93. B. Silvestrini, B. Catanese, G. Barillari, E. Iorio and P. Valeri, *Int. J. Tiss. React.* **V**, 217 (1983).
94. J.R. Pfister, W.E. Wymann, J.M. Mahoney and L.D. Waterbury, *J. Med. Chem.*, **23**, 1264 (1980).
95. A. Beyer-Mears, E. Cruz, J. Nicolas-Alexandre and E. Varagiannis, *Arch. Int. Pharmacodyn. Ther.*, **259**, 166 (1982).
96. G.J. White, Agents Actions, **11**, 503 (1981).
97. P.F. Kador, N.E. Sharpless and J.D. Goosey, *Prog. Clin. Biol. Res.*, **114**, 243 (1982).
98. P.F. Kador and N.E. Sharpless, *Biophys. Chem.*, **8**, 81 (1978).
99. P.F. Kador and N.E. Sharpless, *Mol. Pharmacol.*, **24**, 521 (1983).
100. J. Okuda, I. Miwa, K. Inagaki, H. Morie and M. Nakayama, *Biochem. Pharmacol.*, **31**, 3807 (1982).
101. H. Ono and S. Hayano, *Nippon Ganka Gakkai Zasshi*, **86**, 353 (1982).
102. P.F. Kador, J.H. Kinoshita, W.H. Tung and L.T. Chylak Jr., *Invest. Ophthalmol. Vis. Sci.*, *19*, 980 (1980).

Chapter 18. Vitamin D: Metabolism and Mechanism of Action

Hector F. DeLuca and Heinrich K. Schnoes
Department of Biochemistry, University of Wisconsin-Madison
College of Agricultural and Life Sciences
Madison, Wisconsin 53706

INTRODUCTION

This report summarizes the current status of the vitamin D endocrine system, focussing on its biochemical aspects. For more detailed treatments of the subject and for discussions of the older literature and of clinical and chemical results, earlier reviews should be consulted.[1-8]

METABOLISM OF VITAMIN D

Functional Pathways of Vitamin D Metabolism (Figure 1) - The activation of vitamin D_3 to its hormonal form, 1α,25-dihydroxyvitamin D_3 (1,25-$(OH)_2D_3$) involves two sequential hydroxylations - at carbon 25 and carbon 1, respectively (Figure 1). The enzymology of this conversion has been studied extensively and also reviewed previously.[2-4] The principal system for 25-hydroxylation is a microsomal P-450-dependent enzyme in liver parenchymal cells.[9-11] A liver mitochondrial 25-hydroxylase is also known,[12] but appears unlikely to play a significant role in physiological vitamin D metabolism, because of its high K_m and lack of substrate specificity.

The 1α-hydroxylase of chick kidney, an exclusively mitochondrial enzyme, is a three-component system comprising a flavoprotein, iron-sulfur protein and cytochrome P-450, the iron-sulfur protein of which is identical with beef adrenal ferredoxin.[13,14] Recent reports[15,16] of a rat cytochrome P-450 functioning as a 25-OH-D_3-1α-hydroxylase in the absence of the flavoprotein and the iron-sulfur protein require further confirmation by definitive product characterization.

There have been several reports of extrarenal 1α-hydroxylation. Removal of kidneys in pregnant rats, for example, does not prevent formation of 1α,25-$(OH)_2D_3$ *in vivo*, and placenta has been identified as an extrarenal site of 1α-hydroxylation.[17,18] Reports that isolated bone cells in culture and intestinal cells are capable of 1α-hydroxylation,[19-21] are inconclusive since it is not clear whether cells in culture represent expression *in vivo*. Results bearing on this question have been obtained by two groups who reexamined the *in vivo* synthesis of 1,25-$(OH)_2D_3$ in anephric rats using highly sensitive methods.[22-23] No 1,25-$(OH)_2D_3$ could be detected in the absence of renal tissue, whereas animals made uremic by ureteric ligation but retaining kidneys produced massive amounts. Although there may be some *in vitro* synthesis of 1α,25-$(OH)_2D_3$ by cells in culture, it is unlikely that this has any physiologic significance.

ISBN 0-12-040519-9

$1,25-(OH)_2D_3$ is metabolized very rapidly *in vivo*.[24,25] A major pathway is side-chain oxidative cleavage to yield a C-23 acid, calcitroic acid,[26,27] a likely excretory product of $1,25-(OH)_2D_3$ formed in intestine, liver and perhaps elsewhere. Intermediates between $1,25-(OH)_2D_3$ and calcitroic acid have not yet been elucidated, although compounds such as $1,25-(OH)_2$-24-oxo-D_3[28-30] and $1,23,25-(OH)_3$-24-oxo-D_3[30,31] (Figure 2), generated in various *in vitro* systems, may represent intermediary stages of side-chain metabolism to the acid. These 24-oxo-derivatives presumably result from the further metabolism of $1,24,25-(OH)_3D_3$ (Figure 1), a long known[1-4] metabolite of $1,25-(OH)_2D_3$, with apparently no functional role. Excretion of vitamin D compounds occurs predominantly via the bile into the small intestine; other than calcitroic acid and some dehydration products of $1,25-(OH)_2D_3$,[32] the excretory metabolites in bile remain unknown.

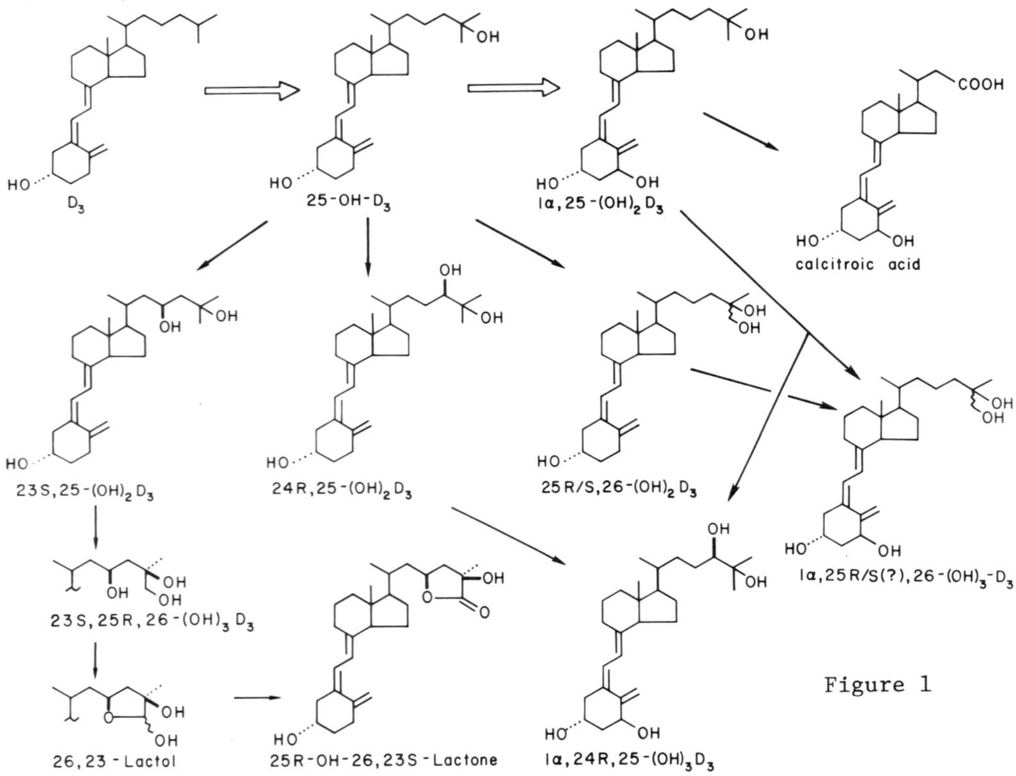

Figure 1

Pathways of Vitamin D Metabolism to Non-functional Metabolites - All metabolism thus far studied courses through 25-OH-D_3 (Figure 1). An early result was the demonstration of 26-hydroxylation, in kidney and perhaps elsewhere, to form $25,26-(OH)_2D_3$.[1-5] Both the *in vivo* and the *in vitro* product of 26-hydroxylation is a (ca. 1:1) mixture of C-25-epimers,[33] implying a non-stereospecific 26- and 27-hydroxylase, or hydroxylation by two different enzymes. $25,26-(OH)_2D_3$ is considerably less active than its precursor and likely does not represent an activation form or an intermediate to functional metabolites. It can serve, *in vitro*, as a substrate for the 1α-hydroxylase, but the product, $1\alpha,25,26-(OH)_3D_3$,[34,35] also obtainable by 26-hydroxylation of $1\alpha,25-(OH)_2D_3$ (Figure 1), has no known function.

More recently, the 23-hydroxylation of 25-OH-D_3 to form $23S,25-(OH)_2D_3$,[36] a precursor[37] to 25-OH-D_3-26,23-lactone[38,39] was discovered. This pathway appears to proceed via 26-oxidation to the triol,[40,41] then to a lactol[42] and finally oxidation to the lactone.[42] The intermediates of this sequence (Figure 1) are biologically inactive, and the in vivo significance of the pathway which apparently occurs predominantly, but not exclusively, in the kidney,[37,43,44] is far from clear.

The major metabolic fate of 25-OH-D_3 other than 1α-hydroxylation is 24-hydroxylation to produce 24R,25-dihydroxyvitamin D_3.[1-4] For this compound functional roles (in mineralization of bone, in regulation of parathyroid secretion, in chick embryonic development, and in cartilage growth and proliferation) reviewed in more detail elsewhere,[3-5] have been proposed, but the experimental evidence in support of a specific function for 24,25-$(OH)_2D_3$ is inconclusive.

A very strong argument against a functional role for 24,25-$(OH)_2D_3$ comes from recent experiments with a 24-difluoro-analog of 25-OH-D_3 (24,24-F_2-25-OH-D_3),[45,46] a compound that is 1α-hydroxylated in vivo, but does not undergo 24-hydroxylation.[47-49] In vitamin D-responsive systems, this synthetic analog proved to be at least as active as 25-OH-D_3.[50-52] Of particular importance is that this difluoro compound can support rats through two generations as the sole source of vitamin D.[53,54] These animals are completely normal in every respect, with normal growth and normal bone mineralization.[54] These results demonstrate that 24-hydroxylation is not required for any significant function of vitamin D. Likewise, the difluoro analog has been shown to be fully active and effective in promoting normal egg hatchability and normal chick embryonic development, again demonstrating that 24-hydroxylation is not required for embryonic development in birds[55,56] (Hart & DeLuca, submitted). In the face of these results, a functional role of the 24-hydroxylated metabolites in the regulation of calcium homeostasis appears highly improbable. The metabolic fate of 24,25-$(OH)_2D_3$ is not known in detail, although 1α-hydroxylation to 1,24,25-$(OH)_3D_3$[1-4] (Figure 1), and side-chain cleavage to the C-24-acid (cholacalcioic acid,[57] Figure 2) have been demonstrated.

Recently, a synthetic 26,27-hexafluoro derivative[58,59] of 25-OH-D_3 has been used to test whether 26-hydroxylation or lactone formation might be of functional importance. Again, the hexafluoro compound is as active as 25-OH-D_3 in the vitamin D-responsive systems.[60,61] As far as can be determined today, therefore, the sequence of 25-hydroxylation followed by 1α-hydroxylation to form 1,25-$(OH)_2D_3$ is the sole functional pathway of vitamin D metabolism.

Pharmacological Metabolism of Vitamin D (Figure 2) - The techniques of metabolite isolation and identification being by now well-defined, there has been a notable recent surge of publications describing the characterization of new metabolites. Figure 2 offers a summary of the compounds identified. Most of these metabolites have been generated either in tissue homogenates in vitro, often at very high substrate levels, or in animals dosed with pharmacological amounts of vitamin D; hence, their physiological significance is obscure. For example, from the plasma of chicks dosed with massive quantities of D_3, 24R-OH-D_3, 25-OH-Δ^{23}-D_3, 25-OH-24-oxo-D_3 as well as 24,25,26-$(OH)_3D_3$ and 23,24,25-$(OH)_3D_3$ (Figure 2) have been isolated,[62] whereas the peroxy-lactone was obtained from the serum of rats given large vitamin D_3 doses.[63] The 25-hydroxy-24-oxo-D_3 compound has also been generated in chick kidney homogenates,[64] and further incubation of it with chick kidney homogenates leads to either 1α,25-$(OH)_2$-24-oxo-D_3, or 23,25-$(OH)_2$-24-oxo-D_3 and the 23,24,25-triol depending on the D-status of animals used as a kidney source.[28] Similarly, incubation of 23S,25-$(OH)_2D_3$ with kidney homogenates of a D-deficient animal generates 1α,23S,25-$(OH)_3D_3$,[65] whereas the same homogenates from D-replete chicks produce 25-OH-23-oxo-D_3,[66] and the latter, incubated with a D-deficient kidney homogenate is 1-hydroxylated as expected to 1α,25-$(OH)_2$-23-oxo-D_3.[67] Incubation of the 26,23-lactone (Figure 1) under similar conditions gives the corresponding 1α-hydroxylated derivative[43] (Figure 2), and the 1α,23,25-$(OH)_3$-24-oxo-D_3 as well as 1α-hydroxylated lactone (Figure 2) is produced from various 1α-hydroxylated precursors in appropriately-treated intestinal or kidney homogenates, or mucosal cells.[30,31-68]Obviously these variations on the same theme can be continued endlessly. Conceivably, some of these compounds may be metabolic intermediates (i.e. to side-chain degraded excretion products) but, given their highly non-physiological mode of production, it is equally, and perhaps much more valid to regard these "metabolites" simply as interesting structural analogs generated by biochemical means.

The 24- and 25-dehydro compounds (Figure 2) were obtained as minor 1,25-$(OH)_2D_3$-degradation products from rat bile,[32] and the 10-oxo-19-nor-5,6-*trans*-derviative of D_2, D_3 and 25-OH-D_3 (structures not shown) resulting from the incubation of bovine rumen flora with the respective

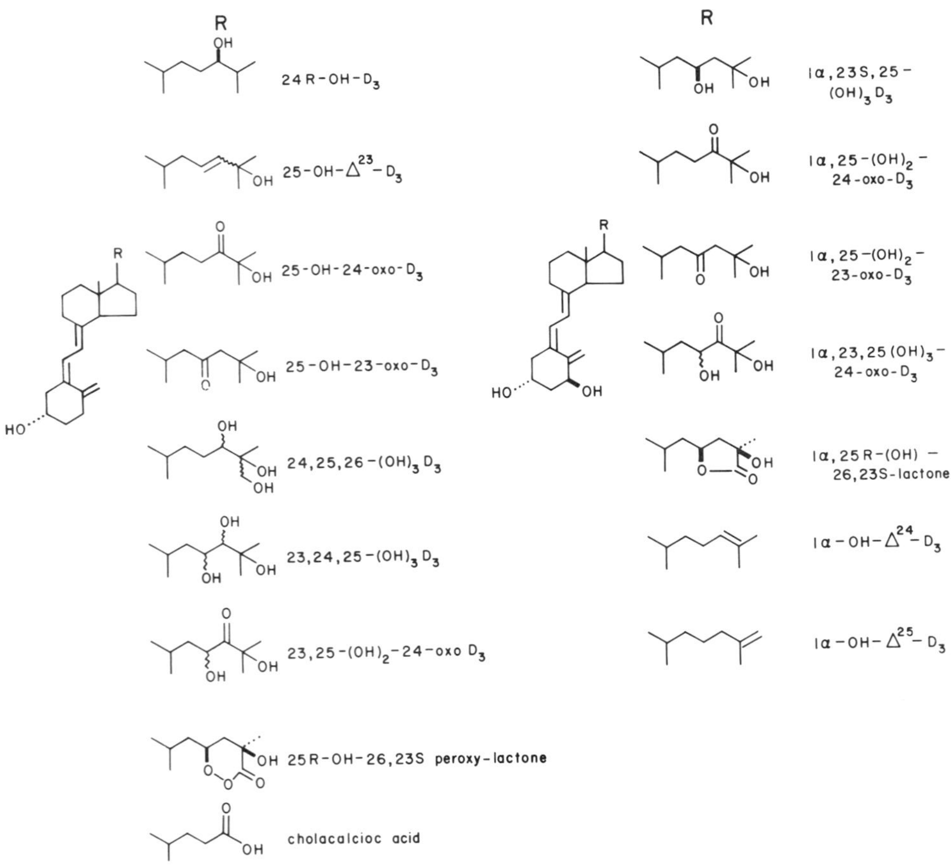

Figure 2

vitamin D substrates.[69] 25-Hydroxy-10-oxo-19-nor-vitamin D_3 and its 5,6-trans isomer have also been obtained from kidney cell cultures treated with 25-OH-D_3.[70]

Regulation of Vitamin D Metabolism - It is well established that hypocalcemia markedly stimulates 1α-hydroxylation of 25-OH-D_3 to form 1,25-$(OH)_2D_3$, and that this effect is mediated through parathyroid hormone secreted by the parathyroid glands in response to lowered serum calcium levels.[2-4] Several investigators have now developed methods allowing direct measurement of mammalian 25-OH-D-1α-hydroxylase.[71-74] Using these techniques, the role of the parathyroid glands in regulating the 1α-hydroxylase has been reaffirmed.[75] Most important, however, is the demonstration that parathyroidectomy results in a rapid change in serum calcium concentration with a much delayed response in reduction of the 25-OH-D-1α-hydroxylase and an increase in the 25-OH-D-24-hydroxylase.[72,75] The change in serum calcium occurs within 2-3 hours, whereas the 1α-hydroxylase requires some 12 hours to fall to low levels. Injection of parathyroid hormone results in rapid changes in serum calcium and phosphorus, whereas 12-24 hours are required for the 1α-hydroxylase to respond. These results demonstrate that the 25-OH-D-1α-hydroxylase is not involved in the minute-to-minute regulation of serum calcium concentration. Instead this function must be relegated to the parathyroid hormone working on the systems rendered sensitive by previously circulating 1,25-$(OH)_2D_3$. There is reasonable evidence to suggest that parathyroid hormone stimulation of the 25-OH-D-1α-hydroxylase is mediated by cAMP, since cAMP added to cultures of chick kidney cells or kidney perfusates will increase 1α-hydroxylase activity,[76,77] but the exact regulating mechanism remains unknown.

Phosphate deprivation increases 25-OH-D-1α-hydroxylase by approximately 5-fold,[2-4,75] yet this manipulation increases serum 1,25-$(OH)_2D_3$ by substantial amounts.[2-4,75] It is likely, but not yet established, that phosphate deprivation modulates circulating 1,25-$(OH)_2D_3$ levels both by increasing 1α-hydroxylase activity and by affecting degradation of 1,25-$(OH)_2D_3$.

One of the major regulators of the 1α-hydroxylase is 1,25-$(OH)_2D_3$ itself. Administration of this compound to vitamin D-deficient animals brings about a suppression of the 1α-hydroxylase and a stimulation of the 25-OH-D-24-hydroxylase.[78] This can be demonstrated clearly in cultures[79,80] and in perfused organs.[81] Furthermore, agents that block transcription appear to block this response, implying a transcriptional event in the regulation of the 25-OH-D-1α-hydroxylase.[82] Under conditions of hypocalcemia, however, increasing dosages of vitamin D or of 1,25-$(OH)_2D_3$ do not inhibit the 1α-hydroxylase but actually increase the activity of this system.[83] In fact, circulating levels of 750 pg/ml of 1,25-$(OH)_2D_3$ can be achieved under hypocalcemic conditions with high doses of vitamin D. On the other hand, in animals on an adequate calcium, or on a low phosphorus diet, increasing doses of vitamin D suppress the 1α-hydroxylase. Thus, 1,25-$(OH)_2D_3$ can either stimulate the 1α-hydroxylase or suppress it depending upon the calcium status of the animals. The complex molecular mechanism of regulation of the 25-OH-D-1α-hydroxylase remains a major area of investigation.

THE FUNCTIONS OF 1,25-$(OH)_2D_3$

Certainly one of the major functions of 1,25-$(OH)_2D_3$ is the stimulation of intestinal calcium and intestinal phosphate transport.[2-5]

Of particular importance is the recent observation that 1,25-$(OH)_2D_3$ stimulates intestinal calcium transport in a complex biphasic manner.[84] An injection of 1,25-$(OH)_2D_3$ causes a rapid rise in calcium transport to a peak value at 6 hours, followed by a decline to a low value at 12 hours, and a rise to a second maximum at 24 hours, which is sustained for several days. A second injection of 1,25-$(OH)_2D_3$ after the 24-hour period results in a super-induction of the initial response in addition to the second response. These findings suggest that there are at least two mechanisms of intestinal calcium transport responsive to 1,25-$(OH)_2D_3$. The first response undoubtedly represents the transport activity of existing villus cells, whereas the second response likely results from an effect of 1,25-$(OH)_2D_3$ on the crypt cells that then differentiate and migrate up the villus region to promote intestinal calcium transport. A similar biphasic response has been observed for phosphate transport.[85]

Another well-known function of 1,25-$(OH)_2D_3$ is the mobilization of calcium from bone so as to maintain normal plasma calcium concentration.[2-6] This 1,25-$(OH)_2D_3$-mediated process, however, cannot occur unless parathyroid hormone is present,[86] and it is probable that parathyroid hormone is responsible for bone mobilization, but requires for its action certain 1,25-$(OH)_2D_3$-induced cellular events.[3,4,6]

The role of 1,25-$(OH)_2D_3$ in the kidney is even less well understood, although its localization in the distal renal tubule cells, specifically in the nuclei, is known.[87] No clear role for 1,25-$(OH)_2D_3$ in renal tubular reabsorption of phosphorus has been established, and this area remains controversial. There is evidence that 1,25-$(OH)_2D_3$ stimulates renal reabsorption of calcium in the distal tubule but little else is known concerning that mechanism.[88]

The long-standing question of the role of vitamin D metabolites in the synthesis of collagen matrix elaborated by osteoblasts or chondrocytes of the epiphyseal growth plate, and in the subsequent mineralization,[3-6] was recently examined directly in completely vitamin D-deficient animals.[89,90] These animals were then infused with calcium and phosphorus in the jugular veins to maintain plasma calcium and phosphorus in the normal range despite vitamin D deficiency.[89] Animals so maintained showed entirely normal bone growth and normal mineralization of both cartilage and bone, a result also confirmed by histomorphometric measurements.[90] These results demonstrate that vitamin D plays no direct role in the synthesis of the organic matrix of bone, in epiphyseal plate cartilage growth and mineralization, nor in the mineralization process of bone. Of considerable interest is that the vitamin D-deficient animals infused with calcium and phosphorus accumulated calcium and phosphorus in their bones to a much greater extent than animals given vitamin D.[90] This probably is the result of a failure of the resorption process in vitamin D deficiency, and suggests that vitamin D must play an important role in the bone modeling and remodeling process.

1,25-$(OH)_2D_3$ may play a role in tissues other than those involved in calcium transport. In particular, Stumpf and coworkers have shown that 1,25-$(OH)_2D_3$ specifically localizes in the nuclei of skin cells, especially the malpighian layer,[91] in endocrine cells of the stomach,[91] in the islet cells of the pancreas,[92] parathyroid gland cells,[93] certain cells of the pituitary,[91] and of the brain,[94] in addition to the expected target sites, such as osteoblasts, osteocytes, intestinal villus cells, and renal tubule cells.[95] Since these tissues also contain the 1,25-$(OH)_2D_3$ receptor,[96] a possible functional role for 1,25-$(OH)_2D_3$ is suggested. It has been shown, for example, that glucose-stimulated insulin secretion is blunted in vitamin D deficiency and restored by 1,25-$(OH)_2D_3$.[97] It is unclear, however, how much of this response is the result of a change in plasma calcium concentration and of changes in food consumption.[98] An increase in 7-dehydrocholesterol levels in skin in response to 1,25-$(OH)_2D_3$ has also been demonstrated.[99] A significant and physiologic role of 1,25-$(OH)_2D_3$ in parathyroid hormone secretion or synthesis has been difficult to establish,[100,101] and definitive methods remain to be developed to examine this question.

The recent demonstration that 1,25-$(OH)_2D_3$ stimulates the differentiation of myeloid leukemia cells (MI and human HL-60 cells) into monocytes in a dose-dependent manner,[102,103] suggests a potential role for vitamin D metabolites in cellular differentiation. Somewhat higher doses of 1,25-$(OH)_2D_3$ are required than those effective in cultures of embryonic bone or chick embryonic intestine, but the compounds that show maximum effectiveness in terms of calcium action also show maximum effect in stimulating the differentiation process. Low calcium media or calcium channel blockers increase the responsiveness of these cells to 1,25-$(OH)_2D_3$.[105] These cells contain receptors for 1,25-$(OH)_2D_3$, and mutants low in receptor, are unresponsive to 1,25-$(OH)_2D_3$ treatment.[105] These exciting results have focussed attention on the possibility that some vitamin D compound or analog may be effective in treating certain acute leukemias by inducing the differentiation of malignant cells to non-proliferative monocytes. This work has been extended to show the aggregation of monocytes into polynuclear cells in response to 1,25-$(OH)_2D_3$, a process postulated, but not yet proven, to represent osteoclastic formation.[106,107]

Receptors for 1,25-$(OH)_2D_3$ have also been found in 60% of the breast carcinoma cell lines available.[108-110] Addition of 1,25-$(OH)_2D_3$ to mammary carcinoma or melanoma at low concentrations stimulates proliferation, but at higher concentrations, proliferation is markedly inhibited.[111,112] In mutant cell lines lacking receptor, these responses do not occur.

Another interesting finding is the possible effect of 1,25-$(OH)_2D_3$ on the carcinogenesis process of skin. The compound is reported to suppress papilloma development in skin induced by agents such as methylcholanthrene and promoters such as phorbol esters;[113] however, 1,25-$(OH)_2D_3$ has also been shown to substitute for phorbol esters in promoting methylcholanthrene-induced transformation of Balb 3T3 fibroblasts.[114] These interactions of 1,25-$(OH)_2D_3$ and its analogs with malignant cells will certainly receive great attention in the next few years.

MOLECULAR MECHANISM OF ACTION OF 1,25-$(OH)_2D_3$

This section will be restricted to intestinal calcium transport, since it represents the focus of current efforts to understand the mechanism of action. Much work has been done to demonstrate the existence of a macromolecule that binds 1,25-$(OH)_2D_3$ in the target tissues. Both chick and mammalian intestines contain a protein sedimenting at 3.2 to 3.7 S that binds 1,25-$(OH)_2D_3$ with a K_d of 5 x 10^{-11} M.[96] The association and dissociation rate constants have been determined for chick and man and for preparations from different tissues. The receptor from chick intestine has been purified to homogeneity[115,116] and monoclonal antibodies have been obtained.[117] The chick receptor appears to be a single polypeptide chain of about 63,000 kD.[115,116] Little is known about its chemistry, except that SH groups appear to be involved in hormone binding,[115-119] and that it has a DNA- as well as a hormone-binding site.[115,116,120] Extensive work has been carried out on the selectivity of this receptor for 1,25-$(OH)_2D_3$,[2-4,96] and it has been used, in fact, as a method of assay of the vitamin D hormone because of its marked sensitivity and specificity for 1,25-$(OH)_2D_3$.[121] There is some question about the intracellular location of this receptor, but current evidence, based on cell fractionation work from two different groups, supports the idea that the 1,25-$(OH)_2D_3$ receptor is predominantly in the nucleus even without ligand.[122,123] Addition of ligand apparently causes a structural alteration and a tighter binding to chromatin.[124,125] In animals given adequate amounts of vitamin D, approximately 20% of this receptor exists as the hormone-receptor complex, the remainder being in the free form,[126,127] and the administration of large amounts of 1,25-$(OH)_2D_3$ is required to convert the bulk of the free receptor to receptor-hormone complex.[127] A full discussion of the voluminous receptor literature is beyond the scope of this review, and readers are directed elsewhere for a more detailed treatment.[3-5,96,128]

That interaction of 1,25-$(OH)_2D_3$ with the receptor is required for function is demonstrated by the disease, vitamin D-dependency rickets Type II, an autosomal recessive disorder, which has been shown to be, at least in part, a receptor defect.[129,130] In addition, it has been shown that neonatal rats have no active transport of calcium and are unresponsive to 1,25-$(OH)_2D_3$ until 14-16 days postpartum.[131] However at 14-16 days postpartum, the responsiveness to 1,25-$(OH)_2D_3$ appears and correlates exactly with the appearance of receptor in intestinal tissue.[132] Adrenalectomy delays this response, whereas hydrocortisone injections cause precocious appearance of receptor and intestinal calcium transport.[133] In fact, incubation of tissue explants at 14 days postpartum with hydrocortisone causes <u>in vitro</u> appearance of the receptor.[134] These results argue that intestinal calcium transport requires interaction of 1,25-$(OH)_2D_3$ with the receptor.[135]

There has been considerable debate as to whether 1,25-$(OH)_2D_3$ functions in intestinal calcium transport through a nuclear-mediated mechanism.[2-4] This question has been addressed in recent experiments using intestinal organ cultures which show that cycloheximide or actinomycin D can block intestinal calcium transport response to 1,25-$(OH)_2D_3$.[136] When the inhibitor is removed by placing the tissue in fresh medium, the responsiveness to 1,25-$(OH)_2D_3$ reappears. Thus, transcription and translation appear to be involved in the intestinal response to 1,25-$(OH)_2D_3$. Rasmussen and colleagues have postulated a liponomic action of 1,25-$(OH)_2D_3$ in promoting intestinal calcium transport,[137] based on results with isolated vesicles from brush borders of intestinal villus cells in which 1,25-$(OH)_2D_3$ caused increased calcium transport,[138] and the finding that cholesterol esters and phosphatidylcholine and phosphatidylethylolamine levels in membranes are altered by 1,25-$(OH)_2D_3$ treatment.[139] The significance of these results remains undetermined at the present time.

A major problem in explaining intestinal calcium transport responses to 1,25-$(OH)_2D_3$ is the lack of information on the proteins or gene products of 1,25-$(OH)_2D_3$ action. Only one protein induced by 1,25-$(OH)_2D_3$ has been identified. This is the 27,000 kD calcium-binding protein in the chicken[140] and the 8,000-12,000 kD protein found in rat intestine.[141] Unfortunately, intestinal calcium transport response does not correlate well with the appearance and disappearance of these calcium-binding proteins.[142] Thus, it appears that more is involved in the intestinal calcium transport response than simply the induction of calcium binding protein. However, from experiments with embryonic chick organ cultures, using two-dimensional gel electrophoresis and computer-based spot analysis, it is clear that 1,25-$(OH)_2D_3$-induced calcium-binding protein appears in that tissue sufficiently early to play a role in calcium transport.[143,144] Nevertheless, additional gene products of vitamin D expression are being sought. There have been reports of 1,25-$(OH)_2D_3$-stimulated biosynthesis of actin and other labeled proteins, but without a clear indication that any of these proteins are responsible for calcium transport.[145] The development of two-dimensional gel electrophoresis, double-label techniques, and sensitive computer analysis of these gels should soon yield more detailed information on the 1,25-$(OH)_2D_3$-induced gene products.

Current information on the role of 1,25-$(OH)_2D_3$ in eliciting intestinal calcium transport is fragmentary and incomplete. It is unknown how 1,25-$(OH)_2D_3$ enters the villus cell, or how it translocates to the nucleus where it interacts with its receptor. Hormone binding undoubtedly induces a change in the receptor, since it can be shown that hormone-receptor complex elutes at a higher salt concentration from DNA cellulose than does the free receptor.[124,125] This receptor-hormone complex somehow interacts with the chromatin to elicit expression of specific genes that code for calcium and phosphorus transport proteins. One of these proteins is the calcium-binding protein discussed above; the others remain to be determined. They may function at the brush border membrane, in the cytoplasm, in the Golgi apparatus, and at the basal-lateral membrane to promote the transfer of calcium. The final expulsion of calcium at the basal-lateral membrane involves a sodium exchange mechanism[146] or a calcium ATPase pump mechanism.[147]

References

1. H.F. DeLuca, H.E. Paaren and H.K. Schnoes, in "Annual Reports of Medicinal Chemistry," Vol. 15, H.-J. Hess, Ed., Academic Press, New York, 1980, p. 288.
2. H.F. DeLuca, H.E. Paaren and H.K. Schnoes, in "Topics in Current Chemistry," Vol. 83, M.J.S. Dewar, K. Hafner, E. Heilbronner, S. Ito, J.-M. Lehn, K. Niedenzu, C.W. Rees, K. Schäfer, G. Wittig and F.L. Boschke, Eds., Springer-Verlag, Berlin, 1979, p. 1.
3. H.F. DeLuca and H.K. Schnoes, Ann. Rev. Biochem., 45, 631 (1976); 52, 411-439 (1983).
4. H.F. DeLuca, in "Bone and Mineral Research," Annual 1, W.A. Peck, Ed., Excerpta Medica, Amsterdam, 1983, p. 7
5. A.W. Norman, J. Roth and L. Orci, Endocrine Rev., 3, 331 (1982).
6. P.H. Stern, Pharmacological Rev., 32, 47 (1980).
7. R.I. Yakhimovich, Russian Chem. Rev. 49, 371 (1980).
8. B. Lythgoe, Chem. Soc. Rev., 9, 449 (1980).
9. T.C. Madhok and H.F. DeLuca, Biochem. J., 184, 491 (1979).
10. P.S. Yoon, P.S. and H.F. DeLuca, Arch. Biochem. Biophys., 203, 529 (1980).
11. S. Anderson, I. Holmberg and K. Wikvall, J. Biol. Chem., 258, 6777 (1983).
12. I. Bjorkhem, I. Holmberg, H. Oftebro and J.I. Pedersen, J. Biol. Chem., 255, 5244 (1980).
13. P.S. Yoon, and H.F. DeLuca, Biochemistry, 19, 2165 (1980).
14. P.S. Yoon, J. Rawlings, W.H. Orme-Johnson and H.F. DeLuca, Biochemistry, 19, 2172 (1980).
15. M. Warner, J. Biol. Chem., 257, 12995 (1982).
16. M. Warner, J. Biol. Chem., 258, 11590 (1983).
17. Y. Tanaka, B. Halloran, H.K. Schnoes and H.F. DeLuca, Proc. Natl. Acad. Sci. USA, 76, 5033 (1979).

18. Y. Weisman, A. Harell, S. Edelstein, M. David, Z. Spirer and A. Golander, Nature, 281, 317 (1979).
19. P.W. Lambert, P. Stern, R.C. Avioli, N.C. Brockett, R.P. Turner, A. Green, I.Y. Fu and N.H. Bell, J. Clin. Invest., 69, 722 (1982).
20. R.T. Turner, J.E. Puzas, M.D. Forte, G.E. Lester, T.K. Gray, G.A. Howard and D.J. Baylink, Proc. Natl. Acad. Sci. USA, 77, 5720 (1980).
21. G.A. Howard, R.T. Turner, D.J. Sherrard and D.J. Baylink, J. Biol. Chem., 256, 7738 (1981).
22. L. Reeve, Y. Tanaka and H.F. DeLuca, J. Biol. Chem., 258, 3615 (1983).
23. T.D. Shultz, J. Fox, H. Heath, III and R. Kumar, Proc. Natl. Acad. Sci. USA, 80, 1746 (1983).
24. R.W. Gray, A.E. Caldas, D.R. Wilz, J. Lemann, Jr., G.A. Smith and H.F. DeLuca, J. Clin. Endocrinol. Metab., 46, 756 (1978).
25. J.W. Coburn, G.F. Bryce, B.S. Levine, J.P. Mallon, F. Singer and O.N. Miller, in "Osteoporosis: Recent Advances in Pathogenesis and Treatment," H.F. DeLuca, H.M. Frost, W.S.S. Jee, C.C. Johnston, Jr. and A.M. Parfitt, Eds., University Park Press, Baltimore, 1981, p. 485 (abstract).
26. R.P. Esvelt, H.K. Schnoes and H.F. DeLuca, Biochemistry, 18, 3977 (1979).
27. R.P. Esvelt, M.A. Fivizzani, H.F. Paaren, H.K. Schnoes and H.F. DeLuca, J. Org. Chem., 46, 456 (1981).
28. S. Yamada, M. Ohmori, H. Takayama, Y. Takasaki and T. Suda, J. Biol. Chem., 258, 457 (1983).
29. J.L. Napoli, C. Pramanik, P.M. Royal, T.A. Reinhardt and R.L. Horst, J. Biol. Chem., 258, 9100 (1983).
30. E. Mayer, J.E. Bishop, R.A.S. Chantraratna, W.H. Okamura, J.R. Kruse, G. Popjak, N. Ohnuma and A.W. Norman, J. Biol. Chem., 258, 13458 (1983).
31. J.L. Napoli, and R.L. Horst, Biochemistry, 22, 5848 (1983).
32. B.L. Onisko, R.P. Esvelt, H.K. Schnoes and H.F. DeLuca, Biochemistry, 19, 4124 (1980).
33. N. Ikekawa, N., N. Koizumi, E. Ohshima, S. Ishizuka, T. Takeshita, Y. Tanaka and H.F. DeLuca, Proc. Natl. Acad. Sci. USA, 80, 5286 (1983).
34. Y. Tanaka, H.K. Schnoes, C.M. Smith and H.F. DeLuca, Arch. Biochem. Biophys., 210, 104 (1981).
35. T.A. Reinhardt, J.L. Napoli, B. Praminik, E.T. Littledike, D.C. Beitz, J.J. Partridge, M.R. Uskokovic and R.L. Horst, Biochemistry, 20, 6230 (1981).
36. Y. Tanaka, J.K. Wichmann, H.K. Schnoes and H.F. DeLuca, Biochemistry, 20, 3875 (1981).
37. Y. Tanaka, H.F. DeLuca, H.K. Schnoes, N. Ikekawa and T. Eguchi, Proc. Natl. Acad. Sci. USA, 78, 4805 (1981).
38. J.K. Wichmann, H.F. DeLuca, H.K. Schnoes, R.L. Horst, R.M. Shepard and N.A. Jorgensen, Biochemistry, 18, 4775 (1979).
39. T. Eguchi, S. Takatsuto, M. Ishiguro, N. Ikekawa, Y. Tanaka and H.F. DeLuca, Proc. Natl. Acad. Sci. USA, 78, 6579 (1981).
40. S. Ishizuka, S. Ishimoto and A.W. Norman, Arch. Biochem. Biophys., 217, 264 (1982).
41. J.L. Napoli and R.L. Horst, Biochem. J., 206, 173 (1982).
42. S. Yamada, K. Nakayama, H. Takayama, T. Shinki, Y. Takasaki and T. Suda, J. Biol. Chem., 259, 884 (1984).
43. Y. Tanaka, J.K. Wichmann, H.E. Paaren, H.K. Schnoes and H.F. DeLuca, Proc. Natl. Acad. Sci. USA, 77, 6411 (1980).
44. J.L. Napoli, B.C. Pramanik, J.J. Partridge, M.R. Uskokovic and R.L. Horst, J. Biol. Chem., 257, 9634 (1982).
45. Y. Kobayashi, T. Taguchi, T. Terada, J. Oshida, M. Morisaki and N. Ikekawa, JCS Perkin, I, 85 (1982).
46. S. Yamada, M. Ohmori and H. Takayama, Tetrahedron Lett., 21, 1859 (1979).
47. B.P. Halloran, H.F. DeLuca, E. Barthell, S. Yamada, M. Ohmori and H. Takayama, Endocrinology, 108, 2067 (1981).
48. Y. Tanaka, J.K. Wichmann, H.F. DeLuca, Y. Kobayashi and N. Ikekawa, Arch. Biochem. Biophys., 225, 649 (1983).
49. Y. Tanaka, H.F. DeLuca, H.K. Schnoes, N. Ikekawa and Y. Kobayashi, Arch. Biochem. Biophys., 199, 473 (1980).
50. Y. Tanaka, H.F. DeLuca, Y. Kobayashi, T. Taguchi, N. Ikekawa and M. Morisaki, J. Biol. Chem., 254, 7163 (1979).
51. S. Okamoto, Y. Tanaka, H.F. DeLuca, S. Yamada and H. Takayama, Arch. Biochem. Biophys., 206, 8 (1981).
52. S. Okamoto, C. Smith, H.F. DeLuca, S. Yamada and H. Takayama, J. Nutr., 113, 1607 (1983).
53. K. Jarnagin, R. Brommage, H.F. DeLuca, S. Yamada and H. Takayama, Am. J. Physiol., 244, E290 (1983).
54. R. Brommage, K. Jarnagin, H.F. DeLuca, S. Yamada and H. Takayama, Am. J. Physiol., 244, E98 (1983).

55. S. Ameenuddin, M. Sunde, H.F. DeLuca, N. Ikekawa and Y. Kobayashi, Science, 217, 451 (1982).
56. S. Ameenuddin, M.L. Sunde, H.F. DeLuca, N. Ikekawa and Y. Kobayashi, Arch. Biochem. Biophys., 226, 666 (1983).
57. H.F. DeLuca, and H.K. Schnoes, in "Vitamin D: Basic Research and Its Clinical Application," A.W. Norman, K. Schaefer, D.v. Herrath, H.G. Grigoleit, J.W. Coburn, H.F. DeLuca, E.B. Mawer and T. Suda, Eds., Walter de Gruyter, Berlin, 1979, p. 445.
58. Y. Kobayashi, T. Taguchi, N. Kanuma, N. Ikekawa and J.-I. Oshida, Tetrahedron Lett., 22, 4309 (1981).
59. Y. Kobayashi, T. Taguchi, N. Kanuma, N. Ikekawa and J-I. Oshida, J. Chem. Sco. Chem. Commun., 10, 459 (1980).
60. Y. Tanaka, D.N Pahuja, J.K. Wichmann, H.F. DeLuca, Y. Kobayashi, T. Taguchi and N. Ikekawa, Arch. Biochem. Biophys., 218, 134 (1982).
61. Y. Tanaka, H.F. DeLuca, Y. Kobayashi and N. Ikekawa, Arch. Biochem. Biophys., 229, 348 (1984).
62. J.K. Wichmann, H.K. Schnoes and H.F. DeLuca, Biochemistry, 20, 7385 (1981).
63. S. Ishizuka, S. Ishimoto and A.W. Norman, J. Biol. Chem., 257, 14708 (1982).
64. Y. Takasaki, T. Suda, S. Yamada, H. Takayama and Y. Nishii, Biochemistry, 20, 1681 (1981).
65. J.L. Napoli, and R.L. Horst, Biochem. J., 214, 261 (1983).
66. R.L. Horst, T.A. Reinhardt, B.C. Pramanik and J.L. Napoli, Biochemistry, 22, 245 (1983).
67. R.L. Horst, T.A. Reinhardt and J.L. Napoli, Biochem. Biophys. Res. Commun., 107, 1319 (1982).
68. N. Ohnuma and A.W. Norman, Arch. Biochem. Biophys., 213, 139 (1982).
69. J.L. Napoli, J.L. Sommerfeld, B.C. Pramanik, R. Gardner, A.D. Sherry, J.J. Partridge, M.R. Uskokovic and R.L. Horst, Biochemistry, 22, 3636 (1983).
70. R.U. Simpson, J.K. Wichmann, H.E. Paaren, H.K. Schnoes and H.F. DeLuca, Arch. Biochem. Biophys., 230, 21 (1984).
71. Y. Tanaka and H.F. DeLuca, Anal. Biochem., 110, 102 (1981).
72. Y. Tanaka and H.F. DeLuca, Proc. Natl. Acad. Sci. USA, 78, 196 (1981).
73. R. Vieth and D. Fraser, J. Biol. Chem., 254, 12455 (1979).
74. H.J. Armbrecht, T.V. Zenser and B.B. Davis, J. Clin. Invest., 66, 1118 (1980).
75. Y. Tanaka and H.F. DeLuca, Am. J. Physiol., 246, E168 (1984).
76. N. Horiuchi, T. Suda, H. Takahashi, E. Shimazawa and E. Ogata, Endocrinology, 101, 969 (1977).
77. H.L. Henry, Min. Elect. Metab., 8, 179 (1982).
78. Y. Tanaka, R.S. Lorenc and H.F. DeLuca, Arch. Biochem. Biophys., 171, 521 (1975).
79. H.L. Henry, J. Biol. Chem., 254, 2722 (1979).
80. U. Trechsel, J-P. Bonjour and H. Fleisch, J. Clin. Invest., 64, 206 (1979).
81. A.M. Rosenthal, G. Jones, S.W. Hooh and D. Fraser, Am. J. Physiol., 239, E12 (1980).
82. R.G. Larkins, S.J. MacAuley and I. MacIntrye, Nature, 252, 412 (1974).
83. Y. Tanaka and H.F. DeLuca, Biochem. J., 214, 893 (1983).
84. B.P. Halloran, and H.F.DeLuca, Arch. Biochem. Biophys. 208, 477 (1981).
85. B. Kabakoff, N.C. Kendrick and H.F. DeLuca, Am. J. Physiol. 6, E470 (1982).
86. M. Garabedian, Y. Tanaka, M.F. Holick and H.F. DeLuca, Endocrinology 94, 1022 (1974).
87. W.E. Stumpf, M. Sar, R. Narbaitz, F.A. Reid, H.F. DeLuca and Y. Tanaka, Proc. Natl. Acad. Sci. USA 77, 1149 (1980).
88. R.A.L. Sutton, and J.H. Dirks, Fed. Proc. 37, 2112 (1978).
89. J.L. Underwood, and H.F. DeLuca, Am. J. Physiol., in press (1984).
90. R.S. Weinstein, J.L. Underwood, M.S. Hutson and H.F. DeLuca, Am. J. Physiol., in press (1984).
91. W.E. Stumpf, M. Sar, F.A. Reid, Y. Tanaka and H.F. DeLuca, Science 206, 1188 (1979).
92. S.A. Clark, W.E. Stumpf, M. Sar, H.F. DeLuca and Y. Tanaka, Cell Tissue Res. 209, 515 (1980).
93. W.E. Stumpf, M. Sar, F.A. Reid, S. Huang, R. Narbaitz and H.F. DeLuca, Cell Tissue Res. 221, 333 (1981).
94. W.E. Stumpf, M. Sar, S.A. Clark and H.F. DeLuca, Science 215, 1403 (1982).
95. W.E. Stumpf, M. Sar and H.F. DeLuca, in "Hormonal Control of Calcium Metabolism" (D.V. Cohn, R.V. Talmage and J.L. Matthews, eds.), Excerpta Medica, Amsterdam, p. 1981, p. 222.
96. R.T. Franceschi, R.U. Simpson and H.F. DeLuca, Arch. Biochem. Biophys. 210, 1 (1981).
97. A.W. Norman, B.J. Frankel, A.M. Heldt and G.M. Grodsky, Science 209, 823 (1980).
98. B.S. Chertow, W.I. Sivitz, N.G. Baranetsky, S.A. Clark, A. Waite and H.F. DeLuca, Endocrinology 113, 1511 (1983).
99. R.P. Esvelt, H.F. DeLuca, J.K. Wichmann, S. Yoshizawa, J. Zurcher, M. Sar and W.E. Stumpf, Biochemistry 19, 6158 (1980).
100. P. Golden, A. Greenwalt, K. Martin, E. Bellorin-Font, R. Mazey, S. Klahr and E. Slatopolsky, Endocrinology 107, 602 (1980).

101. P. Golden, R. Mazey, A. Greenwalt, K. Martin and E. Slatopolsky, Min. Elect. Matab. 2, 1 (1979).
102. E. Abe, C. Miyaura, H. Sakagami, M. Takeda, K. Konno, T. Yamazaki, S. Yoshiki and T. Suda, Proc. Natl. Acad. Sci. USA 78, 4990 (1981).
103. H. Tanaka, E. Abe, C. Miyaura, T. Kuribayashi, K. Konno, Y. Nishii and T. Suda, Biochem. J. 204, 713 (1982).
104. Y. Shiina, E. Abe, C. Miyaura, H. Tanaka, S. Yamada, M. Ohmori, K. Nakayama, H. Takayama, I. Matsunaga, Y. Nishii, H.F. DeLuca and T. Suda, Arch. Biochem. Biophys. 220, 90 (1982).
105. T. Suda, in "Proceedings from the VIIIth Intl. Conference on Calcium Regulating Hormones, October 16-24, 1983.
106. E. Abe, C. Miyaura, H. Tanaka, Y. Shiina, T. Kuribayashi, S. Suda, Y. Nishii, H.F. DeLuca and T. Suda, Proc. Natl. Acad. Sci. USA 80, 5583 (1983).
107. Z. Bar-Shavit, S. Teitelbaum, P. Reitsma, A. Hall, L.E. Pegg, J. Trial and A.J. Kahn, Proc. Natl. Acad. Sci. USA 80, 5907 (1983).
108. J.A. Eisman, I. MacIntrye, T.J. Martin and J.M. Moseley, Lancet, 1335 (1979).
109. J.A. Eisman, T.J. Martin and I. MacIntrye, Lancet, 1188 (1980).
110. M.R. Haussler, D.J. Mangelsdorf, H.P. Koeffler, C.A. Donaldson and J.W. Pike, Abstract at the VIII Intl. Conf. on Calcium Regulating Hormones, October 22-24, 1983, E-11, p. 82.
111. N.C. Partridge, R.J. Frampton, J.A. Eisman, V.P. Michelangeli, E. Elms, T.R. Bradley and T.J. Martin, FEBS Lett. 115, 139 (1980).
112. R.J. Frampton, S.A. Omond and J.A. Eisman, Cancer Res. 43, 4443 (1983).
113. A.W. Wood, R.L. Chang, M-T. Huang, M. Uskokovic and A.H. Conney, Biochem. Biophys. Res. Commun. 116, 605 (1983).
114. T. Kuroki, K. Sasaki, K. Chida, E. Abe and T. Suda, Gann 74, 611 (1983).
115. R.U. Simpson and H.F. DeLuca, Proc. Natl. Acad. Sci. USA 79, 16 (1982).
116. R.U. Simpson, A. Hamstra, N.C. Kendrick and H.F. DeLuca, Biochemistry 22, 2586 (1983).
117. J.W. Pike, J.W., S.L. Marion, C.A. Donaldson and M.R. Haussler, J. Biol. Chem. 258, 1289 (1983).
118. W.S. Mellon, R.T. Franceschi and H.F. DeLuca, Arch. Biochem. Biophys. 202, 83 (1980).
119. M.R. Walters, W. Hunziker, D. Konami and A.W. Norman, J. Receptor Res. 2, 331 (1982).
120. J.W. Pike, and M.R. Haussler, Proc. Natl. Acad. Sci. USA 76, 5485 (1979).
121. J.A. Eisman, A.J. Hamstra, B.E. Kream and H.F. DeLuca, Science 193, 1021 (1976).
122. B.E. Kream, R.D. Reynolds, J.C. Knutson, J.A. Eisman and H.F. DeLuca, Arch. Biochem. Biophys. 176, 779 (1976).
123. M.R. Walters, W. Hunziker and A.W. Norman, J. Biol. Chem. 255, 6799 (1980).
124. J.W. Pike, and M.R. Haussler, J. Biol. Chem. 258, 8554 (1983).
125. W. Hunziker, M.R. Walters, J.E. Bishop and A.W. Norman, J. Biol. Chem. 258, 8642 (1983).
126. E.R. Massaro, R.U. Simpson and H.F. DeLuca, Proc. Natl. Acad. Sci. USA 80, 2549 (1983).
127. W. Hunziker, M.R. Walters, J.E. Bishop and A.W. Norman, J. Clin. Invest. 69, 826 (1982).
128. R. Link and H.F. DeLuca, in "The Receptors" (P.M. Conn, ed.), Academic Press, in press (1984).
129. C. Eil, U.A. Liberman, J.F. Rosen and S.J. Marx, New Engl. J. Med. 304, 1588 (1981).
130. U.A. Liberman, C. Eil, P. Holst, J.F. Rosen and S.J. Marx, J. Clin. Endocrinol. Metab. 57, 958 (1983).
131. B.P. Halloran, and H.F. DeLuca, Am. J. Physiol. 239, G473 (1980).
132. B.P. Halloran and H.F. DeLuca, J. Biol. Chem. 256, 7338 (1981).
133. E.R. Massaro, R.U. Simpson and H.F. DeLuca, Am. J. Physiol. 244, E230 (1983).
134. E. Massaro, R.U. Simpson and H.F. DeLuca, J. Biol. Chem. 257, 13736 (1982).
135. H.F. DeLuca, R.T. Franceschi, B.P. Halloran and E.R. Massaro, Fed. Proc. 41, 66 (1982).
136. R.T. Franceschi and H.F. DeLuca, J. Biol. Chem. 256, 3848 (1981).
137. H. Rasmussen, O. Fontaine and T. Matsumoto, Ann. NY Acad. Sci. 372, 518 (1981).
138. H. Rasmussen, O. Fontaine, E. Max and D.B.P. Goodman, J. Biol. Chem. 254, 2993 (1979).
139. H. Rasmussen, T. Matsumoto, O. Fontaine, and D.B.P. Goodman, Fed. Proc. 41, 72 (1982).
140. R.H. Wasserman and A.N. Taylor, Science 152, 791 (1966).
141. R.H. Wasserman and J.J. Feher, in "Calcium Binding Proteins and Calcium Function" (R.H. Wasserman, R.A. Corradino, E. Carafoli, R.H. Kretsinger, D.H. MacLennan, and S.L. Siegel, eds.), Elsevier, New York, 1977, p. 292.
142. R. Spencer, M. Charman, P.W. Wilson and D.E.M. Lawson, Biochem. J. 170, 93 (1978).
143. C.W. Bishop, N.C. Kendrick and H.F. DeLuca, J. Biol. Chem. 258, 1305 (1982).
144. C.W. Bishop, N.C. Kendrick and H.F. DeLuca, J. Biol. Chem., in press (1984).
145. A.N. Hobden, M. Harding and D.E.M. Lawson, Nature 288, 718 (1980).
146. D.L. Martin and H.F. DeLuca, Am. J. Physiol. 216, 1351 (1969).
147. R.H. Wasserman, C.S. Fullmer, S.A. Meyer, F. Shimura and H.J. Armbrecht, Proceedings of the VIIIth Intl. Conf. on Calcium Regulating Hormones, October 16-24, 1983.

Chapter 19. Interleukin 2

John J. Farrar, William R. Benjamin, Lorraine Cheng and Beverly A. Pawson
Roche Research Center, Hoffmann-La Roche, Inc.
Nutley, New Jersey 07110

INTRODUCTION

In the past ten years it has become clear that the induction phase of the immune response of both B and T lymphocytes is regulated by macrophages and subpopulations of T lymphocytes which serve to either enhance or suppress the immune response. Studies on the mechanisms of action of these regulatory processes have shown that, in many cases, the regulatory functions are mediated by soluble factors. These mediators (e.g. Interleukins 1 and 2, Interferon-γ, B cell growth factor and B cell differentiation factor) are produced by the regulatory cells and exert their enhancing or suppressing effects on the responding precursor cells of the immune system.[1] These soluble factors have been designated lymphokines and in most cases have been found to be either proteins or glycoproteins ranging in molecular weight from 10,000 to approximately 150,000.[1] The lymphokines have been found to be extremely potent *in vitro* in inducing very dramatic enhancing or suppressing effects in immune functions at extremely low concentrations (between 10^{-9} and 10^{-12}M). Recently, there have been a number of demonstrations of *in vivo* biological effects of administered lymphokines; thereby, confirming that the biological effects observed in *in vitro* culture systems can, in part, be duplicated in intact animals.

BIOLOGY OF INTERLEUKIN 2

Cellular Interactions in the Production of IL-2. One of the best characterized lymphokines is Interleukin 2 (IL-2). This lymphokine is produced by helper T lymphocytes following stimulation with either T cell mitogens or various antigens.[2,3] As with other T cell-derived lymphokines, the production of IL-2 has been shown to be dependent upon a cellular interaction between the lymphokine-producing T lymphocyte and accessory macrophages. These macrophages serve to provide the inductive signals to the T lymphocyte via antigen presentation and stimulation of the T lymphocyte via the secretion of Interleukin 1.[4] This cellular interaction between the macrophage and the T cell culminates in the induction of the lymphocyte to produce and secrete IL-2. Extensive studies using lymphocyte populations from both mouse and man have delineated the subpopulation of T lymphocytes producing IL-2 as the helper T cell population (Lyt 1^+23^- in mouse and T 4^+ in man) although in some instances T 8^+ cells have been reported to produce IL-2.[5]

Biological Activities of IL-2. The most commonly used *in vitro* biological assay for IL-2 is the stimulation of IL-2-dependent T cells to proliferate.[6] Incorporation of tritiated thymidine (^{3}H-Tdr) by an IL-2-dependent, cloned murine cytolytic T-cell line (CTLL) is used as a measure of proliferation and is related to the amount of IL-2 present in the sample compared to that incorporated by a standard.[6] Due to laboratory variations in the IL-2 standard used (no international IL-2

ISBN 0-12-040519-9

standard has been adopted), interlaboratory comparisons of IL-2 activity are difficult.[7] In some laboratories, the standard, assigned a value of 1 U/ml, is that amount of IL-2 that will induce 50% of the maximum stimulation at a dilution of about 1:10.[6,7] IL-2 of human as well as murine, rat and ape origin can be measured in this assay.[6] Although the original experiments utilized IL-2-dependent cytotoxic T cell lines, more recent experiments have shown that other subpopulations of T cells including suppressor T cells[8] and helper T cells[9] are also responsive to IL-2. Therefore, it appears that IL-2 may serve as a growth factor for a variety of subpopulations of T lymphocytes.

In addition to those studies which have examined the induction of growth of T lymphocyte populations by IL-2, a variety of other immunological activities have been observed. Numerous studies have demonstrated that IL-2 is able to induce resting cytotoxic T lymphocyte (CTL) precursors to proliferate in response to allogeneic stimulation and culminate in the generation of cytotoxic T effector cells capable of lysing radiolabelled target cells.[10] This property has generally been considered to be one of the premier physiological activities of IL-2 and has promoted interest in the restorative effects of IL-2 in in vivo models of immune deficiency (see below). Although the mechanism by which IL-2 stimulates cytotoxic T effector cell precursors is unknown, recent evidence indicates that antigen-activated cytotoxic T cell precursors express IL-2 receptors on their surface, and that the proliferation-inducing signal to these effector cell precursors is mediated through an interaction between IL-2 and the specific cell surface receptor.[7] Recent information indicates that other protein factors distinct from Interleukin 2 may induce IL-2-responsive proliferating T lymphocytes to differentiate into cytotoxic effector cells.[11]

In addition to the capacity of IL-2 to augment the induction and growth of antigen-specific cytotoxic T cells, IL-2 has also been implicated as a growth and activation factor for antigen-nonspecific natural killer (NK) cells.[12] These cells have been implicated in immune functions associated with tumor surveillance.

In the early studies conducted in 1978, the potential role of IL-2 in the induction of antibody-forming B cell responses was shown.[13] These experiments demonstrated that IL-2 was inseparable from the lymphokine capable of restoring antibody responses of B cells in populations of lymphocytes depleted of T helper cells. Although the exact target cell of the IL-2 in these studies was not determined, it seems likely that IL-2 was either enhancing antibody synthesis indirectly by the activation of a particular subset of helper T lymphocytes or directly by activation of the antibody forming B cell precursors. Even though the prevailing notion is that the IL-2 enhances antibody synthesis indirectly, recent experiments indicate that activated B lymphocytes may express receptors for IL-2. Therefore, the mechanism by which IL-2 enhances antibody production remains essentially unknown.

In addition to the above biological activation processes, which measure either the proliferative response of the particular target cell or the induction of some terminal effector cell function such as antibody formation or cytotoxic T lymphocyte development, IL-2 has also been implicated in the induction of synthetic processes within the target populations. For example, IL-2 has been shown to induce certain subsets of T lymphocyte populations to produce gamma interferon.[14,15] The significance of this observation may be broader than initially expected in that gamma interferon, which is another glycoprotein lymphokine, probably mediates important regulatory effects in the induction of both NK and antigen-specific CTL populations as well as antibody-forming cells.[14,16] These latter results raise the possibility that IL-2, in part, mediates its enhancing effect on these subpopulations of cells indirectly through the induction of gamma interferon. Finally, recent evidence indicates that IL-2 may induce gamma interferon via an induction of the cyclic nucleotide pathway and arachidonic acid cascade intermediates.[17] In these studies, it was demonstrated that inducers of cyclic GMP as well as leukotrienes were able to mimic IL-2 in the induction of gamma interferon.

In summary, although the induction of T cell growth probably represents one of the primary functions of IL-2, there is mounting evidence that IL-2 is also involved in the induction of a number of cellular events (synthetic and differentiative) that are separable from cell division.

CHEMISTRY AND GENE CLONING

Purification and Biochemical Properties of IL-2. Early approaches to the purification of IL-2 involved classical preparative and chromatographic methods coupled with a search for T cell lines which did not produce other factors (such as colony stimulating factor, B cell growth factors, lymphocyte activating factor). IL-2 from the cloned human leukemia T-cell line JURKAT has been used for purification and characterization,[7,18-21] monoclonal antibody formation[22,23] and receptor binding studies with radiolabeled material.[7]

Purification of JURKAT-derived IL-2 by Sephadex gel filtration and isoelectric focusing gave a single major peak of activity at pI 8.2. Sodium dodecyl sulfate polyacrylamide gel electrophoresis (SDS-PAGE) of that peak yielded a single band at 15,500 daltons, which was found to be free of other factors and the lectin used to induce the production of IL-2.[7] Stern et al.[24] reported a method by which JURKAT-derived IL-2 was purified to homogeneity using reverse phase HPLC. A single band corresponding to the reported molecular weight of IL-2 was obtained by SDS gel electrophoresis analysis.[24] Biochemical heterogeneity in IL-2 obtained from other sources was shown by neuraminidase and glycosidase treatment to be due to O-linked post-translational sialylation and glycosylation.[7] Evidence suggests that the threonine at position 3 (see Figure I) is glycosylated.[25] The recent development of a monoclonal antibody immunoaffinity column has afforded single step purification of large quantities of JURKAT IL-2. The homogeneity of this material was demonstrated by SDS-PAGE and reverse phase HPLC.[22] The amino acid sequence that was obtained was identical to the sequence predicted by the nucleotide sequence determined from gene cloning experiments (see below).[23,26,27]

Murine IL-2 from the EL-4 thymoma and the cloned T-cell leukemia LBRM-33 cell line was found to exhibit an apparent molecular weight of 30,000 daltons as determined by gel filtration.[28-30] Isoelectric focusing of the material has demonstrated a spectrum of pI's between 4.2 and 5.0. Protein bands retaining IL-2 activity were found in the 21,000-25,000-dalton range following SDS-PAGE separation. The molecular heterogeneity in mouse IL-2 is apparently due to variable amounts of glycosylation since treatment with neuraminidase results in quantitative recovery of essentially a single species of IL-2 with pI 5.1.[31]

Cloning of the IL-2 Gene. Taniguchi and co-workers[26] cloned the cDNA coding for human IL-2 via cDNA libraries originally prepared from JURKAT IL-2 mRNA. Repetition of that process yielded a clone containing a plasmid with an 880 base pair cDNA insert. Approximately 2500 colonies

Figure I. Nucleotide and Deduced Amino Acid Sequence of Human Interleukin 2[a,b]

```
                                                        1
                                                        Met Tyr Arg Met Gln Leu Leu Ser Cys Ile Ala
                                                        ATG TAC AGG ATG CAA CTC CTG TCT TGC ATT GCA

                                20
Leu Ser Leu Ala Leu Val Thr Asn Ser Ala Pro Thr Ser Ser Ser Thr Lys Lys Thr Gln Leu Gln Leu
CTA AGT CTT GCA CTT GTC ACA AAC AGT GCA CCT ACT TCA AGT TCT ACA AAG AAA ACA CAG CTA CAA CTG

                    40
Glu His Leu Leu Leu Asp Leu Gln Met Ile Leu Asn Gly Ile Asn Asn Tyr Lys Asn Pro Lys Leu Thr
GAG CAT TTA CTG CTG GAT TTA CAG ATG ATT TTG AAT GGA ATT AAT AAT TAC AAG AAT CCC AAA CTC ACC

        60                                                                              80
Arg Met Leu Thr Phe Lys Phe Tyr Met Pro Lys Lys Ala Thr Glu Leu Lys His Leu Gln Cys Leu Glu
AGG ATG CTC ACA TTT AAG TTT TAC ATG CCC AAG AAG GCC ACA GAA CTG AAA CAT CTT CAG TGT CTA GAA

                                                                            100
Glu Glu Leu Lys Pro Leu Glu Glu Val Leu Asn Leu Ala Gln Ser Lys Asn Phe His Leu Arg Pro Arg
GAA GAA CTC AAA CCT CTG GAG GAA GTG CTA AAT TTA GCT CAA AGC AAA AAC TTT CAC TTA AGA CCC AGG

                                                                120
Asp Leu Ile Ser Asn Ile Asn Val Ile Val Leu Glu Leu Lys Gly Ser Glu Thr Thr Phe Met Cys Glu
GAC TTA ATC AGC AAT ATC AAC GTA ATA GTT CTG GAA CTA AAG GGA TCT GAA ACA ACA TTC ATG TGT GAA

                                                    140
Tyr Ala Asp Glu Thr Ala Thr Ile Val Glu Phe Leu Asn Arg Trp Ile Thr Phe Cys Gln Ser Ile Ile
TAT GCT GAT GAG ACA GCA ACC ATT GTA GAA TTT CTG AAC AGA TGG ATT ACC TTT TGT CAA AGC ATC ATC

            153
Ser Thr Leu Thr
TCA ACA CTA ACT
```

a. ref. 26

b. Cleavage of the secretion protein between Ser^{20}-Ala^{21} gives mature human IL-2 with 133 amino acids.

were screened to identify that clone. Restriction endonuclease cleavage and nucleotide sequencing allowed the deduction of the amino acid sequence as shown in Figure I. The human IL-2 protein consists of 153 amino acids with a calculated MW 17631.7. The hydrophobic N-terminal region of the deduced sequence is thought to be a signal peptide which is cleaved (possibly between ^{20}Ser and ^{21}Ala) during the secretion process to yield mature IL-2 containing 133 amino acids with MW 15,420.5 (consistent with MW 15,000 reported for JURKAT-derived human IL-2.[18]) The cloned gene was introduced into a plasmid appropriate to mammalian cells and was transfected into a monkey cell line COS-7.[26] The IL-2 activity thus obtained promoted T-cell growth in the IL-2 bioassay, was neutralized by an anti-IL-2 monoclonal antibody and, when purified by Sephadex gel chromatography, gave material of MW 15,000 which was indistinguishable from IL-2 obtained from JURKAT cells.[26] DuPont workers[32], also working with JURKAT cell mRNA, obtained a cDNA insert whose sequence essentially matched that reported by Taniguchi et al.[26] for mature IL-2 but lacked those bases coding for pre-IL-2. Other groups have cloned IL-2 cDNA from human splenocyte mRNAs[27] and human tonsillar cell mRNAs[33,34] and conclude that only one IL-2 gene is present in the human genome.

IL-2 Receptor Studies. Studies of the IL-2 receptor have been performed with ^{35}S-met-IL-2 and ^{3}H-leu, ^{3}H-lys-IL-2 prepared by stimulation of

JURKAT cells in the presence of the radiolabeled amino acids.[7] Binding of labeled IL-2 with murine CTLL cells was rapid and dissociation was slow. Further, degradation of the receptor-bound labeled IL-2 suggests it is adsorbed and internalized by the cells. Binding of ^{35}S-met-IL-2 with CTLL cells indicated a single set of approximately 15,000 receptor sites per cell with a $K_d \cong 10^{-11}$. The number of receptor sites was increased in stimulated compared to unstimulated lymphocytes. Binding of labeled IL-2 to human cell lines has been related to the dependence of those cells on exogenous or endogenous IL-2.[7] No competition for ^{35}S-met-IL-2 binding was observed with a number of factors or mitogens including α- and γ-interferon, lymphocyte activating factor, epidermal and nerve growth factors, colony stimulating factor, phytohemaglutinin (PHA), concanavalin A (Con-A) and phorbol myristate acetate (PMA). Complete inhibition of binding was observed with IL-2 derived from JURKAT and human tonsillar cell sources and with the monoclonal antibody obtained from immunization of BALB/c mice with microgram quantities of IL-2.[7,22] In addition, the biological response for IL-2 in the standard IL-2 proliferation bioassay and ^{35}S-met-IL-2 binding were found to have similar dose response curves.[7]

IN VIVO ACTIVITY IN ANIMAL MODELS

Altered Production of IL-2. As described above, IL-2 appears to be a key lymphokine in the immunoregulatory circuits of immune responses. Thus, an increased or decreased production of IL-2 can cause alterations in immune response control mechanisms. In fact it has been suggested that a deficiency in IL-2 production may result in an impairment of those immunologic control mechanisms which may lead to autoimmune diseases such as systemic lupus erythematosus.[35] Using murine models, evidence has been obtained that a deficiency in IL-2 production may be implicated in the decreased immune response associated with aging[36] and infections due to *Trypanosoma cruzi*[37] and *Mycobacterium lepraemurium*.[38] In general, a reduction in the *in vitro* capacity of cultured spleen cells from mice with the above mentioned conditions to produce IL-2 in response to stimulation by the T cell mitogen Con A has been noted. However, conclusions based on net amounts of IL-2 activity in stimulated culture supernatants bear the caveat that the levels of activity can be affected not only by the amount of IL-2 produced but also by the amount utilized. In addition to the apparent decreased production of IL-2, the impaired *in vitro* proliferative response of these spleen cells to Con A could not be normalized by the addition of a source of exogenous IL-2 to the culture, suggesting a defect in responsiveness to IL-2.[37,38]

The Effects of Murine IL-2 on Cell-Mediated Immunity in Nude Mice in vivo. Some of the early experiments on the *in vivo* effects of IL-2 were conducted in athymic nude mice (nu/nu) which lack immunocompetent T cells. In a series of experiments conducted by Wagner *et al*.[39-41], an alloantigen (irradiated C57BL/6 mouse splenic lymphocytes) was injected subcutaneously (s.c.) into BALB/c nu/nu mice. Simultaneously, semi-purified IL-2 was injected either s.c. at the same site or intravenously (i.v.). Within 5 days the spleen and lymph node cells from these mice were harvested and shown *in vitro* to be able to mount an antigen-specific cytolytic response against target cells. Thus, IL-2 appeared to be able to restore alloreactive CTL responses in nude mice *in vivo*. Similar experiments have been performed to demonstrate that IL-2 also restores humoral immune responses of athymic mice.[42] Treatment of nude mice with IL-2 restored the IgM, IgG, and IgA responses to sheep red blood cells.[42] Similarly, nude mice injected i.v. with sheep red blood cells and treated with IL-2 daily developed antigen-specific T helper cells[43], suggesting that IL-2 restored the humoral response indirectly via its effects on T cells. IL-2 has also been shown to augment NK cell reactivity in nude mice.[44]

The Effects of Murine IL-2 on Cell-Mediated Immunity in Normal Mice in vivo. Scientists at the Fred Hutchinson Cancer Research Center have conducted a series of *in vivo* experiments in normal mice utilizing "highly purified" murine IL-2 preparations.[18,44-46] In these studies, IL-2 was obtained from PHA-stimulated LBRM murine lymphoma cells. The IL-2 in the culture supernatant was then purified by ammonium sulfate precipitation, gel filtration chromatography, and preparative isoelectric focusing. The IL-2 preparations obtained contained no detectable contamination with other lymphokines. In the *in vivo* studies conducted with this preparation of IL-2, CBA/J mice were given a single intraperitoneal (i.p.) injection of IL-2 in the absence of alloantigen. A significant augmentation of NK cell activity was observed in both splenocytes and peritoneal exudate cells (PEC). On the other hand, IL-2 treatment did not result in splenic cytolytic activity against NK-insensitive target cells. The i.v. administration of IL-2 produced similar results.

To study whether IL-2 would augment alloreactive CTL generation *in vivo*, BALB/c mice were immunized with allogeneic EL-4 tumor cells and treated with IL-2. A two-fold increase in cytolytic activity against EL-4 target cells in both PEC and spleen cells was found, whereas no effect on an antigenically distinct tumor cell target (SL-3) could be observed. Mice receiving IL-2 alone in the absence of alloantigen failed to develop CTL activity against EL-4 tumor cells. The rate of development of alloreactive CTL activity in both PEC and spleen cells was not significantly influenced by the *in vivo* administration of IL-2. These experiments demonstrate that the IL-2-augmented CTL reactivity was specific for the immunizing antigen. The augmentation of CTL activity of PEC induced by the i.p. administration of IL-2 and alloantigen was dose-dependent with a plateau of enhanced CTL activity observed at the highest dose (100 U/ml) of IL-2 administered.[18] These authors also noted that the *in vivo* administration of IL-2 completely abolished the induction of tolerance to the hapten dinitrofluorobenzene.

In a recent study, Rosenberg *et al*.[47] examined two preparations of IL-2 (spleen cell-derived and EL-4 thymoma-derived) for their effects on CTL generation in C57BL/6 and DBA/2 mice. The i.p., s.c., or i.v. administration of IL-2 for 3 days to normal C57BL/6 mice or those primed with irradiated P815 mastocytoma tumor cells increased the cytotoxicity of spleen cells against the specific allogeneic P815 target. Along similar lines, with both C57BL/6 mice immunized with DBA/2 lymphocytes and DBA/2 mice immunized with C57BL/6 lymphocytes, the injection of IL-2 over a three-day period markedly enhanced the generation of specific splenic CTL against the immunizing antigen.[47]

The Effects of Human IL-2 on Cell-Mediated Immunity in Mice. In a recent study, 48 highly purified human IL-2 administered s.c. to C57BL/6 mice, that had been injected with L1210 tumor cells and then immunosuppressed with a high dose of cyclophosphamide, restore the lymph node cell cytotoxic activity against L1210 leukemia target cells. However, IL-2 did not enhance CTL activity in normal C57BL/6 mice.[48] These results differ from those of Hefeneider *et al*.[49] who found that murine IL-2 augumented the cytotoxicity of normal spleen cells *in vivo*. The reason for this apparent discrepancy in the *in vivo* effect of IL-2 has not been established.

Antitumor Effects of IL-2 in Mice. Inoculation of either the FBL-3, a Friend virus-induced leukemia, or EL-4(G-), a chemically induced leukemia, into syngeneic C57BL/6 mice is lethal. Cyclophosphamide treatment of the mice extended survival time, but all animals eventually died. Adoptive chemoimmunotherapy experiments[50,51] with tumor sensitized spleen cells grown in the presence of IL-2 were performed to determine if increased survival time resulted. Spleen cells, obtained from C57BL/6 mice immunized with irradiated FBL-3 or EL-4(G-) cells *in vivo*, were secondarily immunized to the respective irradiated tumor antigen *in vitro* in the presence of IL-2-rich supernatants. These conditions allowed clonal expansion of antigen-specific CTL. Treatment of tumor-bearing mice with cyclophosphamide and those *in vitro*-propagated CTL increased survival. This type of therapy was specifically effective against the immunizing tumor in terms of prolonging the median survival time, however, no mice were cured. Other workers have also demonstrated varying degrees of effectiveness with the use of adoptive chemoimmunotherapy in different murine tumor models and employing different preparations of human and rat IL-2.[52,53] Antitumor effects have also been shown with the use of adoptive immunotherapy without chemotherapeutic agents.[54-57]

The antitumor effects of exogenously administered IL-2 have also been examined.[51,58] In these studies, using a purified IL-2 preparation from the lymphoma cell line LBRM-33, no antitumor effect of daily IL-2 therapy, either by itself or in combination with cyclophosphamide was observed in C57BL/6 mice injected with FBL-3 leukemia cells. However, IL-2 administration significantly potentiated the effects of adoptive immunotherapy utilizing CTL with 11 of 16 mice being completely cured.[50] Palladino *et al*.[59] have found that IL-2 in combination with adoptive immunotherapy was more effective than IL-2 alone in prolonging the survival of $CB6F_1$ mice injected with RL♂1 leukemia cells. Other workers have found varying degrees of antitumor effect with the use of mouse,[60] rat[61] or EL-4 cell line[57] derived IL-2-containing culture supernatant as the only therapeutic agent.

IL-2 and Graft Rejection. An immunoregulatory effect of IL-2 has also been demonstrated in graft rejection models. Skin from B10.BR mice grafted on to B6AF1 recipient mice was rejected more rapidly if the recipient mice were injected i.v. with $B6AF_1$ lymphocytes sensitized *in vitro* to irradiated B10.BR stimulator cells and allowed to proliferate in IL-2 rich spleen cell culture supernatants.[62] In another experiment, the survival of cardiac allografts in rats was decreased if the recipient rats were given sensitized lymphocytes plus IL-2-containing Con A-induced spleen cell culture supernatants.[63] These studies indicate that IL-2 is involved in graft rejection and suggest that inhibitors of IL-2 or agents which block IL-2 receptors may be of clinical significance in transplantation.

Pharmacokinetics in Animals. The clearance of murine IL-2 prepared from the EL-4 cell line has been studied in DBA/2 mice.[57] Partially purified IL-2 administered i.v. was rapidly cleared from the serum, (t 1/2=3-5min), whereas a more sustained serum level of IL-2 was maintained after i.p. administration. An even more sustained and prolonged serum level of IL-2 was achieved after s.c. administration.[47,64,65] The short half-life of i.v. administered IL-2 was also demonstrated in nude mice, in irradiated mice, and in splenectomized mice, suggesting that the binding of IL-2 to T cells *in vivo* did not account for its short serum half-life. The main site of IL-2 clearance appeared to be the kidney. In another experiment, a partially purified preparation of murine IL-2, was cleared equally rapidly from the blood of normal and T cell-depleted BALB/c mice.[66] These results also indicated that IL-2 was not cleared by adsorption to T cells.

CLINICAL STUDIES

In Vitro Production and Responsiveness. The production of IL-2 by peripheral blood mononuclear cells or lymphocytes in response to *in vitro* stimulation with mitogen or alloantigen was shown to be defective in patients with systemic lupus erythematosus,[67-70] lepromatous leprosy,[71] bone marrow transplantation,[72] solid tumors,[73] T cell immunodeficiency,[74-76] and rheumatoid arthritis.[70] In some instances, a restoration of the lymphocyte proliferative response in the presence of IL-2 has been demonstrated.[71,75] These studies suggest that IL-2 is implicated in the impaired immune function in certain disease states and point out the possible clinical application of this lymphokine.

It has been demonstrated that highly purified human IL-2 restored the marked decrease in NK cell activity and cytomegalovirus (CMV)-specific cytotoxicity of peripheral blood lymphocytes obtained from six patients with acquired immunodeficiency syndrome (AIDS) and CMV infection.[77]

Increased production of IL-2 has been observed in other clinical conditions. For example, lung cells (presumably containing T lymphocytes) from patients with sarcoidosis and high-intensity alveolitis spontaneously release IL-2.[78]

In Vivo Administration. Recently, Phase I clinical trials utilizing both purified natural and recombinant IL-2 were initiated with patients having immunodeficiency diseases including acquired immune deficiency syndrome (AIDS). In one such clinical trial for which data has been published, human IL-2 purified to apparent homogeneity was administered s.c. daily to a child with Nezelof's T cell deficiency who was retrospectively suspected to have AIDS.[76] He received 5 escalating doses of IL-2. However, his clinical condition deteriorated and he died 8 days after initiation of IL-2 therapy. At present there are not enough clinical data available to evaluate the effectiveness of IL-2 in treating immunodeficiency disease states.

Pharmacokinetics in Humans. The pharmacokinetics of partially purified human IL-2 given i.v. have been studied in two patients with metastatic melanoma.[79] The clearance of IL-2 from the circulation was exponential over one hour and the initial IL-2 blood level was related to the dose given. The half-life of IL-2 was $\sim$22.5 minutes. Side effects included transient rigor, pyrexia, tachycardia, hypotension, hypoglycemia, nausea and vomiting, increased cortisol levels, lymphocytopenia, and signs of mild intravascular coagulation. There was no significant change in liver enzyme levels, and no long-term effects were observed. Since the IL-2 preparation was not purified to homogeneity it is not known if the side effects were due to IL-2 or to other contaminating components. It should be noted that similar side effects using more highly purified preparations of IL-2 have not been observed in mice administered IL-2.

CONCLUSIONS

Thus, in little more than 5 years, IL-2 research has progressed from the initial descriptions of its *in vitro* biological activities, to its purification to molecular homogeneity, gene cloning, demonstrations of *in vivo* activity in mice, and into early efforts in the area of immunotherapy in man. It is expected that with the availability of recombinant IL-2, numerous phase I clinical trials will begin in earnest in patients with various forms of acquired immunodeficiency and malignancy.

REFERENCES

1. "Interleukins, Lymphokines, and Cytokines", J.J. Oppenheim, S. Cohen, Eds., Academic Press, New York, 1983.
2. J. Watson, S. Gillis, J. Manbrook, D. Mochizuki, K. Smith, J. Exp. Med. 150, 849 (1979).
3. E.L. Lattime, S. Gillis, C. David, O. Stutman, Eur. J. Immunol. 11, 67 (1981).
4. E.L. Larson, N.N. Iscove, A. Coutinho, Nature 283, 664 (1980).
5. S. Meuer, S.F. Schlossman, E.L. Reinherz, Cell. Immunol. 70, 405 (1982).
6. S. Gillis, M.M. Ferm, W. Ou, K.A. Smith, J. Immunol. 120, 2027 (1978).
7. R.J. Robb, Immunobiol. 161, 21 (1982).
8. D.E. Parks, Fed. Proc. Fed. Am. Soc. Exp. Biol. 41, 592 (1982).
9. A. Lanzavecchia, M. Ferrarini, F. Celada, Eur. J. Immunol. 12, 468 (1982).
10. D. Zagury, D. Morgan, G. Lenoir, M. Fouchard, M. Feldman, Int. J. Cancer 31, 427 (1983).
11. D.N. Männel, W. Falk, W. Dröjc, J. Immunol. 130, 2508 (1983).
12. T. Timonen, J.R. Ortaldo, B.M. Stadler, G. Bonnard, S.O. Sharrow, R.B. Herberman, Cell. Immunol. 72, 178 (1982).
13. J.J. Farrar, P.L. Simon, W.J. Koopman, J. Fuller-Bonar, J. Immunol. 121, 1353 (1978)
14. W.L. Farrar, H.M. Johnson, J.J. Farrar, J. Immunol. 126, 1120 (1981).
15. T. Kasahara, J.J. Hooks, S.F. Dougherty, J.J. Oppenheim, J. Immunol. 130, 1784 (1983).
16. J.Y. Ojen, J.A. Heinbaugh, H.T. Holden, R.B. Herberman, J. Immunol. 122, 175 (1979).
17. H.M. Johnson, B.A. Torres, D.L. Archer, W.L. Farrar, Cell. Immunol. 70, 382 (1982).
18. S. Gillis, D.Y. Mochizuki, P.J. Conlon, S.H. Hefeneider, C.A. Ramthun, A.E. Gillis, M.B. Frank, C.S. Henney, J.D. Watson, Immunological Rev. 63, 167 (1982).
19. J.D. Watson, D. Mochizuki, M.B. Frank, S. Gillis, in "Immunopharmacology and the Regulation of Leukocyte Function", Vol. 19, D.R. Webb, Ed., Marcel Dekker Inc., New York/Basle, 1982, p. 39.
20. J. Watson, M.B. Frank, D. Mochizuki, S. Gillis, in "Lymphokines", Vol. 6, E. Pick, Ed., Academic Press, New York, 1982, p. 95.
21. J.W. Meir, R.C. Gallo, in "Lymphokines", Vol. 6, E. Pick, Ed., Academic Press, New York, 1982, p. 137.
22. K.A. Smith, M.F. Favata, S. Oroszlan, J. Immunol. 131, 1808 (1983).
23. R.J. Robb, R.M. Kutny, V. Chowdhry, Proc. Natl. Acad. Sci. 80, 5990 (1983).
24. A.S. Stern, Y.-C.E. Pan, D.L. Urdal, D.Y. Mochizuki, S. DeChlara, R. Blacher, J. Wideman, S. Gillis, Proc. Natl. Acad. Sci. 81, 871 (1984).
25. R.J. Robb, R.M. Kutny, M. Panico, H. Morris, W.F. DeGrado, V. Chowdhry, Biochem. Biophys. Res. Commun. 116, 1049 (1983).
26. T. Taniguchi, H. Matsui, T. Fujita, C. Takaoka, N. Kashima, R. Yoshimoto, J. Hamuro, Nature 302, 305 (1983).
27. R. Devos, B. Plaetinck, H. Cherovtre, G. Simons, W. Degrave, J. Tavernier, E. Remaut W. Fiers, Nucleic Acids Res. 11, 4307 (1983).
28. J.J. Farrar, W.R. Benjamin, M.L. Hilfiker, M. Howard, W.L. Farrar, J. Fuller-Farrar, Immunological Rev. 63, 129 (1982).
29. J.D. Watson, D.Y. Mochizuki, S. Gillis, Fed. Proc. Fed. Am. Soc. Exp. Biol. 42, 2747 (1983).
30. K.A. Smith, Immunobiol. 161, 1 (1982).
31. J.J. Farrar, J. Fuller-Farrar, P.L. Simon, M.L. Hilfiker, B.M. Stadler, W.L. Farrar, J. Immunol. 125, 2555 (1980).
32. K.J. Livak, L.A. Ivanoff, Y. DeVries, R. Arentzen, S.R. Petteway, J. Cell. Biol. 97, 420a (1983).
33. S. Maeda, N. Nishino, K. Obaru, S. Mita, H. Nomiyama, K. Shimada, K. Fujimoto, T. Teranishi, T. Hiruno, K. Onoue, Biochem. Biophys. Res. Commun. 115, 1040 (1983).
34. S. Mita, S. Maeda, K. Obaru, N. Nishino, K. Shimada, T. Hirano, K. Onoue, T. Ogawa, H. Ogawa, Biochem. Biophys. Res. Commun. 117, 114 (1983).
35. N. Talal, M. Dauphinec, D. Wofsy, Arthritis Rheum. 25, 838 (1982).
36. F. Joncourt, Y. Wang, F. Kristensen, A.L. deWeck, Immunobiol. 163, 521 (1982).
37. A. Harel-Bellan, M. Joskowicz, D. Fradelizi, H. Eisen, Proc. Natl. Acad. Sci. 80, 3466 (1983).
38. A. Hoffenbach, P.H. Lagrange, M.A. Bach, Infect. Immun. 39, 109 (1983).
39. H. Wagner, C. Hardt, K. Heeg, K. Pfizenmaier, W. Solbach, R. Bartlett, H. Stockinger M. Röllinghoff, Immunol. Rev. 51, 215 (1980).
40. H. Wagner, C. Hardt, K. Heeg, M. Röllinghoff, K. Pfizenmaier, Nature 284, 278 (1980).
41. H. Wagner, C. Hardt, K. Heeg, K. Pfizenmaier, H. Stötter, M. Röllinghoff, Immunobiol. 161, 139 (1982).

42. L.S. English, A.M. Smith, Fed. Proc. Fed. Soc. Exp. Biol. 41, 303 (1983).
43. H. Stötter, E. Rüde, H. Wagner, Eur. J. Immunol. 10, 719 (1980).
44. P.J. Conlon, S. Hefeneider, S. Gillis, Fed. Proc. Fed. Am. Soc. Exp. Biol. 41, 369 (1982).
45. P.J. Conlon, S.H. Hefeneider, C.S. Henney, S. Gillis, in "The Potential Role of T Cells in Cancer Therapy", (Prog. Cancer Res. Ther.), Vol. 22, A. Fefer, A.L. Goldstein, Eds., Raven Press, New York, 1982, p. 113.
46. S.H. Hefeneider, C.S. Henney, S. Gillis, in "NK Cells and Other Natural Effector Cells", R.B. Herberman, Ed., Academic Press, New York, 1982, p. 421.
47. S.A. Rosenberg, P.J. Spiess, S. Schwarz, Transplantation 35, 631 (1983).
48. V.J. Merluzzi, K. Welte, D.M. Savage, K. Lust-Barney, R. Mertelsmann, J. Immunol. 131, 806 (1983).
49. S.H. Hefeneider, P.J. Conlon, C.S. Henney, S. Gillis, J. Immunol. 130, 222 (1983).
50. M.A. Cheever, P.D. Greenberg, A. Fefer, J. Immunol. 126, 1318 (1981).
51. M.A. Cheever, P.D. Greenberg, A. Fefer, S. Gillis, J. Exp. Med. 155, 968 (1982).
52. E. Kedar, R.B. Herberman, in "NK Cells and Other Natural Effector Cells, R.B. Herberman, Ed., Academic Press, New York, 1982, p. 859.
53. E. Kedar, B.L. Ikejiri, E. Gorelik, R.B. Herberman, Cancer Immunol. Immunother. 13, 14 (1982).
54. C. Ting, S.S. Yang, Int. J. Cancer 30, 625 (1982).
55. T.J. Eberlein, M. Rosenstein, S.A. Rosenberg, J. Exp. Med. 156, 385 (1982).
56. G. Forni, R. Preta, L. Fava, M. Glovarelli, Fed. Proc. Fed. Am. Soc. Exp. Biol. 41, 406 (1982).
57. V. Paetkau, G.B. Mills, R.C. Bleackley, in "The Potential Role of T Cells in Cancer Therapy", (Prog. Cancer Res. Ther.), Vol. 22, A. Fefer, A.L. Goldstein, Eds., Raven Press, New York, 1982, p. 147.
58. M.A. Cheever, P.D. Greenberg, S. Gillis, A. Fefer, in "The Potential Role of T Cell in Cancer Therapy", (Prog. Cancer Res. Ther.), Vol. 22, A. Fefer, A.L. Goldstein, Eds., Raven Press, New York, 1982, p. 127.
59. M.A. Palladino, L.J. Old, H.F. Oettgen, Fed. Proc. Fed. Am. Soc. Exp. Biol. 41, 406 (1982).
60. E.A. Grimm, E. Gorelik, M.M. Rosenstein, S.A. Rosenberg, Cell. Immunol. 70, 409 (1982).
61. J. Bubenik, P. Perlmann, M. Indrová, J. Símová, T. Jandlova, J. Neuwirt, Cancer Immunol. Immunother. 14, 205 (1983).
62. M. Rosenstein, T. Eberlein, M.M. Kemeny, P.H. Sugarbaker, S.A. Rosenberg, J. Immunol. 127, 566 (1981).
63. A.E. Clason, A.J. Duarte, J.W. Kupiec-Weglinski, J.N. Williams, B.S. Wang, T.B. Strom, N.L. Tilney, J. Immunol. 129, 252 (1982).
64. J.H. Donohue, S.A. Rosenberg, Fed. Proc. Fed. Am. Soc. Exp. Biol. 42, 1071 (1983).
65. J.H. Donohue, S.A. Rosenberg, J. Immunol. 130, 2203 (1983).
66. P.F. Muhlradt, H.G. Opitz, Eur. J. Immunol. 12, 983 (1982).
67. J. Alcocer-Varela, D. Alarcon-Segovia, J. Clin. Invest. 69, 1388 (1982).
68. D.A. Horwitz, M. Linker-Israeli, A.C. Bakke, R.C. Kitridou, S. Gendler, C.M. Lemoine, S. Gillis, Arthritis Rheum. 25 (Suppl.), 528 (1982).
69. M. Linker-Israeli, A.C. Bakke, R. Kitridou, S. Gendler, S. Gillis, D.A. Horwitz, J. Immunol. 130, 2651 (1983).
70. N. Talal, M. Dauphinée, P. Christadoss, G. Fernandes, I. Russell, N. Miyasaka, T. Nakamura, Adv. Nephrol. 12, 239 (1983).
71. A. Haragewoin, T. Godal, A.S. Mustafa, A. Belehu, T. Yemaneberhan, Nature 303, 342 (1983).
72. O. Azogui, E. Gluckman, D. Fradelizi, J. Immunol. 131, 1205 (1983).
73. A. Rey, B. Klein, D. Zagury, C. Thierry, B. Serrou, Immunol. Lett. 6, 175 (1983).
74. R. Pagenelli, F. Aiuti, P.C. Beverley, R.J. Levinsky, Clin. Exp. Immunol. 51, 338 (1983).
75. S. Tsuchiya, M. Imaizumi, M. Minegishi, T. Konno, K. Tada, N. Engl. J. Med. 308, 1294 (1983).
76. N. Flomenberg, K. Welte, R. Mertelsmann, N. Kernan, N. Ciobanu, S. Venuta, S. Feldman, G. Kruger, D. Kirkpatrick, B. Dupont, R. Reilly, J. Immunol. 130, 2644 (1983).
77. A.H. Rook, H. Masur, H.C. Lane, W. Fredrick, T. Kasahara, A.M. Macher, J.Y. Djeu, J.F. Manischewitz, L. Jackson, A.S. Fauci, G.V. Quinnan, J. Clin. Invest. 72, 398 (1983).
78. P. Pinkston, P.B. Bitterman, R.G. Crystal, N. Engl. J. Med. 308, 793 (1983).
79. C. Blindon, M. Czerniecki, P. Ruell, A. Edwards, W.H. McCarthy, R. Harris, P. Hersey, Br. J. Cancer 47, 123 (1983).

Section V - Topics in Biology

Editor: Robert W. Egan, Merck Institute for
Therapeutic Research, Rahway, New Jersey 07065

Chapter 20. The Inactivation of Cytochrome P-450

Paul R. Ortiz de Montellano, Department of Pharmaceutical
Chemistry, School of Pharmacy University of California
San Francisco, California 94143

Introduction - The inactivation of cytochrome P-450 by enzyme-specific agents can alter the metabolism of concommitantly administered xenobiotics or can upset essential biosynthetic or catabolic pathways if isozymes devoted to the metabolism of endogenous substrates are affected. These changes in metabolic activity may be toxicologically significant but may, if intentional, be exploited for pharmacological or therapeutic purposes. From a heuristic point of view, catalysis-based inactivation can be used to elucidate the catalytic mechanism of cytochrome P-450 enzymes. Three classes of agents are known that specifically inactivate cytochrome P-450 enzymes by catalysis-dependent mechanisms: (a) agents that bind covalently to the protein, (b) agents that quasi-irreversibly coordinate with the heme iron atom, and (c) agents that specifically alkylate or degrade the prosthetic heme group.[1-7] This review summarizes recent work on agents that irreversibly N-alkylate (Scheme 1) or otherwise alter the prosthetic heme group of cytochrome P-450 and then outlines progress in applying this knowledge to the design of isozyme-specific irreversible inhibitors.

Scheme 1

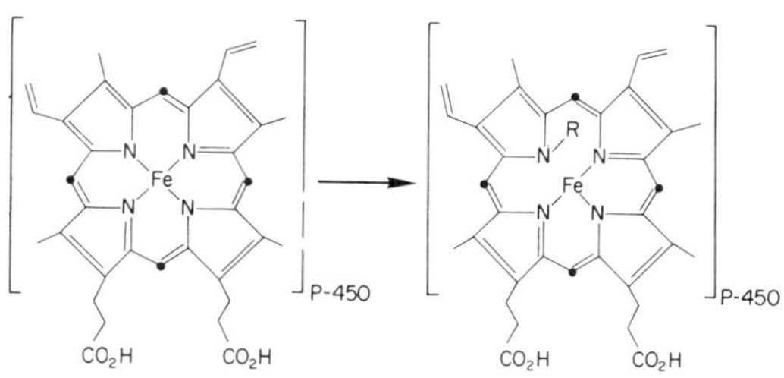

Heme-Destructive Functionalities

Terminal Olefins and Acetylenes - The first recognized and most studied of the suicide substrates for cytochrome P-450 is 2-isopropyl-4-pentenamide (AIA).[6,8] The π-bond of AIA is normally oxidized to the epoxide but approximately once in every 200 catalytic events oxidation of the double bond results in covalent attachment of the substrate to the prosthetic heme group.[6,7] Early work showed that this alkylation yields protoporphyrin IX with AIA bound to one of the nitrogen atoms.[9,10] Chemical and mass spectrometric evidence identified the N-alkyl group as the five- or six-membered lactone expected from intramolecular cyclization of, respectively, the γ- or δ-hydroxylamide obtained if an oxygen atom adds to one end and a porphyrin nitrogen to the other end of the π-bond

ISBN 0-12-040519-9

(Scheme 2).[10] The five-membered lactone structure has recently been confirmed by an NMR spectroscopic analysis.[11] Novonal (2,2-diethyl-4-pentenamide), a structurally related sedative-hypnotic, similarly inactivates cytochrome P-450 and gives rise to a protoporphyrin IX adduct with the corresponding five-membered lactone attached to one of the nitrogens.[11-12]

Scheme 2

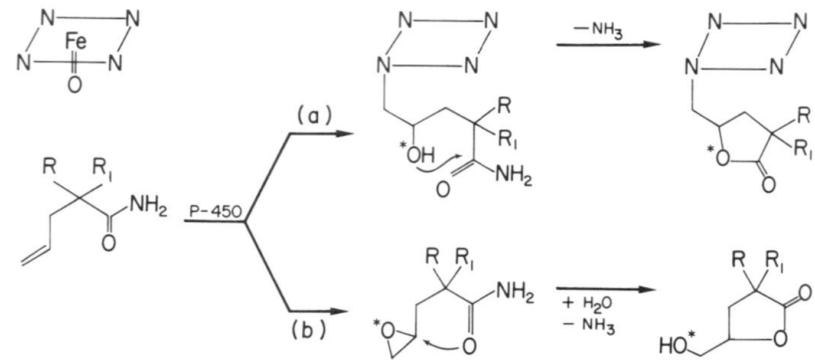

AIA and novonal are oxidized by cytochrome P-450 to lactone metabolites that are structurally analogous to the N-alkyl group in the corresponding heme adducts.[13,14] The lactone metabolite of AIA, as shown by ^{18}O-studies, arises by intramolecular attack of the amide carbonyl on an initially formed epoxide metabolite.[15] The oxygen incorporated into the lactone function in the heme adduct, however, derives from molecular oxygen and thus must be introduced by the catalytic action of the enzyme.[11] This rules out interception by the amide of a catalytically-activated π-bond as a step in heme alkylation because the lactone oxygen would be expected to derive from the amide carbonyl group.

Alkylation of the prosthetic heme of cytochrome P-450 on oxidation of a double bond is not limited to structures that resemble AIA but is generally observed with terminal π-bonds.[6,7] In addition to the 47 olefins and acetylenes listed in a recent review,[6] allyl chloride, allyl alcohol, and several vinylcycloalkanes have been reported to similarly inactivate cytochrome P-450.[16-19] Heme alkylation by all terminal olefins and acetylenes conforms in general to the pattern set by AIA and novonal except that lactonization does not occur in the absence of a properly positioned carbonyl group. The porphyrin N-alkyl group in the heme-adducts that have been characterized to date (ethylene, propene, octene, acetylene, fluroxene, vinyl fluoride, propyne, octyne, ethchlorvynol) is obtained by adding the porphyrin nitrogen to the terminal carbon and the oxygen to the internal carbon of the π-bond.[6] Heme alkylation during the oxidation of ethylene thus gives rise to N-(2-hydroxyethyl)protoporphyrin IX. The hydroxyl group again derives from molecular oxygen.[20] Secondary transformations modify the structure of the heme adduct, however, if the destructive substrate is an acetylene or bears a leaving group on the carbon to which the oxygen is added. The N-alkyl enol function obtained in the reaction with an acetylene tautomerizes to the more stable carbonyl structure. Propyne thus yields N-(2-oxopropyl)protoporphyrin IX.[21] The hydroxyls in the heme adducts with fluroxene (2,2,2-trifluoroethyl vinyl ether) and vinyl fluoride readily eliminate 2,2,2-trifluoroethanol or hydrogen fluoride, respectively, to give the same carbonyl product.[22] It is assumed but still unconfirmed that vinyl chloride, vinylidene chloride, and trichloroethylene analogously alkylate the prosthetic heme of cytochrome P-450.[23,24]

Alkylation of prosthetic heme by the epoxide metabolites can be excluded because (a) epoxides of active olefins do not inactivate the enzyme,[9,10,25] (b) the nitrogen and the oxygen add across the π-bond in a *cis* rather than *trans* fashion,[26] and (c) the heme nitrogen adds to the terminal carbon of vinyl ethers whereas the internal carbon is by far more reactive in the corresponding epoxides.[22] The evidence requires activation of the terminal carbon of the π-bond during cytochrome P-450-catalyzed transfer of oxygen to the internal carbon. The requirement for an *acyclic* alkylating species argues persuasively against exclusively concerted epoxidation mechanisms. The nature of the reactive intermediate, proposed to be a radical, is under continuing investigation.[7,22]

4-Alkyl-1,4-Dihydropyridines - DDC (1, R=Me) causes the accumulation of a green pigment that competitively inhibits the heme biosynthetic enzyme that inserts iron into protoporphyrin IX.[27-29] The green pigment, unambiguously identified as N-methylprotoporphyrin IX,[30] is generated by transfer of the 4-methyl group from DDC to the prosthetic heme moiety upon oxidation of the substrate by the enzyme.[31-33] The *in vivo* inhibition of heme biosynthesis caused by DDC is thus closely linked to its inactivation of cytochrome P-450. Cytochrome P-450 is inactivated by DDC analogues with an ethyl, propyl, isopropyl, isobutyl, or benzyl moiety in place of a 4-methyl group but not by those with a 4-aryl group. N-alkylated porphyrins, however, do not appear to be formed with the 4-isopropyl or 4-benzyl analogues.[33,34] Oxidation of the 4-ethyl analogue (1, R = Et) by microsomal cytochrome P-450 has been shown by spin trapping experiments to release the 4-ethyl group as a free radical.[33] The fact that the 4-ethyl group appears in the N-ethylprotoporphyrin IX heme adduct clearly suggests that the catalytically-released radical reacts with the prosthetic heme group and inactivates the enzyme.

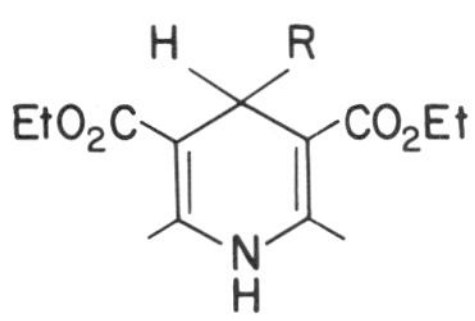

1

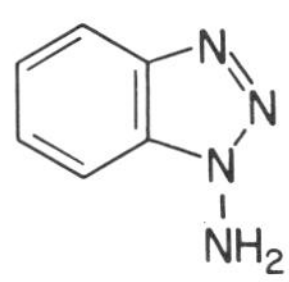

2

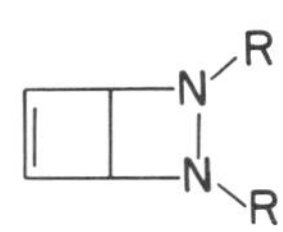

3

Hydrazines - Phenelzine, isoniazid, phenylhydrazine and other hydrazines inhibit a variety of metabolic pathways. A now classical example is the lethal potentiation of meperidine that accompanies inhibition of its metabolism by phenelzine.[35,36] A recent study of the kinetics of inhibition of microsomal dealkylation reactions by phenelzine has been interpreted in terms of reversible inhibitory processes but an earlier, more definitive study demonstrated that phenelzine irreversibly inactives cytochrome P-450.[37,38] The inactivation of cytochrome P-450 is accompanied by the hepatic accumulation of N-(2-phenylethyl)protoporphyrin IX and, as shown by spin trapping experiments, by oxidative formation of the 2-phenylethyl free radical.[39] Phenelzine is thus also oxidized by cytochrome P-450 to a free radical that presumably reacts with the prosthetic heme group.

Phenylhydrazine inactivates cytochrome P-450 in a reaction characterized by time-dependent formation of a transient complex with an absorbance maximum at 480 nm and by eventual loss of heme.[40] The nature of the transient complex and the final fate of the heme group remain ambiguous but results with other hemoproteins (see below) suggest that the complex may involve coordination of the phenyl group with the heme iron and heme loss may result from eventual formation of N-phenylprotoporphyrin IX.

The recently confirmed inhibition of cytochrome P-450 by isoniazid and hydralazine does not result from heme alkylation but rather from the formation of a transient complex with an absorption maximum at 449 nm.[41-44] The complex and its inhibitory effect are dissipated by ferricyanide. The inhibition by isoniazid thus resembles that mediated by 1,1-disubstituted hydrazines, which also form reversible cytochrome P-450 complexes absorbing at 449 nm.[45] Studies with metalloporphyrins indicate that 1,1-dialkylhydrazines are oxidized to aminonitrenes that coordinate to the iron atom.[46] The coordination of diazene metabolites with the heme of cytochrome P-450 also yields complexes with an absorbance maximum at 446 nm.[47]

The reactions of hydrazines with hemoproteins other than cytochrome P-450 are more clearly understood because of their greater accessibility. The inactivation of hemoglobin, myoglobin, and catalase by substituted phenylhydrazines converts their prosthetic heme groups into N-arylprotoporphyrin IX derivatives.[48-50] The key discovery made with these hemoproteins, however, is that the aryl moiety prior to denaturation of the hemoproteins is coordinated to the heme _iron_ atom rather than to the nitrogen on which it is found in the isolated adducts.[51-53] The evidence for this formulation includes a high resolution crystallographic structure of the hemoprotein complex generated from phenylhydrazine and myoglobin.[53] The iron-to-nitrogen shift that yields the isolated N-arylporphyrins takes place as the hemoprotein complexes denature.[52,54,55]

Horseradish peroxidase does not give an iron-phenyl complex with phenylhydrazine but does react with cyclopropanone and nitromethane to yield _meso_ alkylated products (the _meso_ positions are marked by dots in Scheme 1).[50,56,57] No _meso_-alkylated heme adducts have been reported for cytochrome P-450. The factors that favor _meso_ alkylation in horseradish peroxidase but N-alkylation in myoglobin, hemoglobin, catalase, and cytochrome P-450 remain to be elucidated.

Nitrosamines - Nitrosamines depress _in vivo_ cytochrome P-450 concentrations and inhibit xenobiotic metabolism.[58] Dimethyl-, diethyl-, and dipropylnitrosamines have recently been shown to cause the _in vivo_ accumulation of abnormal hepatic pigments.[59] The pigment obtained with diethylnitrosamine, identified as N-(2-hydroxyethyl)protoporphyrin IX, did not detectably contain N-ethylprotoporphyrin IX. The demonstration that N-ethylprotoporphyrin IX is not a precursor of the N-(2-hydroxyethyl) derivative implies that the heme reacts with a preformed 2-hydroxyethyl moiety (β-hydroxylation of diethylnitrosamine is known to occur) or is alkylated during the oxidation of diethylnitrosamine-derived ethylene. The adduct obtained with diethylnitrosamine is the same as that obtained with ethylene.[60] Ethylene formation from diethylnitrosamine has not been reported but alkyl alcohols, which derive from the cationic intermediates required for olefin formation, are known as metabolites of dialkylnitrosamines.[61] Heme alkylation by dimethylnitrosamine must, in any case, involve a species other than an olefin metabolite.[59]

Benzyne and Cyclobutadiene: Stereoelectronic Activation - The catalytic interaction of cytochrome P-450 with 1-aminobenzotriazole (2), a compound that releases benzyne when chemically oxidized, results in time-dependent enzyme inactivation.[62,63] An unusual prosthetic heme adduct is formed in which two vicinal nitrogens of protoporphyrin IX are bridged by an ortho-substituted phenyl ring.[64] The inactivation reaction occurs even when substituents are placed on the exocyclic amino group of 1-aminobenzotriazole, although enzyme inactivation by substituted analogues has not been well characterized. The catalytic role of the enzyme and the structure of the heme adduct indicate that inactivation follows oxidative release of benzyne or a benzyne-like species within the active site of the enzyme.

The chemical decomposition of (3, R = H) yields cyclobutadiene.[65] The cytochrome P-450-catalyzed oxidation of N,N-bis-carbethoxy-2,3-diazabicyclo[2,2,0]hex-5-ene (3, R = CO_2Et), a relatively stable derivative, results in time-dependent inactivation of the enzyme.[66] The inactivation is paralleled by the accumulation of a heme adduct identified as N-(2-cyclobutenyl)protoporphyrin IX, a finding that points to a catalytically unmasked cyclobutadienoid species as the heme-alkylating agent.

Disubstituted Acetylenes and Allenes - The unequivocal demonstration that cytochrome P-450 is destroyed by prosthetic heme alkylation requires isolation and characterization of the heme adduct. These criteria have been met for the inhibitors described above. Cytochrome P-450, however, also catalyzes its own inactivation by mechanisms that destroy the heme group but do not yield detectable heme adducts. The destructive reaction in these instances may give unstable adducts or may degrade the heme by entirely distinct mechanisms. The need for caution in attributing cytochrome P-450 inactivation to heme alkylation is emphasized by the finding that terminal allenes and disubstituted acetylenes inactivate cytochrome P-450 when catalytically acted upon by the enzyme but, in contrast to terminal olefins and acetylenes, do not give detectable heme adducts even though the prosthetic heme moiety is lost.[63,67,68] The reason for this difference is not known but may reflect stringent steric requirements for the N-alkylation of prosthetic heme.

Cyclopropylamines - Benzylcyclopropylamines but not benzylisopropylamines are catalysis-dependent irreversible inhibitors of cytochrome P-450.[69,70] The retention of full destructive activity in N-benzyl-1-(methyl)cyclopropylamine argues that enzyme inactivation does not follow oxidation of the cyclopropylamine to an exocyclic imine because this transformation is blocked by the methyl group. The proposal has therefore been made that oxidation of the cyclopropyl amine to the imine is accompanied by cleavage of the cyclopropyl ring to give a primary homoallylic radical. The carbon radical is envisioned to react with the prosthetic heme group or the protein to inactivate the enzyme.[70,71] Covalent binding of radiolabeled agent to the hemoprotein has been reported and the formation of a "green" pigment mentioned in a footnote, but actual evidence that the cyclopropyl ring is cleaved or that heme alkylation occurs is not available.[71] The mechanism proposed for the inactivation of cytochrome P-450 by cyclopropyl amines closely parallels that formulated by Silverman to explain inactivation of monoamine oxidase by the same substrates.[72] The relevance of mechanism-based inactivation of cytochrome P-450 by cyclopropylamines to inhibition of drug metabolism by the MAO inhibitor tranylcypromine (2-phenylcyclopropylamine) remains unknown.[73]

Aldehydes - The destruction of cytochrome P-450 by aromatic aldehydes is accompanied by equimolar loss of microsomal heme.[74,75] Aliphatic aldehydes also destroy cytochrome P-450 but, unlike the aromatic analogues, only appear to be active in vitro.[77] Enzyme destruction by these mono-aldehydes is distinguished from that mediated by phthalaldehyde by a requirement for NADPH.[75-77] The incubation of octanal with hepatic microsomes from rats pretreated with radiolabeled levulinic acid to tag the heme groups causes the formation of a radiolabeled "green" pigment, but the electronic spectrum of the pigment lacks the features that characterize N-alkylprotoporphyrin IX derivatives.[76]

Halogenated Hydrocarbons - The destruction of cytochrome P-450 by CCl_4,[78] first attributed to lipid peroxidation, has been shown to occur even under conditions where lipid peroxidation is not detectable.[79-82] One possible explanation for this inactivation is that the trichloromethyl radical or a related species obtained by reduction of the halocarbon reacts with the heme moiety or the apoprotein. The ill-defined radio-labeled porphyrins reported in incubations of labeled CCl_4 with hepatic microsomes would provide support for a heme alkylation mechanism were it not for the conflicting report that fluorescent N-alkylated porphyrins similar to those obtained with AIA are not isolated from CCl_4-incubated microsomes by procedures that result in isolation of the AIA adducts.[83,84]

The mechanism of cytochrome P-450 destruction by CCl_4 or the related anesthetic halothane must take into account the reductive formation of a reversible complex with an absorbance maximum at 454 or 469 nm, respectively.[84-88] It is believed, but has yet to be clearly demonstrated, that these complexes involve coordination of the prosthetic heme iron with halocarbon-derived carbenes or anions.

1,2,3-Benzothiadiazoles - The synergistic activity of 1,2,3-benzothiadiazoles with insecticides and drugs has been traced to suicidal inactivation of cytochrome P-450 paralleled by loss of microsomal heme but not by detectable heme adduct formation or enhanced lipid peroxidation.[63,89-91] Covalent binding of a radiolabeled 1,2,3-benzothiadiazole to microsomal proteins has been claimed but no supporting data was presented.[92] The mechanism of action of 1,2,3-benzothiadiazoles remains unknown.

Thiocarbonyls and Thiophosphates - A variety of substances with C=S or P=S functionalities, including diethyldithiocarbamate and parathion, inactivate cytochrome P-450 and result in microsomal heme loss when catalytically acted upon by the enzyme.[93-97] The most informative mechanistic data is that obtained for parathion which suggests that several sulfur atoms are covalently bound but actual loss of catalytic activity correlates with loss of the prosthetic heme group.[96,97] It has been postulated that catalytically-released sulfur atoms react with protein sulfhydryl groups to give hydrodisulfides, one of which subsequently reacts with the heme group. This postulate is supported by the demonstration that benzyl hydrodisulfide destroys cytochrome P-450 and its prosthetic group without a requirement for catalytic turnover of the enzyme but the nature of the products formed in the reaction is not known.[98]

Peroxides - The destruction of cytochrome P-450 during lipid peroxidation is generally assumed to result from nonspecific degradation of microsomal proteins. Lipid peroxidation indeed causes general protein damage and thus falls outside the scope of this review. Evidence nevertheless exists that lipid peroxides can inactivate cytochrome P-450 enzymes with some specificity: (a) lipids extracted from peroxidized microsomes destroy the cytochrome P-450 complement of fresh microsomes,[99] (b) linoleic acid hydroperoxide causes equimolar, concentration-dependent, loss of microsomal cytochrome P-450 and heme, and (c) cytochrome b_5 and cytochrome P-450 reductase are not affected except by relatively high concentrations of linoleic acid hydroperoxide.[100,101]

The Physiological Importance of Isozyme-Specific Inactivation

Isozyme Specificity - Work with purified cytochrome P-450 enzymes has shown that AIA inactivates the major isozyme from phenobarbital-pretreated but not from 3-methylcholanthrene-pretreated rats.[63] The purified phenobarbital-induced enzyme fraction has more recently been fractionated into several individual isozymes and one of the two major isozymes thus obtained ("PB-4") has been shown to be ten times more susceptible to inactivation by AIA than the other ("PB-5").[102] The 3-methylcholanthrene-inducible isozyme is not inactivated by AIA because it does not accept AIA as a substrate but it is readily inactivated by other agents (fluoroxene, 1-aminobenzotriazole, 5,6-dichloro-1,2,3-benzothiadiazole) and thus is not inherently resistant to inactivation.[63] A study of the inactivation of eight purified cytochrome P-450 isozymes by 1,1-dichloroethylene has established that one phenobarbital-inducible isozyme ("PB-B", probably identical to "PB-4") is effectively inactivated, and the other five isozymes from variously induced rats are not detectably inactivated.[23] A clear relationship between inactivation of an isozyme and its ability to oxidize 1,1-dichloroethylene was not found. Resistance to inactivation thus can be conveyed by failure to oxidize the substrate or by factors intrinsic to the active site that suppress the destructive process.

Cytochrome P-450$_{scc}$ - The demonstration of isozyme selectivity and the availability of structure-activity data has triggered efforts to construct irreversible inhibitors for biosynthetic cytochrome P-450 isozymes. Two types of mechanism-based inhibitors have been reported for cytochrome P-450$_{scc}$, the adrenal enzyme that oxidizes cholesterol to pregnenolone.[103,104] Cholesterol analogue (4) with a triple bond between carbons 22 and 23 of the sidechain (the 20,22-bond is that which is normally cleaved) inactivates the enzyme by a catalysis-dependent mechanism.[103] Replacement of carbon 24 and the attached sidechain carbons with a trimethylsilyl group (5) also results in inactivation of the enzyme.[104]

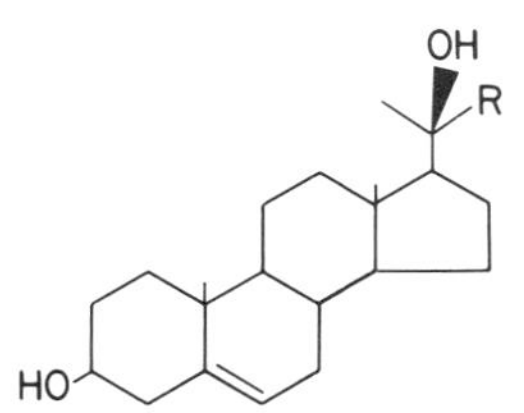

4 a. $R=C\equiv CCH_3$

b. $R=C\equiv CCH_2CH_2CH_2CH_3$

c. $R=C\equiv CCH_2CH_2C\equiv CH$

5 $R=CH_2CH_2SiMe_3$

Aromatase - Inhibition of aromatase, the cytochrome P-450 enzyme that oxidizes testosterone to estradiol, offers a means for the control of estrogen-dependent tumors. A number of mechanism-based inhibitors that may act through heme-alkylation have been described for this enzyme. Three separate laboratories have attached an acetylenic group to the C-19 methyl that is normally hydroxylated by aromatase in the demethylation-aromatization sequence and have shown that the resulting sterols inactivate the placental enzyme.[105-107] It has also been reported that norethisterone, a 17-ethinyl sterol shown earlier to alkylate the prosthetic heme of hepatic cytochrome P-450,[67] inactivates placental aromatase.[108,109] Aromatase is furthermore irreversibly inactivated by the analogue of androst-4-ene-3,17-dione with an allene moiety instead of a methyl group at C-10.[105] Other catalysis-dependent irreversible inhibitors of aromatase act by mechanisms which probably do not involve the prosthetic heme.[107,110-113]

Fatty Acid ω-Hydroxylase - Fatty acids, including prostaglandins, leukotrienes, and other arachidonic acid derivatives, are hydroxylated at the ω-position by a cytochrome P-450 isozyme that, at least for some fatty acids, is induced by clofibrate but not by classical cytochrome P-450 inducers.[114] This ω-hydroxylase is inactivated by 11-dodecynoic and 10-undecynoic acids, fatty acid analogues with a triple bond between the ω and ω-1 carbon atoms.[115] Suppression of fatty acid ω-hydroxylation occurs without measurable loss of total microsomal cytochrome P-450 and thus reflects inactivation of a quantitatively insignificant fraction of the total microsomal cytochrome P-450 that nevertheless is responsible for all the ω-hydroxylation.

Cinnamic Acid 4-Hydroxylase - The p-hydroxylation of cinnamic acid in the biosynthesis of lignins by plants is mediated by a cytochrome P-450 enzyme.[116] The cinnamic acid 4-hydroxylase of Jerusalem artichoke tubers is preferentially inactivated by 1-aminobenzotriazole in a time- and catalysis-dependent manner.[117] The specific inhibition of cytochrome P-450 enzymes in plants provides a possible avenue for the design of new classes of herbicides and other crop-protection agents.

Summary - The inactivation of cytochrome P-450 during catalytic turnover of 2-isopropyl-4-pentenamide, initially thought to be a property of the homoallylic amide functionality, is now known to be intrinsic to the metabolism of several common functionalities. These include olefins, acetylenes, aldehydes, and a growing list of more complex moieties. The toxicological consequences of cytochrome P-450 inactivation, notably inhibition of the metabolism of co-administered agents and disruption of the heme biosynthetic pathway, have been reviewed recently and therefore have not been extensively discussed here.[6] The relatively recent evidence that cytochrome P-450 isozymes involved in the biosynthesis and catabolism of endogenous substrates can be specifically inactivated, however, has been summarized because it implies that specific, nontoxic, inhibitors of important (and pharmacologically interesting) physiological processes may be developed. These processes include the biosynthesis of cholesterol, bile acids, and sterol hormones; the activation of vitamin D; the catabolism of fatty acids, prostaglandins, leukotrienes, and prostacyclins; the possible biosynthesis of biologically active ω-hydroxylated arachidonic acid derivatives; and the biosynthesis of essential structures and hormones in plants and insects.

References

1. B. Testa and P. Jenner, Drug Metab. Disp., 12, 1 (1981).
2. K. J. Netter, Pharmacol. Ther., 10, 515 (1980).
3. M. R. Franklin, Pharmacol. Ther. A, 2, 227 (1977).
4. E. Hodgson and R. M. Philpot, Drug Metab. Rev., 3, 231 (1974).
5. J. Halpert and R. A. Neal, Drug Metab. Rev., 12, 239 (1981).
6. P. R. Ortiz de Montellano and M. A. Correia, Ann. Rev. Pharmacol., 23, 481 (1983).
7. P. R. Ortiz de Montellano in "Reviews in Biochemical Toxicology", Vol. 6, E. Hodgson, J. Bend, and R. M. Philpot, Eds., Elsevier North-Holland Inc., New York, N. Y. 1984, in press.
8. M. W. E. Teunissen, M. Van Graft, N. P. E. Vermeulen, and D. D. Breimer, Xenobiotica, 13, 497 (1983).
9. P. R. Ortiz de Montellano, G. S. Yost, B. A. Mico, S. E. Dinizo, M. A. Correia, and H. Kambara, Arch. Biochem. Biophys., 197, 524 (1979).
10. P. R. Ortiz de Montellano, B. A. Mico, H. S. Beilan, and K. L. Kunze, in "Molecular Basis of Drug Action", T. Singer and R. Ondarza, Eds., Elsevier, New York, N. Y., 1981, p. 151.
11. P. R. Ortiz de Montellano, R. A. Stearns, and K. C. Langry, Molec. Pharmacol., in press (1984).
12. M. Brinkschulte-Freitas and H. Uehleke, Arch. Toxicol., 142, 137 (1979).
13. A. Smith, Biochem. Pharmacol., 25, 2429 (1976).
14. H. Uehleke and M. Brinkschulte-Freitas, Naunyn-Schmied. Arch. Pharmacol., 302, 11 (1978).
15. K. S. Prickett and T. A. Baillie, Biomed. Mass Spectrom., in press (1984).
16. P. G. Gervasi, L. Citti, G. Fassina, E. Testai, and G. Turchi, Toxicol. Letters, 16, 217 (1983).
17. E. Testai, L. Citti, P. Gervasi, G. Turchi, Biochem. Biophys. Res. Commun., 107, 633 (1982).
18. K. M. Ivanetich, S. Lucas, J. A. Marsh, M. R. Ziman, I. D. Katz, and J. J. Bradshaw, Drug Metab. Disp., 6, 218 (1978).
19. J. M. Patel, E. Ortiz, and K. C. Leibman, Fed. Proc., 40, 636 (1981).
20. K. L. Kunze, B. L. K. Mangold, C. Wheeler, H. S. Beilan, and P. R. Ortiz de Montellano, J. Biol. Chem., 258, 4202 (1983).
21. P. R. Ortiz de Montellano and K. L. Kunze, Biochemistry, 20, 7266 (1981).
22. P. R. Ortiz de Montellano, K. L. Kunze, H. S. Beilan, and C. Wheeler, Biochemistry, 21, 1331 (1982).
23. R. E. Miller and F. P. Guengerich, Biochemistry, 21, 1090 (1982).
24. D. C. Liebler and F. P. Guengerich, Biochemistry, 22, 5482 (1983).
25. P. R. Ortiz de Montellano and B. A. Mico, Molec. Pharmacol., 18, 128 (1980).
26. P R. Ortiz de Montellano, B. L. K. Mangold, C. Wheeler, K. L. Kunze, and N. O. Reich, J. Biol. Chem., 258, 4208 (1983).
27. T. R. Tephly, A. H. Gibbs, and F. De Matteis, Biochem. J., 180, 241 (1979).
28. F. De Matteis, A. H. Gibbs, and T. R. Tephly, Biochem. J., 188, 145 (1980).
29. H. A. Dailey and J. E. Fleming, J. Biol. Chem., 258, 11453 (1983).
30. P. R. Ortiz de Montellano, H. S. Beilan, and K. L. Kunze, Proc. Natl. Acad. Sci. U.S.A., 78, 1490 (1981).
31. F. De Matteis, A. H. Gibbs, P. B. Farmer, and J. H. Lamb, FEBS Letters, 129, 328 (1981).
32. T. R. Tephly, B.L. Coffman, G. Ingall, M. S. Abou Zeit-Har, H. M. Goff, H. D. Tabba, and K. M. Smith, Arch. Biochem. Biophys., 212, 120 (1981).
33. O. Augusto, H. S. Beilan, and P. R. Ortiz de Montellano, J. Biol. Chem., 257, 11288 (1982).
34. F. De Matteis, C. Hollands, A. H. Gibbs, N. de Sa, and M. Rizzardini, FEBS Letters, 145, 87 (1982).
35. N. R. Eade and K. W. Renton, J. Pharmacol. Exp. Therap., 173, 31 (1970).
36. D. D. H. Craig, Lancet, 559 (1962).
37. P. M. Belanger and A. Atitse-Gbeassor, Can. J. Physiol. Pharmacol., 61, 524 (1983).
38. S. F. Muakkassah and W. C. T. Yang, J. Pharmacol. Exp. Therap., 219, 147 (1981).
39. P. R. Ortiz de Montellano, O. Augusto, F. Viola, and K. L. Kunze, J. Biol. Chem., 258, 8623 (1983).
40. H. G. Jonen, J. Werringloer, R. A. Prough, and R. W. Estabrook, J. Biol. Chem., 257, 4404 (1982).
41. P. R. Powell-Jackson, J. M. Tredger, H. M. Smith, M. Davis, and R. Williams, Biochem. Pharmacol., 31, 4031 (1982).
42. S. Hara, T. Satoh, and H. Kitagawa, Res. Commun. Chem. Path. Pharmacol., 36, 260 (1982).
43. S. F. Muakkassah, W. R. Bidlack, and W. C. T. Yang, Biochem. Pharmacol., 30, 1651 (1981).
44. S. F. Muakkassah, W. R. Bidlack, and W. C. T. Yang, Biochem. Pharmacol, 31, 249 (1982).
45. R. N. Hines and R. A. Prough, J. Pharmacol. Exp. Therap., 214, 80 (1980).
46. D. Mansuy, P. Battioni, and J.-P. Mahy, J. Am. Chem. Soc., 104, 4487 (1982).

47. P. Battioni, J.-P. Mahy, M. Delaforge, and D. Mansuy, Eur. J. Biochem., 134, 241 (1983).
48. P. R. Ortiz de Montellano and K. L. Kunze, J. Am. Chem. Soc., 103, 6534 (1981).
49. S. Saito and H. A. Itano, Proc. Natl. Acad. Sci. U.S.A., 78, 5508 (1981).
50. P. R. Ortiz de Montellano and D. E. Kerr, J. Biol. Chem., 258, 10558 (1983).
51. O. Augusto, K. L. Kunze, and P. R. Ortiz de Montellano, J. Biol. Chem., 257, 6231 (1982).
52. K. L. Kunze and P. R. Ortiz de Montellano, J. Am. Chem. Soc., 105, 1380 (1983).
53. D. Ringe, G. A. Petsko, D. E. Kerr, and P. R. Ortiz de Montellano, Biochemistry, 23, 2 (1984).
54. D. Mansuy, J.-P. Battioni, D. Dupre, E. Sartori, and G. Chottard, J. Am. Chem. Soc., 104, 6159 (1982).
55. H. J. Callot and F. Metz, J. Chem. Soc., Chem. Comm., 947 (1982).
56. J. S. Wiseman, J. S. Nichols, and M. X. Kolpak, J. Biol. Chem., 257, 6328 (1982).
57. D. J. T. Porter and H. J. Bright, J. Biol. Chem., 258, 9913 (1983).
58. E. A. Smuckler, E. Arrhenius, and T. Hultin, Biochem. J., 103, 55 (1967).
59. I. N. H. White, A. G. Smith, and P. B. Farmer, Biochem. J., 212, 599 (1983).
60. P. R. Ortiz de Montellano, H. S. Beilan, K. L. Kunze, and B. A. Mico, J. Biol. Chem., 256, 4395 (1981).
61. K. K. Park, J. S. Wishnok, and M. C. Archer, Chem.-Biol. Interact., 18, 349 (1977).
62. P. R. Ortiz de Montellano and J. M. Mathews, Biochem. J., 195, 761 (1981).
63. P. R. Ortiz de Montellano, B. A. Mico, J. M. Mathews, K. L. Kunze, G. T. Miwa, and A. Y. H. Lu, Arch. Biochem. Biophys., 210, 717 (1981).
64. P. R. Ortiz de Montellano, J. M. Mathews, and K. C. Langry, Tetrahedron, 40, 511 (1983).
65. S. Masamune, N. Nakamura, and J. Spadaro, J. Am. Chem. Soc., 97, 918 (1975).
66. R. A. Stearns and P. R. Ortiz de Montellano, Biochemistry, 22, 11A (1983).
67. P. R. Ortiz de Montellano and K. L. Kunze, J. Biol. Chem., 255, 5578 (1980).
68. P. R. Ortiz de Montellano and K. L. Kunze, Biochem. Biophys. Res. Commun., 94, 443 (1980).
69. R. P. Hanzlik, V. Kishore, and R. Tullman, J. Med. Chem., 22, 759 (1979).
70. T. L. Macdonald, K. Zirvi, L. T. Burka, P. Peyman, and F. P. Guengerich, J. Am. Chem. Soc., 104, 2050 (1982).
71. R. P. Hanzlik and R. H. Tullman, J. Am. Chem. Soc., 104, 2048 (1982).
72. R. B. Silverman and S. J. Hoffman, Biochem. Biophys. Res. Commun., 101, 1396 (1981).
73. P.-M. Belanger and A. Atitse-Gbeassor, Biochem. Pharmacol., 31, 2679 (1982).
74. J. M. Patel, C. Harper, and R. T. Drew, Drug Metab. Disp., 6, 368 (1978).
75. J. M. Patel, C. R. Wolf, and R. M. Philpot, Biochem. Pharmacol., 28, 2031 (1979).
76. I. N. H. White, Chem.-Biol. Interact., 39, 231 (1982).
77. J. M. Patel, Toxicol. Appl. Pharmacol., 48, 337 (1979).
78. R. O. Recknagel, E. A. Glende, and A. M. Hruszkewycz in "Free Radicals in Biology", Vol. 3, W. A. Pryor, Ed., Academic Press, New York, New York, 1977, p. 97.
79. H. De Groot and W. Haas, Biochem. Parmacol., 30, 2343 (1981).
80. G. Poli, K. Cheeseman, T. F. Slater, and M. U. Dianzani, Chem.-Biol. Interact., 37, 13 (1981).
81. H. D. Colby, W. C. B. Brogan, and P. R. Miles, Toxicol. Appl. Pharmacol., 60, 492 (1981).
82. R. L. Waller, E. A. Glende, and R. O. Recknagel, Biochem. Pharmacol., 32, 1613 (1983).
83. G. Fernandez, M. C. Villarruel, E. G. D. de Toranzo, and J. A. Castro, Res. Chem. Commun. Pathol. Pharmacol., 35, 283 (1982).
84. P. A. Krieter and R. A. Van Dyke, Chem.-Biol. Interact., 44, 219 (1983).
85. O. Reiner and H. Uehleke, Hoppe-Seyler's Z. Physiol. Chem., 352, 1048 (1971).
86. P. J. Cox, L. J. King, and D. V. Parke, Xenobiotica, 6, 363 (1976).
87. H. de Groot, U. Harnisch, and T. Noll, Biochem. Biophys. Res. Commun., 107, 885 (1982).
88. H. J. Ahr, L. J. King, W. Nastainczyk, and V. Ullrich, Biochem. Pharmacol., 31, 383 (1982).
89. J. C. Felton, D. W. Jenner, and P. Kirby, J. Agr. Fd. Chem., 18, 671 (1970).
90. M. Skrinjaric-Spoljar, H. B. Mathews, J. L. Engel, and J. E. Casida, Biochem. Pharmacol., 20, 1607 (1971).
91. P. R. Ortiz de Montellano and J. M. Mathews, Biochem. Pharmacol., 30, 1138 (1981).
92. J. Pedemonte, J. Olate, P. Cervantes, C. Oberti, and L. Gil, Biochem. Pharmacol., 30, 1483 (1981).
93. J. Jarvisalo, A. H. Gibbs, and F. De Matteis, Molec. Pharmacol., 14, 1099 (1978).
94. M. J. Obrebska, P. Kentish, and D. V. Parke, Biochem. J., 188, 107 (1980).
95. G. E. Miller, M. A. Zemaitis, and F. E. Greene, Biochem. Pharmacol., 32, 2433 (1983).
96. J. Halpert, D. Hammond, and R. A. Neal, J. Biol. Chem., 255, 1080 (1980).
97. J. Halpert and R. A. Neal, Drug Metab. Rev., 12, 239 (1981).
98. T. Sawahata and R. A. Neal, Molec. Pharmacol., 21, 464 (1982).
99. A. M. Hruszkewycz, E. A. Glende, and R. O. Recknagel, Toxicol. Appl. Pharmacol., 46, 695 (1978).

100. E. Jeffery, A. Kotake, R. El Azhary, and G. J. Mannering, Molec. Pharmacol., 13, 415 (1977).
101. D. E. Nerland, M. M. Iba, G. J. Mannering, Molec. Pharmacol., 19, 162 (1981).
102. D. J. Waxman and C. Walsh, J. Biol. Chem., 257, 10446 (1982).
103. A. Nagahisa, R. W. Spencer, and W. H. Orme-Johnson," J. Biol. Chem. 258, 6721 (1983).
104. A. Nagahisa and W. H. Orme-Johnson, J. Am. Chem. Soc., in press.
105. B. W. Metcalf, C. L. Wright, J. P. Burkhart, and J. O. Johnston, J. Am. Chem. Soc., 103, 3221 (1981).
106. D. G. Covey, W. F. Hood, and V. D. Parikh, J. Biol. Chem., 256, 1076 (1981).
107. P. A. Marcotte and C. H. Robinson, Steroids, 39, 325 (1982).
108. Y. Osawa, C. Yarborough, and Y. Osawa, Science, 215, 1249 (1982).
109. Y. Osawa, Y. Osawa, C. Yarborough, and L. Borzynski, Biochem. Soc. Trans., 11, 656 (1983).
110. G. A. FLynn, J. O. Johnston, C. L. Wright, and B. W. Metcalf, Biochem. Biophys. Res. Commun., 103, 913 (1981).
111. P. A. Marcotte and C. H. Robinson, Biochemistry, 21, 2773 (1982).
112. A. M. H. Brodie, W. M. Garrett, J. R. Hendrickson, C.-H. Tsai-Morris, P. A. Marcotte, and C. H. Robinson, Steroids, 38, 693 (1981).
113. D. F. Covey and W. F. Hood, Molec. Pharmacol., 21, 173 (1982).
114. D. Kupfer, Pharmac. Ther., 11, 469 (1980).
115. P. R. Ortiz de Montellano and N. O. Reich, J. Biol. Chem., 256, in press (1984).
116. H. A. Stafford, Ann. Rev. Plant. Physiol., 25, 459 (1974).
117. D. Reichhart, A. Simon, F. Durst, J. M. Mathews, and P. R. Ortiz de Montellano, Arch. Biochem. Biophys., 216 522 (1982).

Chapter 21. Phospholipases

Eduardo G. Lapetina, Department of Molecular Biology, The Wellcome Research Laboratories, Burroughs Wellcome Co., Research Triangle Park, NC 27709

Introduction - Phospholipases hydrolyze ester bonds in glycerophospholipids, and are readily subdivided according to the particular bond being attacked (Fig. 1). Hydrolysis of the fatty acylester bond is catalyzed by phospholipases of the A-type.[1-3] Removal of the fatty acid at the sn-1-position is produced by phospholipase A_1 (EC 3.1.1.32) while removal of the fatty acid at the sn-2-position is due to phospholipase A_2 (EC 3.1.1.4). Both phospholipases induce the formation of a lysophospholipid. The subsequent hydrolysis of the fatty acylester bond in lysophospholipids is catalyzed by lysophospholipases (EC 3.4.1.5), which are now recognized as phospholipases of the B type.[1-3]

Hydrolysis of the phosphodiester bonds in phospholipids can be catalyzed by phospholipase C (EC 3.1.4.3)[1] and phospholipase D (EC 3.1.4.4).[1] Mammalian phospholipase C mainly attacks the inositol

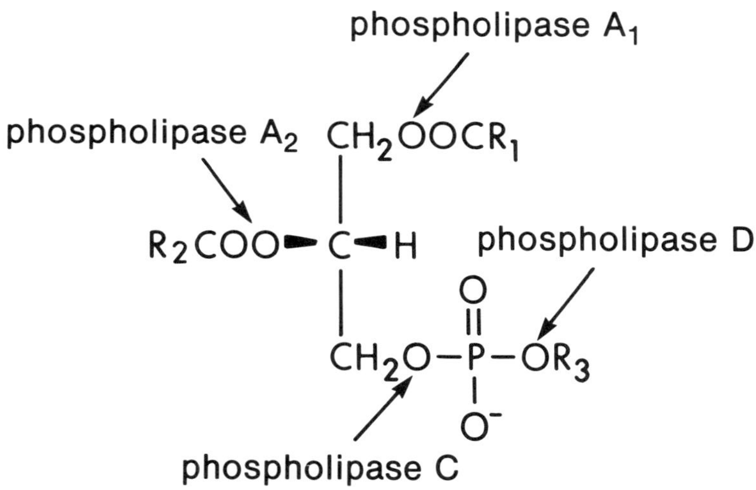

Figure 1: Hydrolysis of ester linkages in glycerophospholipids by phospholipases A_1, A_2, C and D. The fatty acid at the sn-1-position (R_1) is usually saturated while that at the sn-2-position (R_2) is unsaturated. R_3 varies in different phospholipids and could be choline (phosphatidylcholine), ethanolamine (phosphatidylethanolamine), serine (phosphatidylserine), myoinositol (phosphatidylinositol), myoinositol-4-monophosphate (phosphatidylinositol-4-monophosphate) or myoinositol-4,5-bisphosphate (phosphatidylinositol-4,5-bisphosphate). In phosphatidic acid, R_3 is a hydrogen atom.

ISBN 0-12-040519-9

phospholipids such as phosphatidylinositol, phosphatidylinositol-4-monophosphate and phosphatidylinositol-4,5-bisphosphate.[4,5] This enzymatic hydrolysis leads to formation of 1,2-diacylglycerol and, according to the particular inositol phospholipid that is being degraded, myoinositol 1-monophosphate (plus myoinositol 1,2-cyclic phosphate), myoinositol 1,4-bisphosphate or myoinositol 1,4,5-trisphosphate. The hydrolysis of phospholipids by phospholipase D yields phosphatidic acid; this enzyme is predominantly found in higher plants.[1]

Various reviews on phospholipase A or C have emphasized the properties, purification and subcellular distribution of these enzymes.[1-5] The purpose of this chapter is mainly directed to the increasing importance of these enzymes in the regulation of cellular functions.

Acylation of Phospholipids - The esterification of long-chain, unsaturated fatty acids in the sn-2-position could be explained by different enzyme activities such as specific long-chain acyl-CoA synthetases and CoA-dependent or CoA-independent transacylases.[6-11] For example, a long-chain acyl-CoA synthetase has been demonstrated in isolated platelet membranes that is specific for arachidonate and other long-chain unsaturated fatty acids.[6,7] Transfer of arachidonic acid between phospholipids has been observed by the action of CoA-dependent and CoA-independent transacylases. The CoA-dependent transacylases were demonstrated in lymphocytes,[8] pancreatic acini[9] and liver microsomes.[10] A CoA-independent transacylase, recently observed in platelets, catalyzes the synthesis of arachidonoyl-plasmenylethanolamine by acylation of lysoplasmenylethanolamine with arachidonic acid derived from phosphatidylcholine.[11]

Phospholipases A_1 and A_2 - Phospholipases A_1 and A_2, respectively, catalyze the liberation of fatty acids from the sn-1- and sn-2-positions of the diacylglycerolphospholipids. The sn-1-position of the diacylglycerolphospholipids is commonly esterified by saturated fatty acids while the sn-2-position is predominantly esterified by unsaturated fatty acids.

Recently, a great deal of attention has been focused on the liberation of unsaturated fatty acids from the sn-2-position of various phospholipids by the action of phospholipase A_2, based on the fact that the liberation of unsaturated fatty acids such as dihomogammalinolenic acid, arachidonic acid or eicosapentaenoic acid initiates the formation of a wide range of physiologically active metabolites. These three fatty acids are referred to as eicosanoid precursors since they are substrates for cyclooxygenase and lipoxygenases.[12,13] The eicosanoid precursors are not found as free substances in most cells and liberation from the sn-2-position of phospholipids must precede metabolic oxygenation.[14] The generation of the free eicosanoids from phospholipids is thus a rate-limiting process and normally occurs as a consequence of the stimulation of specific cell-surface receptors.[14] Various cell types release eicosanoid precursors from their phospholipids in response to specific agonists such as bradykinin, angiotensin II, vasopressin, thrombin, collagen, epinephrine, ADP, platelet-activating factor, chemotactic peptides, IgE-Ag, C5a, concanavalin A, histamine and phorbol esters.[14] Phosphatidylcholine, phosphatidylethanolamine, phosphatidylinositol, phosphatidic acid, plasmenylethanolamine and 1-O-alkyl-2-arachidonoyl-sn-glycero-3-phosphocholine are all known to release arachidonic acid in different cells.[14-16]

Phospholipase A_2 has been detected in almost any cell in which its presence has been investigated.[1] This activity has extensively been studied in liver cells where it is present in different cell organelles such as plasma membranes, microsomes, golgi membranes and lysosomes.[1] With the exception of the lysosomal enzyme, all phospholipase A_2 activities require an alkaline pH (7.8-9.5) and are stimulated by Ca^{2+} ions. The lysosomal activity requires acidic pH (4.0-5.0) and is inhibited by Ca^{2+} ions.

In platelets it seems that two types of phospholipases A_2 are operative.[17-19] Phosphatidylinositol and phosphatidic acid are degraded by a phospholipase A_2 that requires a neutral pH and micromolar concentration of Ca^{2+}.[18,19] On the other hand, phosphatidylcholine and phosphatidylethanolamine are degraded by a phospholipase A_2 that acts optimally at pH 9.5 and millimolar concentration of Ca^{2+}.[17] These two types of phospholipases A_2 might be acting sequentially in stimulated platelets. The liberation of arachidonic acid from phosphatidylinositol and phosphatidic acid might precede that from phosphatidylcholine and phosphatidylethanolamine.[14] All four lysoderivatives have been detected in stimulated platelets.[17-19,22]

Phospholipase A_2 Activation - The agonist-induced stimulation of phospholipase A_2 does not seem to be regulated in the same way as the pancreatic phospholipase A_2.[3] The inactive zymogen of pancreatic phospholipase A_2 is converted into an active enzyme by the removal of a heptapeptide from the N-terminus by trypsin.[3] Such a mechanism may only apply for digestive enzymes. However, this mechanism cannot be completely disregarded in relation to membrane-associated phospholipases A_2. Trypsin or thrombin stimulate phospholipase A_2 in platelets,[23,24] transformed mouse fibroblasts[25] and endothelial cells[26] but it is not known if this stimulation is due to proteolysis of an inactive proenzyme. A serine-protease might be involved in the stimulation of platelet phospholipase A_2 since a protease inhibitor phenylmethanesulphonyl fluoride blocks phospholipase A_2 activation in platelets stimulated by thrombin.[27]

It seems that mobilization of cellular Ca^{2+} is the most effective way for activation of phospholipase A_2. Most of the cell activators that catalyze the release of eicosanoid precursors also induce Ca^{2+} mobilization. Moreover, the extent of stimulation of phospholipase A_2 by Ca^{2+}-ionophore A23187 in platelets is comparable to the action of thrombin or collagen.[14,15,27] This would indicate that mobilization of Ca^{2+} from intracellular stores to the cytosol would account for the increased phospholipase A_2 activity.

It has also been shown that phospholipase A_2 of intact platelets requires calmodulin for stimulus activation.[28] This calmodulin requirement of phospholipase A_2 from platelets differs from the pancreatic phospholipase A_2. This enzyme requires Ca^{2+} but not calmodulin for activity.[29] A role for calmodulin has also been postulated in Ca^{2+} activation of phospholipase A_2 in renal medulla.[30]

It has been indicated that the phosphodiesteratic cleavage of the inositol phospholipids (phospholipase C activity) precedes and might trigger the activation of phospholipase A_2 in stimulated platelets.[14,20,21] It seems that some of the products formed during the phosphodiesteratic cleavage of the inositides could be related to Ca^{2+} mobilization. One of these products, myoinositol 1,4,5-trisphosphate, has been postulated as the second messenger for Ca^{2+} mobilization.[31] Addition of myoinositol-1,4,5-trisphosphate to pancreatic acinar cells that have previously been made permeable with saponine causes a transient rise in extracellular Ca^{2+}. This increase is not blocked by inhibitors of Ca^{2+} uptake into mitochondria. These permeable cells still respond to carbamylcholine in an effect which is completely inhibited by exogenous added myoinositol-1,4,5-trisphosphate, indicating that the same Ca^{2+} deposit is the target of both agents.[31]

1,2-Diacylglycerol is also formed during the phosphodiesteratic cleavage of the inositol phospholipids[32,33] and recently it has been suggested to play a role in the stimulation of phospholipases.[34] A small percentage of 1,2-diacylglycerol might also be deacylated by the consecutive actions of 1,2-diacylglycerol lipase and 2-monoacylglycerol lipase for the liberation of arachidonic acid.[35] However, most of the 1,2-diacylglycerol is further phosphorylated to phosphatidic acid in stimulated cells.[14] Phosphatidic acid has properties of a Ca^{2+}-ionophore[36-44] and a fusogenic at low concentrations of Ca^{2+}.[45,46] Phosphatidic acid also has a direct stimulatory effect on phospholipase A_2 purified from platelets.[47]

<u>Phospholipase A_2 Inhibition</u> - The liberation of arachidonic acid in stimulated platelets is inhibited by cyclic-AMP.[14,24] This action of cAMP does not seem to be a direct action on platelet phospholipase A_2. Cyclic-AMP decreases the levels of free cytoplasmic Ca^{2+} and reduces in this way the availability of Ca^{2+} for phospholipase A_2 activity. Trifluoperazine, which is a calmodulin antagonist, also inhibits the thrombin- or platelet activating factor-induced mobilization of arachidonic acid from platelet phospholipids.[20,48]

In lungs, macrophages, leukocytes, renal papillary tissue, and some cell culture systems, phospholipase A_2 is inhibited by anti-inflammatory steroids.[49-52] The inhibitory action of the glucocorticoids depends upon unimpaired RNA and protein synthesis.[53,54] It is now clear that the steroids induce the biosynthesis and release of an inhibitor of phospholipase A_2.[49] These phospholipase inhibitor proteins have been isolated from glucocorticoid-treated rabbit neutrophils (lipomodulin with a molecular weight of 40,000[55,56]) and from rat macrophages (macrocortin with a molecular weight of 15,000[56]). Partially purified preparations of these proteins have similar activities with respect to inhibition of arachidonate release from many cells and tissues, and exert anti-inflammatory activity on carrageenan induced paw edema and pleurisy.[57,58] Recent evidence suggests that macrocortin and lipomodulin are closely related.[56] A radioimmunoassay for lipomodulin demonstrated three species of phospholipase inhibitory proteins with molecular weights of 40,000, 30,000 and 15,000 in material from glucocorticoid-treated rats.[58] The 40,000 molecular weight (lipomodulin) protein has the greatest inhibitory effect on phospholipase A_2. Although most of the immunoreactive material was eluted with the 16,000 molecular weight protein (macrocortin), no inhibitory activity could be detected until after dephosphorylation of this protein by alkaline phosphatase treatment.[56] It has been suggested that macrocortin might be the phosphorylated fragment of lipomodulin.[56] Autoantibodies against lipomodulin have been found in patients with rheumatic diseases such as systemic lupus erythematosus.[59]

A recent report indicates that platelets also contain an

endogenous inhibitor of phospholipase A_2.[60] This inhibitor has not yet been identified but is quite different from the activity that is induced by glucocorticoids in macrophages and neutrophils. The platelet inhibitor is heat-stable, trypsin-insensitive and extractable by chloroform/methanol, suggesting that it could be a lipid. It does not decrease phospholipase A_2 by chelating Ca^{2+}. It is not yet known if the inhibition results from an interaction of the inhibitor with the enzyme or with the substrate. Other reports have also indicated the presence of phospholipase A_2 inhibitors in platelets,[61] rabbit granulocytes[62] and leukocytes granules,[63] but the nature of the inhibitor(s) is not known.

In the last few years much attention has been directed to finding new drugs that will specifically inhibit phospholipase A_2 activity. This search has not yet been fruitful. The antimalarial drug, mepacrine, does not seem to have a direct effect on phospholipase A_2 and requires high concentrations.[64] Bromophenacylbromide seems to be more effective for pancreatic phospholipase A_2 activity than for membrane-associated phospholipase A_2.[65,66] Other compounds used as phospholipase A_2 inhibitors such as local anesthetics, chloropromazine and propanolol are also not very effective.[67]

<u>Phospholipase C (Phosphodiesteratic Cleavage of the Inositol Phospholipids)</u> - Mammalian phospholipase C catalyzes the hydrolysis of the glycerophosphate ester bond of phosphatidylinositol, phosphatidylinositol-4-monophosphate and phosphatidylinositol-4,5-bisphosphate (Fig. 2).[4,5,20] This hydrolysis yields 1,2-diacylglycerol and the corresponding mono-, bis- or trisphosphate myoinositols (Fig. 2). The enzyme is mainly present in the soluble fraction, requires Ca^{2+} ions and an acidic pH (5.0-6.0) for maximal activity.[17,18,32,68-83] Evidence from platelets and heart indicate that a single enzyme may be attacking the three inositol phospholipids.[84,85] Phosphodiesteratic cleavage of the inositol phospholipids is induced by a wide range of cellular activators acting on cell surface receptors.[4,86] Recently, a number of investigators have found that the initial receptor activated event may be a breakdown of phosphatidylinositol-4,5-bisphosphate rather than phosphatidylinositol.[87-91] In platelets, it seems that the cytosolic activity degrades the membrane-bound substrates which are only made available to the enzyme during cellular activation.[83,84] The regulatory action for degradation of the inositol phospholipids may then not involve the enzyme directly, but rather the "unmasking" of the substrate that might be closely associated to the specific receptors in the cell surface.[83]

<u>Role of Phospholipase C in Stimulated Cells</u> - Cell activators that induce the phosphodiesteratic cleavage of the inositol phospholipids also evoke cellular Ca^{2+} mobilization.[4] In recent years it has been suggested that some of the products that derive from this phosphodiesterase activity might mobilize Ca^{2+}. 1,2-Diacylglycerol is rapidly phosphorylated to phosphatidic acid and the latter has been shown to have Ca^{2+}-ionophoretic properties.[36-46] However, the ionophoretic properties of phosphatidic acid might not be generalized to all cell systems. Recently a report indicates the failure of phosphatidic acid to facilitate Ca^{2+} fluxes across liposomal membranes.[92]

More recently, myoinositol-1,4,5-trisphosphate has been postulated as a second messenger for mobilizing intracellular calcium.[31] The effect of myoinositol-1,4,5-trisphosphate on Ca^{2+} mobilization has

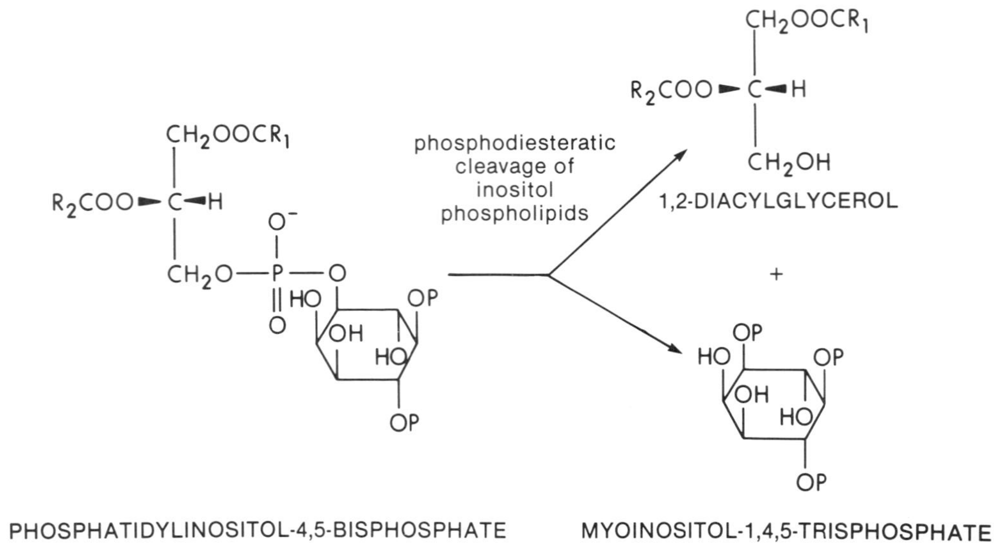

Figure 2: Phosphodiesteratic cleavage (phospholipase C) of phosphatidylinositol-4,5-bisphosphate. Phosphatidylinositol and phosphatidylinositol-4-monophosphate are also hydrolyzed in a similar manner yielding 1,2-diacylglycerol and the corresponding myoinositol phosphates such as myoinositol-1-monophosphate (plus myoinositol-1,2-cyclic phosphate) and myoinositol-4-monophosphate. P indicates phosphate groups.

been studied using pancreatic and liver cells that have been made permeable in various ways.[31,93] Upon addition of micromolar concentrations of myoinositol-1,4,5-trisphosphate, there is a large increase in calcium that seems to originate from an ATP-dependent pool believed to be endoplasmic reticulum.[31,93] There was no release in response to myoinositol-1,4-bisphosphate, myoinositol-1-monophosphate or myoinositol-1,2-cyclic phosphate.[31]

1,2-Diacylglycerol is another product of the phosphodiesteratic cleavage of the inositides that has recently acquired enhanced importance. 1,2-Diacylglycerol activates a phospholipid- and Ca^{2+}-dependent protein kinase C which is implicated in transmembrane signalling, tumor promotion and cellular differentiation.[95,96] 1,2-Diacylglycerol dramatically increases the affinity of protein kinase C for Ca^{2+}-ions, and thereby activates the enzyme without a net increase in the Ca^{2+} concentration.[95,96] Phorbol esters and exogenous 1,2-dioleyl-diacylglycerol or 1-oleoyl-2-acetyl-glycerol activate protein kinase directly without interaction with cell surface receptors.[97] In human platelets, it has been demonstrated that there exists a strict temporal and quantitative correlation between the activation of protein kinase C (detected by the phosphorylation of a 40,000 molecular weight protein), phosphatidic acid production (index of activation of the phosphodiesteratic cleavage of the inositol phospholipids) and shape change following stimulation with platelet-activating factor,[98] arachidonic acid[99] or thrombin.[21] This might reflect the stimulation of kinase C by 1,2-diacylglycerol formed by the activation of phospholipase C by these agents. That would indicate that phospholipase C is the initial response, whilst phosphorylation of the 40,000 molecular weight protein is an epiphenomenon of the formation of 1,2-diacylglycerol.

Phospholipase D - Phospholipase D cleaves the phosphoester bond between phosphatidic acid and the alcoholic moiety of different phospholipids. It is widely present in plants but few reports have indicated the presence of phospholipase D in mammalian tissues.[1] In rat brain a phospholipase D that converts phosphatidylcholine to phosphatidic acid has been studied.[100] Eosinophils also contain a phospholipase D activity.[101] A lysophospholipase D activity seems to have an absolute specificity for 1-alkyl-sn-glycero-3-phosphoethanolamine or -choline[102] and it is present in microsomes of rat brain,[103] kidney, intestine, lung, testes and liver.[104] A recent report indicates that a phospholipase D might be activated in human and rabbit neutrophils stimulated with chemotactic peptides.[105] This activity would induce a net loss of phosphatidylinositol with a net gain in phosphatidic acid. However, this activity has not yet been characterized and its function is not known.

Summary - In the last few years, it has become apparent that products derived from the action of phospholipases A_2 and C are involved in the regulation of important cellular functions. Activation of phospholipase A_2 catalyzes the liberation from phospholipids of eicosanoid precursors which lead to the formation of a wide range of physiologically active metabolites. Products derived from phospholipase C activity, which cleaves only the inositol phospholipids, are related to transmembrane signalling and can modulate reactions such as protein phosphorylation and the intracellular mobilization of Ca^{2+}-ions.

References

1. H. Van den Bosch in "Phospholipids," J.N. Hawthorne and G.B. Ansell, Eds., Elsevier Biomedical Press, Amsterdam, 1982, p. 313.
2. H. van den Bosch, Ann.Rev.Biochem., 43, 243 (1974).
3. A.J. Slotboom, H.M. Verheij and G.H. de Hass in "Phospholipids," J.N. Hathorne and G.B. Ansell, Eds., Elsevier Biomedical Press, Amsterdam, 1982, p. 359.
4. R.H. Michell, Biochem.Biophys.Acta, 415, 81 (1975).
5. S.D. Shukla, Life Sci., 30, 1323 (1982).
6. D.B. Wilson, S.M. Prescott and P.W. Majerus, J.Biol.Chem., 257, 3510 (1982).
7. E.J. Neufeld, D.B. Wilson, H. Sprecher and P.W. Majerus, J.Clin.Invest., 72, 214 (1983).
8. J. Trotter, I. Flesch, B. Schmidt and E. Ferber, J.Biol.Chem., 257, 1816 (1982).
9. R.P. Rubin, Biochem.Biophys.Res.Comm., 112, 502 (1983).
10. R.F. Irvine and R.M.C. Dawson, Biochem.Biophys.Res.Comm., 91, 1399 (1979).
11. R.M. Kramer and D. Deykin, J.Biol.Chem., 258, 13806 (1983).
12. M. Hamberg, J. Svensson and B. Samuelsson, Proc.Natl.Acad.Sci.U.S.A., 72, 2994 (1975).
13. G.W. Taylor and H.R. Morris, Brit.Med.Bull., 39, 219 (1983).
14. E.G. Lapetina, Trends Pharmacol.Sci., 3, 115 (1982).
15. S. Rittenhouse-Simmons and D. Deykin, J.Clin.Invest., 60, 495 (1977).
16. C.L. Swendsen, J.M. Ellis, F.H. Chilton, III, J.T. O'Flaherty and R.L. Wykle, Biochem.Biophys.Res.Comm., 113, 72 (1983).
17. M.M. Billah, E.G. Lapetina and P. Cuatrecasas, J.Biol.Chem., 255, 10227 (1980).
18. M.M. Billah, E.G. Lapetina and P. Cuatrecasas, J.Biol.Chem., 256, 5399 (1981).
19. M.M. Billah and E.G. Lapetina, J.Biol.Chem., 257, 5196 (1982).
20. E.G. Lapetina, Life Sci., 32, 2069 (1983).
21. E.G. Lapetina and W. Siess, Life Sci., 33, 1011 (1983).
22. E.G. Lapetina, M.M. Billah and P. Cuatrecasas, Nature(Lond.), 292, 367 (1981).
23. E.G. Lapetina and P. Cuatrecasas, Proc.Natl.Acad.Sci.U.S.A., 76, 121 (1979).
24. E.G. Lapetina and P. Cuatrecasas, Biochem.Biophys.Acta, 573, 394 (1979).
25. S.L. Hong, R. Polsky-Cynkin and L. Levine, J.Biol.Chem., 251, 776 (1976).
26. B.B. Weksler, C.W. Ley and E.A. Jaffe, J.Clin.Invest., 62, 923 (1978).

27. M.B. Feinstein, E.L. Becker and C. Fraser, Prostaglandins, 14, 1075 (1977).
28. P.Y.-K. Wong and W.Y. Cheung, Biochem.Biophys.Res.Comm., 90, 473 (1979).
29. M.T. Withnall and T.J. Brown, Biochem.Biophys.Res.Comm., 106, 1049 (1982).
30. P.A. Craven and F.R. DeRubertis, J.Biol.Chem., 258, 4814 (1983).
31. H. Streb, R.F. Irvine, M.J. Berridge and I. Schulz, Nature(Lond.), 306, 67 (1983).
32. S. Rittenhouse-Simmons, J.Clin.Invest., 63, 580 (1979).
33. W. Siess, F.L. Siegel and E.G. Lapetina, J.Biol.Chem., 258, 11236 (1983).
34. R.M.C. Dawson, N.L. Hemington and R.F. Irvine, Biochem.Biophys.Res.Comm., 117, 196 (1983).
35. S.M. Prescott and P.W. Majerus, J.Biol.Chem., 258, 764 (1983).
36. J.M. Gerrard, A.M. Butler, D.A. Peterson and J.G. White, Prostaglandins and Medicine, 1, 387 (1978).
37. C. Serhan, P. Anderson, E. Goodman, P. Dunham and G. Weissman, J.Biol.Chem., 256, 2736 (1981).
38. J.M. Gerrard, S.E. Kindom, D.A. Peterson, J. Peller, K.E. Krantz and J.G. White, Am.J.Path., 96, 423 (1979).
39. J.M. Gerrard, S.E. Kindom, D.A. Peterson and J.G. White, Am.J.Path., 97, 531 (1979).
40. J.M. Gerrard, C.C. Clawson and J.G. White, Am.J.Path., 100, 609 (1980).
41. C.J. Limas, Biochem.Biophys.Res.Commun., 95, 541 (1980).
42. D.M. Salmon and T.W. Honeyman, Nature(Lond.), 284, 344 (1980).
43. J.W. Putney, Jr., S.J. Weiss, C.M. Van de Walle and R.A. Haddas, Nature(Lond.), 284, 345 (1980).
44. A. Tokumura, K. Fukuzawa, J. Isobe and H. Tsukatani, Biochem.Biophys.Res.Commun., 99, 391 (1981).
45. R. Sundler and D. Papahadjopoulos, Biochim.Biophys.Acta, 649, 743 (1981).
46. R. Sundler and J. Wijkander, Biochem.Biophy.Acta, 730, 391 (1983).
47. R. Apitz-Castro, M. Cruz, M. Mas and M.K. Jain, Thromb.Res., 23, 347 (1981).
48. E.G. Lapetina, J.Biol.Chem., 257, 7314 (1982).
49. R. Flower, Trends Pharmacol.Sci., 2, 186 (1981).
50. S.L. Hong and L. Levine, Proc.Natl.Acad.Sci.U.S.A., 73, 1730 (1976).
51. R.J. Gryglewski, B. Panczenko, R. Korbut, L. Grodzinska and A. Ocetkiewicz, Prostaglandins, 10, 343 (1975).
52. G.J. Blackwell, R.J. Flower, F.P. Nijkamp and J.R. Vane, Br.J.Pharmacol., 62, 79 (1978).
53. A. Danon and G. Assouline, Nature(Lond.), 273, 552 (1978).
54. R.J. Flower and G.J. Blackwell, Nature(Lond.), 278, 456 (1979).
55. F. Hirata, E. Schiffman, K. Venkatasubramanian, D. Salomon and J. Axelrod, Proc.Natl.Acad.Sci.U.S.A., 77, 2533 (1980).
56. F. Hirata, Y. Notsu, M. Iwata, L. Parente, M. DiRosa and R.J. Flower, Biochem.Biophys.Res.Comm., 109, 223 (1982).
57. F. Hirata in "Advances in Prostaglandin, Thromboxane and Leukotriene Research," Vol. 11, B. Samuelson, R. Paoletti and P.W. Ramwell, Eds., Raven Press, New York, 1983, p. 73.
58. G.J. Blackwell, R. Carnuccio, M. DiRosa, R.J. Flower, C.S.J. Langham, L. Parente, P. Persico, N.C. Russell-Smith and D. Stone, British J.Pharmacol., 76, 185 (1982).
59. F. Hirata, R. del Carmine, C.A. Nelson, J. Axelrod, E. Schiffmann, A. Warabi, A.L. DeBlas, M. Nirenberg, V. Manganiello, M. Vaughan, S. Kumagai, I. Green, J.L. Decker and A.D. Steinberg, Proc.Natl.Acad.Sci.U.S.A., 78, 3190 (1981).
60. L.R. Ballou and W.Y. Cheung, Proc.Natl.Acad.Sci.U.S.A., 80, 5203 (1983).
61. A. Imai, Y. Ishizuka, K. Kawai and Y. Nozawa, Biochem.Biophys.Res.Commun., 108, 752 (1982).
62. P. Patriarca, P. Beckerdite, P. Pettis and P. Elsbach, Biochim.Biophys.Acta, 280, 45 (1972).
63. R. Franson, P. Patriarca and P. Elsbach, J.Lipid Res., 15, 380 (1974).
64. C.A. Dise, J.W. Burch and D.B.P. Goodman, J.Biol.Chem., 257, 4701 (1982).
65. J.J. Volwerk, W.A. Pieterson and G.H. de Haas, Biochemistry, 13, 1446 (1974).
66. E.G. Lapetina, K.A. Chandrabose and P. Cuatrecasas, Proc.Natl.Acad.Sci.U.S.A., 75, 818 (1978).
67. J.P. Cazenave, J. Benveniste and J.F. Mustard, Lab.Invest., 41, 275 (1979).
68. S.L. Hoffman and P.W. Majerus, J.Biol.Chem., 257, 6461 (1982).
69. P. Kemp, G. Hübscher and J.N. Hawthorne, Biochem.J., 79, 193 (1961).
70. W. Thompson, Can.J.Biochem., 45, 853 (1967).
71. R.S. Atherton and J.N. Hawthorne, Eur.J.Biochem., 4, 68 (1968).
72. D. Allan and R.H. Michell, Biochem.J., 142, 599 (1974).
73. A.A. Abdel-Latif, B. Luke and J.P. Smith, Biochim.Biophys.Acta, 614, 425 (1980).
74. M.K. Wassef and M.I. Horowitz, Biochim.Biophys.Acta, 665, 234 (1981).
75. T. Takenawa and Y. Nagai, J.Biol.Chem., 256, 6769 (1981).
76. R.F. Irvine and R.M.C. Dawson, J.Neurochem., 31, 1427 (1978).
77. R.O. Friedel, J.D. Brown and J. Durell, J.Neurochem., 16, 371 (1969).
78. E.G. Lapetina and R.H. Michell, Biochem.J., 131, 433 (1973).
79. A.L. Majumder and F. Eisenberg, Jr., Biochem.Biophys.Res.Commun., 60, 133 (1974).
80. E.G. Lapetina, M. Grosman and O. Canessa de Scarnati, Int.J.Biochem., 7, 507 (1976).

81. M.M. Billah, E.G. Lapetina and P. Cuatrecasas, Biochem.Biophys.Res.Commun., 90, 92 (1979).
82. G. Mauco, H. Chap and L. Douste-Blazy, FEBS Lett., 100, 367 (1979).
83. W. Siess and E.G. Lapetina, Biochem.Biophys.Acta, 752, 329 (1983).
84. S.E. Rittenhouse, Proc.Natl.Acad.Sci.U.S.A., 80, 5417 (1983).
85. M.G. Low and W.B. Weglicki, Biochem.J., 215, 325 (1983).
86. E.G. Lapetina and R.H. Michell, FEBS Letters, 31, 1 (1973).
87. M.M. Billah and E.G. Lapetina, J.Biol.Chem., 257, 12705 (1982).
88. M.M. Billah and E.G. Lapetina, Proc.Natl.Acad.Sci.U.S.A., 80, 965 (1983).
89. M.M. Billah and E.G. Lapetina, Biochem.Biophys.Res.Commun., 109, 217 (1982).
90. C.J. Kirk, J.A. Creba, C.P. Downes and R.H. Michell, Biochem.Soc.Trans., 9, 377 (1981).
91. S.J. Weiss, J.S. McKinney and J.W. Putney, Biochem.J., 206, 555 (1982).
92. R.P. Holmes and N.L. Yoss, Nature(Lond.), 305, 637 (1983).
93. S.K. Joseph, A.P. Thomas, R.J. Williams, R.F. Irvine and J.R. Williamson, J.Biol.Chem., 259, 3077 (1984).
94. M.J. Berridge, Biochem.J., 212, 849 (1983).
95. Y. Kawahara, Y. Takai, R. Minakuchi, K. Sano and Y. Nishizuka, Biochem.Biophys.Res.Commun., 97, 309 (1980).
96. Y. Takai, K. Kaibuchi, T. Matsubara and Y. Nishizuka, Biochem.Biophys.Res.Comm., 101, 61 (1981).
97. K. Kaibuchi, Y. Takai, M. Sawamura, M. Hoshijima, T. Fujikura and Y. Nishizuka, J.Biol.Chem., 258, 6701 (1983).
98. E.G. Lapetina and F.L. Siegel, J.Biol.Chem., 258, 7241 (1983).
99. W. Siess, F.L. Siegel and E.G. Lapetina, J.Biol.Chem., 258, 11236 (1983).
100. M. Saito and J. Kanfer, Biochem.Biophys.Res.Commun., 53, 391 (1973).
101. L.A. Kater, E.J. Goetzl and K.F. Austen, J.Clin.Invest., 57, 1173 (1976).
102. R.L. Wykle and J.M. Schremmer, J.Biol.Chem., 249, 1742 (1974).
103. R.L. Wykle, W.F. Kraemer and J.M. Schremmer, Arch.Biochem.Biophys., 184, 149 (1977).
104. R.L. Wykle, W.F. Kraemer and J.M. Schremmer, Biochem.Biophys.Acta, 619, 58 (1980).
105. S. Cockcroft and D. Alan in "Inositol and phosphoinositides," J. Bleasdale, J. Eichberg and G. Hauser, Eds., Humana Press, Clifton, N.J., 1984, in press.

Chapter 22. Vaccine Synthesis by Recombinant DNA Technology

Dennis G. Kleid, Department of Vaccine Development, Genentech Inc., 460 Point San Bruno Blvd., South San Francisco, California 94080

Introduction - The conventional human and animal viral vaccines are of two types, live-attenuated vaccines and killed-inactivated vaccines. The live-attenuated vaccines are produced initially by passaging virus many times in tissue culture cells until a weakened virus is obtained. Vaccination with a live-attenuated vaccine gives a mild or inapparent infection which leads to the induction of virus neutralizing antibodies. These antibodies protect the vaccinated individual from a subsequent infection with the normal pathogen. The protection is generally long lasting and quite effective for many human diseases such as polio, mumps, measles, rubella, yellow fever. The widespread inoculation of a live-attenuated vaccinia virus (that may have originated from cow pox) has caused the eradication of small pox. The second type of vaccine, inactivated virus vaccines, consists of concentrated whole virus preparations derived from growth in either eggs, tissue culture cells or animal tissues. These materials have been inactivated by treatment with chemical agents and combined with adjuvants such as aluminum hydroxide. These products contain antigenic proteins from the surface of the virus which when injected subcutaneously induce virus neutralizing antibodies. Vaccines for viral diseases such as influenzae and bacterial diseases such as diphtheria, pertussis and tetanus have been produced by this approach. Recently, subunit vaccines or "split virus" vaccines have been investigated which contain only parts of the virus. The most important immunogenic proteins from the virus, the viral surface proteins, are included in these preparations.

Many different animal diseases are kept under control by use of vaccines produced with technologies similar to those described above. Over one billion doses of live-attenuated Newcastle disease vaccine are used each year to protect fowl from this disease. Many other veterinary vaccines such as those for Marek's disease, infectious bovine rhinotracheitis, transmissible gastroenteritis, etc., are in common use. Recently, the trend in veterinary medicine has been toward the killed products, especially in the cattle and swine production industries. One reason is that some of the recently introduced vaccine products contain as many as seven antigens from different viruses in a single mixture. These multiple vaccine combination products seem to give more successful results when inactivated virus preparations are used, probably because of interference between viruses. In summary, work on prevention of infectious diseases continues to be a thriving area of research using the traditional technologies. However, with the introduction of new techniques of recombinant DNA or "gene splicing", combined with a resurgence of work on adjuvants, this area has received a great deal more interest from both researchers and commercial enterprises.

Recombinant DNA technology has allowed researchers to employ vastly different technologies to the process of vaccine production.

ISBN 0-12-040519-9

These new methods are currently under development: (A) biosynthesis of immunogenic proteins and antigenic sites in bacteria; (B) genetically modified eukaryotic cells that synthesize and secrete viral surface proteins; (C) live vaccines that have been genetically attenuated using recombinant DNA methods; and (D) genetically engineered live vaccines that possess genes and express antigens from other viruses. A few of these new technologies have already been applied to products that are now commercially available. Table 1 gives some examples and references for each of these new approaches to vaccine production using recombinant DNA technology.

Table 1. Vaccine Developments from Recombinant DNA Technology that have been tested for induction of neutralizing antibodies.

		(references)
A.	Biosynthesis of immunogenic proteins and antigenic sites in bacteria	
1.	Influenza B virus - heamagglutinin	1-4
2.	Hepatitis B virus - surface antigen	5-10
3.	Rabies virus - glycoprotein G	11
4.	Herpes simplex virus - glycoprotein D	12
5.	Vesicular Stomatitis Virus - glycoprotein D	13
6.	Foot-and-Mouth disease virus - VP1 protein	14-16
B.	Genetically modified eukaryotic cells that synthesize and secrete viral surface proteins.	
1.	Hepatitis B virus surface antigen gene in yeast	17-19
2.	Hepatitis B virus surface antigen gene in tissue culture	20-25
3.	Influenza virus heamgglutinin gene in tissue culture	26-28
4.	Vesicular Stomatitis Virus glycoprotein gene in tissue culture	29
6.	Herpes virus glycoprotein D gene in tissue culture	30,31
C.	Live vaccines that have been genetically attenuated using recombinant DNA methods.	
1.	Herpes simplex virus	32
D.	Genetically engineered live vaccines that possess genes and express antigens from other viruses (33-35).	
1.	Vaccinia virus vaccine with incorporated herpes antigen gene	36
2.	Vaccinia virus with incorporated hepatitis surface antigen gene	37
3.	Vaccinia virus vaccine with incorporated influenza antigen gene	38,39

(A) Biosynthesis of immunogenic proteins and antigenic sites in bacteria - The approach that has received the most effort has been the production in bacterial systems of immunogenic proteins that can induce protective antibodies when incorporated into vaccines. This approach involves first the isolation of the genes for viral or bacterial surface proteins. The genes are either copied from the viral RNA or isolated from the genome then "cloned" by incorporating into special E. coli plasmids as described below. The total nucleic acid sequence of the cloned gene is determined by DNA sequencing methods and thus by deduction the amino acid sequence of the surface protein is determined. The gene of interest is then recovered from the E. coli plasmids using restriction endonucleases and reintroduced into other plasmids called expression vectors. The following reviews a project involving the biosynthetic vaccines under development for Foot-and-Mouth Disease. This project illustrates most of the concepts and problems involved with the development of these materials for use in vaccines.

Biosynthetic Foot-and-Mouth-Disease vaccine - Foot-and-mouth disease (FMD) is a highly contagious disease that afflicts primarily cloven-hoofed animals. There are seven FMD types (A, O, C, ASIA1, SAT1, SAT2, and SAT3) as well as numerous subtypes. The disease is controlled by extensive use of inactivated vaccines and by limitations on cattle transport. Export of fresh meat and livestock from countries that have the disease to disease-free countries such as the United States and Australia is severely restricted. The FMD virus (a picornavirus) is composed of a positive-strand RNA genome of about 8,000 nucleotides within an icosahedral particle about 25nm in diameter. The particle surface is composed of four different virus structural proteins called VP1, VP2, VP3, and VP4. One of these proteins, VP1 (formerly designated VP3)*, when purified and injected into animals induces the production of virus-neutralizing antibody.[15] The gene coding for this protein has been identified and is located near the center of the viral genome. This knowledge was used to isolate and to clone the structural gene for VP1 and to subsequently produce the protein biosynthetically. The procedures for synthesizing and cloning copies of a gene from an RNA molecule have been described in detail and are outlined below.[40]

The FMD viral RNA from each strain of interest was isolated and annealed to a synthetic oligonucleotide that contained a sequence that hybridizes to a region of the genome to the 3' side of the VP1 gene. The sequence of the synthetic oligonucleotide was chosen based on nucleotide sequencing experiments involving the entire FMD genome. In the presence of the enzyme reverse transcriptase, the RNA was copied into DNA (cDNA) beginning with the primer and subsequently into double-stranded cDNA (ds-DNA) by treatment with DNA polymerase. The ds-cDNA was treated with the enzyme S1, which digests single-stranded regions, and the ds-cDNA was isolated by polyacrylamide gel electrophoresis. Segments of ds-cDNA greater than 2,000 base pairs

---

*The fourth FMD virus capsid polypeptide translated has been designated VP3, VP1, or VPthr, due to its variable migration in different polyacrylamide gel electrophoresis systems. This protein will be referred to as VP1 as recommended at the 1982 American Society of Virology meeting at Ithaca, NY.

were treated with an enzyme, deoxynucleotidyl transferase, in the presence of dCTP, to create polymers of cytidine on both ends of the cDNA molecules. These segments were annealed to linearized plasmid pBR322 previously treated to give polymers of guanosine on both ends. The plasmid and the ds-cDNA now annealed together were mixed with E. coli in the presence of calcium ions and briefly heated, causing some plasmid to enter the bacteria. The bacteria were plated on media containing tetracycline (Tc). This antibiotic eliminates bacteria not "transformed" to tetracycline resistance (Tc^R) with the plasmid, and upon replication allows growth of selected bacterial colonies, each containing numerous copies of a single plasmid with its ds-cDNA insert. Plasmids with incorporated ds-cDNA inserts, were isolated from individual bacterial colonies and were analyzed by restriction enzyme mapping and nucleotide sequencing. Some of the inserts contained the gene coding for the VP1 protein.[15] Once the gene sequence for the immunogenic protein was determined, the gene was put back into E. coli so that it is efficiently copied into mRNA translated into protein. This was done with another plasmid, called an expression vector, that contains five essential genetic elements: 1) an RNA polymerase binding site called the promoter; 2) a control element for the level of mRNA produced called an operator; 3) a sequence for directing the protein synthesis called the ribosome binding sequence; 4) a codon for the beginning of translation of the mRNA into protein called the initiation codon; and 5) a gene coding for part of an E. coli protein that will comprise a portion of a stable fusion protein. The expression vector used in this study contained these genetic elements from the E. coli tryptophan operon.[15] This operon has a control element (i.e., the operator) that responds to the concentration of tryptophan in the growth medium. In this system the bacteria are grown to a high density before protein synthesis is induced. This eliminates the deleterious effects that some foreign gene products have on bacterial cell growth. The vector also contains an efficient origin of replication (derived from pBR322) so that many copies of the plasmid reproduce in the host cell, and an antibiotic resistance gene (Tc^R) so that transformed E. coli can be selected simply by adding the antibiotic to the growth media.[15]

In the study described by Yansura et al.[16], only the part of the VP1 gene containing the major antigenic site was incorporated into the expression vector. The part of the gene that contains the antigenic region (amino acids 130-157) was modified by introducing synthetic DNA molecules to the ends of the gene fragment. This resulted in a gene fragment that was suitable for introduction into an expression vector designed to biosynthesize a fusion protein. In this plasmid construction the codon reading frame was maintained so that when this plasmid was incorporated into E. coli a fusion protein was produced that was made up of 190 amino acids from the expression vector linked to 27 amino acids from the VP1 protein from FMDV type A24 Cruzeiro. The protein accumulated in the cell in high levels with at least 20 percent of the total protein of the bacteria being the desired protein product. The material was isolated, purified and formulated with incomplete Freund's oil adjuvant. This was shown to induce neutralizing antibodies in cattle, and protect the animals from FMD in a virus challenge experiment.

At present the only successful virus challenge experiments reported using a recombinant DNA derived fusion protein antigen as a vaccine have been with FMD virus.[15,16] It is interesting that only a very small portion of the immunogenic protein was required to induce neutralizing antibody in cattle. These are the only studies that have demonstrated that animals can actually be protected using this approach. At this time it is not known how many other vaccines can benefit from this technique.

Several other types of expression vectors have been reported which express the protein of interest in different ways. There are vectors designed to produce the protein directly, by incorporating the heterologous surface protein gene linked directly to bacterial gene control signals. There are vectors designed to produce special fusion proteins, where the gene coding for the surface protein is linked in the same reading frame with a gene coding for a bacterial protein that is expected to be released from the bacteria and thus protein recovered from the bacterial growth media.

It was noticed by many research groups that some proteins, especially those having protein domains with high hydrophilic or hydrophobic characteristics could not be synthesized in bacterial systems efficiently. The growth rate of the bacteria is dramatically slowed or the protein is extensively degraded. In other experiments the recovered proteins did not contain all of the antigenic sites of the native protein. This may be due to the inability of E. coli to form the required disulfide bonds correctly, or due to lack of glycosylation, or due to insolubility problems. These problems have been overcome for some proteins by using vectors designed for expression and antigen production in eukaryotic cells.

(B) Genetically modified eukaryotic cells that synthesize and secrete viral subunit proteins - In this approach the genes coding for the surface proteins are linked to eukaryotic gene control elements and directly expressed. Introduction of the viral gene under the control of a eukaryotic gene control elements allows for the production of the protein in a eukaryotic host cell environment much more similar to that where it is normally synthesized. Several examples of the production of antigenic viral proteins in transient tissue culture expression systems using a viral vector based on SV40 have been reported (see Table 1). Recently continuous cell lines have also been described that produce viral antigens that are useful as vaccines (see Table 1). An important example of this is the production of the Herpes gD protein in Chinese hamster ovary cells (CHO cells).[30,31]

Biosynthetic Herpes simplex vaccine - Herpes simplex virus, of which there are two types (type 1 and type 2), cause a number of human diseases, including cold sores, eye and genital infections, and encephalitis. The herpes simplex virion is an enveloped DNA virus with a genome of about 100,000 base pairs. Contained on the viral envelope are four distinct glycoproteins designated gAB, gC, gD, and gE. One of these, gD, has been shown to stimulate the production of type-common virus-neutralizing antibodies. The genes coding for the gD proteins from type 1 and 2 have been isolated from viral DNA and cloned into bacterial plasmids. By nucleotide sequencing methods the gene sequences and thus the amino acid sequences of the these proteins have been determined. The proteins contain a hydrophobic amino-terminal leader peptides that are processed from the mature glycoprotein during synthesis and transport to the cellular membrane. The proteins also have hydrophilic and hydrophobic carboxyl-terminal peptide sequences that cause the protein to be attached to the cell membrane and during virus envelope production become embedded in the surface of the virus. These surface proteins induce the production of antibodies which, when combined with complement and other factors in the sera, inactivate the virus and/or infected cells.

The paper of Berman *et al.*[30] describes the construction of a plasmid vector that contains the Herpes gD from type 1. The gD gene is linked to the promoter of the SV40 virus VP1 protein. This gene control element is very efficient and has been often used to express various proteins in eukaryotic systems. Also contained on the vector is a gene coding for dihydrofolate reductase (DHFR) likewise linked to an SV40 promoter. The plasmid contains the necessary genes to allow for replication and selection in *E. coli*, so that quantities of the plasmid can be produced. The plasmid was introduced by calcium phosphate transfection into Chinese hamster ovary cells (CHO cells) that have a mutation in the DHFR gene. Cells containing the plasmid were selected for DHFR+ by growing the cells in media deficient in hypoxanthine, glycine and thymidine. Isolated from this selection was a continuous cell line that produces the gD protein on its surface.[30] In a similar manner a gD gene with the carboxyl-terminal hydrophilic/hydrophobic transmembrane domain removed by a gene deletion was introduced into the same vector.[31] Removal of the nucleotide sequences that encode the carboxyl-terminal 93 amino acids, including the transmembrane domain, causes the protein to be unable to attach to the cell membrane but instead slip through it. A cell line constructed with this plasmid synthesized and exported the protein into the media. The result was a continuous cell line producing a herpes gD glycoprotein in a truncated form. After recovery of the protein from the media and purification, the protein was combined with Freund's adjuvant and shown to induce neutralizing antibodies and protect mice from an otherwise lethal challenge with Herpes simplex type 2 virus.

(C) Live vaccines that have been genetically attenuated using recombinant DNA methods - Another use of recombinant DNA methods has been the creation of novel attenuated viruses or bacteria by the specific removal of genes or gene sequences that cause the virus to be pathogenic. The only example of this has been recently reported by Roizman *et al.* concerning Herpes simplex virus.[32] The production of vaccines from these materials is expected to follow closely that of traditional vaccines.

(D) Genetically engineered live vaccines that possess genes and express antigens from other viruses - A very interesting concept has led to the creation of new virus species that contain genes from other viruses. For example, the vaccinia virus, formerly used as smallpox vaccine, has been modified to contain the hepatitis B virus surface antigen, the influenza B haemagglutinin, and the Herpes type I gD protein (see Table 1). The gene for these antigens were linked to gene control elements from vaccinia genes and incorporated by recombination into the vaccinia virus.

Genetically Engineered Vaccinia Virus - The vaccinia viruses, or pox virus, cause a number of diseases characterized by skin lesions in both humans and animals. With some pox virus types, for example variola (smallpox), the lesions can be considerably disseminated and deadly. The virus is an enveloped DNA virus with a large genome (180,000 base pairs) coding for over a hundred proteins. The virus has its own RNA polymerase and it replicates using its own unique promoters and gene control elements. The virus has its own thymidine kinase (TK) gene. Vaccinia which are TK^- can be selected for in TK^- tissue culture cells by addition of 5-bromodeoxyuridine.[35] It is also known, by the work of Nakano et al., that vaccinia can recombine with DNA transformed into tissue culture cells.[34] If the DNA contains an active TK^+ the characteristic can be reintroduced into vaccinia by recombination. This is a process called marker rescue.[33-36] It is these features, marker rescue and TK^- selection, that have been utilized in the work of Panicali et al.[36] and Mackett et al.[37] The thymidine kinase gene of the vaccine strain of vaccinia (strain WR) has been cloned and characterized by restriction enzyme mapping. Also characterized are other vaccinia virus genes, for example, the gene coding for a highly expressed protein with a molecular weight of 7.4 kilodaltons has been cloned and the nucleotide sequence determined. The promoter for this gene and the ribosome binding sequence adjacent to it were linked to the Hepatitis B surface antigen gene previously isolated, cloned and sequenced from the Hepatitis B virus. The hybrid gene, part vaccinia virus and part hepatitis B surface antigen, were inserted into the middle of the vaccinia thymidine kinase gene thus interrupting the coding sequence. All of this was incorporated into a plasmid that was constructed and produced in E. coli. A sample of the plasmid was used to transfect cells (using the calcium phosphate procedure) that are susceptible to vaccinia virus (CV-1 cells). The CV-1 cells were then infected with vaccinia virus and subsequently virus was recovered. Some of the virus recombined with the plasmid and now contained an interrupted (and thus inactive) thymidine kinase gene. These recombined vaccinia virus were isolated by growth on TK^- 143 cells in the presence of 5-bromodeoxyuridine, which kills cells containing an active thymidine kinase enzyme. In this way genetically engineered vaccinia virus (TK^-) were obtained that contained an expressed Hepatitis B surface antigen in the middle of the vaccinia TK gene.[37] Research groups at both the NIH[33,35,37,39] and at the New York State Department of Health in Albany[34,36,38] have demonstrated that viruses constructed in this way can accept a number of different antigenic proteins from different viruses. The genetically engineered viruses can infect rabbits, mice and guinea pigs and induce protective antibodies from a number of different viruses including influenza, hepatitis B and herpes simplex.

Summary - There are several advantages to vaccines derived from recombinant DNA technology: 1) there is a potential for higher margin of safety; 2) the products may be less expensive due to savings related to production methods, stability, and quality control; 3) it should be possible to produce vaccines for diseases not amenable to current technology; 4) some of these products can be incorporated into formulations not possible with the currently available materials; and 5) since many of the methods required to develop these vaccines were beyond the technology of the time, it should be possible to develop proprietary vaccine products.

Thus the future of this line of research is sure to lead to many important discoveries and products that are useful for the protection of people and animals from pathogenic organisms. At this time it is not possible to speculate as to which of the above described new technologies will ultimately be the most successful, however it is clear that each will likely find a place.

References

1. J.S. Emtage, W.C.A. Tacon, G.H. Catlin, B. Jenkins, A.G. Porter and N.H. Carey, Nature 283, 171 (1980).
2. I. Heiland and M.J. Gething, Nature 292, 851 (1981).
3. A.R. Davis, D.P. Nayak, M. Ueda, A.L. Hiti, D. Dowbenko and D.G. Kleid, Proc. Natl. Acad. Sci. 78, 5376 (1981).
4. A.R. Davis, T. Bos, M. Ueda, D.P. Nayak, D. Dowbenko and R.W. Compans, Gene 21, 273 (1983).
5. M. Pasek, T. Goto, W. Gilbert, B. Zink, H. Schaller, P. MacKay, G. Leadbetter and K. Murray, Nature 282, 575 (1979).
6. P. Valenzuela, P. Gray, M. Quiroga, J. Zaldivar, H.M. Goodman and W.J. Rutter, Nature 280, 815 (1979).
7. P. Charnay, M. Gervais, A. Louise, F. Galibert and P. Tiollais, Nature 286, 893 (1980).
8. K. Hardy, S. Stahl and H. Kupper, Nature 293, 481 (1981).
9. J.C. Edman, R.A. Hallewell, P. Valenzuela, H.M. Goodman and W.J. Rutter, Nature 291, 503 (1981).
10. P. MacKay, M. Pasek, M. Magazin, R.T. Kovacic, B.Allet, S. Stahl, W. Gilbert, H. Schaller, S.A.Bruce and K. Murray, Proc. Natl. Acad. Sci. USA 78, 4510 (1981).
11. E. Yelverton, S. Norton, J.F. Obijeski and D.V. Geoddel, Science 219, 614 (1983).
12. R.J. Watson, J.H. Weis, J.S. Salstrom, and L.W. Enquist, Science 218, 381 (1982).
13. J.K. Rose and A. Shafferman, Proc. Natl. Acad. Sci. USA 78, 6670 (1981).
14. H. Kupper, W. Keller, C. Kurz, S. Forss, H. Schaller, R. Franze, K. Strohmaier, O. Marquardt, V.G. Zaslavsky, and P.H. Hofschneider, Nature 289, 555 (1981).
15. D.G. Kleid, D. Yansura, B. Small, D. Dowbenko, D.M. Moore, M.J. Grubman, P.D. McKercher, D.O. Morgan, B.H. Robertson and H.L. Bachrach, Science 214, 1125 (1981).
16. D.G. Yansura, D. Dowbenko, G.N. Weddell, M.E. Hoatlin, S.J. Shire, L.A. Bock, E.J. Patzer, D.G. Kleid, D.M. Moore, B.H. Roberston, M.J. Grubman, D.O. Morgan and P.D. McKercher, Miami Winter Symp. 20, 479 (1983).
17. P. Valenzuela, A Medina, W.J. Rutter, G. Ammerer and B.D. Hall, Nature 298, 347 (1982).
18. R.A. Hitzeman, C.Y. Chen, F.E. Hagie, E.J. Patzer, C.C. Liu, D.A. Estell, J.V. Miller. A. Yaffe, D.G. Kleid, A.D. Levinson and H. Oppermann, Nucl. Acids. Res. 11, 2745 (1983).
19. W.J. McAleer, E.B. Buynak, R.Z. Maigetter, D.E. Wampler, W.J. Miller and M.R. Hilleman, Nature 307, 178 (1984).
20. M.F. Dubois, C. Pourcel, S. Rousset, C. Chany and P. Tiollais, Proc. Natl. Acad. Sci. USA 77, 4549 (1980).
21. A.M. Moriarty, B.H. Hoyer, J.W. Shih, J.L. Gerin and D.H. Hamer, Proc. Natl. Acad. Sci. USA 78, 2606 (1981).
22. C-C. Liu, D. Yansura and A.D. Levinson, DNA 1, 213-221 (1982).
23. J.K. Christman, M. Gerber, P.M. Price, C. Flordellis, J. Edelman, and G. Acs, Proc. Natl. Acad. Sci. USA 79, 1815 (1982).
24. C. Pourcel, A. Louise, M. Gervais, N. Chenciner, M-F. Dubois and P. Tiollais, J. Virol. 42, 100 (1982).
25. C.W. Crowley, C-C. Liu and A.D. Levinson, Molecular and Cellular Biology 3, 44 (1983).
26. M.M. Sveda and C-J. Lai, Proc. Natl. Acad. Sci. USA 78, 5488 (1981).
27. M-J. Gething and J. Sambrook, Nature 293, 620 (1981) and Nature 300, 598 (1982).
28. J.R. Hartman, D.P. Nayak and G. C. Fareed, Proc., Natl. Acad. Sci. USA 79, 233 (1982).
29. J.K. Rose and J.E. Bergmann, Cell 30, 753 (1982).
30. P.W. Berman, D. Dowbenko, L.A. Lasky and C.C. Simonsen, Science 222, 524 (1983).
31. L. Lasky, D. Dowbenko, C. Simonsen and P. Berman in: "Modern Approaches to Vaccines", eds. R. Chanock and R. Lerner, Cold Spring Harbor Laboratory, Cold Spring Harbor, N.Y., Chapter 27 (1984).
32. B. Roizman et al. in: "Modern Approaches to Vaccines," eds. R. Chanock and R. Lerner, Cold Spring Harbor Laboratory, Cold Spring Harbor, N.Y. (1984).
33. J.P. Weir, G. Bajszar and B. Moss, Proc. Natl. Acad. Sci. USA 79, 1210 (1982).
34. E. Nakano, D. Panicali and E. Paoletti, Proc. Natl. Acad. Sci. USA 79, 1593 (1982).
35. M. Mackett, G.L. Smith and B. Moss, Proc. Natl. Acad. Sci. USA 79, 7415 (1982).
36. D. Panicali and E. Paoletti, Proc. Natl. Acad. Sci. USA 79, 4927 (1982).
37. G.L. Smith, M. Mackett and B. Moss, Nature 302, 490 (1983).
38. D. Panicali, S.W. Davis, R.L. Weinberg and E. Paoletti, Proc. Natl. Acad. Sci. USA 80, 5364 (1983).
39. G.L. Smith, B.R. Murphy and B. Moss, Proc. Natl. Acad. Sci. USA 80, 7155 (1983).
40. T. Maniatis, E.F. Fritsch and J. Sambrook, Molecular Cloning, A Laboratory Manual, (Cold Spring Harbor Laboratory, Cold Spring Harbor, N.Y.), (1982).

Chapter 23. Collagenases in Tumor Cell Extravasation

T. Turpeenniemi-Hujanen, M.D., Ph.D., U.P. Thorgeirsson, M.D., and L.A. Liotta, M.D., Ph.D.
Laboratory of Pathology, NIH, NCI, Bethesda, MD 20205

INTRODUCTION - Metastasis is a complex multistep process in which tumor cells escape from the primary tumor, enter the hematogenous and lymphatic vascular channels, and are transported to distant organs. (Figure 1)

The circulating tumor cells arrest in the precapillary venules by passive embolization due to clumping or by active attachment to the lumenal vascular surface. In order to form metastases the tumor cells must penetrate the vascular wall to enter the interstitial stroma. Such penetration is thought to be an active process rather than a passive bursting of the vessel due to growth pressure[1]. The two components of the vascular wall which are mechanical barriers to tumor cell penetration are the endothelium and the continuous basement membrane (BM). Ultrastructural studies indicate that tumor cells may traverse the endothelium by a) moving between endothelial cell bodies[2] or b) by destroying the entire endothelial cell[3]. The tumor cell is then held up at the basement membrane for a period of time ranging from 2 to 8 hours[2,3]. The next step is local dissolution of the basement membrane which occurs at the point of tumor cell contact[2,4]. Such dissolution may be mediated by degradative enzymes which lyse the components of the basement membrane.

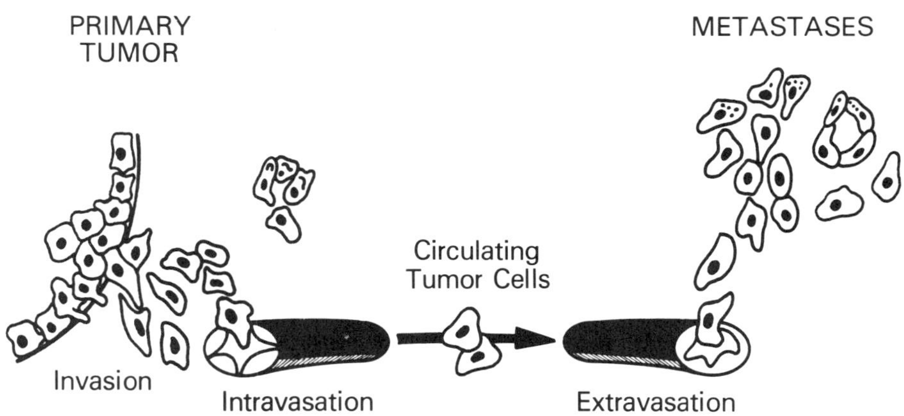

Figure 1: Diagram of the metastatic cascade. Tumor cells invade at the primary tumor site and enter the interstitial stroma. They thereby gain access to blood vessels for further dissemination. Tumor cells invade the vascular wall and are dislodged into the circulation in single cells and clumps. Circulating tumor cells arrest in the precapillary venules of the target organ by adherence or mechanical wedging. They must then exit the circulation to initiate a secondary tumor colony called a metastasis.

ISBN 0-12-040519-9

The basement membrane is an insoluble meshwork with type IV collagen comprising the major structural supporting scaffold. Thus, a limiting factor in the degradation of the basement membrane is lysis of type IV collagen. Classic collagenases which cleave interstitial collagens fail to degrade basement membrane type IV collagen. Type V collagen, which is associated with the basement membrane interface with the stroma, is also resistant to classic interstitial collagenase. Therefore, separate classes of proteases are required to degrade basement membrane associated collagens. One main objective of this review will be to summarize the current biochemical information with regard to type specific collagenolysis. A second objective will be to extrapolate this information into diagnostic and therapeutic applications for patients with metastatic cancer.

Two types of cells, inflammatory cells and malignant cells, are known to possess the biochemical machinery for actively migrating through blood vessel walls. Cellular extravasation is a complex cascade of events involving attachment to endothelium or basement membrane followed by active invasion and migration of cells through vascular BM and adjacent perivascular connective tissue stroma. The mechanism for extravasation of inflammatory cells and tumor cells may be similar. The present review will focus on tumor cell extravasation.

Active penetration of tumor cells into the extracellular matrix is facilitated by enzymatic degradation of various matrix components. One of many proteolytic enzymes involved in the metastatic process degrades collagen, the major structural component of the extracellular matrix. Collagenases are metalloproteases that digest native collagen at neutral pH. Separate collagenases have been found to degrade different types of collagen, i.e. types I-III, collagen type IV[6], and type V collagen[7] (Table I). They control mechanism for production of collagenases is still unknown but the presence of specific collagenase inhibitors in tissues possibly plays an important role in regulating the collagenolytic activity[8]. The concept that collagenases are involved in tumor cell extravasation is supported by a number of studies that show elevated collagenase activi-ties in various malignant human and animal tumors as well as in cultured tumor cells[9-23].

The extracellular matrix and cell-matrix interactions - The extracellular matrix is a dense meshwork of collagenous and noncollagenous glycoproteins, elastin and proteoglycans[24-31]. The matrix composition varies greatly according to the special function of the tissue[24-31]. The most abundant matrix protein, collagen, provides the major physical structure to connective tissues. At least seven genetically different types of collagen are known that differ in their primary structure and tissue distribution[32]. Type IV collagen is specifically found in basement membranes which are dense linear structures produced by parenchymal and endothelial cells at the interface with the connective tissue stoma[25,27-30,33,34]. Type IV collagen is organized into a network of fine filamentous aggregates of molecules that are highly crosslinked through disulfide and lysine derived covalent bonds[32]. The main noncollagenous matrix glycoproteins are fibronectin[35] and laminin[28]. They act as anchorage for the resident matrix cells to the structural components, such as collagen[35-37]. Fibronectin serves as an adhesive protein for a variety of cells but laminin specifically mediates attachment of epithelial and endothelial cells to BM collagen[36]. Fibronectin and laminin greatly influence growth and differentiation of the matrix cells. Other matrix proteins include entactin[27] and elastin[38], different glycosaminoglycans[39,40] and proteoglycans[24].

Table I. Tissue Distribution of the Different Collagen Types

Collagen Type	Tissue Location	Cell Types that Synthesize Collagen
Type I	bone, tendon, skin, dentin, cornea, fascia	fibroblasts, osteoblasts, smooth muscle cells, epithelial cells
Type II	cartilage, cornea, vitreous body	chrondrocytes, neural retinal cells
Type III	skin, lung, liver	fibroblasts, myoblasts
Type IV	basement membrane	epithelial cells, endothelial cells
Type V	all types of connective tissue except hyaline cartilage.	smooth muscle cells, chondrocytes
Type VI	uterus, placental villi, skin	
Type VII	fetal membranes	
1 , 2 , 3 , HMV, LMV, SC	cartilage	

In malignant transformation the molecular composition of the extracellular matrix as well as the cell-matrix interactions are often altered. Malignant cells generally produce less matrix components than their normal counterparts[41-45]. As an example is fibronectin synthesis which is greatly reduced in malignant cells.

The properties of malignant cells that help them move across tissue boundaries include: decreased cell adhesiveness[46], increased cell motility[47], angiogenesis factor(s)[48] and enzymatic activities[49-51]. One of the most insoluble connective tissue barriers that tumor cells must encounter at different stages during the metastatic process is the BM. Tumor cells must traverse the epithelial BM as they progress from in situ to invasive carcinoma as well as during extra- and intravasation. The first step in tumor cell penetration of the vascular BM is laminin mediated attachment to the exposed BM[36,37,52]. Recent studies indicate that the attachment requires a presence of a specific receptor protein for laminin on the cell surface[53-56]. The second step involves focal proteolytic degradation of the BM in the invasion front[2,4,14,57-60]. It has been demonstrated that both inflammatory cells and tumor cells form pseudopodia that contain enough of the degradative enzymes to cause focal defect in the BM[61,62]. The third step is migration of tumor cells through the focal defect in the BM[3,44]. Studies of cells cultured from tumor explants have demonstrated that both tumor cells and host cells may be able to produce type I collagenase. However, the host-derived cells released the enzyme only during limited number of passages while the tumor cells continued to produce collagenase I[22,63]. In some cases cocultures of tumor cells and normal cells have secreted more enzyme activity than each line could produce individually[64,65]. These studies suggest that the tumor cells may stimulate the normal host cells to produce collagenase.

Collagenase I-III - The dense network of collagen fibers of the extracellular matrix is impenetrable to most cells. Invasive cells, such as inflammatory cells and tumor cells produce matrix degrading enzymes in order to cross tissue boundaries. Destruction of the extracellular matrix during tumor invasion has been observed histologically and ultrastructurally [66,67]. Various types of tumor cells have been found to elabaorate collagenolytic enzymes[9-13,15,16,18-20,22,23]. The vertebrate collagenase was first described by Gross and Lapière in 1962[68]. It has neutral pH optimum and cleaves the interstitial collagen types at a single locus near the C-terminal end of the collagen molecule producing 75% and 25% fragments. The collagen fragments are thermally denatured after the initial cleavage and are then susceptible to gelatinases and other nonspecific proteinases[5]. It has also been suggested that the collagenase itself can further degrade the collagen fragments at physiological temperature[69]. The vertebrate collagenases degrade types I, II and III but not types IV and V collagen[5,18,32,70-72]. The rate of degradation for type II collagen is considerably slower than for I and III. The vertebrate collagenases isolated from different sources may also have different preferences in their types of substrates e.g., human granulocyte collagenase degrades type I collagen faster than type III[73].

The level of type I collagenolytic activity associated with malignant tumor cells often correlates with the malignant behavior of these tumors[9,12,13,15,16,18,20,23,74]. Increase of the collagenase I activity has also been demonstrated in experimentally induced tumors by chemical carcinogens[75] or in cells treated with tumor promoters[76]. Tumor cell collagenases that have been partially purified and characterized are similar to the vertebrate collagenase[5,8,10,69,71,73,77-82].

Basement membrane degrading collagenases - Type IV collagenase is inhibited by EDTA, and is thus a metallo protease. It specifically degrades basement membrane collagen, and cleaves procollagen IV into two segments which comprise 1/4 and 3/4 of the total length of the molecule[6,83] (Figure 2). The cleavage site is located near the N-terminal end of the collagen molecule and the cleavage produces a complex containing the highly disulfide-linked "7-S region" joined with the residual parts of the four procollagen IV N-terminal ends. Type IV collagenase has been found in normal migrating cells and metastatic tumor cells[14,17,18,21,47,70,84]. The enzyme is secreted into the culture media in a latent form which can be activated with trypsin and plasmin. Type IV collagenase has been purified from a mouse tumor (Pulmonary metastasis of the T241 sarcoma)[6,83]. The enzyme migrates as a doublet on polyacrylamide gel with a molecular weight of approximately 62,000 and 68,000. Type IV collagenase activity of culture supernatants or cell lysates is assayed by using biosynthetically labelled, acid extracted EHS sarcoma type IV collagen substrate[6,85]. Vascular dissemination of tumor cells is critical in the cascade of events that lead to metastases. Therefore, type IV collagenase might be one of the most important factors although not sufficient, for metastasis formation. Type IV collagenolytic activity of many metastatic tumor cell lines studied has been shown to correclate with their malignancy[17,21,50]. Nevertheless, it is unlikely that there is a strict quantitative relationship between the enzyme level and the metastatic capacity in all types of malignant tumors since there are many other cellular and host factors that are involved in the metastatic process[86,82]. For example, tumor cells that secrete large amounts of collagenase but are also highly susceptible to host defense factors, could be poorly metastatic.

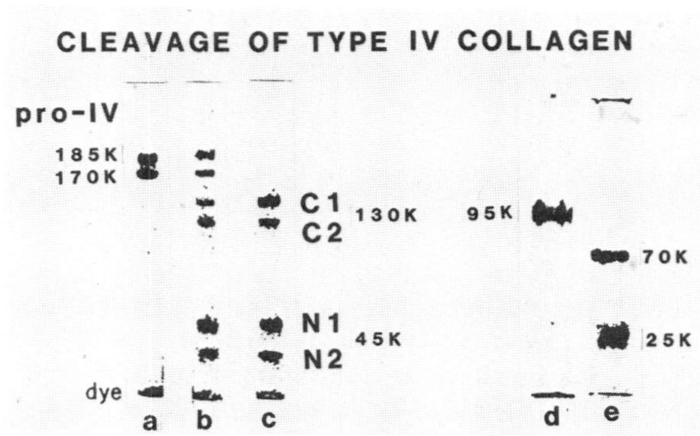

Figure 2: Type IV collagen degradation products identified by sodium doclecyl sulfate polyacrylamide gel electrophoresis (after reduction). The EHS mouse tumor type IV procollagen[88] was incubated with enzyme 5h (a); 18h (b); and at 30°C (c). The cleavage yields peptides corresponding to 2/3 (C_1 and C_2) and 1/3 (N_1 and N_2) of the procollagen molecule. In pepsin digested placenta type IV collagen[6] (d); the placenta type IV collagen plus enzyme (e), two cleavage products are produced, consistent with a single site of cleavage.

Type V collagenase has been identified in alveolar macrophages and in tumor cells[7]. It has been partially purified from culture media of murine reticulum sarcoma cells (M5076) and appears to be a doublet on polyacrylamide gel with a molecular weight of about 80,000. It degrades type V collagen, but not type IV collagen[7].

Regulation of collagenases: The genetic control of collagenases is so far poorly understood. Considering the present knowledge the three possible control mechanisms for collagenases include: 1) secretion of collagenases as latent forms and their activation, 2) production of specific collagenase inhibitors, and 3) factors that stimulate collagenase secretion by cells.

1) Collagenases are found both in activated and latent forms (procollagenases)[16,21,83,89]. Procollagenases can be activated by proteases such as trypsin or plasmin[21,71]. They can also be activated by nonenzymatic methods using ammonium sulfate fractionation, thiol-blocking reagents and ion-exchange chromatography[71,78,90], or by long incubation periods at 37°C (spontaneous activation)[71].

Activation of the collagenase involves a split off of a peptide and yields a lower molecular weight active collagenase[71,83,89,91,92]. Plasmin is proposed to be the most important enzyme in the biological activation of collagenases. It is derived from serum plasminogen by plasminogen activators, which are serine proteases. A varient of tumor cell lines have shown high levels of plasminogen activators, that often correlates with enhanced collagenase activities[19,21,76,93].

2) Some tissues contain tissue collagenase inhibitors, that bind to the collagenases and, thus, locally regulate the rate of collagenolysis. These inhibitors are generally of low molecular weight, diffuse easily through the extracellular matrix and are released into tissue culture supernatants[94,95]. They have been purified from many tissues such as skin[96], aorta[97,98], tendon[99], bone[95] and cartilage[57]. They form an enzyme-inhibitor complex that is often irreversible so that trypsin treatment cannot reactivate the enzyme[95]. Several concepts on the role of collagenase inhibitors in the regulation of collagenolysis have been proposed[8]. One possibility is that the same cell produces both the active collagenase and the inhibitor at different times. Thus, in the absence of collagen breakdown there would be an excess inhibitor and the collagenase production could be initiated on demand. On the other hand, the collagen activating enzymes such as trypsin would also inactivate the inhibitors[8]. Studies of collagenase inhibitors in brain tumor tissues have revealed that less inhibitors are present in the invasive than the non-invasive tumor[23]. In vitro studies have also shown that bovine cartilage derived inhibitors retard tumor cell invasion of a human amnion BM[57]. Plasma contains at least two types of collagenase inhibitors i.e. a non-specific protease inhibitor, α2-macroglobulin which comprises over 90% of the serum collagenase inhibitory capacity, and more specific β1-anticollagenase[100,101].

3) Recent work has demonstrated that tumor cells like inflammatory cells respond to chemotactic factors in vitro. The chemotactic factors discovered to date include: the fifth component of complement[102], N-formylmethionyl-leucyl-phenylalanine (FMLP)[107], bone resorption factor[104], and fibronectin[105]. In response to chemotactic stimuli the migrating epidermal and endothelial cells secrete collagenases[84,106]. Cultured fibroblasts increase their collagenase production when stimulated by interleukin[107]. In contrast to inflammatory cells and endothelial cells that store the enzymes in cytoplasmic granules, tumor cells probably store very little collagenase but produce and secrete on demand[108].

Conclusions:

1. Extravasation of tumor cells is an important step in the metastatic process. It involves various tumor cell factors as well as special properties of extracellular matrix components. Proteolytic enzymes secreted locally at the point of tumor cell penetration degrade the extracellular matrix to make room for invading cells. Although tumor cell invasion is facilitated by their own proteolytic activities, proteases produced by host cells may also play a role in tumor cell invasion.

2. Basement membranes constitute the main structural support to blood vessels and provide a mechanical barrier to tumor cell extravasation. Type IV collagenase produced by various tumor cells in culture is considered to be necessary for proteolytic destruction of the BM during the metastazation. Since many other cellular and host factors are involved in the metastatic process the collagenolytic activities do not always correspond with metastatic behavior although a minimal enzyme activity is necessary to degrade BM for extravasation of tumor cells.

3. Prior to penetration of the vessel wall tumor cells must attach to the BM. For many types of tumor cells, this attachment is mediated through laminin which is one of two major glycoproteins of the extracellular matrix. Laminin binds to a cell surface receptor that was recently purified and characterized.

4. Collagenases usually are secreted as procollagenases which are activated by other proteases such as plasmin and trypsin. Collagenase inhibitors are possibly involved in local regulation of collagenase activities. Collagenolytic activities may therefore be determined by the regional balance between enzyme and inhibitor.

5. In vitro studies have demonstrated that tumor cells are able to produce collagenase and migrate in response to chemotactic factors. This may play in role in the organ selectivity of metastatic spread of different tumors observed in vivo.

Potential Clinical Applications - The major cause of treatment failure in patients with solid tumors is metastasis. The primary tumor can usually be eliminated by current therapeutic techniques. However, half the patients with a newly diagnosed tumor will already have clinically silent micrometastases. Thus major goals of cancer research are to develop improved methods to predict whether an individual patient's tumor has already metastasized, and to provide strategies for prevention or localization of metastases. Biochemical information about tumor collagenases may ultimately be applied to approach these goals.

Antibodies directed against collagenases can be used to localize subpopulations of invasive tumor cells in histologic sections of human tumors [18,45]. Since these proteases may be augmented in metastatic cells, such antibodies could conceivably be used to localize micrometastases in lymph node sections. Primary tumor cell populations are heterogeneous. Subpopulations of highly metastatic tumor cells may preexist in the tumor. These metastatic cells may be selected out during the metastatic cascade. Therefore, the percent of collagenase positive staining tumor cells in a histologic section of a patient's tumor may relate to the clinical aggresiveness of the tumor. Measurement of collagenase by radioimmunoassay may also have prognostic value if applied to tumor samples or serum. A further application of collagenase antibodies could be the localization of clinically silent micrometastases in the patients. Appropriate radionuclide scanning techniques have already been developed to localize monoclonal antibodies injected into patients.

Finally, collagenase inhibitors could also play a future role in the therapy of tumor invasion. Natural collagenase inhibitors have been shown to inhibit tumor invasion in vitro. Synthetic inhibitors may also be developed which block the substrate binding domain, competitively, or irreversibly inhibit the enzyme itself.

REFERENCES

1. U.P. Thorgeirsson, T. Turpeenniemi-Hujanen, L.M. Neckers, D.W. Johnson and L.A. Liotta, Invasion Metastasis, 4, 73 (1984).
2. R.H. Kramer and G.L. Nicolson, Proc. Natl. Acad. Sci., USA, 76 , 5704 (1979).
3. S. Wood Jr., Arch. Pathol., 66, 550 (1958).
4. L.A. Liotta, C.W. Lee and D.J. Morakis, Cancer Lett., 11, 141 (1980).
5. E.D. Harris, Jr. and C.A. Vater, Meth. Enzymol., 82, 423 (1982).
6. L.A. Liotta, K. Tryggvason, S. Garbisa, P. Gehron-Robey and S. Abe, Biochemistry, 20, 100 (1981).
7. L.A. Liotta, W.L. Lanzer and S. Garbisa, Biochem. Biophys. Commun., 98, 184 (1981).
8. A. Sellers and G. Murphy, Int. Rev. Conn. Tiss. Res., 9, 151 (1981).
9. A.C. Taylor, B.M. Levy and J.W. Simpson, Nature, London., 228 366 (1970).
10. M.H. Dresden, S.A. Heilman and J.D. Schmidt, Cancer Res., 32, 933 (1972).
11. E.D. Harris Jr., C.S. Faulkner and S. Wood Jr., Biochem. Biophys. Res. Commun., 48, 1247 (1972).

12. K. Hashimoto, Y. Yamanishi, E. Maeyens, M.K. Dabbous and T. Kanzaki, Cancer Res., 33, 2790 (1973).
13. M. Abramson, R.W. Schilling, C.-C. Huang and R.G. Salome, Ann. Otol. Rhinol. Laryngol., 84, 158 (1975).
14. L.A. Liotta, J. Kleinerman, P. Catanzaro, D. Rynbrandt, J. Natl. Cancer Inst., 58, 1427 (1977).
15. H. Ohyama and K. Hashimoto, J. Biochem., 82, 175 (1977).
16. G. Wirl and J. Frick, Urol. Res., 7, 103 (1979).
17. L.A. Liotta, K. Tryggvason, S. Garbisa, I. Hart, C.M. Foltz and S. Shafie, Nature, London, 284, 67 (1980).
18. D.E. Woolley, L.C. Tetlow, C.J. Mooney and J.M. Evanson, in "Proteinases and tumor invasion", P. Strand, A.J. Barrett and A. Baili, Eds., Raven Press, New York, 1980, p. 97.
19. R.L. O'Grady, L.I. Upfold and R.W. Stephens, Int. J. Cancer, 28, 509 (1981).
20. R.L. O'Grady, P.J. Harrop and D.A. Cameron, Pathology, 14, 135 (1982).
21. T. Salo, L.A. Liotta, J. Keski-Oja, T. Turpeenniemi-Hujanen and K. Tryggvason, Int. J. Cancer, 30, 669 (1982).
22. M.K. Dabbous, M. El-Torky, L. Haney, Sr. B. Brinkley and N. Sobhy, Int. J. Cancer, 31, 357 (1983).
23. A.N. Halaka, R.A.D. Bunning, C.C. Bird, M. Gibson and J.J. Reynolds, J. Neurosurg., 59, 461 (1983).
24. V.C. Hascall and G.K. Hascall in "Cell biology of extracellular matrix", E.C. Hay, Ed., Plenum Press, New York, (1981), p. 39.
25. N.A. Kefalides, R.Alper and C.C. Clark, Int. Rev. Cytol., 61, 167 (1979).
26. J.R. Hassell, P. Gehron-Robey, H.-J. Barrach, J. Wilczek, S.I. Rennard and G.R. Martin, Proc. Natl. Acad. Sci. USA, 77, 4494 (1980).
27. B. Carlin, R. Jaffee, B. Bendon, E.E. Chung, J. Biol. Chem., 256, 5209 (1981).
28. R. Timpl, H. Rhode, P. Gehron-Robey, S.I. Rennard, J.M. Foidart and G.R. Martin, J. Biol. Chem., 254, 9933 (1979).
29. A. Martinez-Hernandez, S. Gay and E.J. Miller, J. Cell. Biol., 92, 343 (1982).
30. R.J. Roll, J.A. Madri and H. Furthmayr, J. Cell. Biol., 85, 597 (1980).
31. E.J. Miller and L. Lunde, Biochemistry, 12, 3153 (1973).
32. P. Bornstein and H. Sage, Ann. Rev. Biochem., 49, 957 (1980).
33. R. Timpl, H. Wiedemann, V. Van Delden, H. Furthmayr and K. Kühn, Eur. J. Biochem., 120, 203 (1981).
34. R. Vracko, Am. J. Pathol., 77, 313 (1974).
35. S. Stenman and A.L. Vaheri, J. Exp. Med., 147, 1054 (1978).
36. V.P. Terranova, D.H. Rohrbach and G.R. Martin, Cell, 22, 719 (1980).
37. C.N. Rao, I.M.K. Margulies, T.S. Tralka, V.P. Terranova, J.A. Madri and L.A. Liotta, J. Biol. Chem., 257, 9740 (1982).
38. C. Franzblau and B. Faris, in "Cell biology of extracellular matrix", E.D. Hay, Ed. Plenum Press, New York, 1981, p 65.
39. Y. Kanwar and M.G. Farguhar, Proc. Natl. Acad. Sci. USA, 76 4493 (1979).
40. V.C. Hascall, in "Biology of Carbohydrates", vol. 1, V. Ginsburg, Ed., John Wiley & Sons, New York, 1981, p 1.
41. M.S. Al-Adnani, J.A. Kirrane and J. O'D. McGee, Br. J. Cancer, 31, 653 (1975).
42. K. Alitalo, J. Keski-Oja and A. Vaheri, Int. J. Cancer, 27, 733 (1981).
43. H.K. Kleinman, R.J. Klebe and G.R. Martin, J. Cell. Biol., 88, 473 (1981).
44. G.P. Siegal, S.H. Barsky, V.P. Terranova and L.A. Liotta, Invasion Metastasis, 1, 54 (1981).
45. S.H. Barsky, G.P. Siegal, F. Jannotta and L.A. Liotta, Lab. Invest., 49, 140 (1983).
46. D.R. Coman, Cancer Res., 4, 625 (1944).
47. S. Wood, Jr., R.R. Baker and B. Marzocchi, in "The Proliferation and Spread of Neoplastic Cells", Williams & Wilkins, Baltimore, 1968, p 495.
48. J. Folkman, Adv. Cancer Res., 19, 331 (1974).
49. A.D. Recklies, K.J. Tiltman, T.A.M. Stoker and A.R. Poole, Cancer Res., 40, 550 (1980).
50. L.A. Liotta, U.P. Thorgeirsson and S. Garbisa, Cancer Metast. Rev., 1, 277 (1982).
51. Ng. Raymond and J.A. Kellen, Med. Hypoth., 10, 291 (1983).
52. V.P. Terranova, L.A. Liotta, R.G. Russo and G.R. Martin, Cancer Res., 42, 2265 (1982).
53. V.P. Terranova, C.N. Rao, T. Kalebic, I.M. Margulies and L.A. Liotta, Proc. Natl. Acad. Sci., 80, 444 (1983).
54. C.N. Rao, S.H. Barsky, V.P. Terranova and L.A. Liotta, Biochem. Biophys. Res. Commun., 111, 804 (1983).
55. H.L. Malinoff and M.S. Wicha, J. Cell. Biol. 96, 1475 (1983).
56. S.H. Barsky, C.N. Rao, D. Hyams and L.A. Liotta, Breast Cancer Res. and Treat., in Press.
57. U.P. Thorgeirsson, L.A. Liotta, T. Kalebic, I.M. Margulies, K. Thomas, M. Riescandelore and R.G. Russo, J. Natl. Cancer Inst., 1049 (1982).
58. P.A. Jones and Y.A. DeClark, Cancer Res., 40, 3222 (1980).
59. R.H. Kramer, R. Gonzalez and G.L. Nicolson, Int. J. Cancer, 26, 639 (1980).
60. G.L. Nicolson, Biochim, Biophys. Acta. Rev. Cancer, 659, 113 (1982).

61. R.G. Russo, L.A. Liotta, U.P. Thorgeirsson, R. Brundage and E. Schiffmann, J. Cell Biol., 91, 459 (1981).
62. T. Kawaguchi, S. Tobai and K. Nakamura, Invasion Metastasis, 2, 40 (1982).
63. M.K. Dabbous, M. El-Torky, L. Haney, N. Sobhy and Sr. B. Brinkley, Exp. Mol. Pathol., 38, 1 (1983).
64. C. Biswas, Biochem. Biophys. Res. Commun., 109, 1026 (1982).
65. N. Henry, Y. Eeckhout, A.-L. von Lamsweerde and G. Vaes, FEBS Lett., In Press.
66. E.A. Bauer, J.M. Gordon, M.E. Reddick and A.Z. Eisen, J. Invest. Dermatol., 69, 363, (1977).
67. M.S.C. Birbeck and D.N. Wheatley, Cancer Res., 25, 490 (1965).
68. J. Gross and C.M. Lapiere, Proc. Natl. Acad. Sci. USA, 48, 1014 (1962).
69. D.E. Woolley, R.W. Glanville, D.R. Roberts and J.M. Evanson, Biochem. J., 169, 265 (1978).
70. L.A. Liotta, S. Abe, P. Gehron-Robey and G.R. Martin, Proc. Natl. Acad. Sci. USA, 76, 2268 (1979).
71. S. Krane, Conn. Tiss. Res., 10, 51 (1982).
72. H.G. Welgus, J.J. Jeffrey and A.E. Eisen, J. Biol. Chem., 256, 9511 (1981).
73. A.L. Horwitz, A.L. Hance and R.G. Grystal, Proc. Natl. Acad. Sci. USA, 74, 897 (1977).
74. D. Tarin, B.J. Hoyt and D.J. Evans, British J. Cancer, 46, 266 (1982).
75. G. Wirl, Conn. Tiss. Res., 5, 171 (1977).
76. J. Gross, D. Moscatelli, E. Jaffe and D. Rifkin, J. Cell. Biol., 95, 974 (1982).
77. Y. Yamanishi, M.K. Dabbous, K. Hashimoto, Cancer Res., 32, 2551 (1972).
78. P.A. McCroskery, J.F. Richards, E.D. Harris, Jr., Biochem. J., 152, 131 (1975).
79. Y. Yamanishi, E. Maeyens, M.K. Dabbous, H. Ohyama and K. Hashimoto, Cancer Res., 33, 2507 (1973).
80. T.E. Cawston and G. Murphy, Meth. Enzymol., 80, 711 (1981).
81. J. Gross and Y. Nagai, Proc. Natl. Acad. Sci. USA, 54, 1197 (1965).
82. A.Z. Eisen, J.J. Jeffrey and J. Gross, Biochim. Biophys. Acta, 151, 673 (1968).
83. T. Salo, L.A. Liotta and K. Tryggvason, J. Biol. Chem., 258, 3058 (1983).
84. T. Kalebic, S. Garbisa, B. Glaser and L.A. Liotta, Science, 221, 281 (1983).
85. K. Tryggvason, P. Gehron-Robey and G.R. Martin, Biochemistry 19, 1284 (1980).
86. B.D. Pauli, D.E. Schwartz, E.J.-M. Thonar and K.E. Kuettner, Cancer Metast. Rev., 2, 129 (1983).
87. R.L. Carter, J. Clin. Pathol., 35, 1041 (1982).
88. L.L. Mainardi, J.M. Seyer and A.H. Kang, Biochem. Biophys. Res. Commun., 97, 1108 (1980).
89. G.P. Stricklin, E.A. Bauer, J.J. Jeffrey and A.Z. Eisen, Biochemistry, 16, 1607 (1977).
90. A. Sellers, E. Cartwright, G. Murphy and J.J. Reynolds, Biochem. J., 167 353 (1977)
91. G. Vaes, Biochem. J., 126, 275 (1972).
92. E. Harper, J. Gross, Biochem. Biophys. Res. Commun., 48, 1147 (1972).
93. M. Paranjpe, L. Engel, N. Young and L.A. Liotta, Life Sci., 26, 1223 (1980).
94. G. Murphy, E.C. Cartwright, A. Sellers and J.J. Reynolds, Biochim. Biophys. Acta, 483, 493 (1977).
95. T.E. Cawston, W.A. Galloway, E. Mercer, G. Murphy and J.J. Reynolds, Biochem. J., 195 159 (1981).
96. H.G. Welgus, G.P. Stricklin, A.Z. Eisen, E.A. Bauer and J.J. Jeffrey, J. Biol. Chem. 254, 1938 (1979).
97. J.C. Nolan, S.C. Ridge, A.L. Oronsky and S.S. Kerwar, Atherosclerosis 35, 93, (1980).
98. S.S. Kerwar, J.C. Nolan, S. Ridge, A.L. Oronsky and L.L. Slakey, Biochim. Biophys. Acta 632, 183 (1980).
99. C.A. Vater, C.L. Mainardi and E.D. Harris Jr., J. Biol. Chem. 254, 3045 (1979).
100. A. Sellers and G. Murphy, Int. Rev. Conn. Tiss. Res., 9, 151 (1981).
101. M. Steinbuch and R. Audran, in "Protease Inhibitors", H. Fritz, H. Tschesche, L.J. Grune and E. Truscheit, Eds., Bayer Symp. Springer-Verlag, Berlin and New York, (1974) p. 78.
102. P.A. Ward, C.T. Cochrane and H.J. Müller-Eberhart, J. Exp. Med. 122, 327 (1965).
103. E. Schiffmann, B.A. Corcoran and S.M. Wahl, Proc. Natl. Acad. Sci. USA, 72, 1059 (1975).
104. W. Orr, J. Varani, M.D. Gondek, P.A. Ward and G.R. Mundy, Science, 203, 176 (1979).
105. J.C. Boversox and N. Sorgente, Cancer Res., 42, 2547 (1982).
106. D. Woodley, L.A. Liotta and R. Brundage, Clinical Res., 30, 266A (Abstr.) (1982).
107. A.E. Postlethwaite, L.B. Lachman, C.L. Mainardi and A.H. Kang, J. Exp. Med., 157, 801 (1983).
108. K.-J. Valle and E.A. Bauer, J. Biol. Chem., 254, 10115 (1979).

Chapter 24. Biology of Leukotrienes

William Kreutner and Marvin I. Siegel
Schering-Plough Corporation
Bloomfield, New Jersey 07003

Introduction - Leukotrienes are potent mediators of allergic, inflammatory and other pathologic events. Studies with murine mastocytoma cells,[1,2] rat basophilic leukemia cells,[3,4] guinea pig lung[5] and human lung[6] have elucidated the biosynthetic pathway of slow-reacting substance of anaphylaxis (SRS-A), and identified it as leukotriene C_4 (LTC_4, 1) and/or leukotriene D_4 (LTD_4, 2).

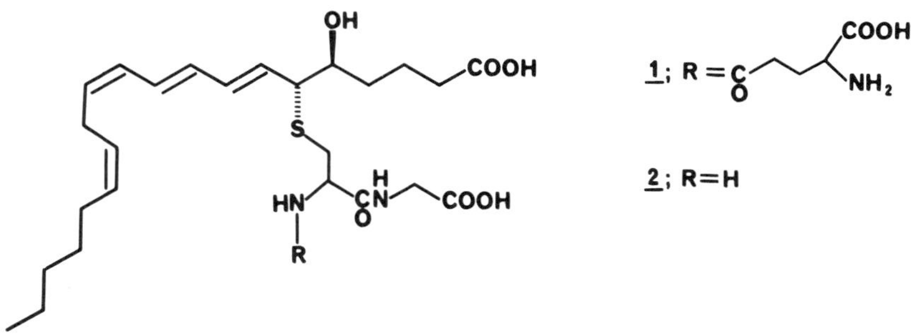

Principal steps in sulfidopeptide leukotriene biosynthesis involve the release of arachidonic acid (AA) from membrane phospholipids in response to various cellular signals, the addition of molecular oxygen to AA to form 5-hydroperoxyeicosatetraenoic acid (5-HPETE) by a 5-lipoxygenase, dehydration of 5-HPETE to an unstable epoxytriene, LTA_4, and addition of the thiol group of glutathione across the epoxide ring to give LTC_4. LTC_4 is further metabolized to LTD_4 by removal of glutamate which, in turn, yields LTE_4 by a loss of glycine. LTA_4 can also be enzymatically hydrated to LTB_4 (5S,12R-dihydroxy-6Z,8E,10E,14Z-eicosatetraenoic acid) or hydrolyzed non-enzymatically to the diastereoisomers, 5S,12R,-dihydroxy-6E,8E,10E,14Z-eicosatetraenoic acids (5,12-diHETEs). Other important lipoxygenase pathways yield 15-HPETE in neutrophils, eosinophils and reticulocytes [7-9] and 12-HPETE in platelets and lymphocytes.[10,11] These hydroperoxides are reduced by peroxidases to the corresponding hydroxy fatty acids (HETEs).

Various inflammatory cells including leukocytes,[12,13] eosinophils,[14,15] macrophages[16,17] and mast cells[18,19] from several species, including humans, produce leukotrienes from either added AA or from endogenous AA in the presence of specific stimulants such as the chemotactic peptide, F-met-leu-phe,[20] platelet-activating factor (PAF),[21] phorbol myristyl acetate (PMA)[22] and IgE directed antigens.[23,24] Leukotrienes then produce a physiological response by interacting with specific, high affinity, receptors. Anti-inflammatory steroids inhibit AA release while aspirin and other non-steroidal anti-inflammatory drugs (NSAID) block the prostaglandin pathway.[25] As yet, there is no approved drug, which is an effective inhibitor of leukotriene biosynthesis or an antagonist of leukotriene action.

ISBN 0-12-040519-9

Effects of Leukotrienes on Respiratory Function

1) In Vivo Actions - The *in vivo* pulmonary response to leukotrienes varies among species. Rats are often resistant to the effects of leukotrienes, but aerosolized LTD_4 causes a bronchospasm in an inbred strain with bronchial hyperresponsiveness.[26] Basenji-Greyhound dogs, an inbred strain with hyperreactive airways, develop an atropine-inhibitable bronchospasm to inhaled LTD_4 while there is no response in mongrel dogs.[27] Furthermore, aerosolized LTD_4 is 300-900 times more potent than histamine in causing a bronchospasm in Ascaris-sensitive rhesus monkeys, whereas non-allergic monkeys are relatively non-responsive.[28] The bronchial response in monkeys to aerosol LTD_4 is inhibited by inhaled FPL-55712 while intravenous FPL-55712 is inactive.[29] Administered into the right atria of monkeys, LTC_4 causes a bronchospasm and a transient increase in atrial pressure followed by a prolonged hypotension.[30]

In healthy humans, inhaled LTC_4 and LTD_4 are more potent than histamine at causing a decrease in maximum expiratory flow rate.[31,32] The effect of LTD_4 is independent of the production of cyclooxygenase products. In contrast to airway hypersensitivity to histamine among human asthmatics, there is no increased sensitivity to inhaled LTD_4.[33]

The importance of cyclooxygenase products, particularly thromboxane A_2, in the bronchospastic response to leukotrienes depends on the species and route of administration. The bronchospasm resulting from intravenous LTC_4 in guinea pigs[34] and LTD_4 in cats[35] is blocked by cyclooxygenase inhibitors. The response to inhaled LTC_4 in guinea pigs is either not inhibited or is enhanced by cyclooxygenase blockade.[34,36]

2) In Vitro Actions - Studies using a superfusion technique suggest that a major portion of the contractile response of guinea pig lung parenchymal strips to LTC_4 and LTD_4 is due to the generation of thromboxane A_2.[37] In contrast, rabbit and rat parenchymal strips do not develop a marked contractile response or synthesize thromboxane A_2 in response to LTC_4 or LTD_4. More recent work has shown that under non-flow conditions, the contractile responses of guinea pig parenchyma to LTC_4 and LTD_4 are independent of thromboxane A_2 generation.[38,39] LTB_4 is 5-10 times more potent than histamine in contracting guinea pig lung strips.[40] Intravenous or aerosol administration of LTB_4 to guinea pigs produces a bronchospasm.[41] These *in vitro* and *in vivo* effects of LTB_4 are secondary to the generation of cyclooxygenase products.

Conversion of LTC_4 to LTD_4 and subsequent blockade of the contractile response by FPL-55712 during incubation with guinea pig trachea has been reported.[42] This LTD_4-induced contraction may utilize predominantly intracellular calcium stores because it is inhibited by TMB-8, an inhibitor of intracellular calcium mobilization, but not by nifedipine.[43] In contrast, on the guinea pig ileum, the slowness of the contractile response to LTD_4, compared to histamine, and the sensitivity to blockade by a slow calcium channel blocker, D-600, suggests that LTD_4 utilizes extracellular calcium.[44]

The Merck-Frosst group prepared the sulfones of LTC_4, LTD_4 and LTE_4 and found that they are almost equipotent to the sulfides on contracting guinea pig trachea and intravenous administration to guinea pigs produces an indomethacin and FPL-55712 inhibitable bronchospasm.[45,46] The contractile activity of LTC_4 and LTD_4 analogs has been examined. The rank order of potency is similar on guinea pig trachea, lung parenchyma and ileum where the 5R, 6S isomers have 0.01-times the potency of the natural

5S, 6R isomers.[47,48] Eight biosynthetically formed sulfidopeptide leukotrienes (LTC_3, 8,9-LTC_3, LTC_4, 11-trans LTC_4, LTC_5, LTD_4, LTE_4 and 11-trans LTE_4) were studied for contractile activity on guinea pig lung parenchyma.[49] All compounds are full agonists and 5000-times more potent than histamine. There is no potency difference between 11-cis and 11-trans isomers of LTC_4 or LTE_4. Incubation of LTE_4 with glutathione and gamma-glutamyltranspeptidase results in the formation of LTF_4.[50] LTF_4 is equipotent to LTD_4 on guinea pig trachea, but only 0.01-times as potent on ileum and *in vivo*.[51]

Purified human lung mast cells release SRS-A upon stimulation with anti-IgE.[52] Eosinophils (horse) release leukotrienes upon stimulation with ionophore A-23187 [53] and platelets (rabbit) release leukotrienes on stimulation with thrombin or PAF.[54] Rat alveolar macrophages [55] and human peritoneal macrophages [56] stimulated with A-23187 release LTC_4 and LTB_4, but not LTD_4. The production of LTB_4 by human alveolar macrophages has also been demonstrated.[57] IgE mediated release of SRS-A occurs with alveolar but not peritoneal rat macrophages.[55] Sensitized human lung challenged with antigen, releases sufficient amounts of LTC_4, D_4 and E_4 to cause contraction of human bronchi *in vitro*.[58]

A major finding has been the importance of lipoxygenase products in the secretion of airway mucus. Mucus glycoprotein secretion from human bronchi in culture is stimulated by low concentrations (1-10 nM) of various HETEs, with 12- and 15-HETE most active.[59,60] Antigen-provoked mucus release is inhibited by the lipoxygenase inhibitors ETYA, nordihydroguiaretic acid and alpha naphthol.[59] A recent report claims a significant decrease in mucociliary clearance after oral ingestion of aspirin by healthy volunteers.[61] This effect presumably is due to an augmented production of lipoxygenase products after cyclooxygenase inhibition. Other reports have shown increased secretion of mucus after intraarterial injection of LTC_4 or LTD_4 into dogs,[62-64] intratracheal administration of LTC_4 into cats [65] or *in vitro* addition of LTC_4 or LTD_4 to human bronchial mucosa.[66]

Leukotriene Actions in the Gastrointestinal System - Large species differences have been found in the response of gastrointestinal tissue to sulfidopeptide leukotrienes. Human gastrointestinal muscle (ileum, stomach, jejunum, colon) does not contract to SRS-A from guinea pig lung.[67] In contrast, both LTC_4 and LTD_4 contract the rat stomach and colon but not the duodenum or ileum.[68] LTC_4, D_4 and E_4 produce cyclooxygenase independent contractions of the guinea pig gall bladder.[69] At high concentrations (10^{-5}M), LTC_4 and LTD_4 also provoke acid secretion from isolated rabbit gastric parietal cells.[70] A role for leukotrienes in inflammatory bowel disease is suggested by the release of SRS-A from antigen challenged sensitized colonic mucosa.[71]

Lipoxygenase products may also be essential for insulin release by pancreatic islets. Glucose-induced insulin release is inhibited by lipoxygenase inhibitors, including BW-755C, nordihydroguairetic acid and 15-HETE.[72,73] Also, 5-HETE augments insulin release triggered by low concentrations of glucose.[73] Rat pancreatic islets produce 5-, 12- and 15-HETE from arachidonic acid.[73]

Actions of Leukotrienes on Cardiovascular Tissue - Purified SRS-A generated from guinea pig lung has been shown to increase vascular permeability in guinea pig skin, especially if injected together with a vasodilator like PGE_2.[74] Intradermal LTC_4 and LTD_4 cause a wheal and flare[75,76] and increase microvascular blood flow [77] in human skin. In a hamster cheek pouch preparation, low concentrations of LTC_4 and LTD_4 cause intense arteriolar constriction followed by a marked increase in vascular permeability.[78] Airway edema associated with an anaphylactic reaction may be a leukotriene dependent response, since local application of LTC_4, D_4 or E_4 to guinea pig trachea _in vivo_ produces an increase in microvascular permeability.[79]

Low concentrations of LTD_4 (1-10nM) cause a reversible cessation of beating of rat heart cells in culture.[80] Evidence was presented by Burke _et al_. for a role of leukotrienes in cardiac anaphylaxis.[81] The cardiac dysfunction associated with anaphylaxis is partially duplicated by addition of LTC_4, D_4 or E_4. All leukotrienes produce a negative inotropic effect, decrease coronary flow of isolated guinea pig heart, and decrease the contractile force of pectinate muscle from human heart. LTD_4 also potentiates the tachycardia and arrhythmias caused by histamine. Sulfidopeptide leukotrienes also appear to be potent contractile agents on canine renal arteries[82] and rat and cat coronary arteries.[83] These effects are independent of thromboxane A_2 generation.

The _in vivo_ effects of leukotrienes on blood pressure and heart rate are quite complex. Generally, there is a biphasic response consisting of an initial transient increase in systemic blood pressure followed by a sustained systemic hypotension after a bolus of 1-10 µg/kg of LTC_4 or LTD_4.[84-86] Intravenous LTC_4 decreases cardiac output and rate while increasing total peripheral resistance.[85] Pretreatment with indomethacin potentiates the initial increase in blood pressure and attenuates the sustained hypotension and changes in cardiac function due to LTC_4. Low doses of LTD_4 injected into the left circumflex coronary artery of sheep result in coronary vasoconstriction and impaired ventricular function.[87]

Leukotrienes may also play a pathogenic role in hypoxic pulmonary hypertension. Hypoxic pulmonary vasoconstriction in sheep is prevented or reversed by FPL-55712.[88] LTC_4 and LTD_4 are present in lung lavage fluid obtained from human newborns with a clinical diagnosis of persistent pulmonary hypertension and hypoxemia.[89]

<u>Leukotrienes in Inflammatory Disorders</u> - Leukotrienes are produced by cells (PMNs and macrophages) that are present in large numbers at inflammatory sites.[90] If stimulated with the calcium ionophore A-23187, PMNs produce LTB_4, which has potent chemokinetic activity and is also an aggregating agent for neutrophils.[91] The potency of LTB_4 as a chemotactic agent is similar to that of the complement component C_{5a} or the synthetic peptide F-met-leu-phe.[92-94]

In a hamster cheek pouch preparation, LTB_4 causes adhesion of leukocytes to vascular endothelium.[76,78] Evidence has also been presented that LTB_4 is eosinophil chemotactic factor (ECF).[95] However, ECF stimulates the migration of eosinophils, but not neutrophils while LTB_4 is equally chemotactic for both cell types.[96] The neutrophil chemotactic activity of LTB_4 is blocked competitively by acetyl LTB_4.[97] Intradermal injection of LTB_4 (0.2-1.5 nmole) into human skin causes an immediate erythema and wheal followed by a delayed reaction of erythema and induration at 1-4 hr with perivascular neutrophil infiltrates.[76]

LTB_4 has been shown to cause the release of enzymes from human PMNs.[98,99] Compared to C_{5a} and F-met-leu-phe, LTB_4 is a weak secretogogue and is active only at concentrations 10-fold greater than needed to stimulate chemotaxis or adherence of neutrophils. Neutrophil degranulation caused by LTB_4 is inhibited by LTB_4-dimethylamide (K_D = 0.2 μM).[100] In addition, it has recently been reported that LTB_4 may have immunomodulatory activity.[101-105] For example, LTB_4 appears to cause specific suppression of human lymphocyte function, possibly by inducing suppressor cells. Moreover, LTB_4 augments natural cytotoxic cell activity. These latter activities may play important roles in chronic inflammatory disease. Injection of LTB_4 into rabbit skin,[106] rabbit eye,[107] guinea pig peritoneal cavity[94] and human skin [108] results in leukocyte accumulation at the injection site. In a hamster cheek pouch preparation, the movement of leukocytes through the vascular endothelium as a result of the local application of LTB_4 has been directly observed.[106] The increase in vascular permeability due to LTB_4 is augmented by the co-administration of a vasdilator prostaglandin and is dependent upon the presence of neutrophils.[109]

The role of LTB_4 as an inflammatory mediator is supported by studies showing high amounts of LTB_4 in the skin chamber fluid of involved skin from psoriatics [110] and in gouty effusions.[111] In contrast, LTC_4 and LTD_4, but not LTB_4, are detected by bioassay and HPLC in carrageenan-induced pleurisy in rats.[112] Leukotrienes may also be involved in the modulation of pain in inflammatory lesions. In the rat paw, LTB_4 causes slight hyperalgesia.[113] However, the mechanism of action of this response is not understood.

Leukotrienes C_4, D_4 and E_4 may also be important mediators of inflammatory processes. LTD_4 and LTE_4 influence vasopermeability in guinea pig skin and their activities are potentiated by the presence of prostaglandin vasodilators.[114] In addition, LTC_4 and LTD_4 increase neutrophil adherance and may also affect release reactions of macrophages.[115]

Elevated levels of both LTB_4 and LTC_4 are found in psoriatic lesions in man.[114,116] Ulcerative colitis has among its characteristic features an accumulation of neutrophils in mucosal sites. Samples of tissue from inflamed sites in such patients contain elevated levels of LTB_4.[117] In addition, these samples have an increased ability to synthesize 5-lipoxygenase products compared to normal tissue. Synovial fluids from rheumatoid arthritic or gouty patients also contain large numbers of neutrophils and elevated levels of LTB_4.[90,111]

Receptor Binding of Leukotrienes - Initial reports suggested the presence of tissue specific heterogeneity of LTD_4 receptors.[118] This is supported by more recent experiments on guinea pig trachea utilizing FPL-55712.[119] The availability of high specific activity radiolabeled LTC_4 and LTD_4 has permitted the detection of specific binding sites in lung for these ligands. [^{3}H]-LTD_4 exhibits high affinity (K_D= 10^{-10}M), saturable binding to membranes of guinea pig lung.[120] Binding is enhanced by divalent cations (Ca^{2+}, Mg^{2+}, Mn^{2+}) and inhibited by Na^+. LTE_4, but not LTC_4, has high affinity for the binding site suggesting that LTC_4 and LTD_4 interact at different sites. High affinity binding of [^{3}H]-LTC_4 was also demonstrated on lungs from rat [121] and guinea pig [122,123] and on an intact smooth muscle cell line, DDT_1 MF-2, derived from syrian hamster vas deferens.[124] The LTC_4 binding is biochemically distinct from LTD_4 binding sites.[120,122] The K_D for FPL-55712 is greater against LTC_4 than LTD_4. Binding of LTD_4, but not LTC_4 is altered in the presence of guanine nucleotides.

Evidence has also been presented that LTB_4 interacts with a specific cell receptor.[93-125] The difference in chemotactic and secretogogue activity of LTB_4 has been explained in terms of the former being associated with high affinity receptors and the latter with low affinity receptors.[126]

Pharmacology of Leukotriene Biosynthesis Inhibitors and Antagonists

1) Biosynthesis Inhibitors - BW-755C (3) and phenidone (4) are inhibitors of the lipoxygenase and cyclooxygenase pathways[127-128] and inhibit *in vitro* anaphylactic contractions of airway smooth muscle from guinea pigs.[127-129] BW-755C inhibits SRS-A, but not histamine, release from antigen challenged sensitized human lung[130] and when given by inhalation, attenuates antigen-induced bronchospasm in Ascaris-sensitive monkeys.[131]

BW-755C inhibits prostaglandin synthesis in inflammatory exudates but does not inhibit PGI_2 synthesis by gastric mucosa.[132] BW-755C also reduces the concentration of LTB_4, thromboxane B_2 and PGE_2 in exudates derived from the subcutaneous implantation of carrageenan impregnated sponges in rats.[133] PMN migration into the inflammatory exudate is also decreased. BW-755C, but not indomethacin, reduces the size of a myocardial infarct produced in dogs by coronary occlusion followed by reperfusion.[134] The mechanism may involve inhibition of lipoxygenase to reduce the production of LTC_4 and LTD_4 which constrict coronary arteries and of LTB_4 which is chemotactic for inflammatory cells.

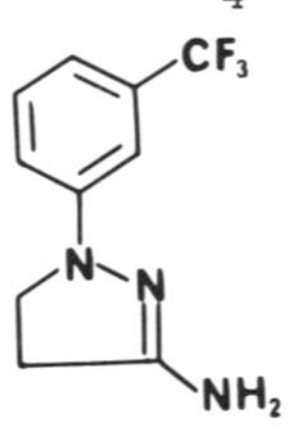

3

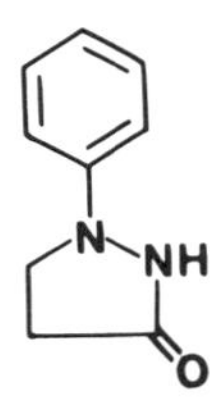

4

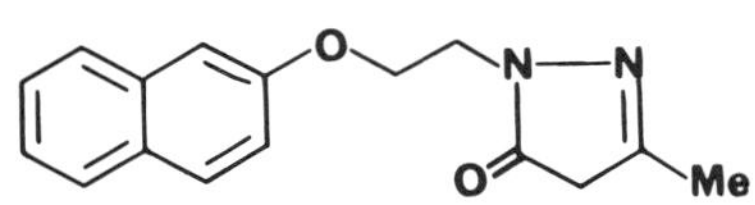

5

At concentrations lower than those needed to inhibit either lipoxygenase or cyclooxygenase, BW-755C enhances lymphocyte activation by mitogens.[135] BW-755C and phenidone, applied topically to mouse skin, prevent the induction of epidermal ornithine decarboxylase caused by application of the tumor promoter 12-0-tetra-decanoyl phorphol-13-acetate.[136]

Nafazatrom (5) inhibits 5- and 12-lipoxygenases but not the cyclooxygenase of B16a tumor cells and also inhibits tumor proliferation.[137] Other studies have shown nafazatrom increases PGI_2 synthesis by aortic strips and has antithrombotic activity.[138]

Benoxaprofen (6) has antiinflammatory activity and also inhibits the release of SRS-A.[139] At a concentration of 100 μM, benoxaprofen inhibits a 5-lipoxygenase in guinea pig peritoneal cells and HL-60 cells with no effect on the 12-lipoxygenase of human platelets or a soybean 15-lipoxygenase.[140] In rat PMNs, benoxaprofen is a better inhibitor of cyclooxygenase than 5-lipoxygenase. Its *in vivo* activity in animal models is consistent with these results, because inhibition of edema occurs at lower doses than inhibition of cellular influx to inflamed sites. Benoxaprofen is claimed to be ineffective clinically to inhibit aspirin-induced bronchospasm[141] but has shown beneficial activity in the treatment of psoriasis,[142-143] a disease associated with the presence of increased levels of lipoxygenase products in epidermal lesions.

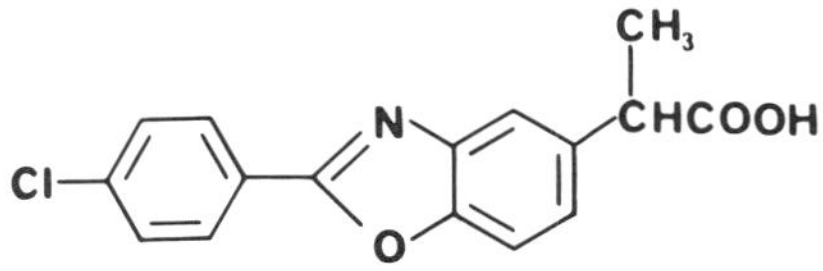

6

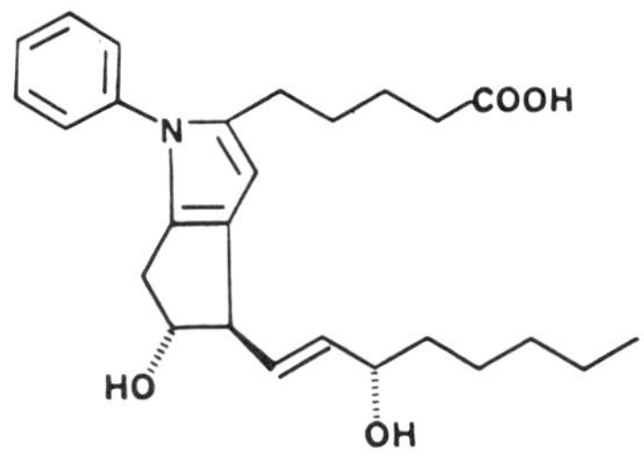

7

U-60,257 (7) is an inhibitor of glutathione-S-transferase[144] but its principal action seems to be inhibition of a 5-lipoxygenase.[145] U-60,257 inhibits SRS, but not histamine, release from human lung, competitively antagonizes the contactile effects of LTC_4 and LTD_4 on guinea pig ileum and inhibits lysosomal enzyme release from human PMNs. In Ascaris sensitive monkeys, U-60,257 given by aerosol (0.05-1%) or intravenously (0.01-5 mg/kg) inhibits antigen-induced bronchoconstriction.[146]

2) Leukotriene Antagonists - FPL-55712 (8) consistently inhibits contractile responses to LTD_4, but its effect on responses to LTC_4 are less clear. Krell et al. claim that FPL-55712 does not antagonize LTC_4 on guinea pig parenchymal strips.[147] On trachea and bronchus, FPL-55712 antagonizes responses to low concentrations of LTC_4, but potentiates responses to high concentrations of LTC_4. Krell also showed that antigen-induced bronchospasm in dogs is not inhibited by i.v. or aerosol FPL-55712.[148]

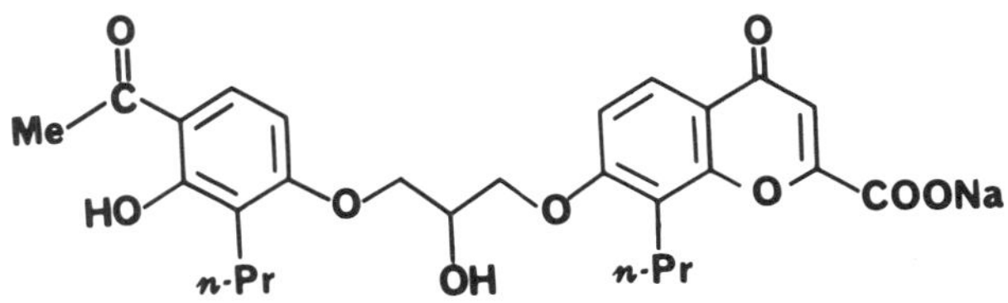

8

Limited studies with FPL-55712 have been conducted in humans. Inhaled FPL-55712 in 4 allergic asthmatics gave inconclusive results.[149] Two of the four patients showed improved FEV_1. Tracheal mucous velocity is decreased in asthmatics challenged with antigen.[150] Inhalation of 0.5-1% FPL-55712 prevents the decrease in mucous velocity but does not inhibit the bronchospasm. Inhalation of LTC_4 or LTD_4 by normal subjects induces coughing which is blocked by aerosol FPL-55712.[151] In Basenji-Greyhound dogs, the bronchospasm to inhaled citric acid is associated with increased plasma levels of SRS (but not histamine) and is partly blocked by FPL-55712.[152]

Much of the pharmacology of FPL-55712 is ascribed to its antagonism of SRS-A; however, at concentrations only slightly higher than those needed to inhibit SRS-A, FPL-55712 also inhibits thromboxane synthetase[153] and lipoxygenase in broken cell, but not intact cell preparations.[154] FPL-55712 also inhibits the extensive necrosis of liver parenchymal cells and death in animals injected with endotoxin and D-galactosamine.[155]

4R,5S,6Z-2-nor LTD_1 (4R-hydroxyl-5S-cysteinylglycyl-6Z-nonadecenoic acid) is an analog of LTD_4 which at a concentration of 100 μM, antagonizes the contractile response to LTD_4, LTC_4 and LTE_4 on guinea pig airways.[156,157] The effects of histamine are not blocked. The vasoconstrictor effects of LTD_4 on guinea pig pulmonary artery are also blocked. In vivo, an intravenous dose of 5 mg/kg inhibits the bronchoconstrictor response to LTD_4 given 1 min later. 4R,5S,6Z-nor LTE_1 also has antagonist activity while 4R,5S,6Z-nor LTC_1 and 4S,5R,6E-nor LTD_1 are agonists.

Summary - The leukotrienes are a group of biologically active mediators derived from arachidonic acid. While LTB_4 is a potent chemotactic agent and may mediate inflammatory reactions, LTC_4 and LTD_4 are predominantly smooth muscle contractile agents with an implied rolé in allergic diseases. The availability of pharmacological agents with in vivo activity that inhibit the biosynthesis or actions of leukotrienes (Chapter 11, Pulmonary and Antiallergy Agents) will greatly facilitate our understanding of the role of leukotrienes in pathological conditions. Steroids with antiinflammatory activity prevent the release of arachidonic acid and, thereby, inhibit the subsequent generation of both cyclooxygenase and lipoxygenase products. Some of the effects of steroids not shared by aspirin-like drugs may, therefore, be due to an inhibition of leukotriene formation.

References

1. R.C. Murphy, S. Hammarstrom and B. Samuelsson, Proc. Natl. Acad. Sci. USA, 76, 4275 (1979).
2. S. Hammarstron, B. Samuelsson, D.A. Clark, G. Goto, A. Marfat, C. Miokowski and E.J. Corey, Biochem. Biophys. Res. Commun., 92, 946 (1980).
3. H.R. Morris, G.W. Taylor, P.J. Piper, M.N. Samhoun and J.R. Tippins, Prostaglandins, 19, 185 (1980).
4. L. Orning, S. Hammarstrom and B. Samuelsson, Proc. Natl. Acad. Sci. USA 77, 2014 (1980).
5. H.R. Morris, G.W. Taylor, P.J. Piper and J.R. Tippins, Nature, 285, 104 (1980).
6. R.A. Lewis, K.F. Austen, J.M. Drazen, D.A. Clark, A. Marfat and E.J. Corey, Proc. Natl. Acad. Sci. USA 77, 3710 (1980).
7. J.Y. Vanderhoek, R.W. Bryant and J.M. Bailey, J. BIol. Chem. 255, 10065 (1980).
8. J. Turk, R.L. Maas, A.R. Brash, L.T. Roberts and J.A. Oates, J. Biol. Chem., 257, 7068 (1982).
9. R.W. Bryant, J.M. Bailey, T. Schewe and S.M. Rappoport, J. Biol. Chem., 257, 6050 (1982).
10. D.H. Nugteren, Biochem. Biophys. Acta, 380, 299 (1975).
11. J.M. Bailey, R.W. Bryant, C.E. Low, M.B. Pupillo and J.Y. Vanderhoek, Cell Immunol., 67, 112 (1982).
12. B. Samuelsson and S. Hammarstrom, Prostaglandins, 19, 645 (1980).
13. P. Borgeat and B. Samuelsson, Proc. Natl. Acad. Sci. USA, 17, 785 (1979).
14. R.L. Mass, A.R. Brash and J.A. Oates, Proc,. Natl. Acad. Sci. USA, 78, 5523 (1981).
15. H.J. Ziltener, P.A. Chavaillaz and A. Jorg, Hoppe-Seyler's Z. Physiol. Chem., 364, 1029 (1983).
16. C.A. Rouzer, W.A. Scott, A.L. Hamill and Z.A. Cohn, J. Exp. Med., 155, 720 (1982).
17. M. Damon, C. Chavis, G. Ph., F.B. Michel and A. Crastes de Paulet, Biochem. Biophys. Res., Commun., 111, 518 (1983).
18. E. Razin, J.M. Mencia-Huerta, R.A. Lewis, E.J. Corey and K.F. Austen, Proc. Natl. Acad. Sci. USA, 79, 4665 (1981).
19. D.W. MacGlashan, R.P. Schleimer, S.P. Peters, E.S. Schulman, G.K. Adams, H.H. Newball and L.M. Lichtenstein, J. Clin. Invest., 70, 747 (1982).
20. W. Jubiz, O. Radmark, C. Malmsten, G. Hansson, J.A. Lindgren, J. Palmbald, A.M. Uden and B. Samuelsson, J. Biol. Chem., 257, 6106 (1982).
21. N.F. Voelkel, S. Worthen, J.T. Reeves, P.M. Henson and R.C. Murphy, Science, 218, 286 (1982).
22. J.L. Humes, S. Sadowski, M. Galavage, M. Goldenberg, E. Subers, R.J. Bonney and F.A. Kuehl, J. Biol. Chem. 257, 1591 (1982).
23. C.A. Rouzer, W.A. Scott, A.L. Hamill, F.T. Liu, D.H. Katz and Z.A. Cohn, J. Exp. Med., 156, 1077 (1982).
24. E. Razin, J.M. Mencia-Huerta, R.L. Stevens, R.A. Lewis, F.T. Liu, E.J. Corey and K.F. Austen, J. Exp. Med.,157, 189 (1983).
25. B. Samuelsson, Science, 220, 568 (1983).
26. G. Burnet, H. Piechuta, R. Hamel, G. Holme and A.W. Ford-Hutchinson, J. Immunol., 131, 434 (1983).

27. C.A. Hirshman, M. Darnell, T. Brugman and J. Peters, Prostaglandins, 25, 481 (1983).
28. R. Patterson, K.E. Harris, L.J. Smith, P.A. Greenberger, M.A. Shaughnessy, P.R. Bernstein and R.D. Krell, Int. Arch. Allergy Appl. Immunol., 71, 156 (1983).
29. J.C. Williams, The Pharmacologist, 25, 234 (1983).
30. G. Smedegard, P. Hedqvist, S.E. Dahlen, B. Revenas, S. Hammarstrom and B. Samuelsson, Nature, 295, 327 (1982).
31. J.W. Weiss, J.M. Drazen, N. Coles, E.R. McFadden, P.F. Weller, E.J. Corey, R.A. Lewis and K.F. Austen, Science 216, 196 (1982).
32. J.W. Weiss, J.M. Drazen, E.R. McFadden, P. Weller, E.J. Corey, R.A. Lewis and K.F. Austen, J. Am. Med. Assoc., 249, 2814 (1983).
33. M. Griffin, J.W. Weiss, A.G. Leitch, E.R. McFadden, E.J. Corey, K.F. Austen and J.M. Drazen, N. Eng. J. Med., 308, 436 (1983).
34. S.E. Dahlen, Acta Physiol. Scand., 118, 415 (1983).
35. G.B. Graybar, K.H. Cowen, E.W. Spannhake, A.L. Hyman and P.J. Kadowitz, Circulation, 68, 400 (1983).
36. A.G. Leitch, E.J. Corey, K.F. Austen and J.M. Drazen, Am. Rev. Resp. Dis., 128, 639 (1983).
37. P.J. Piper and M.N. Samhoun, Prostaglandins, 21, 793 (1981)
38. K.F. Austen, E.J. Corey, J.M. Drazen and A.G. Leitch, Brit. J. Pharmac., 80, 47 (1983).
39. S.E. Dahlen, P. Hedqvist, P. Westlund, E. Granstrom, S. Hammarstrom, J.A. Lindgren and O. Radmark, Acta Physiol. Scand., 118, 393 (1983).
40. P. Sirois, P. Borgeat, H. Jeanson, S. Roy and G. Girard, Prostaglandins and Medicine, 5, 429 (1980).
41. R. Hamel and A.W. Ford-Hutchinson, Prostaglandins, 25, 405 (1983).
42. D.W. Snyder, The Pharmacologist, 25, 541 (1983).
43. S.S. Tucker, The Pharmacologist, 25, 110 (1983).
44. S.R. Findlay, T.M. Leary, C.W. Parker, E.I. Bloomquist and C.R. Scheid, Am. J. Physiol. 243, C133 (1982).
45. Y. Girard, M. Larue, T.R. Jones and J. Rokach, Tetrahedron Letters, 23, 1023 (1982).
46. T. Jones, P. Masson, R. Hamel, G. Brunet, G. Holme, Y. Girard, M. Larue and J. Rokach, Prostaglandins, 24, 279 (1982).
47. S. Okuyana, S. Miyamoto, K. Shimoji, Y. Konishi, D. Fukushima, H. Niwa, Y. Arai, M. Toda and M. Hayashi, Chem. Pharm. Bull., 30, 2453 (1982).
48. S.R. Baker, J.R. Boot, W.B. Jamieson, D.J. Osborne and W.J.F. Sweatman, Biochem. Biophys. Res. Commun., 103, 1258 (1981).
49. S.E. Dahlen, P. Hedqvist and S. Hammarstrom, Eur. J. Pharmac., 86, 207 (1983).
50. K. Bernstroem and S. Hammarstrom, Biochem. Biophys. Res. Commun., 109, 800 (1982).
51. D. Denis, S. Charleston, A. Rackham, T.R. Jones, A.W. Ford-Hutchinson, A. Lord, M. Cirino, Y. Girard, M. Larue and J. Rokach, Prostaglandins, 24, 801 (1982).
52. D.W. MacGlashan, R.P. Schleimer, S.P. Peters, E.S. Schulman, G.K. Adams, H.H. Newball and L.M. Lichtenstein, J. Clin. Invest., 70, 747 (1982).
53. A. Jorg, W.R. Henderson, R.C. Murphy and S.J. Klebanoff, J. Exp. Med., 155, 390 (1982).
54. J.M. Mencia-Huerta, L. Hadji and J. Benveniste, J. Clin. Invest., 68, 1586 (1981).
55. J.A. Rankin, M. Hitchcock, W. Merrill, M.K. Bach, J.R. Brashler and P.W. Askenase, Nature, 297, 329 (1982).
56. J.T. Du, M. Foesh, Y. Maddox and P.W. Ramwell, Biochem. Biophys. Acta, 753, 159 (1983).
57. A.O.S. Fels, N.A. Pawlowski, E.B. Cramer, T.K.C. King, Z.A. Cohn and W.A. Scott, Proc. Natl. Acad. Sci. USA, 79, 7866 (1982).
58. S.E. Dahlen, G. Hansson, P. Hedqvist, T. Bjorck, E. Granstrom and B. Dahlen, Proc. Natl. Acad. Sci. USA, 80, 1712 (1983).
59. Z. Marom, J.H. Shelhamer and M. Kaliner, J. Clin. Invest., 67, 1695 (1981).
60. Z. Marom, J.H. Shelhamer, F. Sun and M. Kaliner, J. Clin. Invest. 72, 122 (1983).
61. T.R. Gerrity, E. Cotromanes, C.S. Garrard, D.B. Yeates and R.V. Lourenco, N. Eng. J. Med., 308, 139 (1983).
62. H.G. Johnson and M.L. McNee, Prostaglandins, 25, 237 (1983).
63. H.G. Johnson, M.L. McNee, M.A. Johnson and M.D. Miller and J.A. Nadel, Int. Arch. Allergy Appl. Immunol., 71, 214 (1983).
64. H.G. Johnson, R.A. Chinn, D.R. Morton, M.L. McNee, M.D. Miller and J.A. Nadel, Agents and Actions, 13, 1 (1983).
65. A.C. Peatfield, P.J. Piper and P.S. Richardson, Brit. J. Pharmac., 77, 391 (1982).
66. S.J. Coles, K.H. Neill, L.M. Reid, K.F. Austen, Y. Nii, E.J. Corey and R.A. Lewis, Prostaglandins, 25, 155 (1983).
67. A. Ghelani and M.C. Holroyde, Methods and Findings Exp. Clin. Pharmac., 3, 385 (1981).
68. M.M. Goldenberg and E.M. Subers, Eur. J. Pharmac., 78, 463 (1982).
69. P. Yusko, R.A. Hall and A.W. Ford-Hutchinson, Prostaglandins, 25, 397 (1983).

70. R. Magous, J.P. Bali, J.C. Rossi and J.P. Girard, Biochem. Biophys. Res. Commun., 114, 897 (1983).
71. R.H. Wolbling, U. Aehringhaus, B.A. Peskar, K. Morgenroth and B.M. Peskar, Prostaglandins, 25, 809 (1983).
72. S.A. Metz, W.Y. Fujimoto and R.P. Robertson, Life Sci., 32, 903 (1983).
73. S. Yamamoto, M. Ishii, T. Nakadate, T. Nakaki and R. Kato, J. Biol. Chem., 258, 12149 (1983).
74. T.J. Williams and P.J. Piper, Prostaglandins, 19, 779 (1980).
75. R.A. Lewis, N.A. Soter, E.J. Corey and K.F. Austen, Clin. Res., 29, 492A (1981).
76. R.D.R. Camp, A.A. Coutts, M.W. Greaves, A.B. Kay and M.J. Walport, Brit. J. Pharmac., 80, 497 (1983).
77. H. Bisgaard, J. Kristensen and J. Sondergaard, Prostaglandins, 23, 797 (1982).
78. S.E. Dahlen, J. Bjork, P. Hedqvist, K.E. Arfors, S. Hammarstrom, J.A. Lindgren and B. Samuelsson, Proc. Natl. Acad. Sci. USA, 78, 3887 (1981).
79. D.F. Woodward, M.A. Wasserman and B.M. Weichman, Eur. J. Pharmac., 93, 9 (1983).
80. M. Clark, The Pharmacologist, 25, 220 (1983).
81. J.A. Burke, R. Levi, Z.G. Guo and E.J. Corey, J. Pharmac. Exp. Ther., 221, 235 (1982).
82. B.M. Altura, The Pharmacologist, 25, 193 (1983).
83. D.M. Roth, The Pharmacologist, 25, 216 (1983).
84. G. Feuerstein, Z. Zukowska-Grojec and I.J. Kopin, Eur. J. Pharmac., 76, 107 (1981).
85. V.J. Iacopino, T.M. Fitzpatrick, P.W. Ramwell, J.C. Rose and P.A. Kot, J. Pharm. Exp. Ther., 227, 244 (1983).
86. E.B. Cramer, L. Polose, N.A. Pawlowski, Z.A. Cohn and W.A. Scott, Proc. Natl. Acad. Sci. USA, 80, 4109 (1983).
87. F. Michelassi, L. Landa, R.D. Hill, E. Lowenstein, W.D. Watkins, A.J. Petkau and W.M. Zapol, Science, 217, 841 (1982).
88. S.T. Ahmed and W. Oliver Jr., Am. Rev. Resp. Dis., 127, 566 (1983).
89. K.R. Stenmark, L. James, N.F. Voelkel, W.H. Toews, J.T. Reaves and R.C. Murphy, N. Eng. J. Med., 309, 77 (1983).
90. L.B. Klickstein, C. Shapleigh and E.J. Goetzl, J. Clin. Invest., 66, 1166 (1980).
91. A.W. Ford-Hutchinson, M.A. Bray, M.V. Doig, M.E. Shipley and M.J.H. Smith, Nature, 286, 264 (1980).
92. E.J. Goetzl and W.C. Pickett, J. Immunol., 125, 1789 (1980).
93. D.W. Goldman and E.J. Goetzl, J. Immunol., 129, 1600 (1982).
94. M.J.H. Smith, A.W. Ford-Hutchinson and M.A. Bray, J. Pharm. Pharmac., 32, 517 (1980).
95. B.M. Czarnetski, Int. Arch. Allergy Appl. Immunol. 67, 181 (1982).
96. A.M. Uden, J. Palmblad, J.A. Lindgren and C. Malmsten, Int. Arch. Allergy Appl. Immunol., 72, 91 (1983).
97. E.J. Goetzl and W.C. Pickett, J. Exp. Med., 153, 482 (1981).
98. T.E. Rollins, B. Zanolari, M.S. Springer, Y. Guindon, R. Zamboni, C.K. Lau and J. Rokach, Prostaglandins, 25, 281 (1983).
99. I. Hafstrom, J. Palmblad, C.L. Malmsten, O. Radmark and B. Samuelsson, FEBS Letters, 130, 146 (1981).
100. H.J. Showell, I.G. Otterness, A. Marfat and E.J. Corey, Biochem. Biophys. Res. Commun., 106, 741 (1982).
101. M. Rola-Pleszczynski, L. Gagnon and P. Sirois, Biochem. Biophys. Res. Commun., 113, 531 (1983).
102. D.G. Payan and E.J. Goetzl, J. Immunol., 131, 551 (1983).
103. M. Rola-Pleszczynski, P. Borgeat and P. Sirois, Biochem. Biophys. Res. Commun., 108, 1531 (1982).
104. D.R. Webb, I. Nowowiejski, C. Healy and T.J. Rogers, Biochem. Biophys. Res. Commun., 104, 11617 (1982).
105. D.R. Payan and E.J. Goetzl, J. Clin. Immunol., 1, 266 (1981).
106. M.A. Bray, F.M. Cunningham, A.W. Ford-Hutchinson and M.J.H. Smith, Brit. J. Pharmac., 72, 483 (1981).
107. P. Bhattacherjee, B. Hammond, J.A. Salmon, R. Stepney and K.E. Eakins, Eur. J. Pharmac. 73, 21 (1981).
108. R.D.R. Camp, Brit. J. Pharmac., 75, 168 (1982).
109. C.V. Wedmore and T.J. Williams, Nature, 284, 646 (1980).
110. P.M. Dowd, S.D. Brain, R.D.R. Camp, A. Kobza Black, N.J. Fincham and M.W. Greaves, Brit. J. Dermatol., 108, 225 (1983).
111. S.A. Rae, E.M. Davidson and M.J.H. Smith, The Lancet, Nov. 20, 1982, p. 1122.
112. A. Ueno, K. Tanaka and M. Katori, Prostaglandins, 26, 493 (1983).
113. A. Rackman and A.W. Ford-Hutchinson, Prostaglandins, 25, 193 (1983).
114. A.W. Ford-Hutchinson and A. Rackman, Brit. J. Dermatol., 109, Suppl. 25, 26 (1983).
115. E. Schenkelaars and I.L. Bonta, Eur. J. Pharmac., 86, 477 (1983).
116. M.W. Greaves, Brit. J. Dermatol., 109, 115 (1983).
117. P. Sharon and W.F. Stenson, Gastroenterology, 84, 1306 (1983).
118. J.H. Fleisch, L.E. Rinkema and S.R. Baker, Life Sci., 31, 577 (1981).

119. R.D. Krell, B.S. Tsai, A. Berdoulay, M. Barone and R.E. Giles, Prostaglandins, 25, 171 (1983).
120. S.S. Pong and R.N. DeHaven, Proc. Natl. Acad. Sci. USA, 80, 7415 (1983).
121. S.S. Pong, R.N. DeHaven, F.A. Kuehl and R.W. Egan, J. Biol. Chem., 258, 9616 (1983).
122. G.K. Hogaboom, S. Mong, H.L. An and S.T. Crooke, Biochem. Biophys. Res. Commun., 116, 1136 (1983).
123. R.F. Bruns, W.J. Thomsen and T.A. Pugsley, Life Sci., 33, 645 (1983).
124. S. Krilis, R.A. Lewis, E.J. Corey and K.F. Austen, J. Clin. Invest., 72, 1516 (1983).
125. R.A. Kreisle and C.W. Parker, J. Exp. Med., 157, 628 (1983).
126. D.W. Goldman, L.A. Gifford and E.J. Goetzl, Fed. Proc., 43, 847 (1984).
127. F.P. Nijamp and A.G.M. Ramakers, Eur. J. Pharmac., 62, 121 (1980).
128. J. Baumann, F.V. Bruchhausen and G. Wurm, Prostaglandins, 20, 627 (1980).
129. S.S. Yen and W. Kreutner, Agents and Actions, 10, 274 (1980).
130. C.L. Armour, J.M. Hughes, J.P. Seale and D.M. Temple, Clin. Exp. Pharmac. Physiol., 8, 654 (1981).
131. R. Patterson, J.J. Pruzansky and K.E. Harris, J. Allergy Clin. Immunol., 67, 444 (1981).
132. N.K. Boyghton-Smith and B.J.R. Whittle, Proc. Brit. Pharmac. Soc., Sept. 1981, p. 920 P.
133. J.A. Salmon, P.M. Simmons and S. Moncada, J. Pharm. Pharmac., 35, 808 (1983).
134. K.M. Mullane and S. Moncada, Prostaglandins, 24, 255 (1982).
135. A.T.M. Mobarok and J. Morley, Agents and Actions, 12, 645 (1982).
136. T. Nakadate, S. Yamamoto, M. Ishii and R. Kato, Carcinogenesis, 3, 1411 (1982).
137. K.V. Honn and J.R. Dunn, FEBS Letters, 139, 65 (1982).
138. T.E. Eling, Prostaglandins and Related Lipids, 2, 783 (1982).
139. J.R. Walker, J.R. Boot, B. Cox and W. Dawson, J. Pharm. Pharmac., 32, 866 (1980).
140. J. Harvey, H. Parish, P.P.K. Ho, J.R. Boot and W. Dawson, J. Pharm. Pharmac., 35, 44 (1983).
141. W.R. Lumry, J.G. Curd, W.E. Brocklehurst, R.A. Simon and D.D. Stevenson, J. Allergy Clin. Immunol., 69, 93 (1982).
142. K. Kragballe and T. Herlin, Arch. Dermatol., 119, 548 (1983).
143. J.J. Vorhees, Arch. Dermatol., 119, 541 (1983).
144. M.K. Bach, J.R. Brashler, F.A. Fitzpatrick, R.L. Griffin, S.S. Iden, H.G. Johnson, M.L. McNee, J.C. McGuire, H.W. Smith, R.J. Smith, F.F. Sun and M.A. Wasserman in "Advances in Immunopharmacology 2". J.W. Hadden, L. Chedid, P. Dukor, F. Spreafico and D. Willoughby, Ed., Pergamon Press, Oxford, 1983, p. 549.
145. S.S. Sun and J.C. McGuire, Prostaglandins 26, 211 (1983).
146. H.G. Johnson, M.L. McNee, M.K. Back and H.W. Smith, Int. Arch. Allergy Appl. Immunol., 70, 169 (1983).
147. R.D. Krell, M. O'Donnell, R. Osborn, K. Falcone, L. Vickery, M. Grous, C. Kinzig, D. Bryan and J. Gleason, Fed., Proc., 40, 2586 (1981).
148. R.D. Krell, J. McCoy, R. Osborn and L.W. Chakrin, Int. J. Immunopharmac., 2, 55 (1980).
149. T.H. Lee, M.J. Walport, A.H. Wilkinson, M. Turner-Warwick and A.B. Kay, The Lancet, Aug. 8, 1981, p. 304.
150. T. Ahmed, D.W. Greenblatt, S. Birch, B. Marchette and A. Wanner, Am. Rev. Resp. Dis., 124, 110 (1981).
151. M.C. Holroyde, R.E.C. Altounyan, M. Cole, M. Dixon and E.V. Elliott, Agents and Actions, 11, 573 (1981).
152. C.A. Hirshman, J.E. Peters, J. Butler, J. Hanifin, H. Downes and R.K. Lynn, Am. Rev. Resp. Dis., 125, 65 (1982).
153. A.F. Welton, W.C. Hope, L.D. Tobias and J.G. Hamilton, Biochem. Pharmac., 30, 1378 (1981).
154. F.B. Casey, B.J. Appleby and D.O. Buck, Prostaglandins, 25, 1 (1983).
155. W. Hagmann and De. Keppler, Naturwissenschaften, 69, 594 (1982).
156. B.M. Weichman, The Pharmacologist, 25, 114 (1983).
157. J.G. Gleason, Biochem. Biophys. Res. Commun. 117, 732 (1983).

Chapter 25. Endogenous Natriuretic Agents

Mary Anna Napier and Edward H. Blaine
Merck Sharp and Dohme Research Laboratories
Rahway, NJ 07065 and West Point, PA 19486

Introduction - Endogenous natriuretic factors, which are important in the maintenance of the extracellular fluid volume, have been detected in plasma and urine and have been extracted from brain tissue and from cardiac atria. Expansion of the extracellular fluid volume leads to increased sodium ion excretion (natriuresis) by the kidney, independent of changes in glomerular filtration rate (GFR) and of the renin-angiotensin-aldosterone system.[1] Considerable evidence indicates that volume expansion (VE) in laboratory animals and in man causes the release of a humoral natriuretic factor, which is excreted in the urine. One such agent that appears to be an endogenous inhibitor of the sodium transport system has been referred to as natriuretic hormone (NH). Many studies have implicated NH in the pathogenesis of hypertension in man and experimental animals.[2-6] A second substance, isolated from cardiac atrial tissue and designated atrial natriuretic factor (ANF), does not affect Na^+,K^+-ATPase and is structurally different from NH. The role of ANF in blood pressure control or hypertension may relate to cardiac atrial receptors that are presumed to be sensitive to changes in intraatrial pressure, possibly resulting in release of ANF.[7] This chapter will review the isolation, the characterization of the chemical and biological properties, and the possible roles of these agents in the control of extracellular fluid volume and development of hypertension. More detailed reviews of the natriuretic hormone have appeared recently.[2-6]

Natriuretic Hormone - A humoral factor which causes natriuresis when cross circulated to recipient animals has been demonstrated in VE donor animals.[1,8,9,10] Extracts of blood or urine from VE subjects also causes natriuresis in assay animals, indicating a transferable natriuretic substance.[4] This agent is either less active or undetectable in nonexpanded subjects.[4] A natriuretic substance has been extracted from kidney tissue of VE animals, but not from kidneys of hydropenic animals.[11] The evidence suggests that NH increases sodium ion excretion by inhibiting reabsorption as a result of direct inhibition of the renal Na^+,K^+-ATPase, the energy-requiring sodium ion transport pump.[12] A high concentration of the Na^+,K^+-ATPase inhibitor, digoxin, produces natriuresis by impairing the cellular pump of the renal tubule.[12]

An unknown pressor agent has been observed in the blood of animals and man with VE, low renin, hypertension.[13] Suppressed Na^+,K^+-ATPase activity in cardiovascular tissues of animals with either one-kidney renal, deoxycorticosterone acetate or reduced renal mass hypertension suggests that this pressor substance is an inhibitor of the Na^+,K^+-ATPase. By inhibiting the Na^+,K^+-ATPase of vascular smooth muscle, it is possible that such an agent could contribute to increased vascular resistance or

ISBN 0-12-040519-9

reactivity associated with hypertension.[14] A substantial increase in arterial pressure has been observed following injection of digitalis preparations in intact animals[15] as well as in excised arterial[16,17] and venous[16] tissues.

A theoretical role for a salt-excreting hormone in hypertension was proposed by Dahl et al.[18,19] and expanded by Blaustein,[20] Haddy et al.,[14] and deWardener.[21] Blood pressure is a function of both total volume of the blood and the resistance to its flow through the circulatory system. The resistance to flow is determined by the degree of constriction of the arterioles. A primary defect in the kidney that reduces its ability to excrete sodium results in increased blood volume, release of natriuretic hormone, the inhibition of the renal Na^+,K^+-ATPase and increased urinary sodium excretion. At the same time, however, the vascular tissues would be exposed to the same Na^+,K^+-ATPase inhibitor, causing increased non-specific sensitivity to constrictor agents such as angiotensin, vasopressin or norepinephrine.[22,23] Inhibition of the vascular Na^+,K^+-ATPase leads to increased levels of intracellular Ca^{++} in blood vessels, either by a Na^+-Ca^{++} exchange mechanism[20] or by partial depolarization of the cell membranes increasing the permeability to calcium.[14] This increases the sensitivity to circulating pressor agents and results in enhanced contraction and increased blood pressure.

Another mechanism proposed by Buckalew and Gruber involves a more direct role of the sympathetic nervous system.[4] In nonhypertensive individuals, NH in conjunction with other natriuretic forces, increases renal sodium excretion and maintains normal plasma volume. Since NH can cause natriuresis with no effect on blood pressure, volume can be regulated in normal individuals without causing hypertension[24]. In hypertensive subjects, renal response to NH is blunted, causing NH levels to become higher than in normotensive individuals. Pathologically elevated NH levels lead to increased blood pressure either by activating the sympathetic nervous system, by increasing vascular reactivity, or both. The increased blood pressure adds another natriuretic force helping to overcome the defect in renal sodium excretion. Thus, hypertension is a result of the need to regulate volume in the presence of a defect in renal sodium excretion. Control of NH release by a central site was suggested by impaired natriuresis and impaired secretion of NH in rats with lesions of the anteroventral third ventricle (AV3V) region of the brain.[25]

Evidence that NH may be an endogenous Na^+,K^+-ATPase inhibitor includes the inhibition of sodium transport in an anuran membrane, a model of the renal distal tubule,[4] and in isolated rabbit collecting tubule.[26] Natriuretic urine extracts also displace ^{3}H-ouabain from renal Na^+,K^+-ATPase.[27] Direct inhibition of Na^+,K^+-ATPase in vitro has also been reported with extracts of plasma[28,29] and urine.[30] Hamlyn et al. show a significant correlation between the Na^+,K^+-ATPase inhibitory activity in the plasma and the mean arterial pressure of the donor patient.[29] A cytochemical assay for inhibition of guinea pig renal Na^+,K^+-ATPase[30] has been used to show the presence of a circulating inhibitory activity that is 25-times greater in the plasma of subjects on high salt than those on low salt diets.[31] Plasma from hypertensive patients, plasma with renin values below normal, and plasma from older subjects exhibited higher levels of Na^+,K^+-ATPase inhibitory activity.[32]

A natriuretic extract of urine from normal or salt-loaded[33,34] and uremic[35] humans was also shown to inhibit the Na^+,K^+-ATPase in pig renal tubular membranes. The inhibition is greatest in the proximal and distal tubules and in the thick ascending limb of the loop of Henle,[30] and the extract is antinatriferic on the serosal surface of frog skin.[31] Plasma induced inhibition is similarly distributed within the renal tubule. Urine from VE dogs and uremic patients has also been reported to contain an NH-like activity,[36-38] and a natriuretic factor has been described in the serum of patients with chronic uremia.[39-41]

Buckalew and Gruber hypothesized that antibodies against a Na^+,K^+-ATPase inhibitor, such as digoxin, could be used as probes for natriuretic hormone on the basis that antibodies to drugs that bind to specific receptors might recognize and bind to the endogenous ligand.[28] In two of their studies, cross reactivity of the inhibitor of Na^+,K^+-ATPase with anti-digoxin antibodies was observed.[28,30] Rudd et al. showed that there was significantly more Na^+,K^+-ATPase inhibition and digoxin immunoreactivity in the plasma of VE dogs compared to hydropenic dogs.[42] The plasma extract with digoxin immunoreactivity from VE dogs is also natriuretic in rats. Digoxin immunoreactivity is found in the plasma of Rhesus and African Green monkeys with 2-kidney, 1-clip Goldblatt hypertension. The hypertensive monkeys have two- to three-fold higher levels of digoxin immunoreactivity than the normotensive controls.[43] Increased excretion of a digoxin-like hormone in rats during salt-loading, which is further enhanced during the development of hypertension and adaptation to chronic renal failure, has been reported recently.[44] In contrast to these studies, Hamlyn et al. did not observe any anti-digoxin immunoreactivity in plasma samples from normal or hypertensive patients.[29]

Impaired sodium efflux was measured in leukocytes from patients with essential hypertension. Incubation of normal leukocytes with plasma from hypertensive patients caused impaired sodium transport.[45] Furthermore, partially purified fractions from plasma, as well as urine, inhibit sodium efflux from peripheral blood leukocytes.[46] The severity of the defect in leukocyte cation transport is inversely related to the plasma renin activity and is greatest in patients with essential hypertension in whom the renin response to sodium restriction was atypical.[47] Although these studies support the hypothesis that NH is a digitalis-like substance which inhibits Na^+,K^+-ATPase in numerous tissues including the kidney, it remains to be determined whether the natriuretic effect of NH is due exclusively to inhibition of renal Na^+,K^+-ATPase.

It has been proposed that the hypothalamus might secrete a substance which controls sodium excretion.[48] The AV3V region of the brain is important for regulating blood pressure, since its destruction prevents several forms of hypertension.[49] Furthermore, less Na^+ is excreted, the blood level of Na^+ is elevated, and NH is absent in animals with AV3V lesions.[50,51] Consequently, several groups have prepared hypothalamic or brain extracts and studied their antinatriferic or Na^+,K^+-ATPase inhibitory properties, but natriuretic activity was not reported. On the other hand, an acetone extract of the rat hypothalamus (but not cerebral cortex, pituitary, or other tissues) possessed very potent Na^+,K^+-ATPase inhibitory activity which was increased 150-fold in animals on a high sodium diet.[52] The plasma from these animals also contained Na^+,K^+-ATPase inhibitory activity, with seven-fold more activity in the plasma of the rats on the high sodium diet. Aqueous-acetone extraction of rat brain, in the presence of nitrogen, resulted in a specific Na^+,K^+-ATPase inhibitor.[53] A low molecular weight "ouabain-like" factor, which inhibits in vitro Na^+,K^+-ATPase activity, blocks ^3H-ouabain binding to brain microsomal Na^+,K^+-ATPase or inhibits ^{86}Rb uptake in human erythrocytes, has been partially purified from bovine hypothalamus,[54,55] guinea pig[56] and rat[57] brain.

The chemical structure of NH has not been determined and there is no assurance that the activities ascribed to the various extracts are caused by the same substance. Furthermore, there is no standardized assay which all the investigators agree is best for detecting NH. However, there is general agreement that the substance is low molecular weight ($\lesssim$ 1000 daltons), acidic, water soluble and heat stable.[4] The presence of two natriuretic factors in the plasma and urine of volume expanded subjects has been reported.[58] One factor causes natriuresis in rats after a 20 min delay and appears to be larger than the second factor which produces immediate natriuresis. The low molecular weight factor is antinatriferic (inhibits ion flux in isolated frog skin or toad bladder membrane preparations) while the higher molecular weight factor is not. These results and the finding that NH activity increases in plasma following incubation for 30 minutes at room temperature supports the presence of an NH precursor.[59] Evidence for the peptide nature of NH is indicated by its sensitivity to enzymatic digestion[30,39,40,60,61] and its behavior during chromatography by techniques used for the isolation of small peptides.[62] Buckalew and coworkers have studied a heptapeptide with sequence homology to a fragment of ACTH/αMSH (Met-Glu-His-Phe-Arg-Trp-Gly [or Asp]), which may be related to an endogenous NH since its biological effects mimic those of NH.[63-66] The peptide inhibits Na^+,K^+-ATPase, is natriuretic at low doses, and is hypertensive at high doses.[62,63] Others, however, do not believe the factor they isolate from urine of salt-expanded dogs is a peptide.[4,42,60] In spite of the reported cross reactivity of NH with anti-digoxin antibodies, NH does not appear to be a steroid, since it is isolated in aqueous solutions and is susceptible to acid hydrolysis.[4]

Expansion of the extracellular fluid volume, resulting in a natriuretic response affects proximal tubule sodium resorption.[67,68] However, inhibition of sodium resorption at a more distal nephron site such as the collecting duct, also occurs during expansion of the extracellular fluid volume and is a major determinant of the magnitude of natriuresis.[69,70] Recollection micropuncture techniques were used to show a significant decrease in fractional sodium resorption in the proximal tubules of the superficial nephrons of uremic rats when a natriuretic response occurred following the administration of a serum derived NH.[71] In rabbit, the renal collecting duct (tubule) has been implicated as an important site for the regulation of active sodium transport by the natriuretic factor. A natriuretic sample applied to the peritubular surface of the isolated perfused cortical collecting tubule inhibited the potential difference and decreased the net sodium flux from the lumen to the peritubular surface.[26] This is consistent with results from studies on the isolated toad bladder (a structure with many similarities to the distal nephron) which have shown that NH acts at the serosal surface and inhibits transepithelial transport by reducing sodium movement across the serosal barrier.[44]

Atrial Natriuretic Factor - Evidence has accumulated which suggests that the cardiac atria might function as a sensor for detecting changes in extracellular fluid volume. Indeed, numerous experimental procedures which alter intraatrial pressure or stretch the atrial wall result in significant changes in water and electrolyte excretion. Such alterations in intraatrial pressure should reflect changes in and provide a means for regulating extracellular fluid volume.[7,72]

Changes in water excretion that are secondary to altered intra-atrial pressure can be attributed largely to changes in the secretion of vasopressin by the posterior pituitary gland. This reflex has been documented both in experimental animals and in man, and the relevant data has been reviewed recently by Bie.[73] On the other hand, it has not been possible, until recently, to account for the increased urinary Na^+ excretion associated with changes in atrial pressure.

de Bold et al. have extracted a substance from the cardiac atria of rats which, when injected into other rats, results in a 30-fold increase in urinary Na^+ excretion and a 10-fold increase in urine volume.[74] Similar extracts of cardiac ventricle were without effect. Numerous independent laboratories have now confirmed the initial observation of atrial natriuretic factor (ANF).[75-88] This activity has been observed in the atrium of all mammalian species studied, including man.[83,89,90] The specific activity appears to be highest in rat atria. The distinction between this material and NH rests on several criteria: 1) the minimum molecular weight for ANF is at least 3000 daltons[89] while NH is proported to be less than 1000,[12] 2) NH appears to inhibit Na^+,K^+-ATPase while ANF has no such activity,[4,76] and 3) NH is thought to be hypertensive while ANF clearly lowers blood pressure.[4,74]

"Specific granules"[91] have been described in the myocytes of the cardiac atria which closely resemble the storage granules of other peptide hormone secreting tissues. These granules co-purify with ANF[92,93] and immunocytochemical studies confirm the association between the specific atrial granules and ANF.[94,95] Furthermore, the highest concentrations of antigenic material are localized in the subpericardial area of the atrial appendages.[94] Direct bioassay of dissected atria confirm these high levels of ANF activity.[96]

It is not understood what factors control the synthesis, storage and secretion of ANF, but alterations in body fluid volume appear to change granule density and extractable activity. Granularity increases markedly in response to salt or water deprivation and decreases when extracellular fluid volume is expanded secondary to desoxycorticosterone treatment.[97] However, there is no correlation between the granularity of the atria and the amount of extractable ANF.[98] It is possible that the storage form of ANF under these conditions may not be active, and further processing is required before biological activity can be expressed. Although extractable ANF is reduced in the atria of spontaneously hypertensive rats, it is not known whether this change is related to the elevated blood pressure.[99]

Potent natriuretic activity was evident in a simple phosphate buffered saline extract of whole atria.[74] Subsequently, the material was found to be acid stable.[75,87,88,100,101] While the extract is both acid and heat[75,101] stable, trypsin,[75,89,93] as well as other proteolytic enzymes[76] such as chymotrypsin, aminopeptidase A and carboxypeptidases B and C destroy natriuretic activity. However, treatment with carboxypeptidase A only blunted activity,[76] and concanavalin A had no effect.[75] These observations are consistent with ANF being a small peptide.

Purification of ANF from rat atria by several independent laboratories utilizing different isolation procedures confirmed its peptidic nature.[85,102,103,104] An amino acid composition without sequence was reported from two laboratories[85,102] but the total number of residues differed considerably (49 and 36) and there were differences in composition. An amino acid sequence for rat ANF was reported essentially simultaneously by Flynn *et al.*,[103] Currie *et al.*,[105] Kangawa and Matsuo,[106] Seidah *et al.*[107] and Napier *et al.*[87,88] The longest peptide reported has the 33 amino acid sequence shown below, 1, which contains the Cys^{12}-Cys^{28} disulfide bridge, and is the Tyr acid rather than the amide.[87,88,107]

H-Leu-Ala-Gly-Pro-Arg^{5}-Ser-Leu-Arg-Arg-Ser^{10}-Ser-Cys-Phe-Gly-
Gly^{15}-Arg-Ile-Asp-Arg-Ile^{20}-Gly-Ala-Gln-Ser-Gly^{25}-Leu-Gly-Cys-
Asn-Ser^{30}-Phe-Arg-Tyr-OH

1

Other sequences are N-terminal truncated versions of this peptide (22,28, 31,32 residues).[87,88,103,107] Currie *et al.* have purified two atrial peptides (atriopeptins I and II) of 21 and 23 amino acids.[105] The longer of these is identical to the 10-32 fragment of 1. Kangawa and Matsuo have isolated ANF from human atria and determined the amino acid sequence of a 28 amino acid peptide which differs from the rat ANF sequence only at position 17 where Ile is replaced by Met.[106] Based on the sequence discovered by Seidah *et al.*,[106] synthetic material was made and the important confirmation of the natural material was obtained. The synthetic peptide composed of the 8-33 sequence of Seidah *et al.* has full biological activity.[107]

Initially, the principal biological activity of ANF was thought to be natriuresis. Most investigators reported 30 to 40-fold increases in urinary sodium excretion with lesser effects on urinary K^+ excretion. Although the mechanism of the natriuresis has not been studied extensively, reports using micropuncture and microcatheterization techniques suggest that inhibition of Na^+ transport in the far distal nephron (i.e. collecting duct) is the principal mechanism of action.[77,108] These data must be interpreted cautiously, however, because the magnitude of the natriuretic response would argue for a more proximal site of action. Indeed, a recent report by Seymour *et al.*, in which synthetic ANF was infused directly into the renal artery of anesthetized dogs, indicated that fractional sodium excretion exceeded 10% during maximal natriuresis.[109] Based on what is currently known about sodium delivery to the various nephron segments, this would argue for a site of action at least in the cortical diluting segment (a site similar to thiazide diuretics). Nevertheless, the data do not exclude multiple sites of action, or changes in renal hemodynamics or nephron heterogeneity. Some evidence also suggests that changes in glomerular filtration are important in the natriuretic response, but these studies were conducted in isolated perfused kidneys.[110] Other studies in intact animals have not demonstrated major changes in either glomerular filtration or renal blood flow.[74,78,79]

In addition to the potent natriuretic activity, ANF has significant relaxant effects on vascular and perhaps other smooth muscle tissues.[84,110,113,114] Currie *et al.* observed relaxation of rabbit aorta and chick rectum, which had previously been contracted with epinephrine and carbachol, respectively.[84] They use this spasmolytic action to monitor ANF activity and are the first group to separate smooth muscle relaxant activity of similar peptides isolated from cardiac atria, although both peptides are natriuretic.[105] A recent report by Winquist *et al.* has detailed the vascular smooth muscle relaxing effect of synthetic ANF using a variety of agonists and different vascular smooth muscle preparations.[114] Based on these observations, ANF appears to have spasmolytic properties similar to sodium nitroprusside.

ANF has potent effects on cyclic GMP (cGMP) levels. In vitro, addition of atrial extracts to minced kidney tissue or to primary kidney cell cultures results in increased levels of cGMP.[115] Injection into anesthetized rats resulted in a 4-fold increase in plasma cGMP levels and a 28-fold increase in urinary excretion.[115] Sodium nitroprusside also causes increased cGMP levels in treated tissues, but probably acts directly through stimulation of guanylate cyclase. Further studies must be done to clarify the role of cGMP in the vasorelaxant and the natriuretic actions of ANF.

The potent natriuretic activity of ANF has inevitably led to comparisons with known diuretics. One report by Sonnenberg et al. indicated that probenecid blunted the natriuretic activity of ANF.[111] Because probenecid interferes with the natriuretic activity of most other diuretics by competing with their secretory transport into the nephron lumen, these authors suggest that ANF might be acting as an endogenous diuretic. Furosemide is the diuretic most frequently cited as similar to ANF, with regard to electrolyte excretion. ANF promotes sodium, potassium and calcium excretion with chloride as the principal anion. Although furosemide has been a reference standard in at least one paper,[89] it should be pointed out that rigorous dose-response studies comparing ANF to other diuretics have not been conducted. The most definitive studies suggested a ceiling similar to the thiazide diuretics, a maximum fractional sodium excretion of about 10%.[109] Also, loop diuretics such as furosemide inhibit NaCl transport in bull frog cornea, an analog of the medullary thick ascending limb of the loop of Henle whereas atrial extracts are inactive in this preparation.[112]

Because of the potential for ANF to be involved in blood pressure control, several investigators have studied its effects on the cardiovascular system of normal and hypertensive animals. In anesthetized normotensive and spontaneously hypertensive rats, atrial extract decreased arterial blood pressure in association with a decrease in both total peripheral resistance and cardiac contractility.[116] Also, a negative chronotropic effect has been reported in rats.[117] Dahl salt-sensitive rats appear to have greater amounts of ANF in their atria, but are hyporesponsive when exogenous material is injected.[90] Results from this study also suggest an increase in renal papillary plasma flow and a washout of the medullary osmotic gradient, factors which could contribute significantly to the associated natriuresis and diuresis. Increased medullary and inner cortical blood flow also were reported in normotensive rats.[118]

Summary - In the case of natriuretic hormone, twenty-five years of research have failed to characterize a pure substance that has natriuretic activity and is an inhibitor of Na^+,K^+-ATPase, although, a vast body of circumstantial evidence does favor the existence of such a substance. Atrial natriuretic factor, on the other hand, is a recent discovery which yielded readily to chemical characterization and synthesis. Its biological profile suggests that it may be a component of the various control systems subserving fluid volume and pressure regulation. However, no convincing evidence has yet been offered that atrial natriuretic factor plays an important physiological role in pressure or volume homeostasis.

REFERENCES

1. H.E. deWardener, I.H. Mills, W.F. Clapham and C.J. Hayter, Clin. Sci., 21, 249 (1961).
2. N.G. Levinsky in "Advances in Metabolic Disorders," Vol. 7, R. Levine and R. Luft, Eds., Academic Press, 1974, p 37.
3. J.L. Marx, Science, 212, 1255 (1981).
4. V.M. Buckalew, Jr. and K.A. Gruber, Ann. Rev. Phys., 46, 327 (1984).
5. H.E. deWardener and G.A. MacGregor, Lancet, I, 1450 (1982).
6. M.B. Pamnani, D.L. Clough, S.J. Huot, F.J. Haddy in "Vasodilatation," P.M. Vanhoutte and I. Lensen, Eds., Raven Press, New York, 1981, p 391.
7. O.H. Gauer and J.P. Henry, Physiol. Rev., 43, 423 (1963).
8. C.A. Knock and H.E deWardener, Clin. Sci., 59, 411 (1980).
9. C.A. Knock, Clin. Sci., 59, 423 (1980).
10. H.E. deWardener, I.H. Mills, W.F. Clapham and C.J. Hayter, Clin. Sci., 21, 249 (1961).
11. H.C. Gonick and L.F. Saldanha, J. Clin. Invest., 56, 247 (1975).
12. F.H. Epstein in "Modern Diuretic Therapy in the Treatment of Cardiovascular and Renal Disease," A.F. Lant and G.M. Wilson, Eds., Excerpta Medica, Amsterdam, 1973, p 188.
13. H.E. deWardener and G.A. MacGregor, Kidney Int., 18, 1 (1980).
14. F.J. Haddy, M.B. Pamnani and D.L. Clough, Life Sci., 24, 2105 (1979).
15. J. Ross, Jr., J.A. Waldhausen and E. Braunwald, J. Clin. Invest., 39, 930 (1960).
16. D. Cow, J. Physiol., 42, 125 (1911).
17. E. Leonard, Amer. J. Physiol., 189, 185 (1957).
18. L.K. Dahl and M. Heine, Circ. Res., 36, 692 (1975).
19. L.K. Dahl, K.D. Knudsen and J. Iwai, Circ. Res., (Suppl. I) 24, I-21 (1969).
20. M.P. Blaustein, Amer. J. Physiol., 232, C165 (1977).
21. H.E. deWardener and G.A. MacGregor, J. Chron. Dis., 34, 233 (1981).
22. J.H. Cort, T. Dousa, V. Pliska, B. Lichendus, J. Safarova, M. Vranesic and J. Rudinger, Amer. J. Physiol., 215, 921 (1968).
23. W.C. Plunkett, P.M. Hutchins, K.A. Gruber and V.M. Buckalew, Jr., Hypertension, 4, 581 (1982).
24. M.A. Rudd, K.A. Gruber and V.M. Buckalew, Jr., Fed. Proc., 40, 613 (1981).
25. S. Bealer, J.R. Haywood, A.K. Johnson, K.A. Gruber, V.M. Buckalew and M.J. Brody, Fed. Proc., 38, 1232 (1979).
26. L.G. Fine, J.J. Bourgoignie, K.H. Hwang and N.S. Bricker, J. Clin. Invest., 58, 590 (1976).
27. T. Bohan, L. Potter and J.J. Bourgoignie in "Hormonal Regulation of Sodium Excretion," B. Lichardus, R.W. Schrier and J. Ponec, Eds., Elsevier, New York, N.Y., 1980, p 393.
28. K.A. Gruber, J.M. Whitaker and V.M. Buckalew, Jr., Nature, 287, 743 (1980).
29. J.M. Hamlyn, R. Ringel, J. Schaeffer, P.D. Levinson, B.P. Hamilton, A.A. Kowarski and M.P. Blaustein, Nature, 300, 650 (1982).
30. D. Klingmuller, E. Weiler and H.J. Kramer, Klin. Wochenschi., 60, 1249 (1982).
31. H.E. deWardener, E.M. Clarkson, L. Bitensky, G.A. MacGregor, J. Alaghband-Zadeh and J. Chayen, Lancet, 1, 411 (1981).
32. G.A. MacGregor, S. Fenton, J. Alaghband-Zadeh, N.D. Markandu, J.E. Roulston and H.E. deWardener, Clin. Sci., 61, 17s (1981).
33. E.M. Clarkson, S.M. Raw and H.E. deWardener, Kidney Int., 16, 710 (1979).
34. E.M. Clarkson, D.R. Young, S.M. Raw and H.E. deWardener in "Hormonal Regulation of Sodium Excretion," B. Lichardus, R.W. Schrier and J. Ponec, Eds., Elsevier, New York, 1980, p 333.
35. R. Dzurik, B. Lichardus, V. Spustova, J. Ponec, A. Gerykova and P. Bakos, Physiologia Bohemoslovaca, 31, 573 (1982).
36. A. Licht, S. Stein, C.W. McGregor, J.J. Bourgoignie and N.S. Bricker, Kidney Int., 21, 339 (1982).
37. J.J. Bourgoignie, K.H. Hwang, E. Ipakchi and N.S. Bricker, J. Clin. Invest., 53, 1559 (1974).
38. M.A. Kaplan, J.J. Bourgoignie, J. Rosecan and N.S. Bricker, J. Clin. Invest., 53, 1568 (1974).
39. N.S. Bricker, J.J. Bourgoignie and S. Klahr, Arch. Int. Med., 126, 860 (1970).
40. J. Bourgoignie, S. Klahr and N.S. Bricker, J. Clin. Invest., 50, 303 (1971).
41. J.J. Bourgoignie, K.H. Hwang, C. Espinel, S. Klahr and N.S. Bricker, J. Clin. Invest., 51, 1514 (1972).
42. M.A. Rudd, K.A. Gruber and V.M. Buckalew, Jr., Fed. Proc., 40, 613 (1981).
43. K.A. Gruber, J.M. Whitaker, L.L. Rudel, B.C. Bullock and V.M. Buckalew, Jr., Fed. Proc., 40, 535 (1981).
44. R.A. Kelly, C.S. Wilcox, W.E. Mitch, T.W. Smith and N. Spicer, Clin. Res., 31, 432A (1983).
45. L. Poston, R.B. Sewell, S.P. Wilkinson, P.J. Richardson, R. Williams, E.M. Clarkson,

G.A. MacGregor and H.E. deWardener, Brit. Med. J., 282, 847 (1981).
46. L. Poston, S. Wilkinson, R.B. Sewell and R. Williams, Clin. Sci., 63, 243 (1982).
47. R.P.S. Edmondson and G.A. MacGregor, Brit. Med. J., 282, 1267 (1981).
48. H.W. Smith, Amer. J. Med., 23, 623 (1957).
49. M. Brody and A.K. Johnson in "Frontiers in Neuroendocrinology," L. Martini and W.F. Ganong, Eds., Raven Press, New York, Vol. 6, 1980.
50. A.K. Johnson in "Perspectives in Nephrology and Hypertension," H. Schmitt and P. Meyer, Eds., Wiley, New York, 1978, p 106.
51. S. Bealer, J.R. Haywood, A.K. Johnson, K.A. Gruber, V.M. Buckalew and M.J. Brody, Fed. Proc., 38, 1232 (1979).
52. J. Alaghband-Zadeh, S. Fenton, K. Hancock, J. Millett and H.E. deWardener, J. Endocrinol., 98, 221 (1983).
53. K.R. Whitmer, E.T. Wallick, D.E. Epps, L.K. Lane, J.H. Collins and A. Schwartz, Life Sci., 30, 2261 (1982).
54. G.T. Haupert, Jr., L. Kormos, L. Anderson, S. Ray and L.C. Cantley, Abstracts, 16th Ann. Mtg. Amer. Soc. Nephrology, p 192A (1983).
55. G.T. Haupert, Jr. and J.M. Sancho, Proc. Natl. Acad. Sci., 76, 4658 (1979).
56. M.C. Fishman, Proc. Natl. Acad. Sci., 76, 4661 (1979).
57. D. Lichtstein and S. Samuelov, Biochem. Biophys. Res. Comm., 96, 1518 (1980).
58. H.E. deWardener, Clin. Sci. Mol. Med., 53, 1 (1977).
59. K.A. Gruber and V.M. Buckalew, Jr., Proc. Soc. Exp. Biol. Med., 159, 463 (1978).
60. V.M. Buckalew, Jr. and K.A. Gruber in "The Kidney in Liver Disease," 2nd ed., M. Epstein, Ed., Elsevier, New York, N.Y., 1983, p 479.
61. K.A. Gruber and V.M. Buckalew, Jr. in "Hormonal Regulation of Sodium Excretion," B. Lichardus, R.W. Schrier and J. Ponec, Eds., Elsevier, New York, N.Y., 1980, p 349.
62. A.N. Radhakrishnan, S. Stein, A. Licht, K.A. Gruber and S. Udenfriend, J. Chromatography, 132, 552 (1977).
63. J.F. Hennessy, V.M. Buckalew, Jr., J.R. Lymangrover and K.A. Gruber, Clin. Res., 30, 848A (1982).
64. M.C. Klein, P.M. Hutchins, V.M. Buckalew, Jr., J.F. Hennessy, J.R. Lymangrover and K.A. Gruber, Fed. Proc., 42, 1020 (1983).
65. K.A. Gruber, J.F. Hennessy, J.R. Lymangrover, M.C. Klein, P.M. Hutchins and V.M. Buckalew, Jr., Fed. Proc., 42, 1174 (1983).
66. J.R. Lymangrover, J.F. Hennessy, M.C. Klein, P.M. Hutchins, V.M. Buckalew, Jr. and K.A. Gruber, Fed. Proc., 42, 1174 (1983).
67. B.M. Brenner, K.H. Falchuk, R.I. Keimowitz and R.W. Berliner, J. Clin. Invest., 48, 1519 (1969).
68. J.H. Dirks, W.J. Cirksena and R.W. Berliner, J. Clin. Invest., 44, 1160 (1965).
69. H. Sonnenberg, Amer. J. Physiol., 223, 916 (1972).
70. J.H. Stein, R.W. Osgood, S. Boonjarern and T.F. Ferris, J. Clin. Invest., 52, 2313 (1973).
71. H. Weber, J.J. Bourgoignie and N.S. Bricker, Amer. J. Physiol., 226, 419 (1974).
72. K.L. Goetz, G.C. Bond and D.D. Bloxham, Physiol. Rev., 55, 157 (1975).
73. P. Bie, Physiol. Rev., 60, 961 (1980).
74. A.J. de Bold, H.B. Borenstein, A.T. Veress and H. Sonnenberg, Life Sci., 28, 89 (1981).
75. N.C. Trippodo, A.A. MacPhee, F.E. Cole and H.L. Blakesley, Proc. Soc. Exp. Biol. Med., 170, 502 (1982).
76. G. Thibault, R. Garcia, M. Cantin and J. Genest, Hypertension 5 (Suppl. I), I-75 (1983).
77. J.P. Briggs, B. Steipe, G. Schubert and J. Schnermann, Pfluegers Arch., 395, 271 (1982).
78. R. Keeler, Can. J. Physiol. Pharm., 60, 1078 (1982).
79. D.M. Pollock and R.O. Banks, Clin. Sci., 65, 47 (1983).
80. A. Bonham, K. Barron, M. Allen, D. Pollock, R. Banks and M.J. Brody, Fed. Proc., 42, 494 (1983).
81. S. Hathaway and S. Solomon, Fed. Proc., 42, 475 (1983).
82. M.J.F. Camargo, H.D. Kleinert, J.E. Sealey, J.H. Laragh and T. Maack, Fed. Proc., 42, 475 (1983).
83. M.N. Nemeh and J.P. Gilmore, Circ. Res., 53, 420 (1983).
84. M.G. Currie, D.M. Geller, B.R. Cole, J.G. Boylan, W. YuSheng, S.W. Holmberg and P. Needleman, Science, 221, 71 (1983).
85. R.T. Grammer, H. Fukumi, T. Inagami and K.S. Misono, Biochem. Biophys. Res. Commun., 116, 696 (1983).
86. I. Kitor, D.J. Levenson, O.C. Ohuoha and V.J. Dzau, Abstracts, 16th Ann. Mtg. Amer. Soc. Nephrology, p 167A (1983).
87. M.A. Napier, R.S. Dewey, G. Albers-Schonberg, C.D. Bennett, J.A. Rodkey, E.A. Marsh, M. Whinnery, A.A. Seymour and E.H. Blaine, Fed. Proc., 43, 501 (1984).
88. M.A. Napier, R.S. Dewey, G. Albers-Schonberg, C.D. Bennett, J.A. Rodkey, E.A. Marsh, M. Whinnery, A.A. Seymour and E.H. Blaine, Biochem. Biophys. Res. Commun. (in press).

89. N.C. Trippodo, A.A. MacPhee and F.E. Cole, Hypertension 5 (Supp. I), I-81 (1983).
90. M. Ganquli, Y. Hirata and L. Tobian, Abstracts, 16th Ann. Mtg. Amer. Soc. Nephrology, p 191A (1983).
91. J.D. Jamieson and G.E. Palade, J. Cell Biol., 23, 151 (1964).
92. A.J. de Bold, Can. J. Physiol. Pharm., 60, 324 (1982).
93. R. Garcia, M. Cantin, G. Thibault, H. Ong and J. Genest, Experientia, 38, 1071 (1982).
94. M. Cantin, J. Gutkowski, G. Thibault, R.W. Milne, S. Ledoux, S. Minli, R. Garcia, P. Hamet and J. Genest, Histochem. (in press).
95. M. Cantin, J. Gutkowski, R.W. Milne, G. Thibault, R. Garcia and J. Genest, 37th Ann. Mtg. Council for High Blood Pressure Research, 1983, abst. 39 (1983).
96. J. Briggs, W. Kriz and J. Schnermann, Circulation 68 (Suppl. II), abst. 172 (1983).
97. A.J. de Bold, Proc. Soc. Exp. Biol. Med., 161, 508 (1979).
98. G. Thibault, R. Garcia, M. Cantin and J. Genest, Fed. Proc., 42, 611 (1983).
99. H. Sonnenberg, S. Milojevic, C.K. Chong and A.T. Veress, Hypertension, 5, 672 (1983).
100. A.J. de Bold, Fed. Proc., 40, 554 (1981).
101. A.J. de Bold, Proc. Soc. Exp. Biol. Med., 170, 133 (1982).
102. A.J. de Bold and T.G. Flynn, Life Sci., 33, 297 (1983).
103. T.J. Flynn, M.L. de Bold and A.J. de Bold, Biochem. Biophys. Res. Comm., 117, 859 (1983).
104. G. Thibault, R. Garcia, N.G. Seidah, C. Lazure, M. Cantin, M. Chretien and J. Genest, FEBS Lett., 164, 286 (1983).
105. M.G. Currie, D.M. Geller, B.R. Cole, N.R. Siegel, K.F. Fok, S.P. Adams, S.R. Eubanks, G.R. Galluppi and P. Needleman, Science, 223, 67 (1984).
106. K. Kangawa and H. Matsuo, Biochem. Biophys. Res. Comm., 118, 131 (1984).
107. N.G. Seidah, C. Lazure, M. Chretien, G. Thibault, R. Garcia, M. Cantin, J. Genest, R.F. Nutt, S.F. Brady, T.A. Lyle, W.J. Paleveda, C.D. Colton, T.M. Ciccarone and D.F. Veber, Proc. Natl. Acad. Sci. (in press).
108. H. Sonnenberg, W.A. Cupples, A.J. de Bold and A.T. Veress, Can. J. Physiol. Pharm., 60, 1149 (1982).
109. A.A. Seymour, R. Nutt, E. Mazack and E.H. Blaine, Fed. Proc., 43, 502 (1984).
110. H.D. Kleinert, J.E. Sealey, J.H. Laragh and T. Maack, Physiologist, 26, A-59 (1983).
111. H. Sonnenberg, C.K. Chong and A.T. Veress, Can. J. Physiol. Pharm., 59, 1278 (1981).
112. D.C. Throckmorton and J.P. Gilmore, Fed. Proc., 42, 475 (1983).
113. R.C. Deth, K. Wong, S. Fukozawa, R. Rocco, J.L. Smart, C.J. Lynch and R. Awad, Fed. Proc., 41, 983 (1982).
114. R.J. Winquist, E.P. Faison and R.F. Nutt, Europ. J. Pharmacol. (in press).
115. P. Hamet, J. Tremblay, G. Thibault, R. Garcia, M. Cantin and J. Genest, Abstracts, 65th Ann. Mtg. Endocrine Society, p 289 (1983).
116. U. Ackermann, T. Irizawa and H. Sonnenberg, Fed. Proc., 41, 1353 (1982).
117. U. Ackermann, T.G. Irizawa, S. Milojevic and H. Sonnenberg, Fed. Proc., 42, 1096 (1983).
118. H.B. Borenstein, W.A. Cupples, H. Sonnenberg and A.T. Veress, J. Physiol., 334, 133 (1983).

Section VI - Topics in Chemistry and Drug Design

Editor: Richard C. Allen, Hoechst-Roussel Pharmaceuticals Inc., Somerville, New Jersey 08876

Chapter 26. Enzymic Methods in Organic Synthesis

Mark A. Findeis and George M. Whitesides, Department of Chemistry, Harvard University, Cambridge, Massachusetts 02138

Introduction - Enzyme-based synthetic chemistry has continued to grow rapidly in recent years.[1-6] Technical problems, which have inhibited the widespread use of enzymes as catalysts, have been much reduced by the introduction of new procedures for enzyme immobilization and stabilization and for *in situ* cofactor recycling. A widespread interest in asymmetric synthesis has focussed attention on the demonstrated utility of enzymic catalysis in producing chiral fragments. In fact, the major hindrance to the widespread use of enzymic catalysis is the residual unfamiliarity of many classically trained synthetic chemists in the techniques of enzyme isolation, manipulation, and assay. This last barrier is disappearing as biochemistry and enzymology become an accepted part of the education of an organic chemist.

This review emphasizes procedures which use partially or highly purified enzymes to catalyze organic reactions potentially useful in medicinal chemistry. Biochemical procedures using microbiological transformations or cell culture are not discussed.

General Techniques - More than two thousand enzymes are known,[7] and several hundred can be obtained commercially. Many other enzymic activities are available through straightforward isolations or small-scale fermentations. Increasing attention is being paid to large-scale production of enzymes;[8] Kula has developed efficient methods based on liquid-liquid extractions.[9-11] Given sufficient demand, many enzymes can be produced in quantity by recombinant DNA techniques.

We note that highly purified enzyme preparations are not always necessary in synthetic applications since contaminating enzymes may have no effect on the reactants and products present in the reaction mixture. If an enzyme to be used in synthesis has intrinsically low specific activity (units of catalytic activity per weight of protein), crude preparations can cause practical problems by requiring large volumes of immobilization medium and correspondingly large reactor volumes. The use of microbial cells in enzyme-catalyzed synthesis represents a limiting case in purification; since *no* purification is involved, their manipulation is straightforward. The activity of these preparations may be low. In favorable cases, however, they represent

ISBN 0-12-040519-9

the simplest basis for enzyme-catalyzed reactions, and several successful industrial processes have been developed using immobilized whole cells.[12]

Enzyme Immobilization - Enzymes used in synthetic applications are commonly immobilized in or on insoluble materials because immobilization enhances their stability and allows their recovery and reuse. While bench-scale experiments are most conveniently carried out as batch processes, industrial processes often depend on long-lived immobilized enzymes in continuous processes.

Many immobilization methods have been developed.[12-14] Only a few of these techniques deserve explicit comment. Glutaraldehyde is the most commonly used bifunctional reagent in immobilization,[12] forming covalent linkages of still incompletely defined nature between amino groups. It is used to bind enzymes to solid supports or cross-link enzymes adsorbed on a support, and to cross-link enzymes with carrier proteins or with themselves to form insoluble aggregates. Immobilization procedures for industrial applications based on functionalized ceramics cross-linked with enzyme via glutaraldehyde have been developed.[14,15] Wood _et al_. have developed an immobilization procedure using polyurethane-based membranes.[16] Whitesides _et al_. have developed an immobilization method based on polyacrylamide-_co_-N-acryloxysuccinimide cross-linked with triethylene tetramine in the presence of enzyme.[17] This method is particularly useful with the relatively delicate enzymes useful in organic synthesis. Kula _et al_. have used membrane reactors containing soluble enzymes.[18] The enzymes are not immobilized, but reactor performance has been good.

Enzyme Stabilization - Enzyme stabilization is a concern before, during, and after immobilization. Immobilization often greatly increases the stability of enzymes.[19-22] Reasons for this stabilization are not clearly understood. Enzyme deactivation during the immobilization process is often troublesome. Addition of substrates or inhibitors of the enzyme during immobilization helps to occupy and protect the active site, and increases yields on immobilization.[17] A number of strategies are useful in maintaining activity in soluble and immobilized enzymes during use.[23-26] Thiol reagents (dithiothreitol, β-mercaptoethanol, 1,3-dithiopropan-2-ol) maintain the reduced state of catalytically essential thiols in the enzyme. Chelating agents inhibit metal ion catalyzed oxidations of enzymes.[27] The stability of soluble enzymes can be enhanced by the addition of polyols, salts, and certain polymers, and by chemical modification.[23,28-32] Thermally inactivated immobilized enzymes have been reactivated by thiol reagents and reversible heat treatment.[33]

Cofactor Regeneration - About 70% of enzymes use nucleoside triphosphates, nicotinamide derivatives [NAD(P)(H)], or CoA as cofactors. These enzymes include many of those of greatest interest in the synthesis of fine chemicals. Since these cofactors are too expensive to be used stoichiometrically, it has been necessary to develop recycling systems for them. The problem of recycling the nucleoside triphosphates is essentially solved at the level of laboratory-scale synthesis,[34-38] and that of recycling NAD derivatives is well-advanced toward solution.[39-46] None of these schemes has been tested on a production scale, although they work satisfactorily for syntheses of several moles of products. Recycling of ATP from ADP or AMP rests on the development of practical syntheses of the phosphate donors acetyl phosphate[37] and phosphoenolpyruvate[35] to be used with the enzymes acetate kinase and pyruvate kinase, respectively. Acetate kinase is also applicable to recycling of GTP, UTP, CTP and the corresponding 2′-deoxy-nucleoside triphosphates.[1,34] Regeneration of these species from nucleoside monophosphates is not truly practical since adenylate kinase is specific for AMP, but relatively few reactions generate nucleoside monophosphate. CoA recycling has not been explored. Recycling of S-adenosyl-L-methionine, a cofactor in enzyme-catalyzed transmethylations, is currently difficult.[47] The conversion of reduced nicotinamide cofactors [NAD(P)H] to the oxidized form [NAD(P)], developed by Jones _et al._, is the most widely used procedure for this transformation, but suffers from the need for large amounts of flavin and from slow reaction rates.[39] Oxidative regeneration based on the conversion of α-ketoglutarate to glutamic acid works well and does not require the presence of oxygen.[36] The best system for reductive regeneration of NADH from NAD is based on formate dehydrogenase.[18,46,48] This procedure works very well, although the enzyme does not accept NADP as a substrate and is relatively expensive. The most practical procedure for regeneration of NADPH uses glucose-6-phosphate dehydrogenase.[42]

Equilibrium Manipulation - Enzymes are catalysts and therefore serve only to accelerate attainment of equilibrium. In many instances the equilibrium constant for a given reaction in water does not adequately favor the desired product or water acts as an undesired reactant. Sometimes product can be favored using excess starting material, or by reacting the product in an irreversible manner.[49] Occasionally precipitation of product will drive an unfavorable reaction, but precipitation can foul immobilized enzymes. Kinetic control of a reaction can sometimes be used to maximize yields.[49] In cases where less polar materials are being manipulated, water miscible organic cosolvents have been used, but this technique can reduce enzyme activity.[50-53] Careful selection of cosolvent can, however, maintain enzyme activity and dramatically shift equilibrium.[54] A more general approach for working with less polar compounds is the use of a two-phase system incorporating a water-immiscible organic solvent. This approach has been discussed extensively by Martinek and coworkers.[55-59]

Enzyme Mediated Processes

Simple Hydrolases and Isomerases - These are enzymes requiring no added cofactors, and which catalyze hydrolyses, isomerizations, some condensations, and related reactions. Such enzymes are among the simplest to use and are the most widely used in industry (Table 1). A valuable introduction to processes using these enzymes has been published.[12]

Table 1. Selected Industrial Applications of Enzymes.

			ref.
penicillin-G	⟶	6-aminopenicillanic acid	12,60,61
fumaric acid	⟶	L-aspartic acid	12,62
fumaric acid	⟶	L-malic acid	12,63
starch	⟶	glucose	30,64
glucose	⟶	fructose	30,65,66
N-acetyl-D,L-amino acids	⟶	L-amino acids	12
L-arginine	⟶	L-citrulline	12

Amidases - Acylase is used in continuous production of L-amino acids.[12] It has been used to resolve D,L-phenylalanine[67] and α-formyl-ϵ-acyl-D,L-lysine.[68] Penicillins have been synthesized from 6-aminopenicillanic acid with penicillinase.[69]

Esterases - Esterases from several sources have been used in stereoselective and regioselective hydrolysis of esters in the preparation of chiral esters, acids, and alcohols. Sih and coworkers have developed a valuable theoretical treatment for quantitative analysis of such biochemical kinetic resolutions relating the extent of conversion (c) of racemic substrate, the optical purity (ee) of the starting material and products, and the enantiomeric selectivity of the enzyme.[70] Recycling of enantiomerically enriched substrate can greatly increase ee. Racemic threo ester 1 was hydrolyzed with pig liver esterase (PLE) to (-)-acid 2 (64% ee). Reesterification and incubation to 80% c resulted in 2 with >90% ee. Erythro ester 3 was treated with *Gliocladium roseum* to hydrolyze the 2S,3S isomer. At 70% c the remaining ester had 95% ee.

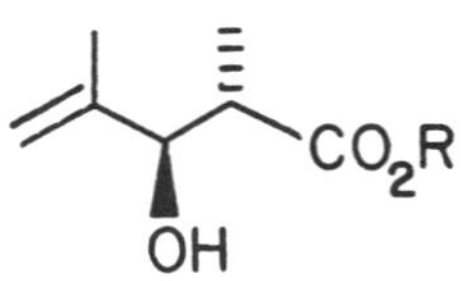

(±) 1, R = CH_3
(-) 2, R = H

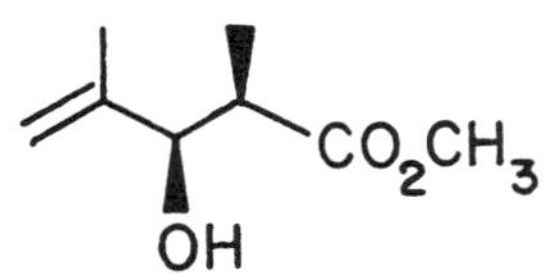

(+) 3

PLE was used in the enantiotopically selective hydrolysis of diethyl-β-methylglutarate 4 to the half-ester 5. Reduction, derivatization and resolution raised the ee of 6 from 69% to 91%.[70]

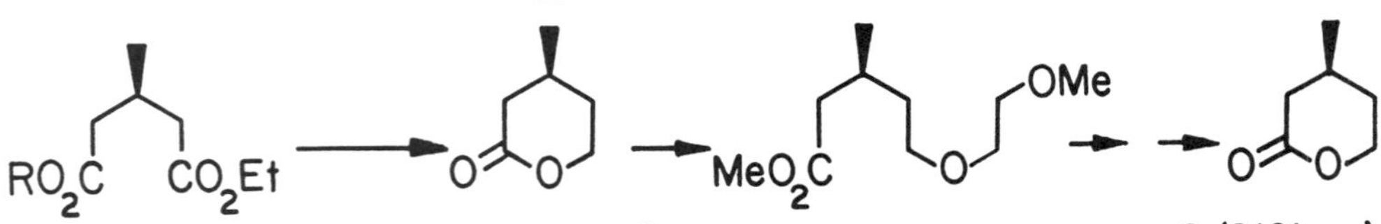

4, R = Et; 5, R = H 6 (69 % ee) 6 (91% ee)

Substituent effects have been studied in PLE-catalyzed hydrolysis of a range of symmetrical dicarboxylic acids and a model has been developed to determine absolute configuration in the monoester products.[71]

Esterases have successfully resolved compounds 7-11. Imidazolone 7 was converted to (+)-biotin.[72] Dimethyl-2,4-dimethylglutarates 8 and 9 were resolved with PLE and *G. roseum*.[73] Dimethyl-β-aminoglutarate 10 was used as starting material for syntheses of R- and S-4-[(methoxycarbonyl)methyl]-2-azetidinones,[74] useful nuclei for carbapenem β-lactam antibiotic synthesis. R-Esters of 3-hydroxy-3-methylalkanoic acids 11 were prepared for testing as inhibitors of 3-hydroxy-3-methylglutarylCoA reductase and in compactin analogue syntheses.[75] Schneider and coworkers have used PLE-catalyzed hydrolyses in asymmetric syntheses of (1R, 3R)-chrysanthemic 12, permethrinic 13 and caronic acid 14 derivatives,[76] and in syntheses of disubstituted monoalkyl malonates[77] and 1,2-cycloalkane dicarboxylate monomethyl esters.[78]

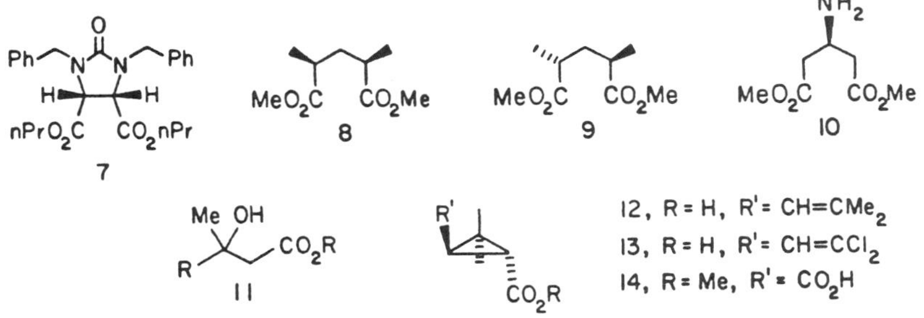

7, 8, 9, 10, 11

12, R = H, R' = CH=CMe2
13, R = H, R' = CH=CCl2
14, R = Me, R' = CO2H

Cambou and Klibanov[79] have used transesterifications catalyzed by PLE and yeast lipase to resolve a variety of alcohols with great effectiveness. Their novel approach employed a biphasic system of aqueous enzyme solution absorbed in a porous solid phase placed in a mixture of "matrix ester" (methyl propionate or tributyrin) and racemic alcohol (Scheme I).

Scheme I.

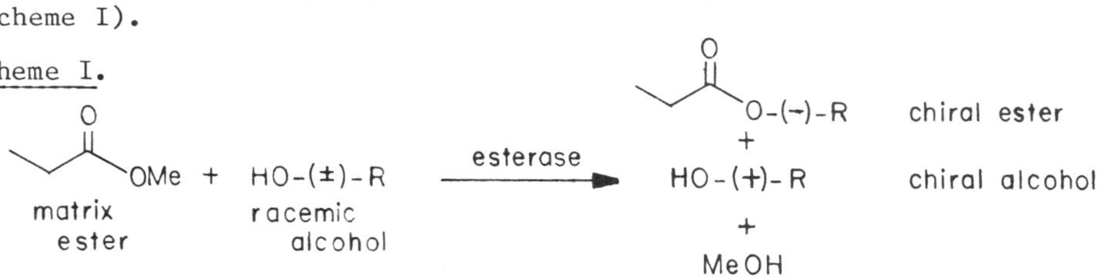

The low water content of this system simplified subsequent purification and effectively suppressed ester hydrolysis. Lipases have also been applied to resolutions of chloroglycerol derivatives,[80,81] 3-acetylthiocycloheptene,[82] and 2-chloromethyl-1-propyl propionate.[83] Cholinesterases have been used to resolve D,L-carnitine.[84]

Proteases - Proteases have been applied to several synthetic purposes the most important of which are peptide bond synthesis and protein semisynthesis. Recent extensive reviews cover this area.[85,86] Proteases have been used in ester synthesis[87] and resolution.[88] Semisynthesis of human insulin has been achieved by enzymic removal and replacement of one amino acid in porcine insulin.[89,90] All peptide bonds in the N-terminal octapeptide of dynorphin [H-Tyr-Gly-Gly-Phe-Leu-Arg-Arg-Ile-OH] have been formed using proteases.[91] A precursor of aspartame has been made by thermolysin-catalyzed condensation of benzyloxycarbonyl-L-aspartic acid and L-phenylalanine methyl ester.[92]

Aldolases - Aldolases catalyze reversible aldol condensations of sugars.[93] A well-studied enzyme is fructose-1,6-diphosphate aldolase from rabbit muscle. This enzyme exhibits a high specificity for dihydroxyacetone phosphate as the nucleophile, but tolerates a range of aldehydes as electrophiles (Scheme II).[94] This broad specificity allows synthesis of sugars such as 6-deoxyfructose[95] and isotopically labeled glucose derivatives.[94] Other aldolases exist with different substrate specificities for possible application to preparative sugar synthesis.

Scheme II.

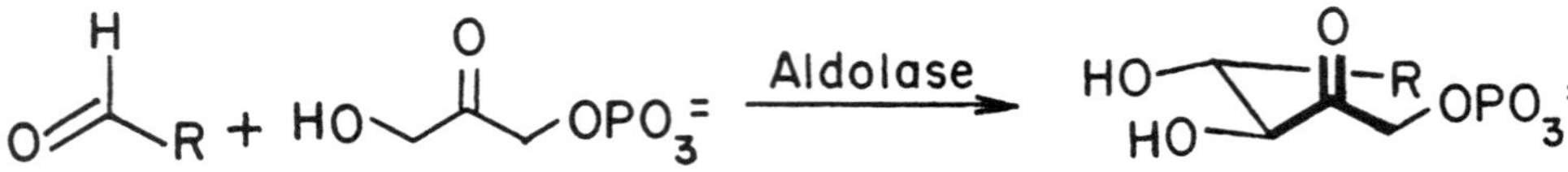

Other - Farnesylpyrophosphate synthetase has been used in asymmetric synthesis of isoprenoids.[96] Potato acid phosphatase has been applied to mild hydrolysis of polyprenyl pyrophosphates.[97] Sulfatase-catalyzed hydrolysis of β-napthol sulfate has been used to separate α- and β-napthols.[98] NAD[99] and flavin adenine dinucleotide[100] have been made by enzymic coupling reactions.

Cofactor Requiring Enzymes

Phosphorylation - Enzymic phosphorylation with coupled in situ ATP regeneration has been used to prepare glucose-6-phosphate,[101] sn-glycerol-3-phosphate,[102] creatine-phosphate,[103] and 5-phosphoribosyl-1-pyrophosphate.[104]

Chiral Redox Chemistry - Nicotinamide cofactor dependent oxidoreductases have been applied in chiral synthesis. Enantioselectivity is often only moderate, but in certain cases excellent results have been obtained. Horse liver alcohol dehydrogenase (HLADH) is the most widely explored enzyme for these purposes with most of the work in this area having been done by Jones and coworkers.[105] HLADH has been studied sufficiently thoroughly to allow modeling of the active site.[106] Cis-3-(2-hydroxyethyl)cyclopentanol 15 was oxidized by HLADH to (+)3S-(2-hydroxyethyl) cyclopentanone 16 with 97% ee; the remaining diol had 70% ee.[107]

OH (±) 15

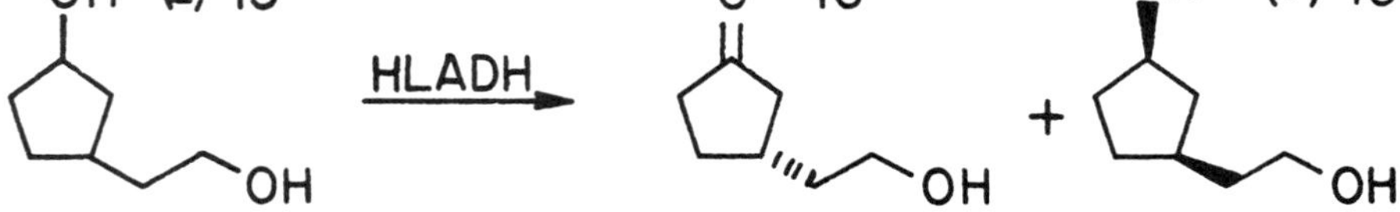

2-Substituted tetrahydropyran-4-ones were reduced to (-)trans alcohols with 100% ee.[108] Remaining (+)ketones had 51-86% ee. HLADH-catalyzed oxidation of monocyclic mesodiols 17 proceeds to bicyclic lactones 18 in

17

18 (100% ee)

high yield with complete stereotopic selectivity.[109] Essentially no selectivity is seen in oxidation of the trans diols. Cis-decalin-2,7-dione 19 is reduced specifically to (-)(7S,9S,10R)-7-hydroxy-cis-decalin-2-one 20[110,111] which can be converted to (+)4R-twistanone 21 in 51% yield from the dione. D- and L-lactic dehydrogenases reduce chloropyruvic acid 22 to D- and L-chlorolactic acids 23; these can be

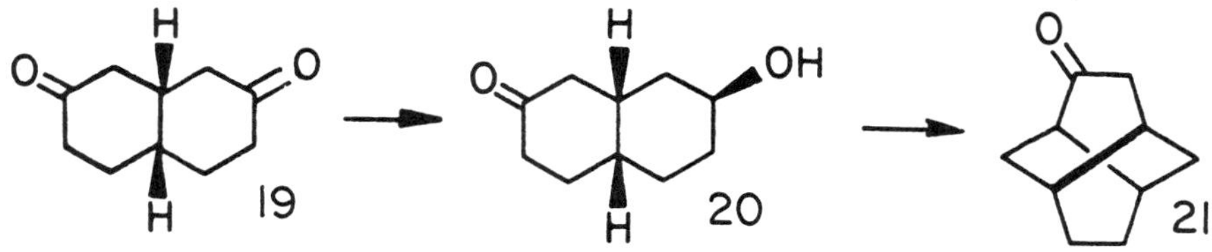

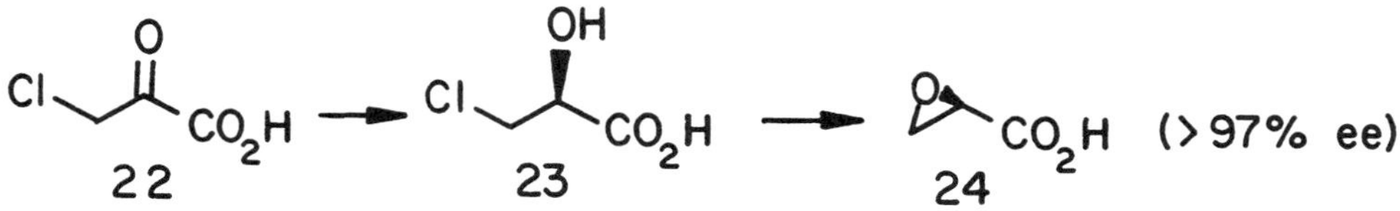

converted to epoxyacrylic acids 24.[112] Progesterone has been reduced to 20-β-hydroxy-pregn-4-ene-3-one with 20-β-hydroxysteroid dehydrogenase using reversed micelles in organic solvent and H_2 as ultimate reductant.[113] Microbial reductions have been used in intermediate steps in syntheses of natural brefeldin-A,[114] (+)compactin[115] and L-carnitine.[116] L-leucine dehydrogenase has been used in the reductive amination of α-ketoisocaproate to L-leucine[18] in a membrane reactor. NAD was covalently modified by attachment of polyethylene glycol to make it unable to cross the membrane.

Multi-Enzyme Cofactor Requiring Processes - A more complex level of applied enzymology is reached in the use of multi-enzyme schemes to synthesize complex molecules. Examples of such syntheses include ribulose-1,5-diphosphate,[36] important in the study of ribulose-diphosphate carboxylase; lactosamine,[117] from the first use of the Leloir pathway enzymes in synthesis; and S-adenosyl-L-methionine.[47]

Oxidations - Enzymes which functionalize inactivated carbon are often difficult to obtain and handle. Many examples exist of oxidations using microbial fermentations,[118] e.g. for steroids, and recently in olefin oxidation.[119] Such preparative transformations have not been achieved with purified enzymes and are unlikely to be amenable to large-scale *in vitro* approaches because of instability and complexity of the enzyme systems. Klibanov and coworkers, however, have developed systems with horseradish peroxidase[120,121] and xanthine oxidase[122] for oxidation of aromatic alcohols and amines, for use in syntheses and waste water treatment. Hydroxyphenyl compounds can be oxidized to dihydroxy derivatives. L-DOPA has been made from L-tyrosine in this manner.[123] Cyclohexanone was oxidized to ε-caprolactone with a bacterial oxygenase.[124]

The Future - Enzymic synthetic methods will see increased use in research and industry. Numerous examples exist of preparations of useful quantities of chiral compounds for use in synthesis, and the use of hydrolytic enzymes for simple synthesis. More importantly, enzymes will allow the facile synthesis of complex molecules important in biological research. Immunology, neurobiology, endocrinology, molecular genetics, membrane biology, and plant and insect biology are areas becoming more molecular in scope. Research in such fields will increasingly depend on biologically active compounds not readily accessible by more conventional chemistry. Molecules that are water soluble, or highly functionalized such as carbohydrates, nucleic acids, lipids, and proteins may prove available by enzymic synthetic methods. Enzymology will be useful in modifications of poly- and oligosaccharides and proteins. Enzymes will also see growth in applications in medical diagnostics and treatment, and in food chemistry.[125]

Recombinant DNA and RNA methods rely on enzymes, and as these methods are developed, the opportunity for enzyme engineering of synthetic catalysts will grow.[126,127] Enzyme-based synthetic methods will be an important part of future organic synthesis, especially in synthesis of new pharmaceutical products.

Acknowledgement - We acknowledge financial support from the National Institutes of Health, Grant GM 30367.

References

1. G.M. Whitesides and C.-H. Wong, Aldrichimica Acta, 16, 27 (1983).
2. J.B. Jones in "Asymmetric Synthesis," J.D. Morrison, Ed., Academic Press, New York, N.Y., 1983.
3. C.J. Suckling and H.C.S. Wood, Chem. Brit., 15, 243 (1979).
4. J.B. Jones, C.J. Sih and D. Perlman, Eds., "Applications of Biochemical Systems in Organic Chemistry," Tech. Chem., 10, Wiley, New York, N.Y., 1976.
5. J.B. Jones, Method. Enzymol., 44, 831 (1976).
6. C.J. Sih, E. Abushanab and J.B. Jones, Annu. Repts. Med. Chem., 12, 298 (1977).
7. International Union of Biochemistry. Nomenclature Committee. "Enzyme Nomenclature, 1978," Academic Press, New York, N.Y. (1979).
8. B.S. Hartley, T. Atkinson and M.D. Lilly, Eds., Phil. Trans. R. Soc. Lond., B 300, 237 (1983).
9. M.R. Kula, K.H. Kroner and H. Hustedt. Adv. Biochem. Eng., 24, 73 (1982).
10. P.-A. Albertsson, Adv. Protein Chem., 24, 309 (1970).
11. P.-A. Albertsson, "Partition of Cell Particles and Macromolecules," 2nd ed., Almquist and Wiksell, Stockholm, Wiley-Interscience, N.Y., 1971.
12. I. Chibata, "Immobilized Enzymes--Research and Development." Halsted Press, New York, N.Y., 1978.
13. K. Mosbach, Ed., "Immobilized Enzymes," Method. Enzymol., 44 (1976).
14. A.M. Klibanov, Science, 219, 722 (1983).
15. R.P. Rohrbach, U.S. 4,268,419 (1979).
16. L.L. Wood, F.J. Hartdegen, P.A. Hahn, U.S. 4,312,946 (1982).
17. A. Pollak, H. Blumenfeld, M. Wax, R.L. Baughn and G.M. Whitesides, J. Am. Chem. Soc., 102, 6324 (1980).
18. C. Wandrey, A.F. Buckmann and M.R. Kula, Biotechnol. Bioeng., 23, 2789 (1981).
19. A.M. Klibanov, Anal. Biochem., 93, 1 (1979).
20. K. Martinek, A.M. Klibanov, V.S. Goldmacher and I.V. Berezin, Biochim. Biophys. Acta, 485, 1 (1977).
21. K. Martinek, A.M. Klibanov, V.S. Goldmacher, A.V. Tchernysheva, V.V. Mozhaev, I.V. Berezin and B.O. Glotov, Biochim. Biophys. Acta, 485, 13 (1977).
22. A.M. Klibanov, N.O. Kaplan and M.D. Kamen, Arch. Biochem. Biophys., 199, 545 (1980).
23. R.D. Schmid, Adv. Biochem. Eng., 12, 41 (1979).
24. A.M. Klibanov, Biochem. Soc. Trans., 11, 19 (1983).
25. A.M. Klibanov, N.O. Kaplan and M.D. Kamen, Enzyme Eng., 5, 135 (1980).
26. A.M. Klibanov, N.O. Kaplan and M.D. Kamen, Proc. Nat. Acad. Sci. U.S.A., 75, 3640 (1978).
27. A.M. Klibanov, N.O. Kaplan and M.D. Kamen, Biochim. Biophys. Acta, 547, 411 (1979).
28. J.B. Jones and D.R. Dodds, Can. J. Chem., 57, 2533 (1979).
29. S. Branner-Jørgensen, Biochem. Soc. Trans., 11, 20 (1983).
30. C. Bucke, Biochem. Soc. Trans., 11, 13 (1983).
31. K. Gekko and S.N. Timasheff, Biochem., 20, 4667; 4677 (1981).
32. T. Arakawa and S.N. Timasheff, Biochem., 21, 6536; 6545 (1982).
33. A.M. Klibanov and V.V. Mozhaev, Biochem. Biophys. Res. Comm., 83, 1012 (1978).
34. C.-H. Wong, S.L. Haynie and G.M. Whitesides, J. Am. Chem. Soc., 105, 115 (1983).
35. B.L. Hirschbein, F.P. Mazenod and G.M. Whitesides, J. Org. Chem., 47, 3765 (1982).
36. C.-H. Wong, A. Pollak, S.D. McCurry, J.M. Sue, J.R. Knowles and G.M. Whitesides, Method. Enzymol., 89, 108 (1982).
37. D. Crans and G.M. Whitesides, J. Org. Chem., 48, 3130 (1983).
38. R.L. Baugh, O. Adalsteinsson, and G.M. Whitesides, J. Am Chem. Soc., 100, 304 (1978).

39. J.B. Jones and K.E. Taylor, Can. J. Chem., 55, 2969 (1976).
40. C.-H. Wong and G.M. Whitesides, J. Am. Chem. Soc., 104, 3542 (1982).
41. C.-H. Wong and G.M. Whitesides, J. Org. Chem., 47, 2816 (1982).
42. C.-H. Wong and G.M. Whitesides, J. Am. Chem. Soc., 103, 4890 (1981).
43. C. Burstein, H. Ounissi, M.D. Legoy, G. Gellf and D. Thomas, Appl. Biochem. Biotechnol., 6, 329 (1981).
44. E. Chave, E. Adamowicz and C. Burstein, Appl. Biochem. Biotechnol., 7, 431 (1982).
45. C.-H. Wong and G.M. Whitesides, J. Am. Chem. Soc., 105, 5012 (1983).
46. Z. Shaked and G.M. Whitesides, J. Am. Chem. Soc., 102, 7104 (1980).
47. A. Gross, S. Geresh and G.M. Whitesides, Appl. Biochem. Biotechnol., 8, 415 (1982).
48. H. Schutte, M.R. Kula and H. Sahm, Method. Enzymol., 89, 527 (1982).
49. K. Martinek and A.N. Semenov, J. Appl. Biochem., 3, 93 (1981).
50. L.G. Butler, Enzyme Microb. Technol., 1, 253 (1979).
51. J.B. Jones and H.M. Schwartz, Can. J. Chem., 60, 335 (1982).
52. J.B. Jones and D.H. Pliura, Can. J. Chem., 59, 2921 (1981).
53. J.B. Jones and M.M. Mehes, Can. J. Chem., 57, 2245 (1979).
54. G.A. Homandberg, J.A. Mattis and M. Laskowski, Jr., Biochem., 17, 5220 (1978).
55. K. Martinek, A.N. Semenov and I.V. Berezin, Biochim. Biophys. Acta, 658, 76 (1981).
56. K. Martinek and A.N. Semenov, Biochim. Biophys. Acta, 658, 90 (1981).
57. K. Martinek, A.N. Semenov and I.V. Berezin, Dokl. Akad. Nauk. SSSR, 253, 358 (1980).
58. K. Martinek, A.N. Semenov and I.V. Berezin, Dokl. Akad. Nauk. SSSR, 252, 394 (1980).
59. A.M. Klibanov, G.P. Samokhin, K. Martinek and I.V. Berezin, Biotechnol. Bioeng., 19, 1351 (1977).
60. B.J. Abbott, Adv. Appl. Microbiol., 20, 203 (1976).
61. E. Lagerlof, L. Nathorst-Westfelt, B. Ekstrom and B. Sjoberg, Method. Enzymol., 44, 759 (1976).
62. I. Chibata, T. Tosa and T. Sato, Method. Enzymol., 44, 739 (1976).
63. I. Chibata, T. Tosa and I. Takata, Trends. Biotechnol., 1, 9 (1983).
64. M.K. Weibel, W.H. McMullen and C.A. Starace, Dev. Ind. Microbiol., 19, 103 (1977).
65. N.B. Havewala and W.H. Pitcher, Jr., Enzyme Eng., 2, 315 (1974).
66. N.H. Mermelstein, Food Technol., 29, 20 (1975).
67. T. Kitahara and S. Asai, Agric. Biol. Chem., 47, 991 (1983).
68. Y. Minematsu, Y. Shimohigashi, M. Waki and N. Izumiya, Bull. Chem. Soc. Jap., 51, 1899 (1978).
69. A.N. Semenov, K. Martinek, V. Yu.-K. Shvyadas, A.L. Margolin and I.V. Berezin, Dokl. Akad. Nauk. SSSR, 258, 1124 (1981).
70. C.-S. Chen, Y. Fujimoto, G. Girdaukas and C.J. Sih, J. Am. Chem. Soc., 104, 7294 (1982).
71. P. Mohr, N. Waespe-Sarcevic, C. Tamm, K. Gawronska and J.K. Gawronska, Helv. Chim. Acta, 66, 2501 (1983).
72. S. Iriuchijima, K. Hasegawa and G. Tsuchihashi, Agric. Biol. Chem., 46, 1907 (1982).
73. C.-S. Chen, Y. Fujimoto, C.J. Sih, J. Am. Chem. Soc., 103, 3580 (1981).
74. M. Ohno, S. Kobayashi, T. Iimori, Y.-F. Wang and T. Izawa, J. Am. Chem. Soc., 103, 2405 (1981).
75. W.K. Wilson, S.B. Baca, Y.J. Barber, T.J. Scallen and C.J. Morrow, J. Org. Chem., 48, 3960 (1983).
76. M. Schneider, N. Engel, and H. Boensmann, Angew. Chem. Int. Ed. Engl., 23, 64 (1984).
77. M. Schneider, N. Engel, and H. Boensmann, Angew. Chem. Int. Ed. Engl., 23, 66 (1984).
78. M. Schneider, N. Engel, P. Honicke, G. Heineman and H. Gorisch, Angew. Chem. Int. Ed. Engl., 23, 67 (1984).
79. B. Cambou and A.M. Klibanov, J. Am. Chem. Soc., 106, in press (1984).
80. S. Iriuchijima and N. Kojima, Agric. Biol. Chem., 46, 1153 (1982).
81. S. Iriuchijima, A. Keiyu and N. Kojima, Agric. Biol. Chem., 46, 1593 (1982).
82. S. Iriuchijima and N. Kojima, J.C.S. Chem. Comm., 185 (1981).
83. J. Lavayre, J. Verrier and J. Baratti, Biotechnol. Bioeng., 24 2175 (1982).
84. E.P. Dropsy and A.M. Klibanov, Biotechnol. Bioeng. 26, in press (1984).
85. J.S. Fruton, Adv. Enzymol., 53, 239 (1982).
86. I.M. Chaiken, A. Komoriya, M. Ohno and F. Widmer, Appl. Biochem. Biotechnol. 7, 385 (1982).
87. D. Tarquis, P. Monsan and G. Durand, Bull. Soc. Chim. Fr. II., 76 (1980).
88. Y. Nishida, H. Ohrui and H. Meguro, Agric. Biol. Chem., 47, 2123 (1983).
89. K. Morihara, T. Oka, H. Tsuzuki, Y. Tochino and T. Kanaya, Biochem. Biophys. Res. Comm., 92, 396 (1980).
90. K. Inouye, K. Watanabe, K. Morihara, Y. Tochino, T. Kanaya, J. Emura and S. Sakakibara, J. Am. Chem. Soc., 101, 751 (1979).
91. W. Kullman, J. Org. Chem., 47, 5300 (1982).
92. K. Oyama, S. Nishimura, Y. Nonaka, K. Kihara and T. Hashimoto, J. Org. Chem., 46, 5241 (1981).

93. P.D. Boyer, Ed., "The Enzymes," 7, Chapters 6-9 (1972).
94. C.-H. Wong and G.M. Whitesides, J. Org. Chem., 48, 3199 (1983).
95. C.-H. Wong, F.P. Mazenod and G.M. Whitesides, J. Org. Chem. 48, 3493 (1983).
96. T. Koyama, A. Saito, K. Ogura and S. Seto, J. Am. Chem. Soc., 102, 3614 (1980).
97. H. Fujii, T. Koyama, K. Ogura, Biochem. Biophys. Acta, 712, 716 (1982).
98. G. Pelsy, A.M. Klibanov, Biotechnol. Bioeng., 25, 919 (1983).
99. D.R. Walt, M.A. Findeis, V.M. Rios-Mercadillo, J. Augé and G.M. Whitesides, J. Am. Chem. Soc., 106, 234 (1984).
100. S. Shimizu, K. Yamane, Y. Tani and H. Yamada, Appl. Biochem. Biotechnol., 8, 237 (1983).
101. A. Pollak, R.L. Baughn and G.M. Whitesides, J. Am. Chem. Soc., 99, 2366 (1977).
102. V.M. Rios-Mercadillo and G.M. Whitesides, J. Am. Chem. Soc., 101, 5828 (1979).
103. Y.-S. Shih and G.M. Whitesides, J. Org. Chem., 42, 4165 (1977).
104. A. Gross, O. Abril, J.M. Lewis, S. Geresh and G.M. Whitesides, J. Am. Chem. Soc., 105, 7428 (1983).
105. T. Takemura and J.B. Jones, J. Org. Chem., 48, 791 (1983), and previous papers.
106. J.B. Jones and I.J. Jakovac, Can. J. Chem., 60, 19 (1982).
107. A.J. Irwin and J.B. Jones, J. Am. Chem. Soc., 99, 1625 (1977).
108. J.A. Haslegrave and J.B. Jones, J. Am. Chem. Soc., 104, 4666 (1982).
109. I.J. Jacovac, H.B. Goodbrand, K.P. Lok and J.B. Jones, J. Am. Chem. Soc., 104, 4659 (1982).
110. D.R. Dodds and J.B. Jones, J.C.S. Chem. Comm., 1080 (1982).
111. M. Nakazaki, H. Chikamatsu and M. Taniguchi, Chem. Lett., 11, 1761 (1982).
112. B.L. Hirschbein and G.M. Whitesides, J. Am. Chem. Soc., 104, 4458 (1982).
113. R. Hilhorst, C. Laane and C. Veeger, FEBS Lett., 159, 225 (1983).
114. C. LeDrian and A.E. Greene, J. Am. Chem. Soc., 104, 5473 (1982).
115. C.T. Hsu, N.Y. Wang, L.H. Latimer and C.J. Sih, J. Am. Chem. Soc., 105, 593 (1983).
116. B. Zhou, A.S. Gopalan, F. Vanmiddlesworth, W.-R. Shieh and C.J. Sih, J. Am. Chem. Soc., 105, 5925 (1983).
117. C.-H. Wong, S.L. Haynie and G.M. Whitesides, J. Org. Chem., 47, 5416 (1982).
118. K. Kieslich, "Microbial Transformations of Non-Steroid Cyclic Compounds," John Wiley & Sons, New York, N.Y., 1976.
119. H. Ohta and H. Tetsukawa, J.C.S. Chem. Comm., 849 (1978).
120. A.M. Klibanov, B.N. Alberti, E.D. Morris and L.M. Felshin, J. Appl. Biochem., 2, 414 (1980).
121. A.M. Klibanov and E.D. Morris, Enzyme Microb. Technol., 3, 119 (1981).
122. G. Pelsy and A.M. Klibanov, Biochim. Biophys. Acta, 742, 352 (1983).
123. A.M. Klibanov, Z. Berman and B.N. Alberti, J. Am. Chem. Soc., 103, 6263 (1981).
124. J.M. Schwab, J. Am. Chem. Soc., 103, 1876 (1981).
125. H. Samejima, K. Kimura and Y. Ado, Biochemie, 62, 299 (1980).
126. W.H. Rastetter, Appl. Biochem. Biotechnol., 8, 423 (1983).
127. T.H. Maugh, II, Science, 223, 154 (1984).

Chapter 27. Stable Isotopes in Drug Metabolism and Disposition

Thomas A. Baillie and Albert W. Rettenmeier, Department of Medicinal Chemistry, University of Washington, Seattle, WA 98195
Lisa A. Peterson and Neal Castagnoli, Jr., Department of Pharmaceutical Chemistry, University of California, San Francisco, CA 94143

Since the 1977 review in Annual Reports,[1] the applications of stable-isotope-labeled compounds to biomedical research have increased dramatically. Three international conferences devoted to the use of stable isotopes in the life sciences[2-4] and one to synthesis and general applications[5] have been held. Review articles have focused on mass spectrometric-based analytical methods,[6,7] on applications of stable isotopes in clinical[8-10] and pharmacological research,[11-13] and on more specific studies concerning problems in drug metabolism,[14,15] pharmacokinetics[16,17] and synthesis.[18,19] The present review covers mainly the literature published since 1981 and focuses on recent developments in the use of stable isotopes for investigations of drug metabolism and disposition.

Metabolite Structure Elucidation

Isotope Clusters - The isotope cluster (isotope doublet, twin ion) technique has proved to be a powerful method for the detection and structural identification of drug metabolites in complex biological matrices.[1,11] An equimolar mixture of [8-^{13}C;1,3-$^{15}N_2$]caffeine and unlabeled caffeine was used to establish that the acetyl group of the metabolite, 4-acetyl-6-amino-3-methyluracil, did not contain the original C-8 carbon atom since it was labeled with two ^{15}N atoms only.[20] In this and a related study,[21] [2-^{14}C]caffeine was co-administered to aid in monitoring the excretion of drug metabolites. Although a useful adjunct to the stable isotope cluster technique, the administration of radiotracers in human research is limited on ethical grounds, particularly in pediatric and obstetric populations. Studies on the biotransformation of theophylline in premature newborns[22] and on the placental transfer and fetal metabolism of this compound in humans[23] relied exclusively on the use of [2-^{13}C;1,3-$^{15}N_2$]theophylline as tracer. The stable isotopes of carbon and nitrogen also have been successfully applied to ion cluster metabolism studies of [^{13}C, $^{15}N_2$]hexobarbital.[24]

Although the use of deuterium in drug metabolism involving the isotope cluster technique may be complicated by isotope effects,[25] both cost and convenience have led most workers to rely on this isotope. Recent examples include reports on aminopyrine,[26] pencycuron,[27] phencyclidine,[28,29] fentanyl,[30] and diethylstilbestrol.[31] In the case of steroids and their derivatives, ion cluster studies have been performed exclusively with deuterium-labeled substrates. Examples include investigations with [$^2H_5$]ethynylestradiol,[32] [$^2H_2$]4-hydroxyandrost-4-ene-3,17-dione,[33] [$^2H_2$]-ursodeoxycholic acid,[34] and [$^2H_8$]budesonide.[35]

ISBN 0-12-040519-9

The isotope cluster technique has proved valuable in the identification of metabolites similar or identical to endogenous compounds present in the biological extracts. Use of a hexadeutero variant of valproic acid, a short-chain fatty acid, led to the identification of a new biotransformation product, 2-*n*-propyl-4-oxopentanoic acid.[36] A similar approach was used to reveal a novel pathway of benzoic acid metabolism in the horse.[37-39] Metabolites of rutin, a flavinol glycoside, were identified in the urine of rats[40] and humans[41] given [2',5',6'-$^{2}H_3$]rutin. Deuterium-labeled phenylephrine[42] has been used in a similar fashion.

Derivatization of samples with an equimolar mixture of labeled and unlabeled reagents to produce ion clusters has been applied recently to the characterization of the bleomycins by FD and FAB mass spectrometry.[43] A variation of this technique utilizes a mixture of labeled and unlabeled reagent gases {pyridine/[$^{2}H_5$]pyridine,[44] trimethylchlorosilane (TMCS)/[$^{2}H_9$]TMCS,[45] tetramethylsilane (TMS)/[$^{2}H_{12}$]TMS,[45] and $NH_3/N^{2}H_3$[46,47]} in conjunction with CI mass spectral analyses. The isotope cluster technique also may be exploited with the newer ionization methods of SIMS[48] and FAB[49] through the addition of appropriate counter-ions (for example, Ag^{+}[49]).

Work continues on the development of computer software for automatic screening of large amounts of mass spectral data for specified isotope clusters.[50,51]

The Shift Technique - The stable isotope shift technique refers to a procedure in which the mass spectrum of an unlabeled compound is compared with that of a stable-isotope-labeled counterpart. The observed differences in masses between corresponding ions in the two spectra provide valuable information on ion composition and hence on molecular structure.

This technique may be used in two distinct ways. The drug may be labeled at a specific site(s) and the mass spectra of metabolites derived from the labeled compound compared with those obtained from the unlabeled drug. Recent examples of this application include reports on the metabolism of oxybutynin,[52] tocainide,[53] and ketamine.[54] Alternatively, a heavy isotope (usually ^{2}H) is introduced into the metabolite of interest by reaction with a labeled derivatizing reagent. For example, perdeuteromethylation results in a mass difference of 3 daltons between labeled and unlabeled derivatives per methyl group. Structures of metabolites obtained from caffeine[55] and phenytoin[56] have been confirmed by this technique. Tri(deuteromethyl)silylation has been employed in the identification of metabolites of carbamazepine[57] and afloqualone.[58]

NMR Studies - The applications of stable isotopes coupled with NMR analysis have multiplied as higher field magnets and sophisticated computer techniques have increased sensitivity. A recent review which discussed NMR studies employing stable isotopes is available.[59]

Natural abundance ^{13}C-NMR spectroscopy has been applied to the structure analysis of several metabolites derived from acetaminophen.[60] The structures of metabolites of prenalterol[61] and *trans*-sobrerol[62] have been studied. ^{13}C-NMR analyses of metabolites derived from ^{13}C-enriched amitriptyline,[63] BHT,[64] propachlor[65] and aminopyrine[66,67] have been reported.

Mechanistic Studies

Cytochrome P-450 Catalyzed Reactions - Studies with $^{18}O_2$ have established that the cytochrome P-450 mediated hydroxylation of camphor by the bacterial enzyme[68] and the enzyme purified from rat liver[69] results in the incorporation of atmospheric oxygen in the 5-exo-hydroxylation product. Cumene [$^{18}O_2$]hydroperoxide will transfer its peripheral oxygen atom to a variety of compounds which serve as substrates for mammalian cytochrome P-450.[70] As expected, $^{18}O_2$ served as the oxygen source for the bacterial cytochrome P-450 catalyzed epoxidation of 5,6-dehydrocamphor.[71] N-Hydroxymethylcarbazole, formed by the cytochrome P-450 catalyzed oxidation of N-methylcarbazole, incorporates ^{18}O exclusively from dioxygen.[72,73] Under anaerobic conditions cytochrome P-450 may catalyze the intramolecular transfer of oxygen present in tertiary amine N-oxides.[70] Mechanistic studies on S-dealkylation and S-oxidation reactions also have used ^{18}O tracer methods.[74]

The oxidation of 1,2-dideuterocyclohexene by both cytochrome P-450 and model chemical systems gave cyclohexenol with various degrees of allylic rearrangement of the deuterium atoms, suggesting a mechanism involving caged radical intermediates.[75] The bioactivation of terminal olefins by cytochrome P-450 leads to the alkylation of the heme prosthetic group. Mechanistic studies using trans-1-[^{2}H]octene as substrate demonstrated that the trans-stereochemistry is retained in both the heme alkylation product and the epoxide. This result is inconsistent with heme alkylation occurring by reaction with octene oxide itself.[76] The structures of heme adducts produced from several olefins now have been determined and it has been established, using oxygen-18, that the adducts contain an atom of oxygen derived from the atmosphere.[77] Mechanistic studies with deuterium-labeled arenes have provided evidence against a proton abstraction or direct insertion mechanism for the hydroxylation of certain benzenoid systems at the meta position.[78]

Finally, ^{15}N-labeling has been employed in an investigation of the process through which metabolism of 3,5-diethoxycarbonyl-1,4-dihydrocollidine by cytochrome P-450 leads to self-destruction of the enzyme and formation of N-methylprotoporphyrin IX.[79]

Metabolic Pathways - Compounds labeled at specific sites with stable isotopes have been employed in the identification of short-lived, potentially toxic metabolites.[80] The metabolic formation of reactive episulfonium ions from the conjugation of 1,2-dihaloethanes with glutathione has been studied with deuterium-labeled substrates.[81,82] The detection of iminocyclophosphamide from cyclophosphamide was aided by deuterium and oxygen-18 labeling techniques.[83] Quantitative studies on the α-hydroxylation bioactivation pathway of carcinogenic nitrosamines were approached by measuring $^{15}N_2$ evolution from doubly ^{15}N-labeled N-nitrosodimethylamine and N-nitrosomethylaniline from rat liver homogenates.[84] The evolution of ^{13}CO from $^{13}CCl_4$ could be distinguished from unlabeled CO derived from the destruction of heme and lipid peroxidation.[85] The two isotopes of chlorine (^{35}Cl and ^{37}Cl) have been used to probe the origin of electrophilic chlorine species generated during the metabolism of CCl_4 in hepatic microsomal preparations.[86] Reductive dehalogenation of CCl_4 to the trichloromethylperoxy radical has been proposed as the initial event which leads to the electrophilic chlorine species.[87] In a study using oxygen-18 as a metabolic tracer, the porphyrinogenic agent allylisopropylacetamide was shown to undergo biotransformation to three different chemically reactive intermediates which appear to alkylate different cellular constituents.[88]

Using $^{18}O_2$ Anderson et al. showed that the two major phenolic metabolites of N-phenyl-2-naphthylamine produced by liver enzymes are likely to be derived via arene oxide intermediates.[89] The mechanism by which bromobenzene is converted to 4-bromocatechol in isolated rat hepatocytes was shown with $^{18}O_2$ to proceed via dehydrogenation of a dihydrodiol intermediate rather than by two successive hydroxylation reactions of the substrate.[90] Similar studies have been performed on chlorpromazine.[91]

Using ^{2}H-labeled drugs, arene oxides have been implicated in the metabolism of phenytoin[92] and propranolol.[93] Deuterium-labeled variants of propranolol also have been used to study stereoselective aspects of its metabolism and disposition.[93,94] Spontaneous decomposition of anti-neoplastic agents such as the 2-haloethylnitrosoureas to give reactive alkylating and carbamoylating intermediates has been examined with deuterium and oxygen-18 labeled compounds.[95-97] Studies of the enterohepatic recycling of ^{13}C- and ^{2}H-labeled mercapturic acid conjugates of propachlor have been reported.[65,98-100]

Deuterium Kinetic Isotope Effects - The analysis of deuterium kinetic isotope effects provides a useful tool for the elucidation of mechanisms of drug metabolism and toxicity.[101-103] Recent studies of deuterium isotope effects on cytochrome P-450 catalyzed reactions have led to a general consensus that these types of oxidations proceed by hydrogen radical abstraction, with the transient formation of a carbon radical prior to recombination with a hydroxyl radical to form the final product.[104-106] Cytochrome P-450 catalyzed oxidative N-dealkylations tend to proceed with modest isotope effects at best.[107-109] Studies with 5-[$^{2}H_5$]phenyl-5-phenylhydantoin showed that no isotope effect was observed for both para- and meta-hydroxylations with rat liver microsomes.[110] Analogous results were obtained with deuterium-labeled warfarin,[78] suggesting that these biotransformations occur by an addition-rearrangement mechanism. In vitro kinetic isotope studies have been reported with monoamine oxidase[111,112] and epoxide hydrolase.[113] The introduction of deuterium at C-6 of penicillanic acid resulted in a significant increase in its β-lactamase inhibiting properties.[114]

In vivo alterations in rates of metabolism and excretion of deuterium-labeled versus unlabeled drugs have been discussed.[115] Such changes in metabolic patterns can influence the pharmacological properties of some drugs such as the increased potency of the hexadeuterated analog of the gastric antisecretory agent N,N-dimethyl-N'-[2-(diisopropylamino)ethyl]-N'-(4,6-dimethyl-2-pyridyl)urea.[116] Substitution of deuterium in the N-methyl groups of imipramine caused a decrease in systemic clearance, increase in half life and, upon oral administration, an increased bioavailability.[117] The uptake, clearance and brain levels of the hallucinogen N,N-dimethyltryptamine were potentiated upon deuterium substitution at the α and β positions of the side chain.[118] The in vivo rate of degradation of the antidepressant phenelzine was found to be retarded in a deuterium-labeled analog,[119] which apparently was responsible for alterations in the observed biochemical,[119] behavioral,[120] and spontaneous motor activity effects of the drug.[121] Kinetic isotope effects also have been used to aid in the elucidation of the in vitro and in vivo metabolic pathways of the anticancer agent 6-mercaptopurine.[122] A variety of isotope effect studies have been conducted in an effort to characterize critical bioactivation steps which may be responsible for the toxicity of xenobiotics. Examples include investigations on the metabolism of allyl alcohol,[123] methoxyflurane,[124] BHT,[125] 1,2-dibromoethane,[126] chloroform[127,128] and carcinogenic nitrosamines.[129-134]

Pharmacokinetic Investigations

Quantitative Applications - The use of stable-isotope-labeled compounds as internal standards for the quantitation of drugs and metabolites in biological fluids offers a unique combination of sensitivity and selectivity of detection for pharmacokinetic studies. The principles of the technique have been outlined, and applications up to 1981 have been compiled.[135] Specific aspects of the isotope dilution method, e.g. its utility as a reference technique[136] and associated procedures for handling the data generated from the use of multiple isotope tracers simultaneously,[137] have been discussed. Examples of isotope dilution methods used in clinical psychopharmacology have also been reviewed.[138]

Deuterium continues to be the most widely used isotope for preparation of labeled standards and appears to be well suited for this purpose. However, certain ^{2}H-labeled compounds have been found to become diluted with the protio forms upon storage, presumably due to isotope exchange phenomena. A case in point is [3,3,4,4-$^{2}H_4$]6-oxoprostaglandin $F_{1\alpha}$.[139] Stable isotopes other than deuterium have been employed with increasing frequency for internal standards. A sensitive assay for glyceryl trinitrate in human plasma by GC-negative ion CIMS made use of two multiply-^{15}N-labeled analogs of the drug. The [$^{15}N_3$] variant was employed to minimize absorption of glyceryl trinitrate on the GC column and a [$^{2}H_5$,$^{15}N_3$] species served as internal standard for quantitative purposes.[140] [^{13}C,$^{15}N_3$]Guanfacine has been used as an internal standard for guanfacine[141,142] and [$^{2}H_3$,$^{13}C_6$]dexamethasone for the corresponding unlabeled corticosteroid.[143]

In certain situations, ^{18}O may be suitable for labeling purposes[144,145] and some exploratory work has been carried out with polychlorinated compounds enriched at > 90% excess with ^{37}Cl.[146,147] In the latter case, greatly simplified molecular ion clusters were evident in the mass spectra of ^{37}Cl-labeled compounds such as heptachlor and related organochlorine insecticides. Although ^{34}S is available commercially, compounds enriched in this isotope do not appear to have been used as yet in the field of drug metabolism.[148] Lithium is important in the treatment of manic depressive illness and a stable isotope dilution assay for ^{6}Li, using ^{7}Li as internal standard, has been published.[149]

Where preparation of stable-isotope-labeled drug metabolites is required, an economical approach is to administer the labeled drug to a suitable animal species and to isolate the corresponding labeled metabolites from urine samples. Deuterium-labeled metabolites of antipyrine were obtained in this manner,[150] while deuterated metabolites of 6-oxo-$PGF_{1\alpha}$ were prepared by incubation of the labeled parent prostaglandin with *Mycobacterium rhodochrous*.[151]

The "soft" ionization MS techniques of FD[152] and FAB[153-155] have been employed successfully for quantitative applications with stable-isotope-labeled internal standards and interesting developments in the use of microwave plasma discharges, either alone or in combination with mass spectrometic detection, promise to extend the utility of stable isotope labeling methods in quantitative studies of drug dispositon.[156,157] New methods have been reported for the measurement of ^{15}N enrichment in ammonia, based on conversion into hexamethylenetetramine,[158] and for the analysis of ^{13}CO by combustion to $^{13}CO_2$.[159,160]

Pharmacokinetics Determined under "Steady-State" Conditions - Administration of a single "pulse dose" of a stable-isotope-labeled analog is an effective method for defining the kinetics of a drug which is being given chronically, as there is no need to interrupt the dosage regimen to conduct the study.[11,16] The value of this technique for use in clinical pharmacology is illustrated convincingly by work on the changes in valproic acid pharmacokinetics which take place during pregnancy. These parameters were determined by following the disappearance of a bolus dose of [1,2-$^{13}C_2$]valproic acid given to a pregnant epileptic patient.[161] For ethical reasons, this type of study could not have been carried out by alternative methodology using radioactive isotopes. A similar investigation of valproic acid pharmacokinetics in non-pregnant humans was performed using a deuterium-labeled variant of the drug.[162]

It is essential that the labeled species chosen for administration be demonstrated to be pharmacokinetically equivalent to its unlabeled counterpart. The likelihood for non-equivalence is remote when ^{13}C or ^{15}N is employed. Reports have appeared in which investigators have validated the use of [^{13}C,^{15}N]phenobarbital[163-165] and [^{13}C,$^{15}N_2$]phenytoin[166] for such applications. Concern over apparent differences *in vivo* between phenytoin and a deuterium-labeled analog led Poupaert *et al.*[167] to synthesize [$^{13}C_6$]phenytoin for pulse dose studies. Nevertheless, when properly validated, drugs labeled with deuterium can be used for administration in kinetic experiments. Recent examples of the latter have involved studies on the influence of long-term infusions of lidocaine on the kinetics of this agent,[168] to investigations of methadone kinetics during maintenance treatment,[169,170] and to stereoselective aspects of the disposition of methadone enantiomers in humans given the racemic drug.[171] The single-dose kinetics of Δ^1-tetrahydrocannabinol were studied in light and heavy cannibis users with the aid of a $^2H_3$-labeled variant of the drug.[172]

Bioavailability Studies - The use of stable isotopes in studies of absolute or relative bioavailability offers a number of important advantages over the conventional "cross-over" approach in which different (unlabeled) formulations are administered on separate occasions.[11,16,173,174] This is especially true for drugs which are subject to extensive first-pass elimination, since these compounds usually exhibit the greatest variation in bioavailability as a function of time.[175] The absolute bioavailabilities of timolol[176,177] and bepridil[178] have been studied by the simultaneous oral and intravenous administration of unlabeled and ^{13}C-labeled forms, respectively. The bioavailability of a standard formulation of clovoxamine fumarate was compared with that of a slow-release preparation using [^{13}C]clovoxamine as the reference dosage form.[179] The absolute bioavailabilities of verapamil[180] and methadone[181] have been investigated with the aid of specifically deuterated analogs, and the results of a pilot study on the oral bioavailability of captopril have been published.[182]

Additional examples of the use of stable isotopes in measurements of drug bioavailability include the study of stereoselective increases in propranolol bioavailability during chronic dosing[183] and the influence of urinary pH and route of administration on meperidine disposition in man.[184] This technique also can be employed to quantify the conversion of prodrugs into the active species *in vivo*, as illustrated in a study on prodrugs of indomethacin.[185]

Conclusions

Stable isotope techniques offer a number of significant advantages over alternative methods for investigations of drug metabolism and disposition. These advantages, some of which have been highlighted in this brief review, are becoming more widely appreciated, and with the growing availability of mass spectrometers suitable for the measurement of isotopically-enriched molecules, the use of stable isotopes in metabolic and pharmacokinetic studies may be expected to increase steadily in the coming years. Although stable isotopes will never replace their radioactive counterparts for many types of application, they have become well established as indispensible tools for use in metabolic investigations.

References

1. S.D. Nelson and L.R. Pohl, Annu. Rep. Med. Chem., 12, 319 (1977).
2. T.A. Baillie, Ed., "Stable Isotopes. Applications in Pharmacology, Toxicology and Clinical Research," The Macmillan Press Ltd., London, 1978.
3. E.R. Klein and P.D. Klein, Eds., "Stable Isotopes. Proceedings of the Third International Conference," Academic Press, New York, 1979.
4. H.-L. Schmidt, H. Förstel and K. Heinzinger, Eds., "Stable Isotopes. Proceedings of the 4th International Conference," Elsevier, Amsterdam, 1982.
5. W.P. Duncan and A.B. Susan, Eds., "Synthesis and Applications of Isotopically Labeled Compounds," Elsevier, Amsterdam, 1983.
6. A.L. Burlingame, T.A. Baillie, P.J. Derrick and O.S. Chizhov, Anal. Chem., 52, 214R (1980).
7. A.L. Burlingame, A. Dell and D.H. Russell, Anal. Chem., 54, 363R (1982).
8. D. Halliday and M.J. Rennie, Clin. Sci., 63, 485 (1982).
9. R.M. Caprioli and D.M. Bier, in "Biochemical Applications of Mass Spectrometry. First Supplementary Volume," G.R. Waller and O.C. Dermer, Eds., Wiley, New York, 1980, p. 895.
10. J. Sjövall, in "Advances in Mass Spectrometry," 8B, A. Quayle, Ed., Heyden and Son, London, 1980, p. 1069.
11. T.A. Baillie, Pharmacol. Rev., 33, 81 (1981).
12. T.A. Baillie, H. Hughes, and D.S. Davies, in "Stable Isotopes," H.-L. Schmidt, H. Förstel, and K. Heinzinger, Eds., Elsevier, Amsterdam, 1982, p. 187.
13. N.J. Haskins, Biomed. Mass Spectrom., 9, 269 (1982).
14. D.R. Hawkins, in "Isotopes: Essential Chemistry and Applications," I.A. Eldridge and I.L. Ioues, Eds., The Chemical Society, London, 1980, p.232.
15. W.J.A. VandenHeuvel, in "Synthesis and Applications of Isotopically Labeled Compounds," W.P. Duncan and A.B. Susan, Eds., Elsevier, Amsterdam, 1983, p. 77.
16. P.J. Murphy and H.R. Sullivan, Annu. Rev. Pharmacol. Toxicol., 20, 609 (1980).
17. M. Eichelbaum, G.E. von Unruh and A. Somogyi, Clin. Pharmacokin., 7, 490 (1982).
18. D.G. Ott, "Syntheses with Stable Isotopes of Carbon, Nitrogen and Oxygen," John Wiley and Sons, New York, 1981.
19. A.P. Tulloch, Prog. Lipid Res., 22, 235 (1983).
20. A.R. Branfman, M.F. McComish, R.J. Bruni, M.M. Callahan, R. Robertson and D.W. Yesair, Drug Metab. Dispos., 11, 206 (1983).
21. M.M. Callahan, R.S. Robertson, A.R. Branfman, M.F. McComish and D.W. Yesair, Drug Metab. Dispos., 11, 211 (1983).
22. J.L. Brazier, B. Ribon, M. Desage and B. Salle, Biomed. Mass Spectrom., 7, 189 (1980).
23. J.L. Brazier, B. Ribon, M. Desage and B. Salle, in "Recent Developments in Mass Spectrometry in Biochemistry, Medicine and Environmental Research," 7, A. Frigerio, Ed., Elsevier, Amsterdam, 1981, p. 27.
24. M. Desage, J.L. Brazier and R. Guilluy, Int. J. Mass Spectrom. Ion Phys., 48, 85 (1983).
25. K. Mimura and S. Baba, Chem. Pharm. Bull. (Tokyo), 29, 2043 (1981).
26. T. Goromaru, T. Furuta, S. Baba, A. Noda and S. Iguchi, Chem. Pharm. Bull. (Tokyo), 29, 1724 (1981).
27. I. Ueyama, S. Kurogochi, I. Kobori, T. Hoshino, Y. Ishii and I. Takase, J. Agric. Food Chem., 30, 1061 (1982).
28. A.K. Cho, R.C. Kammerer and L. Abe, Life Sci., 28, 1075 (1981).
29. G. Hallström, R.C. Kammerer, C.H. Nguyen, D.A. Schmitz, E.W. Di Stefano and A.K. Cho, Drug Metab. Dispos., 11, 47 (1983).
30. T. Goromaru, H. Matsuura, T. Furuta, S. Baba, N. Yoshimura, T. Miyawaki, T. Sameshima, Drug Metab. Dispos., 10, 542 (1982).
31. J. Weidenfeld, P. Carter, V.N. Reinhold, S.B. Tanner IV and L.L. Engel, Biomed. Mass Spectrom., 5, 587 (1978).
32. B.-L. Sahlberg, M. Axelson, D.J. Collins and J. Sjövall, J. Chromatogr., 217, 453 (1981).
33. D.A. Marsh, L. Romanoff, K.I.H. Williams, H.J. Brodie and A.M.H. Brodie, Biochem. Pharmacol., 31, 701 (1982)
34. M. Tohma, Y. Nakata, H. Yamada, T. Kurosawa, I. Makino and S. Nakagawa, Chem. Pharm. Bull. (Tokyo), 29, 137 (1981).

35. S. Edsbäcker, S. Jönsson, C. Lindberg, Å. Ryrfeldt and A. Thalén, Drug Metab. Dispos., 11, 590 (1983).
36. A. Acheampong, F. Abbott and R. Burton, Biomed. Mass Spectrom., 10, 586 (1983).
37. M.V. Marsh, A.J. Hutt, J. Caldwell and R.L. Smith, Biochem. Pharmacol., 30, 1879 (1981).
38. M.V. Marsh, J. Caldwell, A.J. Hutt, R.L. Smith, M.W. Horner, E. Hougton and M.S. Moss, Biochem. Pharmacol., 31, 3225 (1982).
39. M.V. Marsh, J. Caldwell, R.L. Smith, M.W. Horner, E. Houghton and M.S. Moss, Xenobiotica, 11, 655 (1981).
40. S. Baba, T. Furuta, M. Fujioka and T. Goromaru, J. Pharm. Sci., 72, 1155 (1983).
41. S. Baba, T. Furuta, M. Horie and H. Nakagawa, J. Pharm. Sci., 70, 780 (1981).
42. J.R. Crowley, C.M. Williams and M.J. Fregly, J. Pharm. Pharmacol., 35, 264 (1983).
43. A. Dell and H.R. Morris, Biochem. Biophys. Res. Commun., 102, 730 (1981).
44. D.I. Carroll, J.G. Nowlin, R.N. Stillwell and E.C. Horning, Anal. Chem., 53, 2007 (1981).
45. R.N. Stillwell, D.I. Carroll, J.G. Nowlin and E.C. Horning, Anal. Chem., 55, 1313 (1983).
46. M.V. Buchanan, Anal. Chem., 54, 570 (1982).
47. T. Cairns and E.G. Siegmund, Anal. Chem., 54, 2458 (1982).
48. K.L. Busch, B. Hsuing Hsu, Y.-X. Xie and R.G. Cooks, Anal. Chem., 55, 1157 (1983).
49. J. Allison, B. Musselman, G. Dolnikowski and J. Watson, presented at the 31st Annual Conference on Mass Spectrometry and Allied Topics: Boston, Ma., May 8-13, 1983.
50. R.J. Anderegg, Anal. Chem., 53, 2169 (1981).
51. R.J. Anderegg, J. Chromatogr., 275, 154 (1983).
52. B. Lindeke, G. Hallström, C. Johansson, Ö. Ericsson, L.-I. Olsson and S. Strömberg, Biomed. Mass Spectrom., 8, 506 (1981).
53. R. Venkataramanan, F.S. Abbott and J.E. Axelson, J. Pharm. Sci., 71, 491 (1982).
54. J.D. Adams, Jr., T.A. Baillie, A.J. Trevor and N. Castagnoli, Jr., Biomed. Mass Spectrom., 8, 527 (1981).
55. E. Houghton, Biomed. Mass Spectrom., 9, 103 (1982).
56. S.W. Johnson, Drug Metab. Dispos., 10, 510 (1982).
57. K. Lertratanangkoon and M.G. Horning, Drug Metab. Dispos., 10, 1 (1982).
58. S. Furuuchi, M. Otsuka, Y. Miura and S. Harigaya, Drug Metab. Dispos., 11, 371 (1983).
59. D.M. Rackham, in "Isotopes: Essential Chemistry and Applications," J.A. Elvidge and J.R. Jones, Eds., The Chemical Society, London, 1980, p.97.
60. J.A. Hinson, T.J. Monks, M. Hong, R.J. Highet and L.R. Pohl, Drug Metab. Dispos., 10, 47 (1982).
61. K.-J. Hoffmann, A. Arfwidsson and L.O. Borg, Drug Metab. Dispos., 10, 173 (1982).
62. P. Ventura, M. Schiavi and S. Serafini, Xenobiotica, 13, 139 (1983).
63. D.R. Hawkins and I. Midgley, J. Pharm. Pharmacol., 30, 547 (1978).
64. L.I. Wiebe, J.R. Mercer and A.J. Ryan, Drug Metab. Dispos., 6, 296 (1978).
65. V.J. Feil, J.E. Bakke, G.L. Larsen and B.E. Gustafsson, Biomed. Mass Spectrom., 8, 1 (1981).
66. A. Noda, T. Goromaru, N. Tsubone, K. Matouyama and S. Iguchi, Chem. Pharm. Bull. (Tokyo), 24, 1502 (1976).
67. A. Noda, N. Tsubone, M. Mihara, T. Goromaru and S. Iguchi, Chem. Pharm. Bull. (Tokyo), 24, 3229 (1976).
68. D.C. Heimbrook and S.G. Sligar, Biochem. Biophys. Res. Commun., 99, 53 (1981).
69. T.L. Macdonald, L.T. Burka, S.T. Wright and F.P. Guengerich, Biochem. Biophys. Res. Commun., 104, 620 (1982).
70. H. Kexel, B. Limach, J. Möller and H.-L. Schmidt, in "Stable Isotopes," H.-L.Schmidt, H. Förstel and K. Heinzinger, Eds., Elsevier, Amsterdam, 1982, p.229.
71. M.H. Gelb, P. Mälkönen and S.G. Sligar, Biochem. Biophys. Res. Commun., 104, 853 (1982).
72. J.P. Shea, G.L. Valentine and S.D. Nelson, Biochem. Biophys. Res. Commun., 109, 231 (1982).
73. G.L. Kedderis, L.A. Dwyer, D.E. Rickert and P.F. Hollenberg, Mol. Pharmacol., 23, 758 (1983).
74. Y. Watanabe, T. Numata, T. Iyanagi and S. Oae, Bull. Chem. Soc. Japan, 54, 1163 (1981).
75. J.T. Groves, O.F. Akinbote and G.E. Avaria, in "Microsomes, Drug Oxidations, and Chemical Carcinogenesis," M.J. Coon, A.H. Conney, R.W. Estabrook, H.V. Gelboin, J.R. Gillette and P.J. O'Brien, Eds., Academic Press, Inc., 1980, p.253.
76. P.R. Ortiz de Montellano, B.L.K. Mangold, C. Wheeler, K.L. Kunze and N.O. Reich, J. Biol. Chem., 258, 4208 (1983).
77. K.L. Kunze, B.L.K. Mangold, C. Wheeler, H.S. Beilan and P.R. Ortiz de Montellano, J. Biol. Chem., 258, 4202 (1983).
78. E.D. Bush and W.F. Trager, Biochem. Biophys. Res. Commun., 104, 626 (1982).
79. P.R. Ortiz de Montellano, H.S. Beilan and K.L. Kunze, Proc. Natl. Acad. Sci. U.S.A., 78, 1490 (1981).
80. S.D. Nelson, in "Synthesis and Applications of Isotopically Labeled Compounds," W.P. Duncan and A.B. Susan, Eds., Elsevier, Amsterdam, 1983, p. 89.
81. P.J. van Bladeren, D.D. Breimer, J.A.T.C.M. van Huijgevoort, N.P.E. Vermeulen and A. van der Gen, Biochem. Pharmacol., 30, 2499 (1981).
82. J.C. Livesey, M.W. Anders, P.W. Langvardt, C.L. Putzig and R.H. Reitz, Drug Metab. Dispos., 10, 201 (1982).
83. C. Fenselau, J.P. Lehman, A. Myles, J. Brandt, G.S. Yost, O.M. Friedman and O.M. Colvin, Drug Metab. Dispos., 10, 636 (1982).
84. M.B. Kroeger-Koepke, S.R. Koepke, G.A. McClusky, P.N. Magee and C.J. Michejda, Proc. Natl. Acad. Sci. U.S.A., 78, 6489 (1981).
85. R.N. Harris and M.W. Anders, Drug Metab. Dispos., 9, 551 (1981).
86. B.A. Mico and L.R. Pohl, Biochem. Biophys. Res. Commun., 107, 27 (1982).
87. B.A. Mico and L.R. Pohl, Arch. Biochem. Biophys., 225, 596 (1983).
88. K.S. Prickett and T.A. Baillie, Biomed. Mass Spectrom., in press (1984)
89. M.M. Anderson, R.K. Mitchum and F.A. Beland, Xenobiotica, 12, 31 (1982).
90. R.E. Billings, W.G. Stillwell and R. Caprioli, Fed. Proc., 41, 1223 (1982).
91. J.M. Perel, L. O'Brien and A.A. Manian, Fed. Proc., 41, 1223 (1982).

92. M.A.A. Moustafa, M. Claesen, J. Adline, D. Vandervorst and J.H. Poupaert, Drug Metab. Dispos., 11, 574 (1983).
93. T. Walle, J.E. Oatis, Jr., U.K. Walle and D.R. Knapp, Drug Metab. Dispos., 10, 122 (1982).
94. T. Walle, M.J. Wilson, U.K. Walle and S.A. Bai, Drug Metab. Dispos., 11, 544 (1983).
95. J.W. Lown and S.M.S. Chauhan, J. Org. Chem., 46, 2479 (1981).
96. J.W. Lown and S.M.S. Chauhan, J. Org. Chem., 47, 851 (1982).
97. J.W. Lown, R.R. Koganty and A.V. Joshua, J. Org. Chem., 47, 2027 (1982).
98. J.E. Bakke, J. Rafter, G.L. Larsen, J.A. Gustafsson and B.E. Gustafsson, Drug Metab. Dispos., 9, 525 (1981).
99. G.L. Larsen and R. Ryhage, Xenobiotica, 12, 855 (1982).
100. G.L. Larsen and J.E. Bakke, Xenobiotica, 13, 115 (1983).
101. J.P. Klinman, Adv. Enzymol., 46, 415 (1978).
102. D.B. Northrup, Annu. Rev. Biochem., 50, 103 (1981).
103. W.W. Cleland, CRC Crit. Rev. Biochem., 13, 385 (1982).
104. J.T. Groves, S. Krishnan, G.E. Avaria, T.E. Nemo, in "Biomimetic Chemistry," D. Dolphin, C. McKennia, Y. Marakami, I. Tabushi, Eds., American Chemical Society, Washington, D.C., 1980, p. 277.
105. M.H. Gelb, D.C. Heimbrook, P. Mälkönen and S.G. Sligar, Biochemistry, 21, 370 (1982).
106. T. Shono, Y. Ohmizu, T. Toda and N. Oshino, Drug Metab. Dispos., 9, 476 (1981).
107. G.T. Miwa, W.A.Garland, B.J. Hodson and A.Y.H. Lu, J. Biol. Chem., 255, 6049 (1980).
108. G.T. Miwa, J.S. Walsh, G.L. Kedderis and P.F. Hollenberg, J. Biol. Chem., 258, 14445 (1983).
109. T. Shono, T. Toda and N. Oshino, J. Am. Chem. Soc., 104, 2639 (1982).
110. M. Claesen, M.A.A. Moustafa, J. Adline, D. Vandervorst and J.H. Poupaert, Drug Metab. Dispos., 10, 667 (1982).
111. M. Husain, D.E. Edmondson and T.P. Singer, Biochemistry, 21, 595 (1982).
112. P.H. Yu, S. Barclay, B. Davis and A.A. Boulton, Biochem. Pharmacol., 30, 3089 (1981).
113. R.B. Westkaemper and R.P. Hanzlik, Arch. Biochem. Biophys., 208, 195 (1981).
114. D.G. Brenner and J.R. Knowles, Biochemistry, 20, 3680 (1981).
115. A.B. Foster, in "Advances in Drug Research," B. Testa, Ed., Academic Press, London, 1984, in press.
116. J.M. Hoffman, C.N. Habecker, A.M. Pietruszkiewicz, W.A. Bolhofer, E.J. Cragoe, Jr., M. Lou Torchiana and R. Hirschmann, J. Med Chem., 26, 1650 (1983).
117. I.W. Taylor, C. Ioannides, P. Sacra, J.C. Turner and D.V. Parke, Biochem. Pharmacol., 32, 641 (1983).
118. S.A. Barker, J.M. Beaton, S.T. Christian, J.A. Monti and P.E. Morris, Biochem. Pharmacol., 31, 2513 (1982).
119. L.E. Dyck, D.A. Durden, P.H. Yu, B.A. Davis and A.A. Boulton, Biochem. Pharmacol., 32, 1519 (1983).
120. C.T. Dourish, K.M. Dewar, L.E. Dyck and A.A. Boulton, Psychopharmacol.,81, 122 (1983).
121. C.T. Dourish, A.J. Greenshaw and A.A. Boulton, Pharmacol. Biochem. Behav., 19, 471 (1983).
122. M. Jarman, J.H. Kiburis, G.B. Elion, V.C. Knick, G. Lambe, D.J. Nelson and R.L. Tuttle, in "Stable Isotopes," H.-L. Schmidt, H. Förstel and K. Heinzinger, Eds., Elsevier, Amsterdam, 1982, p. 217.
123. J.M. Patel, W.P. Gordon, S.D. Nelson and K.C. Leibman, Drug Metab. Dispos., 11, 164 (1983).
124. J.M. Baden, S.A. Rice and R.I. Mazze, Anesthesiology, 56, 203 (1982).
125. T. Mizutani, K. Yamamoto and K. Tajima, Toxicol. Appl. Pharmacol., 69, 283 (1983).
126. R.D. White, A.J. Gandolfi, G.T. Bowden and I.G. Sipes, Toxicol. Appl. Pharmacol., 69, 170 (1983).
127. R.V. Branchflower and L.R. Pohl, Toxicol. Appl. Pharmacol, 61, 407 (1981).
128. M. Ahmadizadeh, C.-H. Kuo and J.B. Hook, J. Toxicol. Environ. Health, 8, 105 (1981).
129. S.S. Hecht and R. Young, Cancer Res., 41, 5039 (1981).
130. M.S. Rao, D.G. Scarpelli and W. Lijinsky, Carcinogenesis, 2, 731 (1981).
131. W. Lijinsky and M.D. Reuber, Cancer Lett., 16, 273 (1982).
132. W. Lijinsky, M.D. Reuber, T.S. Davies and C.W. Riggs, J. Natl. Cancer Inst., 69, 1127 (1982).
133. G. Farrelly, M.L. Stewart, J.E. Saavedra and W. Lijinsky, Cancer Res., 42, 2105 (1982).
134. S.S. Hecht, D. Lin and A. Castonguay, Carcinogenesis, 4, 305 (1983).
135. W.A. Garland and M.L. Powell, J. Chromatogr. Sci., 10, 392 (1981).
136. I. Björkhem, in "Stable Isotopes," H.-L. Schmidt, H. Förstel and K. Heinzinger, Eds., Elsevier, Amsterdam, 1982, p. 593.
137. D.L. Hachey, K. Nakamura, M.J. Kreek and P.D. Klein, in "Stable Isotopes," H.-L. Schmidt, H. Förstel and K. Heinzinger, Eds., Elsevier, Amsterdam, 1982, p.235.
138. L. Bertilsson, in "Clinical Pharmacology in Psychiatry," E. Usdin, Ed., Elsevier, New York, 1981, p. 59.
139. A.I. Mallet, Prostaglandins Leukotrienes Med., 8, 181 (1982).
140. G. Idzu, M. Ishibashi and H. Miyazaki, J. Chromatogr., 229, 327 (1982).
141. C. Julien-Larose, C. Lange, D. Lavene, J.R. Kiechel and J.J. Basselier, Int. J. Mass Spectrom. Ion Phys., 48, 221 (1983).
142. N.D. Barber and J.L Reid, J. Pharm. Pharmacol., 35, 114 (1983).
143. Y. Kasuya, J.R. Althaus, J.P. Freeman and R.K. Mitchum, in "Synthesis and Applications of Isotopically Labeled Compounds," W.P. Duncan and A.B. Susan, Eds., Elsevier, Amsterdam, 1983, p. 349.
144. W.C. Pickett and R.C. Murphy, Anal. Biochem., 111, 115 (1981).
145. D.M. Desiderio and M. Kai, Biomed. Mass Spectrom., 10, 471 (1983).
146. D.H.T. Chien and J.R. Heys, J. Agric. Food Chem., 30, 396 (1982).
147. N. H. Mahle and L.A. Shadoff, Biomed. Mass Spectrom., 9, 45 (1982).
148. R.H. White, Anal. Biochem., 114, 349 (1981).
149. J.R. Lloyd and F.H. Field, Biomed. Mass Spectrom., 8, 19 (1981).

150. B.K. Tang, H. Uchino, T. Inaba and W. Kalow, Biomed. Mass Spectrom., 9, 425 (1982).
151. P. Falardeau, J.A. Oates and A.R. Brash, Anal. Biochem., 115, 359 (1981).
152. H.-R. Schulten, J. Chromatogr., 251, 105 (1982).
153. S.J. Gaskell, B.G. Brownsey, P.W. Brooks and B.N. Green, Int. J. Mass Spectrom. Ion Phys., 46, 435 (1983).
154. S.J. Gaskell, B.G. Brownsey, P.W. Brooks and B.N. Green, Biomed. Mass Spectrom., 10, 215 (1983).
155. C. Fenselau, P.C.C. Feng, T. Chen and L.P. Johnson, Drug Metab. Dispos., 10, 316 (1982).
156. H.G. Hege and J. Weymann, in "Stable Isotopes," H.-L. Schmidt, H. Förstel, and K. Heinzinger, Eds., Elsevier, Amsterdam, 1982, p. 679.
157. S.P. Markey and F.P. Abramson, in "Synthesis and Applications of Isotopically Labeled Compounds," W.P. Duncan and A.B. Susan, Eds., Elsevier, Amsterdam, 1983, p. 291.
158. R.C. Cottrell, P.J. Young, R. Purchase and S.D. Gangolli, Biomed. Mass Spectrom., 9, 43 (1982).
159. A.R. Swanson and M.W. Anders, J.Chromatogr., 207, 365 (1981).
160. A.R. Swanson and M.W. Anders, J. Chromatogr., 234, 268 (1982).
161. W. Wittfoht, H. Nau, D. Rating and H. Helge, in "Stable Isotopes," H.-L. Schmidt, H. Förstel and K. Heinzinger, Eds., Elsevier, Amsterdam, 1982, p. 265.
162. W. Kochen, B. Tauscher, M. Klemens and E. Depene, in "Stable Isotopes," H.-L. Schmidt, H. Förstel and K. Heinzinger, Eds., Elsevier, Amsterdam, 1982, p. 271.
163. A. van Langenhove, J.E. Biller, K. Biemann and T.R. Browne, Biomed. Mass Spectrom., 9, 201 (1982).
164. T.R. Browne, A. van Langenhove, C.E. Costello, K. Biemann and D.J. Greenblatt, J. Clin. Pharmacol., 22, 309 (1982).
165. I.M. Kapetanović and H.J. Kupferberg, Biomed. Mass Spectrom., 7, 47 (1980).
166. T.R. Browne, A. van Langenhove, C.E. Costello, K. Biemann and D.J. Greenblatt, Clin. Pharm. Ther., 29, 511 (1981).
167. J.H. Poupaert, M. Winand and P. Dumont, J. Labelled Comp. Radiopharm., 18, 1827 (1981).
168. L.A. Bauer, T. Brown, M. Gibaldi, L. Hudson, S. Nelson, V. Raisys and J.P. Shea, Clin. Pharmacol. Ther., 31, 433 (1982).
169. E. Änggård, M.-I. Nilsson, J. Holmstrand and L.-M. Gunne, Eur. J. Clin. Pharmacol., 16, 53 (1979).
170. M.-I. Nilsson, E. Änggård, J. Holmstrand and L.-M. Gunne, Eur. J. Clin. Pharmacol., 22, 343 (1982).
171. K. Nakamura, D.L. Hachey, M.J. Kreek, C.S. Irving and P.D. Klein, J. Pharm. Sci., 71, 40 (1982).
172. A. Ohlsson J.-E. Lindgren, A. Wahlén, S. Agurell, L.E. Hollister and H.K. Gillespie, Biomed. Mass Spectrom., 9, 7 (1982).
173. J. Vink, Mass Spectrom. Rev., 1, 349 (1982).
174. R.L. Wolen and C.M. Gruber, Jr., in "Drug Absorption and Disposition: Statistical Considerations," K.S. Albert, Ed., American Pharmaceutical Association Academy of Pharmaceutical Sciences, Washington D.C., 1980, p. 69.
175. M. Eichelbaum, H.J. Dengler, A. Somogyi and G.E. von Unruh, Eur. J. Clin. Pharmacol., 19, 127 (1981).
176. J.R. Carlin, J.D. Rogers, R.O. Davis, R.K. Ferguson and W.J.A. VandenHeuvel, presented at the 30th Annual Conference on Mass Spectrometry and Allied Topics: Honolulu, Hawaii, June 6-11, 1982.
177. W.J.A. VandenHeuvel, J.R. Carlin and R.W. Walker, J. Chromatogr. Sci., 21, 119 (1983).
178. J. Vink, H.J.M. van Hal, F.M. Kaspersen and H.P. Wijnand, Int. J. Mass Spectrom. Ion Phys., 48, 217 (1983).
179. H. De Bree, D.J.K. van der Stel, J.H.M.A. Kaal and J.B. van der Schoot, in "Stable Isotopes," H.-L. Schmidt, H. Förstel and K. Heinzinger, Eds., Elsevier, Amsterdam, 1982, p. 203.
180. A. Somogyi, M. Albrecht, G. Kliems, K. Schäfer and M. Eichelbaum, Br. J. Clin. Pharmacol., 12, 51 (1981).
181. U. Meresaar, M.-I. Nilsson, J. Holmstrand and E. Änggård, Eur. J. Clin. Pharmacol., 20, 473 (1981).
182. A.I. Cohen, R.G. Devlin, E. Ivashkiv, P.T. Funke and T. McCormick, J. Pharm. Sci., 71, 1251 (1982).
183. S.A. Bai, M.J. Wilson, U.K. Walle and T. Walle, J. Pharmacol. Exp. Ther., 227, 360 (1983).
184. R.K. Verbeeck, R.A. Branch and G.R. Wilkinson, Clin. Pharmacol. Ther., 30, 619 (1981).
185. Y. Matsuki, T. Ito, M. Kojima, H. Katsumura, H. Ono and T. Nambara, Chem. Pharm. Bull. (Tokyo), 31, 2033 (1983).

Chapter 28. Drug Discovery at the Molecular Level: A Decade of Radioligand Binding in Retrospect

Michael Williams and David C. U'Prichard
Nova Pharmaceutical Corporation
P.O. Box 21204, Wade Avenue
Baltimore, MD 21228

Introduction - In the design of new therapeutic entities, the medicinal chemist works from information from known active compounds or from the structure of the putative neuroeffector agent thought to be involved in the etiology of the disease state under study.[1] The search for novel chemical entities which have therapeutic potential has become increasingly more complex;[2] thus any rational strategy that will increase the probability of finding a new drug is of considerable interest to those involved in drug research. One such approach is that utilizing radioligand binding assays,[3] which permits the evaluation of compounds for their direct interaction with cell surface recognition sites (receptors) independent of events distal to the actual recognition process. Structure-activity relationships generated using radioligand binding assays theoretically reflect the actual physicochemical properties required for receptor recognition, and thus provide information directly pertinent to the improvement of specificity and potency. In addition, receptor assays require only small amounts of compound and animal tissue, and are relatively rapid to perform. However, like classical pharmacological screens, these assays have certain drawbacks. They take no account of compounds that require metabolic activation, and generally measure only relative potency rather than efficacy, so that it is difficult to assess whether a compound interacting with a ligand binding site is an agonist or antagonist. These issues, and the advances and insights gained through the technique of radioligand binding, are the subject of this review.

The Binding Technique - The radioligand binding assays developed to date are based on the fact that compounds known to interact with given receptors, when labeled to high specific radioactivity (10 Ci/mmole or greater) with ^{3}H or ^{125}I, bind to such sites in membrane or intact cell preparations from mammalian tissue with high affinity (Kd $\cong 10^{-9}$ M) in a reversible, saturable manner. The difference between total and non-specific binding, the specific binding, can then be used to measure compound interactions with the receptor. Ideally, specific binding should represent 70% or greater of the radioactivity bound. However, 50% specific binding, if not ideal, is acceptable. Binding assays where specific binding is only 20-30% of the total, present problems in 'signal-to-noise' reliability for routine screening purposes. In general for screening purposes, compounds with IC_{50} values greater than 10^{-4} M can be deemed 'inactive,' especially if at other recognition sites, they have IC_{50} values in the 10^{-9} M range.

The large majority of radioligand binding assays make use of brain tissue because of its ease in preparation, and its richness in receptors

ISBN 0-12-040519-9

on a weight per weight basis.[3] Other tissues which have known physiological responses to a given agonist tend not to be good targets for radioligand binding because of the low density of receptors present as compared to CNS tissue.[3] The ease with which CNS tissue can be utilized should not, however, preclude examination of peripheral tissue radioligand binding, since it is highly likely that subtle pharmacological and biochemical nuances exist among receptors of a given type as a function of their tissue source. For instance, despite considerable effort, the nicotinic antagonist, mecamylamine, which is exceedingly potent in autonomic ganglia, has no demonstrable binding in the CNS.[4] Furthermore, ligands specific for dopaminergic receptor subtypes in the CNS have not proven suitable for labeling the subtypes present in peripheral tissues.

Validation of Receptor Binding Assays - There has been considerable concern that because of the non-physiological conditions used to examine radioligand binding, the sites labeled may represent *in vitro* artifacts. Such conditions include the use of hypoosmotic buffers; long incubation periods, which are inconsistent with the millisecond time courses related to the proposed physiological actions of neurotransmitter analogues and antagonists; and the use of non-physiological assay temperatures. These basic criticisms have some validity, but in general, like many *in vitro* test paradigms, reflect the interface between the complexity of nature and mankind's necessarily simplistic approaches.

Several criteria have been proposed[5] to validate a binding assay. These are saturability, indicating a finite number of binding sites; reversibility, consistent with a biological process involved in information transfer; correlation of binding affinity with the demonstrated concentration of the neurotransmitter or drug under physiological conditions; tissue and subcellular distribution of binding consistent with the known localization or target site of the ligand; agonist and antagonist pharmacology of the binding site corresponding to the known properties of the compound from other pharmacological tests; and finally, correlation of binding data with biological dose-response curves in identical tissue preparations. Most binding assays currently used fulfill a good many of these criteria.

With these guidelines in mind, the status of the radioligand binding assays currently in use will be considered. In many cases the initial observations of relatively simple recognition sites has progressed to the delineation of receptor subtypes. While offering exciting possibilities in terms of drug development, many of the assays for receptor subtypes have a weak pharmacological basis and in some instances do not fulfill many, if not all, of the criteria listed above.

Binding and Receptor Subtypes

a) Dopamine receptors - Recognition sites for this monoamine, which is implicated in the etiology of both schizophrenia and Parkinsonism, have been labeled with more than 20 ligands.[6] Five or more distinct receptor subtypes have been described; of these, only those designated D-1 and D-2, which activate and inhibit adenylate cyclase activity, respectively, are considered tangible entities.[7] Apomorphine, spiperone, and domperidone label both D-1 and D-2 recognition sites.[8] While newer antagonists such as SCH 23390 (**1**) appear to be selective for the D-1 receptor *in vitro*,[9] *in vivo* this distinction is not apparent.[10] Newer agonists, such as RDS127[11] (2) and PHNO [12] (3), appear to be selective D-2 ligands. Putative dopamine autoreceptor agonists, such as 3-PPP and TL-99, have generated considerable interest.[8,13] Unfortunately, neither binding nor in vivo pharmacological procedures have provided definitive evidence that either of these compounds has absolute selectivity for presynaptic dopamine receptors.[14,15] Whether pre- and post-synaptic receptors represent distinct pharmacological, as opposed to anatomical, entities is a matter for further investigation.

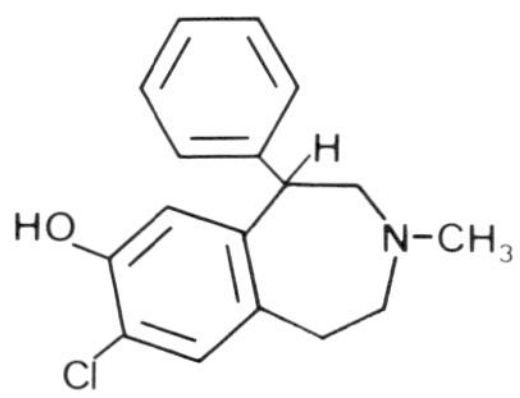

1

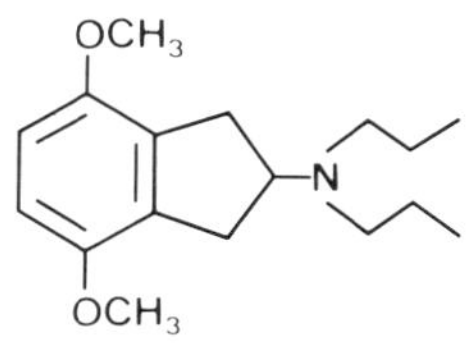

2

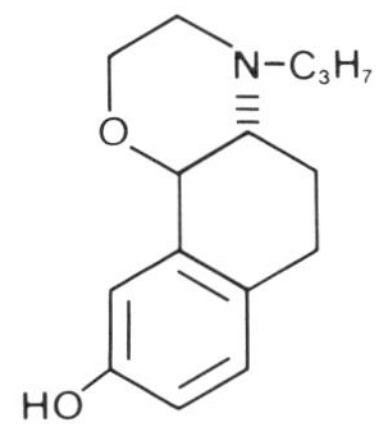

3

b) Alpha-adrenoceptors - A stronger association has been established between radioligand binding sites and functional receptor entities for epinephrine and norepinephrine. Alpha-1 and alpha-2 receptors were originally discriminated by location, being post- and pre-junctional respectively.[16] Subsequently, alpha-2 receptors were shown to be negatively coupled to adenylate cyclase, while alpha-1 receptors stimulate PI turnover and Ca^{2+} entry.[17] Among antagonist radioligands, peptide ergots (e.g. dihydroergocryptine) and phentolamine label both receptors with equal affinity,[18] while prazosin [19] and the aminotetralone, HEAT (BE 2254)[20] are highly potent and selective alpha-1 ligands; WB-4101 is partially selective for the alpha-1 receptor,[21] while yohimbine and its isomer, rauwolscine, are somewhat alpha-2 receptor selective.[22] Alpha-2 antagonists appear to accelerate antidepressant- induced down-regulation of beta- and 5HT-2 receptors, and are currently the subject of intense interest.[23] The selective alpha-2 antagonist, the imidazoline, RX 781094, has been successfully used as a radioligand.[24] Several catecholamines and imidazolines selectively label high affinity states of the alpha-2 receptor.[25] Pre- and postjunctional components of alpha-2 receptor binding have been identified, but no differential pharmacological characteristics have yet been observed.

c) Beta-adrenoceptors - These have been readily labeled with antagonists such as dihydroalprenolol, carazolol, iodohydroxybenzyl-pindolol,[26,27] cyanopindolol and pindolol.[28] Beta receptors were originally subdivided into beta-1 and beta-2 types on the basis of differential agonist and antagonist pharmacology and tissue location.[29] Ligand binding studies corroborate the pharmacological subclassification quite precisely. Since the above mentioned antagonists are not subtype-selective, estimates of subtype relative abundance and competitor preference have relied on the generation of biphasic competition curves, with computer-assisted curve-fitting. These have not, however, been without problems; IPS 339 (4), a beta-1 selective antagonist as judged by computer assisted binding data[30], has little subtype selectivity in other tissue receptor assays.[31]

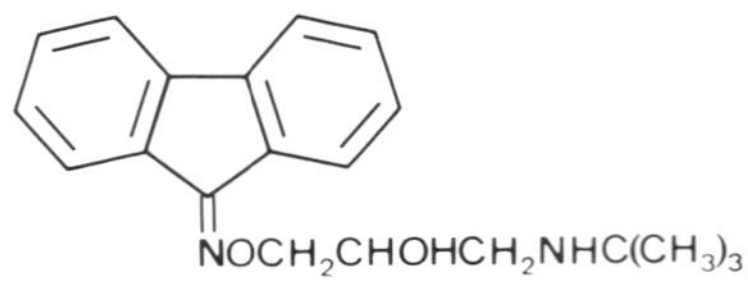

4

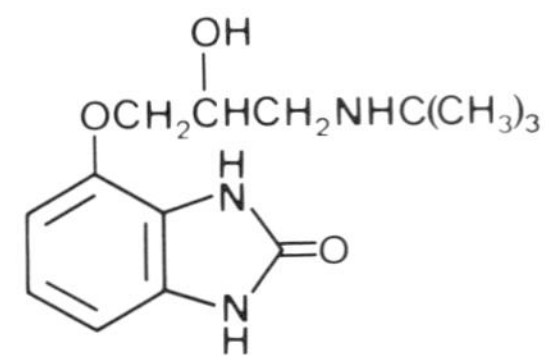

5

The biochemical characterization of beta receptors and their coupling to adenylate cyclase has been intensively studied [32] using agonist radioligands (epinephrine, hydroxybenzylisoproterenol) to label high affinity receptor conformations, especially of the beta-2 receptor;[33] more hydrophilic antagonists (e.g. CGP 12177 (5)) to label cell surface, but not internal, receptor populations (especially useful for examining regulation of beta receptors in cell culture);[34] and photoactivated, irreversibly binding antagonists such as p-azidobenzylcarazolol, to tag the receptor during purification to apparent homogeneity.[35]

d) Serotonin receptors - These have been labeled with serotonin itself and LSD,[36] spiperone[37] using cortical tissue, and more recently, ketanserin.[38] Serotonin receptors in brain tissue appear to exist in two distinct forms termed 5HT-1 and 5HT-2. The former, which can be labeled with serotonin, have been further subdivided into 5-HT-1A and 5-HT-1B sites.[39] Both spiperone and ketanserin label 5-HT-2 receptors in cortical and other brain regions. Interestingly, the pharmacological actions of ketanserin are mediated via alpha-1 receptor blockade.[40] The 5-HT-2 receptor has been implicated in depression, chronic treatment with the tricyclic, amitriptyline, reducing the number of 5-HT-2 binding sites in rat frontal cortex.[41] The physiological relevance of the receptor has, however, been questioned since atypical antidepressants such as mianserin can reduce the density of serotonin-2 sites when given acutely.[42] The relationship of central serotonin binding sites to those designated as M (excitatory; blocked by morphine) and D (inhibitory; blocked by dibenyline) is unclear at the present time.

e) Muscarinic cholinergic receptors - Many highly potent and selective antagonists have been used to label muscarinic receptors in central and peripheral tissues, including N-methylscopolamine, 3-quinuclidinyl benzilate (QNB), N-methyl-4-piperidinyl benzilate, and the covalent ligand, benzylcholine mustard.[43] Excellent correlations have been obtained between affinities at these binding sites and pA values in classical organ preparations such as guinea pig ileum.[43] Agonist competitors at antagonist binding sites exhibit complex, heterogeneous interactions manifested by very "shallow" competition curves; these can be partitioned into "super- high" (SH), "high" (H) and "low" (L) affinity states. Agonist ligands (cis-methyl- dilvasene, acetylcholine, oxotremorine-M) preferentially label, and improve resolution of, the SH state of the receptor.[44] The relative proportion of these three "states" was found to differ markedly in different tissues and brain areas.[44] It has become apparent, however, that there exist two pharmacologically distinct types of muscarinic receptor, designated M-1 and M-2. The agonist McN-A-343 (6),[45] and antagonist pirenzepine (7),[46] are selective for M-1 receptors; no M-2 receptor selective agents are known. Given the apparent tissue specificity for the two receptor types (M-1 in most brain regions and autonomic ganglia;

M-2 in peripheral effector organs such as heart), there is considerable interest in the development of subtype-selective agents.

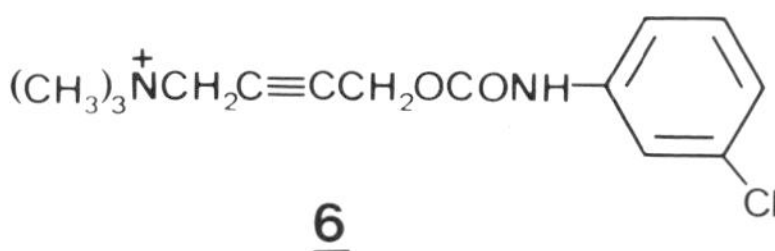

6

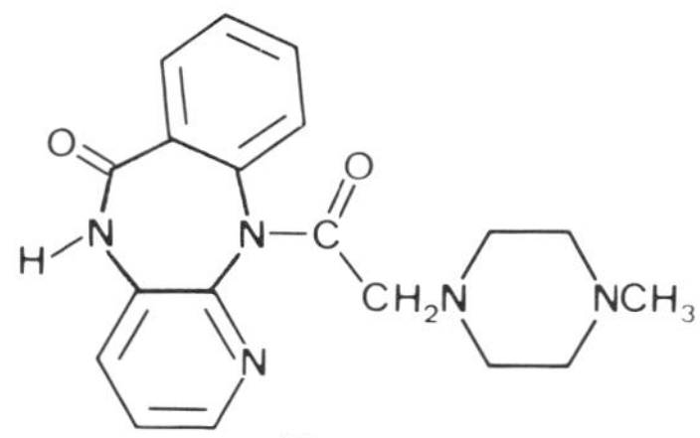

7

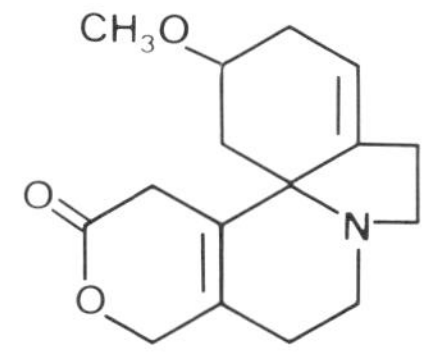

8

f) Nicotinic cholinergic receptors - Acetylcholine (ACh) receptors sensitive to the alkaloid nicotine have been labeled in *Torpedo* tissue and in autonomic and central nervous system using radiolabeled ACh,[47] nicotine[48] and alpha-bungarotoxin (BTX).[49] BTX is, however, not a good ACh antagonist in mammalian nervous tissue.[50] The classical ganglionic blockers, hexamethonium and mecamylamine do not displace these ligands from CNS tissue,[50] a finding which may indicate that agonist and antagonist binding sites are distinct entities. Mecamylamine shows no specific binding in brain tissue.[4] The neuromuscular blocker, dihydro-beta-erythroidine (DBE; (8)) does label central nicotinic recognition sites. DBE binding is insensitive to ganglionic blockers, suggesting that the nicotine recognition site in mammalian brain is more selective for neuromuscular type than ganglionic receptor blockers.[4]

g) GABA receptors - These have been labeled with tritiated GABA[51] and muscimol.[52] Electrophysiological and binding studies in mammalian tissue have led to the discovery of two distinct types of GABA receptor;[53] the GABA-A site, which is labeled by GABA and muscimol and which is sensitive to blockade by bicuculline; and the GABA-B site, which can be selectively labeled by baclofen and is insensitive to blockade by bicuculline. The GABA-B receptor may be linked in a modulatory manner to a membrane bound adenylate cyclase.[54]

h) Excitatory amino acids - As many as four excitatory amino acid receptor subtypes have been described. These are sensitive to the agonists, N-methyl-D-aspartate (NMDA), kainic acid, quisqualic acid and APB (2-amino-4-phosphonobutyric acid).[55] Excitatory amino acids have been termed 'excitotoxins' and can cause cell death in a manner similar to that seen in Huntington's disease.[55] In general, the affinities of the various ligands used to label excitatory amino acid recognition sites in mammalian brain tissue are low compared with other putative neurotransmitter recognition sites, with Kd values around 10^{-6} M.[55]

i) Purinergic receptors - Receptors for the purines may be divided into P-1 and P-2 subtypes.[56] P-1 receptors are adenosine-sensitive and cyclase-linked, while P-2 receptors are ATP-sensitive, affect prosta-glandin synthesis, and have no effect on cyclic AMP production. Two subtypes of the P-1 receptor exist;[57,58] the A-1 or Ri and A-2 or Ra, which, respectively, inhibit or activate adenylate cyclase. Both A-1 and A-2 receptors are sensitive to blockade by xanthines such as caffeine and theophylline. To date, ligand binding assays have only been described for the A-1[57,58] and P-2[59] receptors. Binding studies have led to the description of further subtypes which show species dependence.[60] Functional differences in A-1 receptor function have also been reported.[61,62] The 5'-N-ethylcarboxamide of adenosine, NECA, may show preferential selectivity for A-2 sites.[63]

j) Histamine receptors - Mepyramine has been extensively used to label H-1 receptors,[64] while tiotidine has been shown to demonstrate specific binding consistent with the labeling of an H-2 receptor.[65] The recently described irreversible H-2 blocker, L-643441 (9), may be an ideal ligand for this receptor subclass.[66] In view of the profound effects of histaminergics on CNS tissue function and the search for second generation antiulcer agents, it seems highly appropriate that a reliable H-2 binding assay should be available for drug screening.

k) Opiate receptors - Binding assays for opiates have contributed significantly to the discovery and characterization of the enkephalins and endorphins. At least four, and perhaps five opiate receptor subtypes are amenable to study by radioligand binding techniques.[67] The mu-receptor can be labeled with dihydromorphine (DHM) and by the enkephalin analog, DAGO. The delta receptor, which is involved in spinal analgesia, is labeled with the enkephalin analog, DADLE.[68] The kappa receptor, thought to be involved in the psychotomimetic effects of some opiate partial agonists, especially benzomorphans, can be labeled with ethylketocyclazocine (EKC)[69] or bremazocine (10).[70] The sigma opiate receptor can be labeled with SKF 10,047.[71] The epsilon receptor can be labeled with beta-endorphin, although this is controversial at this time. Meptazinol (11), a newly introduced analgesic agent, has been reported to interact with a subtype of the mu receptor, mu-1 which is responsible for the central analgesic actions of the opiates.[72] By inference, the mu-2 receptor is responsible for the respiratory depression and inhibition of gastric motility associated with opiate action. Radiolabeled meptazinol has not proven to be a satisfactory ligand because of its lipid solubility.

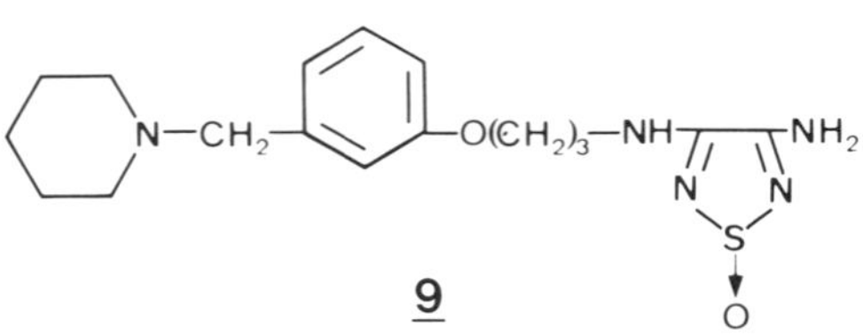

9

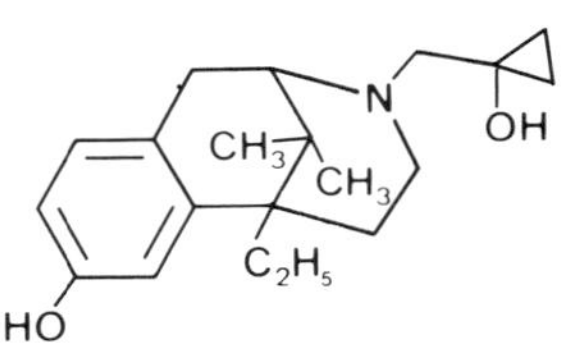

10

l) Benzodiazepine (BZ) receptors - Using ^{3}H-BZs, up to three subclasses of BZ receptors have been described;[73] two central receptors, BZ-1 and BZ-2, which have been reported to mediate the anxiolytic and sedative/ataxic actions of the BZ's, respectively,[74] can be labeled with clonazepam; a third peripheral type receptor, which has low affinity for clonazepam, can be labeled with the BZ, Ro 5-4864.[75] Whether the two central receptors are distinct entities or different forms of a single receptor remains to be determined.[76] Ro 15-1788, the BZ antagonist or 'inverse agonist' can be used to label central BZ receptors.[77] Using the central BZ binding assay, several non-BZs with anxiolytic potential have been reported. These include the pyrazoloquinoline, CGS 9896 [78] and the triazolopyridine, CL 218872,[74] which displace BZ binding; and the pyrazolopyridine, etazolate,[79] which enhances binding. The BZ receptor is somewhat unique in that it exists as a receptor complex involving a GABA recognition site and chloride ionophore, the latter being responsible for the physiological actions of the BZs.[73]

m) Peptide receptors - Many neuromodulatory peptides in addition to the enkephalins and endorphins have been discovered in both the gut and brain which have a common embryological origin. Radioligands have been used to identify the cell surface recognition sites for peptides. The classical peptides first studied were insulin [5] and glucagon; assays for Substance P, cholecystekinin, neurotensin, bombesin, angiotensin, and arginine vasopressin have been described. [80] Although some progress has been made in delineating substance P receptor subtypes using binding criteria, [81] radioligand binding assays for peptides and other related entities such as the leukotrienes [82] should at the present time be described as recognition site, rather than receptor assays. Nonetheless they are exceedingly useful screening tools.

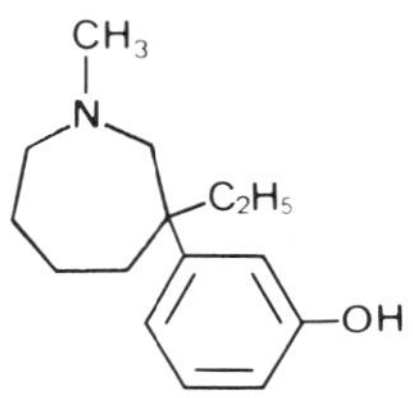

11

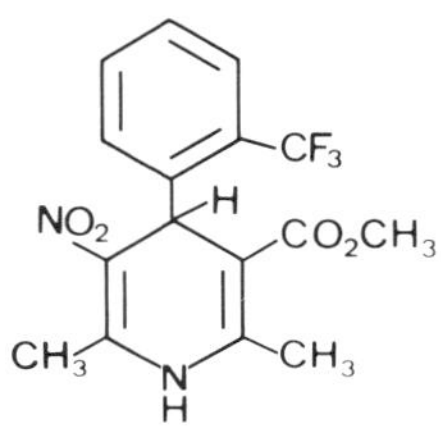

12

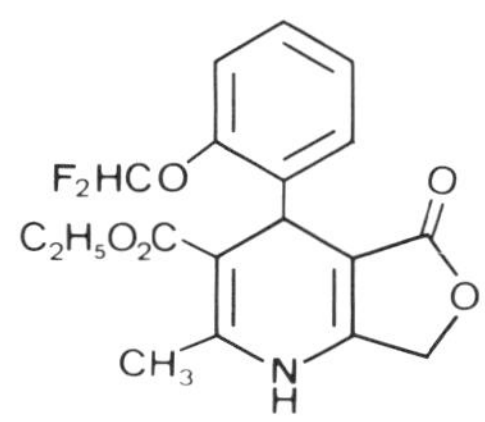

13

n) Receptors for calcium agonists and antagonists - While calcium antagonists bind to ion channels and cannot be considered as receptor antagonists per se, the binding of three classes of such agents has been described: the dihydropyridines [83] (DHP) (e.g. nifedipine and nicardepine); the nitriles[84] (e.g. verapamil and D-600); and the benzothiazepines (e.g. diltiazem). Binding assays for the DHPs have been characterized in brain and visceral tissue. Verapamil, D-600 and diltiazem are weak, partial displacers of DHP binding, which may imply the existence of either distinct calcium channels or a 'receptor' complex with different recognition sites. A binding assay for verapamil has been described recently.[85] Diltiazem can stimulate DHP binding and can reverse the inhibition of binding seen with tiapamil. [86] More recently DHPs have been described which facilitate calcium entry including Bay K 8644 [87] (12) and CGP 28392 (13). [88] While calcium antagonist binding sites are present in brain, their pharmacological significance is unclear since they do not appear to affect calcium fluxes in synaptosomal preparations.[89]

Potpourri - A variety of ligand binding sites are unassociated with defined receptors but are useful in the drug screening process. These include the proteins, labeled by antidepressants, which are associated with reuptake processes for 5-HT (^{3}H-imipramine[90], ^{3}H-paroxetine [91]) and norepinephrine (^{3}H-desmethylimipramine [92]). Other potentially important sites are those for the psychotomimetic phencyclidine (PCP) [93] and the atypical antidepressant, trazodone. [94] The potential significance and usefulness of these and other such sites must remain in limbo until a relevant biochemical or cellular function has been attributed to them. Specific 'drug' recognition sites need not, however, be invoked for each and every novel compound with therapeutic potential. The search for unique binding sites using radiolabeled probes and the extrapolation to a search for the "endogenous ligand" may lead to unnecessary complications.

A potentially useful therapeutic agent may have a distinct 'non-receptor,' but still locus specific, site of action. An example of this in the anticonvulsant/central sympathomimetic/anxiolytic MK 801[95] (14), which has in vivo activity but has no appreciable activity in several ligand binding assays. The pharmacological activity of a compound may also be dependent on activity at more than one recognition site. The example of ketanserin, an in vitro ligand for the 5HT-2 receptor but a alpha-1 antagonist in vivo has been cited.[40] The serotoninomimetic, MK 212 (15), is another instance where the binding profile of the compound and its in vivo activity do not correlate.[96,97]

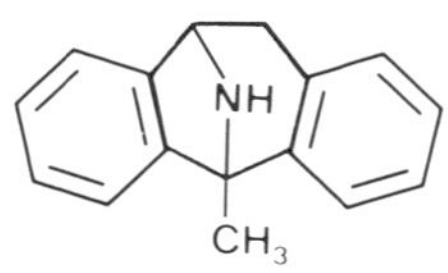

14

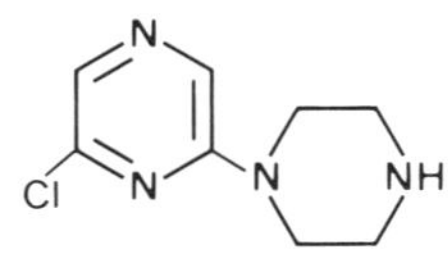

15

Ligand binding methods have facilitated the study of receptor alterations in disease states, whether secondary or presumptively etiological examples incude human postmortem brain tissue obtained from patients with neurodegenerative (Parkinsonism, Huntington's Chorea, Alzheimer's Disease) or psychiatric (Schizophrenia) disorders,[98] and in formed blood elements obtained from patients before, or during, drug treatment (e.g. platelet alpha-2 receptors and H-imipramine binding sites; white cell beta receptors).[99] Especially in the realm of neuropsychiatric disorders, a variety of receptor data has been amassed that is at this stage primarily phenomenological and as yet of little value with regard to etiological issues or rational therapy.

Conclusion - The future of radioligand binding methods for research on receptors and their function can be predicted to follow several directions: (a) the increased use of receptor autoradiography with computer-assisted quantitative imaging to combine biochemical precision with high morphological and spatial resolution;[100] and (b) an increased emphasis on isolation and purification of receptors, facilitated by the use of covalent affinity probes, to allow controlled reconstitution of receptor-effector systems, and to facilitate development of receptor antibodies which in turn will be used as probes to determine specific receptor RNAs and DNAs. This will ultimately lead to a better understanding of receptor expression, and the effects of drugs thereon, and to cloning of receptors in host cells on a scale large enough to provide sufficient material to study in depth the physico-chemical properties of the ligand-binding "active site" of receptor proteins, allowing for truly rational drug design.

References

1. N.C. Cohen, Trends Pharmacol. Sci., 4,503 (1983).
2. F. Gross in "Decision Making in Drug Research," F. Gross, Ed., Raven Press, New York, 1983, p. 1.
3. S.H. Snyder, J. Med. Chem., 26,1667 (1983).
4. M. Williams and J.L. Robinson, J. Neurosci., in press (1984).
5. P. Cuatrecasas and M. Hollenberg, Adv. Protein Chem. 30,251 (1976).
6. P. Seeman, Pharmacol. Rev., 32,229 (1980).
7. J. Kebabian and D. B. Calne, Nature, 277,93 (1979).
8. M. Williams and J. A. Totaro, Eur. J. Pharmacol, 86,35 (1983).
9. J. Hyttel, Eur. J. Pharmacol., 91,153 (1983).
10. R.B. Mailman, D.W. Schulz, D.L. DeHaven, M.H. Lewis and H. Rollema, Science, in press (1984).
11. S.P. Arneric, J.P. Long, M. Williams, D.B. Goodale, J. Mott, J.M. Lakoski and G.F. Gebhart, J. Pharmacol. Exp. Ther., 224,161 (1983).
12. G.E. Martin, M. Williams, G.G. Yarbrough, D.J. Pettibone, B.V. Clineschmidt, P.S. Anderson, J.J. Baldwin and J.H. Jones, Fed. Proc., 43,585 (1984).
13. S. Hjorth, A. Carlsson, D. Clark, K. Svensson, H. Widstrom, D. Sanchez, P. Lindberg, J. Hacksell, L.D. Arvidsson, A. Johansson and J.L.G. Nilsson, Psychopharmacology, 81,89 (1983).
14. K.J. Watling, J.H. Jones and M. Williams, Eur. J. Pharmacol., 82,117 (1982).
15. M. Williams, D.J. Pettibone and G.E. Martin, Life Sci., in press (1984).
16. S.Z. Langer, Biochem. Pharmacol., 23,1793 (1974).
17. J.N. Fain and Garcia-Sainz, J.A. Life Sci., 26,1183 (1980).
18. B.B. Hoffman, A. DeLean, C.L. Wood, D.D. Shocken and R.J. Lefkowitz, Life Sci., 24,1739 (1979).
19. P. Greengrass and R. Bremner, Eur. J. Pharmacol., 55,323 (1979).
20. G. Engel and D. Hoyer, Eur. J. Pharmacol., 73,221 (1981).
21. D.C. U'Prichard, D.A. Greenberg and S.H. Snyder, Mol. Pharmacol., 13,454 (1977).
22. B.D. Perry and D.C. U'Prichard, Eur. J. Pharmacol., 76,461 (1981).
23. D.A. Kendall, R. Duman, J. Slopis and S.J. Enna, J. Pharmacol. Exp. Ther., 222,566 (1982).
24. D.R. Howlett, P. Taylor and D.S. Walter, Br. J. Pharmacol., 76,294P (1982).
25. D.B. Bylund and D.C. U'Prichard, Int. Rev. Neurobiol., 24,343 (1983).
26. K.P. Minneman, R.N. Pittmann and P.B. Molinoff, Annu. Rev. Neurosci., 4,419 (1981).
27. S.R. Nahorski, Trends Pharmacol. Sci., 2,95 (1981).
28. D. Bojanic and S.R. Nahorski, Eur. J. Pharmacol., 93,235 (1983).
29. A.M. Lands, A.A. McAuliff, F.P. Luduena and T.G. Brown, Nature, 214,597 (1967).
30. K.P. Minnemann, A. Hedberg and P.B. Molinoff, J. Pharmacol. Exp. Ther., 211,502 (1979).
31. J.J. Baldwin, D.E. McClure, D.M. Gross and M.J. Williams, J. Med. Chem., 25,931 (1982).
32. B.B. Hoffman and R.J. Lefkowitz, Annu. Rev. Pharmacol. Toxicol., 20,581 (1980).
33. K.A. Heidenreich, G.A. Weiland and P.B. Molinoff, J. Cyclic Nucleotide Res., 6,217 (1980).
34. Unlisted Drugs, 34,119 (1982).
35. T.N. Lavin, S.L. Heald, P.W. Jeffs, R.G.L. Schorr, R.J. Lefkowitz and M.G. Caron, J. Biol. Chem., 256,11944 (1981).
36. J.P. Bennett and S.H. Snyder, Mol. Pharmacol., 12,373 (1976).
37. S.J. Peroutka and S.H. Snyder, Mol. Pharmacol., 16,687 (1979).
38. J.E. Leysen, C.J.E. Niemegeers, J.M. Van Nueten and P.M. Laduron, Mol. Pharmacol., 21,301 (1982).
39. N.W. Pedigo, H.I. Yamamura and D.L. Nelson, J. Neurochem., 36,220 (1981).
40. C. Ennis and B. Cox, J. Pharm. Pharmacol., 32,434 (1980).
41. S.J. Peroutka and S.H. Snyder, J. Pharmacol. Exp. Ther., 215,582 (1980).
42. M.A. Blackshear and E. Sanders-Bush, J. Pharmacol. Exp. Ther., 221,303 (1983).
43. H.I. Yamamura, M. Watson and W.R. Roeske, Adv. Biochem. Psychopharmacol., 37,331 (1983).
44. N.J.M. Birdsall, A.S.V. Burgen and E.C. Hulme, Mol. Pharmacol., 14,723 (1978).
45. R. Hammer and A. Giachetti, Trends Pharmacol. Sci., 5,18 (1984).
46. R. Hammer, C.P. Berne, N.J.M. Birdsall, A.S.V. Burgen and E.C. Hulme, Nature, 283,90 (1980).
47. R.D. Schwartz, R. McGee and K.J. Kellar, Mol. Pharmacol., 22,56 (1982).
48. M.J. Marks and A.C. Collins, Mol. Pharmacol., 22,54 (1982).
49. J. Schmidt, Mol. Pharmacol., 13,283 (1977).
50. S.T. Carbonetto, D.M. Fambrough and K.J. Muller, Proc. Nat. Acad. Sci. U.S.A., 75,1016 (1978).
51. S.J. Enna and S.H. Snyder, Brain Res., 100,81 (1975).
52. M. Williams and E.A. Risley, J. Neurochem., 32,713 (1978).
53. D.R. Hill and N.G. Bowery, Nature 290,149 (1981).
54. E.W. Karbon, R.S. Duman and S.J. Enna, Brain Res., in press (1984).
55. A.C. Foster and G.E. Fagg, Brain Res. Revs., in press (1984).

56. G. Burnstock in "Regulatory Function of Adenosine," R.H. Berne, T.W. Rall and R. Rubio, Eds., Nijhoff, Boston, 1983, p. 49.
57. J.W. Daly, J. Med. Chem., 25,197 (1982).
58. M. Williams, Handbook Neurochem., 6,1 (1984).
59. R.M. Levin, R. Jacoby and A.J. Wein, Mol. Pharmacol., 23,1 (1982).
60. K.M.M. Murphy and S.H. Snyder, Mol. Pharmacol., 22,250 (1982).
61. T.V. Dunwiddie and T. Worth, J. Pharmacol. Exp. Ther., 220,70 (1982).
62. J.J. Katims, Z. Annau and S.H. Snyder, J. Pharmacol. Exp. Ther., 227,167 (1983).
63. S-M. Yeung and R.D. Green, Pharmacologist, 23,184 (1981).
64. R.S.L. Chang, V. Van Tran and S.H. Snyder, Eur. J. Pharmacol., 48,463 (1978).
65. G.A. Gajtkowski, D.B. Norris, T.J. Rising and T.P. Wood, Nature, 304,65 (1983).
66. M.L. Torchiana, R.J. Pendleton, P.G. Cook, C.A. Hanson and B.V. Clineschmidt, J. Pharmacol. Exp. Ther., 224,514 (1983).
67. M.G.C. Gillan and H.W. Kosterlitz, Brit. J. Pharmacol., 77,461 (1982).
68. L.E. Robson and H.W. Kosterlitz, Proc. Roy. Soc. London. Series B, 205,425 (1979).
69. G.W. Pasternak, Proc. Nat. Acad. Sci. U.S.A., 77, 3691 (1980).
70. R.W. Foote and R. Maurer, Eur. J. Pharmacol., 85,99 (1982).
71. G.W. Pasternak, M. Carroll-Buatti and K. Spiegel, J. Pharmacol. Exp. Ther., 219,192 (1981).
72. K. Spiegel and G.W. Pasternak, J. Pharmacol. Exp. Ther., 228,414 (1984).
73. M. Williams, J. Med. Chem., 26,619 (1983).
74. A.S. Lippa, D.J. Critchett, M.C. Sano, C.A. Klepner, D.W. Greenblatt, J. Cooper and B. Beer, Pharmacol. Biochem. Behav., 10,831 (1979).
75. P.J. Marangos, J. Patel, J-P. Boulenger and Clark-Rosenberg, Mol. Pharmacol., 22,26 (1982).
76. I.L. Martin, C.L. Brown and A. Doble, Life Sci., 32,1925 (1983).
77. C. Braestrup, R. Schmiechen, G. Neef, M. Nielsen and E.N. Petersen, Science, 216,1241 (1982).
78. N. Yokiyama and T. Glenn in "Abstracts of Papers," 184th National Meeting of the American Chemical Society, Kansas City, MO., Sept. 12-17, 1982. Abstr. MEDI 58.
79. M. Williams and E.A. Risley, Life Sci., 24,833 (1979).
80. K-J. Chang and P. Cuatrecasas in "Brain Peptides," D.T. Krieger, M.J. Brownstein and J.B. Martin, Eds., Wiley Interscience, New York, 1983, p. 565.
81. C-M. Lee, L.L. Iversen, M.R. Hanley and B.E.B. Sandberg, Naunyn- Schmiedeberg's Arch. Pharmacol., 318,281 (1982).
82. R.F. Bruns, W.J. Thomsen and T.A. Pugsley, Life Sci., 33,645 (1983).
83. R.A. Janis and D.J. Triggle, J. Med. Chem., 26,775 (1983).
84. R.J. Miller, Life Sci., in press (1984).
85. J-P. Galizzi, M. Fosset and M. Lazdunski, Biochem. Biophys. Res. Commun., 118,239 (1984).
86. R.J. Gould, K.M.M. Murphy and S.H. Snyder, Proc. Nat. Acad. Sci. U.S.A., 79,3656 (1983).
87. M. Schramm, G. Thomas, R. Towart and G. Franckoniak, Nature, 303,535 (1983).
88. P. Erne, E. Burgissen, F.R. Buhler, B. Dubache, H. Kuhnis, M. Meier and H. Rogg, Biochem. Biophys. Res. Commun., 118,842 (1984).
89. I. Shallaby, S.A. Freedman and R.J. Miller, Soc. Neurosci. Abstr., 9,565 (1983).
90. R. Raisman, M.S. Briley and S.Z. Langer, Nature, 281,148 (1979).
91. E.T. Mellerup, P. Plenge and M. Engelstoft, Eur. J. Pharmacol., 96,303 (1983).
92. C.M. Lee, J.A. Javitch and S.H. Snyder, J. Neurosci., 2,1515 (1982).
93. R. Quirion and C.B. Pert, Eur. J. Pharmacol., 83,155 (1982).
94. D.A. Kendall, D.P. Taylor and S.J. Enna, Mol. Pharmacol., 23,594 (1983).
95. B.V. Clineschmidt, M. Williams, J.J. Witoslawski, P.R. Bunting, E.A. Risley and J.A. Totaro, Drug Development Res., 2,147 (1982).
96. B.V. Clineschmidt, Gen. Pharmacol., 10,287 (1979).
97. M. Williams and E.A. Risley, Fed. Proc., 41,1330 (1982).
98. R.W. Olsen, P. Van Ness, C. Napias, M. Bergman, and W.W. Tourtelotte, Adv. Biochem. Psychopharmacol., 21,451 (1980).
99. M.S. Briley, S.Z. Langer, R. Raisman, D. Sechter and E. Zarifan, Science, 209,30 (1980)
100. W.S. Young and M.J. Kuhar, Brain Res., 179,255 (1979).

Chapter 29. Forskolin and Adenylate Cyclase: New Opportunities in Drug Design

Kenneth B. Seamon, National Center for Drugs and Biologics, FDA
Bethesda, MD 20205

Cyclic AMP is synthesized by the membrane-bound enzyme adenylate cyclase which is present in almost all mammalian tissues. Hormonal regulation of adenylate cyclase is initiated by the hormone binding to cell surface receptors. The resulting hormone-receptor complex interacts with either stimulatory regulatory proteins, (Ns), or inhibitory regulatory proteins, (Ni), which bind guanine nucleotides and interact with the catalytic subunit (C) of adenylate cyclase to modulate the enzyme activity.

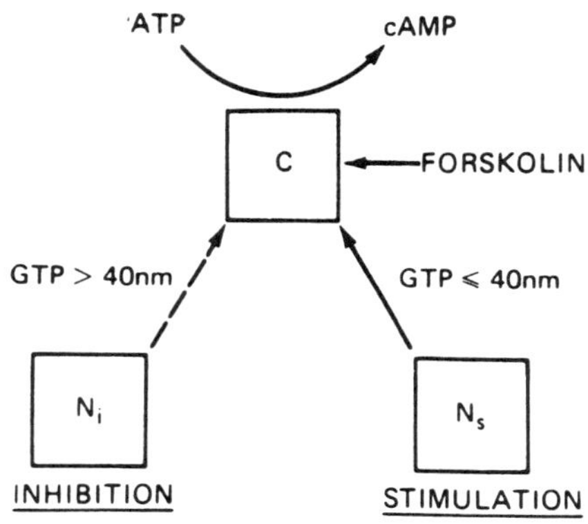

Hormone receptor agonists and antagonists have been used for years as therapeutic agents which affect cyclic AMP levels by acting at receptors. Although there has been much progress in the isolation and purification of the hormone receptors and the regulatory proteins, Ns and Ni, there has been less progress in the purification and characterization of the catalytic subunit of adenylate cyclase. Forskolin (1), the major pharmacologically active diterpene isolated from the roots of Coleus forskohlii,[1-3] can directly activate the

1

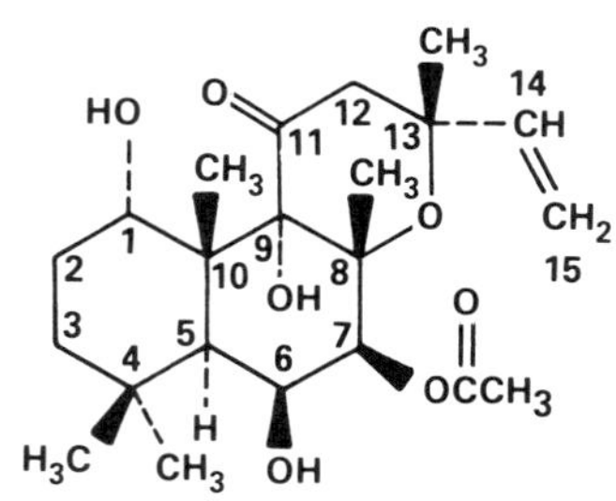

catalytic subunit of adenylate cyclase, and is offering new insights into the regulation of this complex enzyme system. This review will discuss the unique properties of forskolin as an activator of adenylate cyclase which are of potential interest to the medicinal chemist.

ISBN 0-12-040519-9

Activation of Adenylate Cyclase - Metzger and Lindner first demonstrated that forskolin activates adenylate cyclase in rabbit heart and liver membranes.[4,5] Subsequent studies demonstrated that forskolin could activate adenylate cyclase in rat brain membranes,[6] and in membranes from most mammalian tissues.[7] The EC_{50} for forskolin activation is generally between 2μM and 10μM. The maximal extent of activation varies among different tissues and cells. Adenylate cyclase in membranes from rat liver is activated 20-fold,[8] while adenylate cyclase in turkey erythrocyte membranes is activated only 3-fold over basal levels.[9] Forskolin activation of adenylate cyclase is observed in the presence of cyclic AMP phosphodiesterase inhibitors and does not require the presence of calcium or calmodulin.[6] Forskolin activates solubilized adenylate cyclase with similar EC_{50}'s and maximal activations as observed for membrane-bound enzyme.[10-12] Forskolin increases the Vmax of adenylate cyclase with little or no effect on the Km for the substrate, MgATP or MnATP.[11] Forskolin has no effect on the Ka for $MgCl_2$ in activating liver or platelet adenylate cyclase.[8,13]

Forskolin can protect adenylate cyclase in membranes from human platelets, S49 lymphoma cells, and rat brain against thermal denaturation, proteolytic inactivation, and inactivation with N-ethylmaleimide (NEM).[13] Solubilized preparations of adenylate cyclase are also protected against thermal denaturation by forskolin.[11] These include preparations which do not contain functional Ns protein.

Adenylate cyclase from sperm or testis is not stimulated by forskolin.[14,15] These enzymes are not hormonally responsive and are not associated with Ns or Ni proteins. Forskolin can activate sperm membrane adenylate cyclase when extracts from human erythrocyte membranes are added to sperm membranes; however, it is not clear if the reconstitution is revealing adenylate cyclase activity in the human erythrocyte membrane or adding back a factor required for forskolin stimulation of the sperm membrane adenylate cyclase.[15,16]

Adenylate cyclase from invertebrates such as mollusks [17-20] and insects[21,22] is stimulated by forskolin. Adenylate cyclases from *B. Pertussis*,[23] *E. Coli*,[7] *Neurospora*,[24] and *D. Discoidium*[7] are not stimulated by forskolin or Ns.

Interaction with the Catalytic Subunit - Forskolin stimulates adenylate cyclase in membranes from the S49 cyc- lymphoma cell about 20-fold with an EC_{50} of about 25 micromolar.[10,23,25-29] As this mutant of the S49 lymphoma cell line does not contain a functional Ns subunit, it was proposed that forskolin is activating adenylate cyclase by an interaction at the catalytic subunit or other closely associated protein.[10] Forskolin can activate other adenylate cyclase preparations which are unresponsive to hormones or guanine nucleotides and do not contain functional Ns protein. These include detergent solubilized preparations of adenylate cyclase from rat striatum,[11] cortex,[12] heart,[12] and liver.[30] In the absence of a functional Ns protein, it is necessary to include Mn^{2+} in the assay to measure the activity of the catalytic subunit. Forskolin does not require Mn^{2+} to stimulate the catalytic subunit.

The 7-O-hemisuccinate-7-desacetyl derivative of forskolin has been coupled to Sepharose to form an affinity resin.[12] Solubilized adenylate cyclase from rat brain and heart, but not pigeon erythrocytes, binds to this resin and can be eluted from the resin when forskolin is included in the buffer. Preparations of adenylate cyclase devoid of Ns activity are recovered from this column consistent with the proposal that forskolin binds to the catalytic subunit(s). Adenylate cyclase preactivated with guanosine-5',β,γ-imidodiphosphate (GppNHp) also binds to the forskolin-Sepharose column and elutes from the column as a complex of the catalytic subunit(s) and the activated Ns protein.[31]

Potentiation of Forskolin Stimulation by the Ns Protein - Forskolin stimulation of adenylate cyclase can be potentiated by guanine nucleotides and the Ns protein. Forskolin stimulation of adenylate cyclase in wild type S49 lymphoma cell membranes exhibits a time lag and biphasic kinetics indicative of low (Ka=22.0μM) and high (Ka=0.35μM) affinity sites of action.[27,32] Forskolin stimulation of adenylate cyclase in S49 cyc- membranes, which do not contain Ns protein, has kinetics consistent with one site of action with a Ka of 244μM. The high affinity component for activation of the adenylate cyclase in S49 membranes by forskolin requires the Ns protein. The lag time in activation of the wild type S49 adenylate cyclase by forskolin is also associated with the presence of the Ns protein and is not observed in the presence of isoproterenol.

Preparations of the catalytic subunit which do not have functional Ns protein can be stimulated by the addition of Ns and GppNHp and forskolin can potentiate this stimulation.[11] This potentiation requires that forskolin, Ns, GppNHp, and the catalytic subunit be present at the same time. Preactivated Ns (Ns-GppNHp) and forskolin are only additive in their stimulation of the catalytic subunit when they are present in the assay at the same time.

Forskolin can potentiate GppNHp stimulation of adenylate cyclase in membranes from some tissues. This potentiation can be observed when the effect of the Ni protein is suppressed by treatment with pertussis toxin,[33] or by including manganese chloride in the assay medium.[34]

Activation of Cyclic AMP Generation in Intact Cells - Forskolin activates adenylate cyclase in intact cells and tissues with similar characteristics as those observed for activation of the enzyme in membranes and solubilized preparations. These include preparations of rat[35] and human[36] adipocytes, human platelets,[37] tissue slices from brain and other peripheral tissues,[38] and various endocrine and secretory tissues.[39] Forskolin stimulates adenylate cyclase in S49 lymphoma cells,[25,32] rat astrocytomas,[40,41] rat pheochromocytoma cells,[42] cultured pituitary cells,[43-48] cardiomyocytes,[49,50] cultured leydig cells,[51] and cultured kidney cells.[52,53] Forskolin increases intracellular cyclic AMP rapidly with an EC_{50} of about 10μM, and results in elevations of cyclic AMP over basal levels which range between two and fifty-fold, depending on cell type and tissue. Increases in intracellular cyclic AMP by forskolin occur rapidly and are reversible. Increased levels of cyclic AMP are maintained in tissue culture over long periods of time when forskolin is kept in the culture media. Cyclic AMP phosphodiesterase inhibitors augment the increases in intracellular cyclic AMP elicited by forskolin.

Paralleling membrane studies, forskolin elicits increases in intracellular cyclic AMP in the cyc- mutants of the S49 lymphoma cell line which are about ten-fold less than the increase elicited in wild type S49 lymphoma cells.[25] Similar results are seen in the H21a mutant which contains a lesion in Ns impairing its coupling to the catalytic subunit. A maximal stimulation of cyclic AMP synthesis by forskolin in S49 cells requires the presence of a functional Ns protein.

Forskolin stimulation of cyclic AMP synthesis in C6-2B rat astrocytoma cells is decreased after the cells are grown in the presence of cycloheximide, a protein synthesis inhibitor.[40] This occurs without the loss of isoproterenol or cholera toxin stimulated adenylate cyclase. It was suggested that the cycloheximide blocks the synthesis of a protein which has a high turnover rate and is required for forskolin stimulation of adenylate cyclase.[40]

Potentiation of Hormone Action - Low concentrations of forskolin ($<1\mu M$) which elicit only small increases in cyclic AMP can markedly potentiate hormonal responses. This was first observed in rat brain slices, where low concentrations of forskolin markedly potentiated the maximal increase in cyclic AMP and decreased the EC_{50} for stimulation by a number of neurotransmitters including norepinephrine, vasoactive intestinal peptide, PGE_1, and adenosine.[6,38] Receptors coupled to adenylate cyclase which are potentiated by forskolin include β-adrenergic[25,35,36,48,54,55] A_2-adenosine,[38] PGE_1,[37,56] ACTH,[57] and TSH[58-61]. A typical response is illustrated by studies using intact platelets[37] having prostaglandin receptors coupled to adenylate cyclase. PGD_2 increases cyclic AMP levels 10-fold in the absence of forskolin with an EC_{50} of about 5 micromolar. In the presence of 0.1 micromolar forskolin PGD_2 increases cyclic AMP levels about 60-fold with an EC_{50} of $0.5\mu M$. In a similar manner low concentrations of prostaglandins can potentiate the forskolin elicited increases in cyclic AMP.

Potentiation of hormone responses in membranes has been more difficult to observe. Forskolin can potentiate prostaglandin stimulation of adenylate cyclase in human platelet membranes[62] and isoproterenol stimulation of adenylate cyclase in membranes from wild type S49 lymphoma cells[27] and rat adipocytes.[35] These potentiations are not nearly as large as those seen in intact cells and tissues.

Inhibition of Forskolin-Stimulated Activation of Adenylate Cyclase - Inhibition of forskolin-stimulated adenylate cyclase in membranes and cells has been observed with α_2-agonists in platelets,[37,62,63] and adipocytes;[36,64] muscarinic-cholinergic agonists in cardiac cells;[65] A_1-adenosine agonists and insulin in rat adipocytes;[35] somatostatin in S49 lymphoma cells[29,66,67] and pituitary;[47] and dopamine D_2-agonists in pituitary.[68] The inhibition varies between 50 and 75%.

Inhibition of forskolin stimulated adenylate cyclase in membranes by guanine nucleotides can be observed when a functional Ni subunit is present and has been shown in membranes from rat brain,[34] S49 lymphoma cells, [28,66] rat adipocytes,[69] and human platelets.[62] This inhibition can be observed with GTP, GppNHp, or GTP-γS (guanosine 5'-(3-O-thio)triphosphate), is blocked by GDP or GDP-βS (guanosine 5'(2-O-thio)diphosphate), and does not require the presence of an inhibitory hormone. Guanine nucleotide inhibition of adenylate cyclase is not competitive with forskolin and is not observed when membranes are treated with agents that inhibit the action of the Ni protein, NEM and $MnCl_2$. Inhibition of forskolin stimulated adenylate cyclase by GTP is not observed when the Ni protein is ADP-ribosylated by pertussis toxin.[70]

Adenosine analogs inhibit adenylate cyclase by binding to the adenosine P-site associated with the catalytic subunit. Dideoxyadenosine, a potent P-site agonist, inhibits forskolin activated adenylate cyclase in cells,[60,71,72] membranes,[30] and solubilized preparations.[30] This inhibition is not compettitive with forskolin. The affinity of adenylate cyclase for inhibitors at the adenosine P-site is higher when the enzyme is maximally activated, regardless of whether the enzyme is stimulated by forskolin or activated Ns.[30]

Calcium (>50μM) inhibits basal, hormone-stimulated, GppNHp stimulated, and forskolin stimulated adenylate cyclase.[73] This inhibition is not competitive with forskolin and has been suggested to be due to the interaction of the metal ion at a cation binding site on the catalytic subunit.

Forskolin activation of adenylate cyclase can be inhibited by high concentrations of detergents and low molecular weight alcohols.[74,75] Inhibition of forskolin stimulated adenylate cyclase by ethanol can be observed at 0.2M and results in 50% inhibition at 1.0M. Butanol and propanol are more effective at inhibiting forskolin stimulated adenylate cyclase than ethanol. Dimethyl sulfoxide inhibits the stimulation of adenylate cyclase by forskolin much less than ethanol.[75]

Physiological Effects of Forskolin - Forskolin elicits physiological effects consonant with its ability to stimulate adenylate cyclase and increase intracellular cyclic AMP. This has been studied in many different tissues and cells and a complete description of all systems is beyond the scope of this review. Secretory responses which are elicited by forskolin include: chloride-secretion across rat colon descendens,[76] acid and pepsinogen secretion from gastric gland,[77,78] prolactin secretion from pituitary cells,[43,45] luteinizing hormone secretion from pituitary,[79] renin secretion from perfused kidney,[80] insulin release from pancreatic cells,[81] ACTH release from pituitary cells,[46,47] and amylase secretion from pancreatic acinor cells.[82] Other responses which are elicited by forskolin and have been associated with increased intracellular cyclic AMP include lipolysis in adipocytes,[35,36,54] inhibition of platelet aggregation,[37,72,83] relaxation of smooth muscle,[2,3,84] and stimulation of steroidogenesis.[57,85] Forskolin increases cyclic AMP content in oocytes and inhibits progesterone induced meiosis.[86-92] Forskolin increases sodium transport in cultured epithelia[93] and bicarbonate transport in choroid plexus.[94] Reduction of intraocular pressure by forskolin has been observed in rabbit, human, and monkey eye.[95,96] Forskolin stimulates transcription of the prolactin gene in pituitary[97] and increases the biosynthesis of enkephalin in bovine chromafin cells.[98] Forskolin elicited physiological responses are usually potentiated by the natural hormone consistent with the hormonal potentiation of forskolin elicited increases in cyclic AMP.

Forskolin has been shown to affect enzymes other than adenylate cyclase indirectly. Treatment of rabbit cardiac slices with forskolin results in the inhibition of the membrane bound Na^+ K^+ ATPase.[99]

The inhibition of the ATPase is probably due to a cyclic AMP mediated phosphorylation. Tyrosine hydroxylase is activated in striatal slices treated with forskolin,[100] also due to a cyclic AMP mediated phosphorylation.

Structure Activity Relationships for Forskolin Analogs - Forskolin makes up 1% of the dry weight of the roots of *Coleus forskohlii* and is the major diterpene that is isolated from the plant. Other structurally similar diterpenes that occur naturally in the plant include 6-acetyl-7-desacetylforskolin (6), 7-desacetylforskolin (5), 1,9-dideoxyforskolin (7), and 9-deoxyforskolin (8). A number of semisynthetic derivatives of forskolin have been prepared.[101,102] These and the naturally occurring analogs of forskolin have been tested for their ability to activate adenylate cyclase.[103]

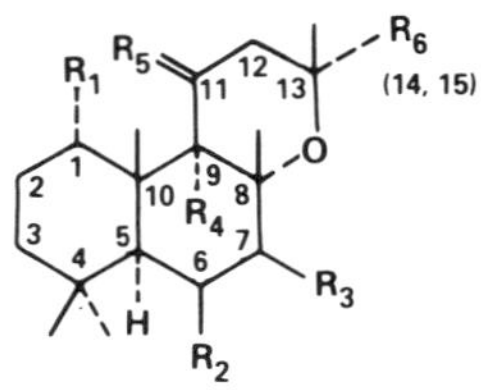

	R_1	R_2	R_3	R_4	R_5	R_6
2	OH	OH	OCOEt	OH	O	$CH{=}CH_2$
3	OH	OH	OCO_2Et	OH	O	$CH{=}CH_2$
4	OH	OH	OSO_2-4-Me-Ph	OH	O	$CH{=}CH_2$
5	OH	OH	OH	OH	O	$CH{=}CH_2$
6	OH	OCOMe	OH	OH	O	$CH{=}CH_2$
7	H	OH	OCOMe	H	O	$CH{=}CH_2$
8	OH	OH	OCOMe	H	O	$CH{=}CH_2$
9	O	O	O	O	O	$CH{=}CH_2$
			CO CO			
10	OH	OH	OCOMe	OH	OH, H	$CH{=}CH_2$
11	OH	OH	OCOMe	OH	O	CH_2CH_3
12	OH	OH	OCOMe	OH	O	$CH{-}CH_2$ (O bridged)

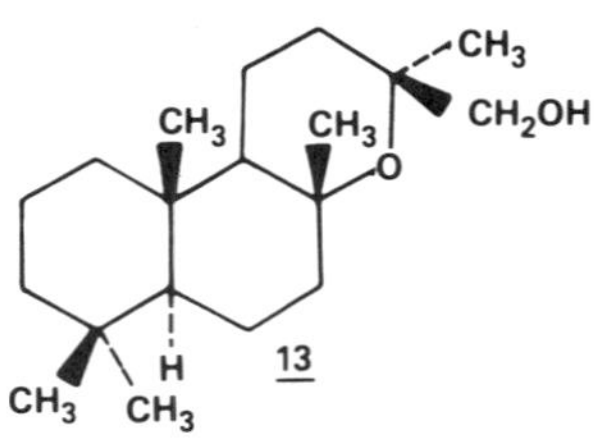

13

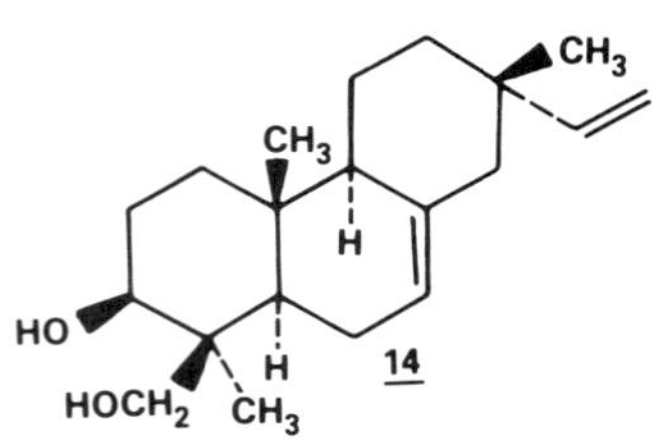

14

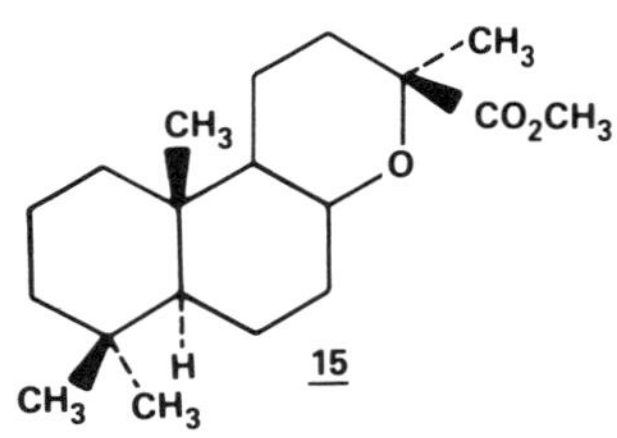

15

The 1-and 9-hydroxyl groups define an important area for forskolin's action. The 1,9-dideoxyderivative (7) is totally inactive and the 1,9,6,7-dicarbonate (9), and 1,9 sulfonate are very weak at activating adenylate cyclase. Other structurally related compounds which are unable to activate adenylate cyclase are 1,6-diketoforskolin, the 14,15-oxide (12), virescenol-B, abietic acid, podocarpic acid, retene, gibberellin, and other diterpenes(13-15).[23,83] None of the inactive diterpenes antagonize forskolin stimulation of adenylate cyclase. Derivatives of forskolin which are partially active include a number of 7-acyl derivatives. The most active of these include the 7-ethylcarbonate (3) and the 7-propionate (2) which are almost equipotent with forskolin. The 7-formyl, 7-desacetyl (5), and 7-hemicsuccinate are less potent than forskolin with EC_{50}'s that range from 50μM to 150μM. The 7-tosyl derivative (4) is much less potent than the other 7-acyl derivatives. The 14,15-dihydroforskolin (11) and 11β-hydroxy-forskolin (10) are less potent than forskolin with EC_{50}'s of about 50 μM. The activity of forskolin analogs does not correlate with lipophilicity.

12-[^3H] forskolin (sp. act. 27 Ci/mmol) has been synthesized and used to determine forskolin binding sites in membranes.[104] 12-[^3H] forskolin binds to rat brain membranes with high affinity, Kd = 27nM, Bmax = 275 fmol/mg protein. The ability of forskolin analogs to compete for these binding sites correlates with their ability to activate adenylate cyclase.

Therapeutic Potential - Forskolin is a potent hypotensive agent due to its peripheral vasodilatory properties. Reduction in blood pressure is observed in dogs (20ug/kg i.v.), renal hypertensive rats (50ug/kg i.v.), and spontaneous hypertensive rats (10 mg/kg, p.o.).[2,105] Forskolin is also a potent positive inotropic agent as observed in isolated guinea pig heart and atrial preparations, and in dog and cat hearts *in vitro*.[2,105] Cardiac output in anesthesized dog was increased at doses ranging from 5ug/kg to 100ug/kg, i.v.[2] Forskolin antagonizes histamine or acetylcholine induced bronchoconstriction making it a potent bronchodilator.[106,107] Forskolin inhibits platelet aggregation *in vitro* and *in vivo*, making it a possible anti-thrombotic agent.[37,72,83] Forskolin produces a rapid and long lasting reduction in intraocular pressure upon topical application to the eye, and would appear to be a valuable agent to treat glaucoma.[95,96] Forskolin has been shown to inhibit pulmonary tumor colonization in mice by B16 murine melanoma cells and is a potential antimetastatic agent.[108]

Summary - Forskolin activates hormone sensitive adenylate cyclase in a completely unique manner. Its ability to potentiate hormonally induced increases in cyclic AMP provides a new mechanism to potentiate hormonal responses *in vitro* and *in vivo*. The exact site of action of forskolin still remains unclear, and it is possible that forskolin acts directly at the catalytic subunit of adenylate cyclase or at another subunit which has not yet been identified. The possibility still remains that an endogenous forskolin compound might exist, which would have the ability to either stimulate adenylate cyclase directly or to put the enzyme in a potentiated state which would be exquisitely sensitive to hormonal stimulation. It also remains to be determined if *in vivo* effects of forskolin are due to the direct stimulation of adenylate cyclase by forskolin or to the potentiation of the action of natural hormones which are in the circulation. Binding sites for forskolin in membranes can now be studied using the radiolabelled forskolin, providing an easy assay to search for forskolin agonists and antagonists. The radiolabelled forskolin should also provide a tool to study the metabolism and disposition of forskolin after *in vivo* administration.

In over 300 articles in the past three years forskolin has been established as a powerful tool to study adenylate cyclase *in vitro* and *in vivo*. It is also anticipated that potential new uses of forskolin as a therapeutic agent will continue to be investigated. Studies with forskolin have thus provided medicinal chemists with a new target in the design of therapeutic agents, the adenylate cyclase enzyme.

Acknowledgement - I would like to acknowledge the cooperation and support of Dr. Noel J. deSouza, Dr. Jurgen Reden, and their colleagues at Hoechst Pharmaceuticals, Bombay during our investigations on forskolin. This review is dedicated to the memory of Dr. Balbir S. Bajwa who made such significant contributions to our understanding of the medicinal chemistry of forskolin.

References

1. S.V. Bhat, B.S. Bajwa, H. Dornauer, and N.J. de Souza, and H.W. Fehlhaber, Tetrahedron Lett., 19, 1669 (1977).
2. E. Linder, A.N. Dohadwalla, and B.K. Bhattacharya, Arzneim.-Forsch., 28, 284 (1978).
3. M.P. Dubey, R.C. Srimal, S. Nityaanand, B.N. Dhawan, J. Ethnopharm., 3, 1 (1981).
4. H. Metzger and E. Lindner, IRCS Med. Sci., 9, 99 (1981).
5. H. Metzger and E. Lindner, Arzneim.-Forsch., 31, 1248 (1981).
6. K.B. Seamon, W. Padgett, and J.W. Daly, Proc. Nat. Acad. Sci. U.S.A., 78, 3363 (1981).
7. K.B. Seamon and J.W. Daly, J. Cyclic Nucleotide Res. 7, 201 (1981).
8. L. Birnbaumer, D. Stengel, M. Desmier, and J. Hanoune, Eur. J. Biochem., 136, 107 (1983).
9. S.A. Morris and J.P. Bilezikian, Arch. Biochem. Biophys., 220, 628 (1983).
10. K.B. Seamon and J.W. Daly, J. Biol. Chem., 256, 9799 (1981).
11. J.L. Bender, and E.J. Neer, J. Biol. Chem., 258, 2432 (1983).
12. T. Pfeuffer, and H. Metzger, FEBS Lett., 146, 369 (1982).
13. J.A. Awad, R.A. Johnson, K.H. Jakobs, and G. Schultz, J. Biol. Chem., 258, 2960 (1983).
14. L.R. Forte, D.B. Bylund and W.L. Zahler, Mol. Pharmacol., 24, 42 (1983).
15. D. Stengel, L. Guenet, M. Desmier, P. Insel and J. Hanoune, Mol. Cell. Endocrinol., 28, 681 (1982).
16. C.R. Schneyer, M.A. Pineyro, and R.I. Gregerman, Life Sci., 33, 275 (1983).
17. L.K. Kaczmarek and J.A. Kauer, J. Neuroscience, 3, 2230 (1983).
18. P. Deterre, D. Paupardintritsch, J. Bockaert, and H.M. Gerschenfeld, Proc. Nat. Acad. Sci. U.S.A., 79, 7934 (1982).
19. A. Eskin and J.S. Takahashi, Science, 220, 82 (1983).
20. F. Strumwasser, L.K. Kaczmarek and K.R. Jennings, Fed. Proc., 41, 2933 (1982).
21. K.R. Jennings, R.W. Steele, and A.N. Starratt, Comp. Biochem. Physiol., 74, 69 (1983).
22. I. Litosch, Y. Saito, and J.N. Fain, Biochem. J., 204, 147 (1982).
23. J.E. Fradkin, W.C. Hardy, and J. Wolff, Endocrinology, 110, 2018 (1981).
24. M.M. Flawia, A.R. Kornblihtt, J.A. Reig, M. Torruella, and H.N. Torres, J. Biol. Chem., 258, 8255 (1983).
25. F.J. Darfler, L.C. Mahan, A.M. Koachman, and P.A. Insel, J. Biol. Chem., 257, 1901 (1982).
26. R.W. Downs and G.D. Aurbach, J. Cyclic Nucleotide Res., 8, 235 (1982).
27. D.A. Green and R.B. Clark, J. Cyclic Nucleotide Res., 8, 337 (1982).
28. J.D. Hildebrandt and L. Birnbaumer, J. Biol. Chem., 258, 3141 (1983).
29. K.H. Jakobs, U. Gehring, B. Gaugler, T. Pfeuffer, and G. Schultz, Eur. J. Biochem., 130, 605 (1983).
30. V.A. Florio and E.M. Ross, Mol. Pharmacol., 24, 195 (1983).
31. T. Pfeuffer, B. Gaugler, and H. Metzger, FEBS Lett., 164, 154 (1983).
32. R.B. Clark, T.J. Goka, D.A. Green, R. Barber, and R.W. Butcher, Mol. Pharmacol., 22, 609 (1982).

33. J.N. Fain, C.J. O'Donnell, I. Mills, and J.A. Garcia-Sainz, Biochem. Biophys. Res. Commun., 116, 651 (1983).
34. K.B. Seamon and J.W. Daly, J. Biol. Chem., 257, 1591 (1982).
35. I. Litosch, T.H. Hudson, I. Mills, S.Y. Li, and J.N. Fain, Mol. Pharmacol., 22, 109 (1982).
36. T.W. Burns, P.E. Langley, B.E. Terry, D.B. Bylund and L.R. Forte, Life Sci.,31, 815 (1982).
37. A.M. Siegl, J.W. Daly, and J.B. Smith, Mol. Pharmacol.,21, 680 (1982).
38. J.W. Daly, W. Padgett, and K.B. Seamon, J. Neurochem.,38, 532 (1982).
39. J.W. Daly, Adv. Cyclic Nucleotide Res., 17, 81 (1984).
40. G. Brooker, C. Pedone, and K. Barovsky, Science,220, 1169 (1983).
41. D.K. Wu and J. Devellis, J. Cyclic Nucleotide Protein Phosphor. Res.,9, 59 (1983).
42. C.S. Rabe, J. Schneider, and R. McGee, J. Cyclic Nucleotide Res.,8, 371 (1982).
43. M.A. Brostrom, C.O. Brostrom, L.A. Brotman and S.S. Green, Mol. Pharmacol.,23, 399 (1983).
44. L.J. Dorflinger and A. Schonbrunn, Endocrinology, 113, 1551 (1983).
45. S. Guild and A.H. Drummond, Biochem. J., 216, 551 (1983).
46. S. Heisler, L. Larose, and J. Morisset, Biochem. Biophys. Res. Commun., 114, 289 (1983).
47. S. Heisler, T.D. Reisine, V.Y.H. Hook, and J. Axelrod, Proc. Nat. Acad. Sci. U.S.A., 79, 6502 (1982).
48. T. Reisine, S. Heisler, V.Y.H. Hook, and J. Axelrod, Biochem. Biophys. Res. Commun., 108, 1251 (1982).
49. M.B. Anandsrivastava and M. Cantin, Arch. Biochem. Biophys., 223, 468 (1983).
50. I.L.O. Buxton and L.L. Brunton, J. Biol. Chem., 258, 233 (1983).
51. T. Lin, Life Sci., 33, 2465 (1983).
52. M. Mettauer, E.M. Giesen, J.L. Imbs, M. Schmidt, and J. Schwartz, IRCS Med. Sci., 11, 887 (1983).
53. H.L. Henry, N.S. Cunningham, and T.A. Noland, Endocrinology, 113, 1942 (1983).
54. R.J. Ho and Q.H. Shi, Biochem. Biophys. Res. Commun., 107, 157 (1982).
55. P. Mauduit, G. Herman, and B. Rossignol, FEBS Lett., 153, 21 (1983).
56. I.L. Karnushina, M. Spatz, and J. Bembry, Life Sci., 32, 1427 (1983).
57. K. Moriwaki, Y. Itoh, S. Iida, and K. Ichihara, Life Sci., 30, 2235 (1982).
58. Y. Totsuka, M.S. Ferdows, T.B. Nielsen, and J.B. Field, Biochim. Biophys. Acta, 756, 319 (1983).
59. Y. Totsuka, T.B. Nielsen, and J.B. Field, Endocrinology, 113, 1088 (1983).
60. J. Vansande, P. Cochaux, and J.E. Dumont, FEBS Lett., 150, 137 (1982).
61. J.E. Fradkin, W.C. Hardy, and J. Wolff, Endrocrinology, 110, 2018 (1981).
62. P.A. Insel, D. Stengel, N. Ferry, and J. Hanoune, J. Biol. Chem., 257, 7485 (1982).
63. K.H. Jakobs, G. Schultz, B. Gaugler, and T. Pfeuffer, Eur. J. Biochem., 134, 351 (1983).
64. K.K. McMahon and R.J. Schimmel, J. Cyclic Nucleotide Res., 8, 39 (1982).
65. R.B. Meeker and T.K. Harden, Mol. Pharmacol., 23, 384 (1983).
66. K.H. Jakobs, K. Aktories, and G. Schultz, Nature, 303, 177 (1983).
67. K.H. Jakobs and G. Schultz, Proc. Nat. Acad. Sci. U.S.A., 80, 3899 (1983).
68. K. Miyazaki, J.M. Saavedra, T.E. Cote, and J.W. Kebabian, Neurochem. Int., 5, 803 (1983).
69. T.H. Hudson and J.N. Fain, J. Biol. Chem., 258, 9755 (1983).
70. J.D. Hildebrandt, R.D. Sekura, J. Codina, R. Iyengar, C.R. Manclark, and L. Birnbaumer, Nature, 302, 706 (1983).
71. J. Vansande, P. Cochaux, J. Mockel, and J.E. Dumont, Mol. Cell Endocrinol., 29, 109 (1983).
72. K.C. Agarwal and R.E. Parks., Biochem. Pharmacol., 31, 3713 (1982).
73. R.D. Lasker, R.W. Downs, and G.D. Aurbach, Arch. Biochem. Biophys., 216, 345 (1982).
74. P. Robberecht, M. Waelbroeck, P. Chatelain, J.C. Camus, and E.J. Christoph, FEBS Lett., 154, 205 (1983).
75. R.D. Huang, M.F. Smith, and W.L. Zahler, J. Cyclic Nucleotide Res., 8, 385 (1982).
76. R.J. Bridges, W. Rummel, and B. Simon, Naunyn-Schmiedebergs Arch. Pharmacol., 323, 355 (1983).
77. C.S. Chew, Am. J. Physiol., 245, 371 (1983).
78. S.J. Hersey, A. Owirodu, and M. Miller, Biochim. Biophys. Acta, 755, 293 (1983).
79. M.J. Cronin, W.S. Evans, E.L. Hewlett, A.D. Rogol, and M.O. Thorner, Neuroendocrinology, 37, 161 (1983).
80. U. Schwertschlag and E. Hackenthal, Eur. J. Pharmacol., 84, 111 (1982).
81. D.E. Wiedenkeller and G.W.G. Sharp, Endocrinology, 113, 2311 (1983).
82. S. Heisler, Can. J. Physiol. Pharmacol., 61, 1168 (1983).
83. S. Adnot, M. Desmier, N. Ferry, and J. Hanoune, Biochem. Pharm., 31, 4071 (1982).
84. M.J. Muller, and H.P. Baer, Naunyn-Schmedebergs, Arch. Pharmacol., 322, 78 (1983).

85. L. Hedin and S. Rosberg, Mol. Cell. Endocrinology, 33, 69 (1983).
86. N. Dekel and I. Sherizly, FEBS Lett., 151, 153 (1983).
87. S.B. Oldham, C.T. Molloy, L.G. Lipson, and T.T. Boggs, Endocrinology, 114, 207 (1984).
88. P.J. Olsiewski and W.H. Beers, Dev. Biol., 100, 287 (1983).
89. S. Schorderetslatkine and E.M. Baulieu, Endocrinology 111, 1385 (1982).
90. R.M. Schultz, R.R. Montgomery, and J.R. Belanoff, Dev. Biol., 97, 264 (1983).
91. R.M. Schultz, R.R. Montgomery, P.F. Wardbailey, and J.J. Eppig, Dev. Biol., 95, 294 (1983).
92. F. Urner, W.L. Herrmann, E.E. Baulieu, and S. Schorderetslatkine, Endocrinology, 113, 1170 (1983).
93. A.W. Cuthbert, and J.A. Spayne, Br. J. Pharmacol., 76, 33 (1982).
94. Y. Saito and E.M. Wright, J. Physiol. (Lond.), 336, 635 (1983).
95. S.P. Bartels, S.R. Lee, and A.H. Neufeld, Curr. Eye Res., 2, 673 (1983).
96. J. Caprioli and M. Sears, Lancet, 1, 958 (1983).
97. G.H. Murdoch, M.G. Rosenfeld, and R.M. Evans, Science, 218, 1315 (1982).
98. L.E. Eiden, and A.J. Hotchkiss, Neuropeptides, 4, 1 (1983).
99. F. Spah, J. Cardiovasc. Pharmacol., 6, 99 (1984).
100. I.R. Katz, D. Smith, and M.H. Makman, Brain Res., 264, 173 (1983).
101. S.V. Bhat, A.N. Dohadwalla, B.S. Bajwa, N.K. Dadkar, H. Dornauer, and N.J. Desouza, J. Med. Chem., 26, 486 (1983).
102. S.V. Bhat, B.S. Bajwa, H. Dornauer, and N.J. de Souza, J. Chem. Soc., Perkin Trans. I, 767 (1982).
103. K.B. Seamon, J.W. Daly, H. Metzger, N.J. de Souza, and J. Reden, J. Med. Chem., 26, 436 (1983).
104. K.B. Seamon, R. Vallancourt, M. Edwards, and J.W. Daly, Proc. Nat. Acad. Sci. U.S.A., in press (1984).
105. N.J. de Souza, A.H. Dohadwalla, and J. Reden, Med. Res. Rev., 3, 201 (1983).
106. J.F. Burka, J. Pharmacol. Exp. Ther., 225, 427 (1983).
107. J.F. Burka, Can. J. Physiol. Pharmacol., 61, 581 (1983).
108. K.C. Agarwal and R.E. Parks, Int. J. Cancer, 32, 801 (1983).

Chapter 30. Recent Progress in the Rational Design of Peptide Hormones and Neurotransmitters

Victor J. Hruby, John L. Krstenansky, and Wayne L. Cody
Department of Chemistry, University of Arizona, Tucson, AZ 85721

Introduction - Any effort to discuss progress in "rational design" of peptide neurotransmitters and hormones (hereafter referred to as peptide hormones) poses certain difficulties. Peptide hormones generally produce their biological effects by interaction with membrane bound receptors, but we know little about the specific chemical and physical (three dimensional) properties of these binding sites. Thus any rational design must utilize our ability to deduce the properties of the receptor system (host) by using well designed peptide hormone analogs (guests), and careful chemical and physical analysis of the structure, conformation and dynamic properties of these analogs in relation to biological activities.

Ultimately hormone-receptor interactions are three dimensional in nature, in which two complementary topological surfaces interact to produce a biological response. These three dimensional features often can be examined by conformational restriction of the peptide hormones.[1-8] For larger peptide hormones (>20 residues), design of extensive secondary structures such as amphiphilic helices can provide important insights into membrane binding, receptor binding, and other aspects of biological recognition. This approach has recently been reviewed[9-11] and will not be discussed here. Rather we will concentrate on smaller peptides (<20 residues) which are conformationally flexible.

Considerable insight into structure-activity relationships for the smaller peptides has been made by classical functional group modifications. This approach has been thoughtfully discussed by Rudinger[12] and Schwyzer[13], and generally is a necessary prelude to rational design based on conformation-biological activity relationships. An important goal of such studies has been the determination of those elements of structure critical for binding (potency), and those necessary for transduction (efficacy). This requires a careful choice of bioassay systems, which allow critical evaluation of potency and efficacy over several orders of magnitude concentration. Rational design must utilize the assay results with a critical understanding of the complexities of receptor systems, lest important aspects of function such as partial agonism or antagonism, superagonism, etc. be missed. In addition, if conformation-activity relationships are to be used for rational design, it is critical that sufficent constraints be incorporated into the peptide so that there can be some expectation that the conformational features will remain when the peptide interacts with its receptor.[14,15] Furthermore, conformational models for biological activity which are postulated must be testable.

Finally, an additional complication must be addressed. Virtually all peptide hormones have multiple receptors of physiological and pharmacological importance, and there is overwhelming evidence that generally each of these receptors utilizes different structural and conformational properties of the hormone. Thus rational design must be directed at specific

ISBN 0-12-040519-9

receptors, distinguishing those features of structure and conformation important to a particular receptor.

Oxytocin - In addition to being the first peptide hormone whose structure was elucidated and then prepared by total synthesis, oxytocin (OT) was also the first peptide hormone for which a conformation-biological activity model was proposed.[16] Though oxytocin has a 20-membered cyclic ring within its structure, H-Cys-Tyr-Ile-Gln-Asn-Cys-Pro-Leu-Gly-NH_2, it has considerable conformational flexibility in aqueous[17] and to a lesser extent in dimethylsulfoxide solutions.[14] Based on a proposed conformation and known structure-function relationships for OT, Walter et al.[16] proposed a model for the relationship of conformation to biological activity at the rat uterine receptor, which was later refined to the "co-operative" model.[18] The basic features of this model were two-fold: the hydrophobic side chains of Ile^3, Gln^4, Pro^7, and Leu^8 were proposed to be at the corners of β turns, primarily important as "binding elements" (address, binding message); the Tyr^2 and Asn^5 side chains interact across the 20-membered ring in a manner which is critical for initiating the oxytocic response (transduction). (For earlier efforts to test this model see reviews by Spatola[19] and Hruby[14,20]). Interestingly, few studies with oxytocin agonists with further conformational restrictions have been reported.

Recently, however, the cycloleucine-2[21] and cycloleucine-8 analogs of OT have been examined. Due to the quaternary α-carbon atom of cycloleucine (cLeu), its conformation is restricted to near the α-helix regions of conformational space.[22] As expected [$cLeu^8$]oxytocin was a full agonist. However, [$cLeu^2$]oxytocin, though a weak agonist (about 1/100 the potency of oxytocin), also was a full agonist.[21] Recently Schwartz and co-workers[23-27] have investigated whether the intramolecular hydrogen bonds proposed in the "cooperative model" were necessary for biological activity, and have found that none are. On the other hand, the very high biological activities of the Δ^3Pro^7-analog of oxytocin, and the biological profiles of the D-Tyr^2, D-Gln^4, Thr^4 and modified Pro^7 analogs[28-32] all support some aspects of the first feature of the co-operative model.

The conformational properties of restricted penicillamine-1 (Pen^1) analogs of oxytocin have been extensively investigated by Hruby et al.[4,14,21,33-37] A proposal for the conformational and structural properties important to antagonist activity at the uterine receptor,[5,14,20] and a proposal for a "dynamic model" for oxytocin agonist and antagonist action have been advanced.[5,14,37] This model is consistent with the "zipper model" of Burgen et al.[38] In the case of OT, this model postulated[5,14,34] that conformational, dynamic, and structural properties of the side chains in positions 2 and 5 are critical for the transduction process and considerable evidence in support of this aspect of the model has been obtained.[14,20-21,33-36,39-41] The model also postulated[5,14] that the conformational and structural properties of the side chains in positions 2, 3, 4, 7, and 8 are important for antagonist binding. Recent studies have provided evidence that Pen-1 antagonists utilize different structural and conformational features than agonists in positions 2 and 8 of the hormone on binding to the uterotonic receptor.[7,21,42]

Vasopressin - Examination of possible conformational features of arginine vasopressin (AVP) led Walter et al.[43] to propose a "biologically active conformation" for vasopressin at the kidney antidiuretic receptor. The key residues for receptor binding were postulated to be the lipophilic side chains of Phe^3, Gln^4, Pro^7, and Arg^8 in vasopressin (H-Cys-Tyr-Phe-Gln-Asn-Cys-Pro-Arg-Gly-NH_2), while the key "active elements" for transduction were postulated to be the Asn^5 side chain and the basic moiety on the side chain of Arg^8. Analogs designed to test this model have not been numerous, though the studies thus far[28,29] are compatible with the Walter model. Recently, a series of excellent papers have appeared in which a large number of vasopressin antagonist analogs were synthesized for the antidiuretic receptor.[44-50] Conformational studies of these analogs have not been reported, though all are conformationally restricted.

Somatostatin - Considerable interest has developed in the use of somatostatin (NH_2-Ala-Gly-Cys-Lys-Asn-Phe-Phe-Trp-Lys-Thr-Phe-Thr-Ser-Cys-COOH) (I) for the control of diabetes. Unfortunately, somatostatin was not biologically stable and therefore not suitable for clinical studies. Therefore rational design of somatostatin analogs has been directed at obtaining small, biologically stable molecules (potency of analogs synthesized will be compared with the *in vitro* inhibition of GH release standardized to somatostatin = 1). Rational design of synthetic somatostatin analogs based on NMR[51-54] and computer-based molecular modeling,[55] has led to the synthesis of small conformationally restricted analogs with high activity (for a review of earlier work see 56).

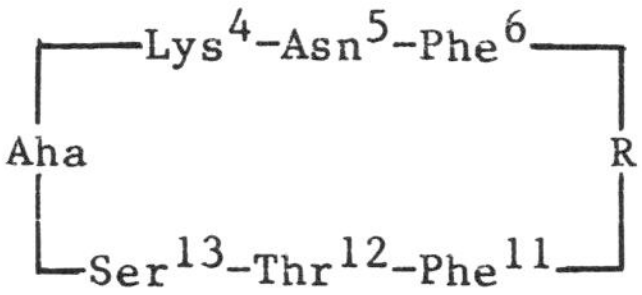

1

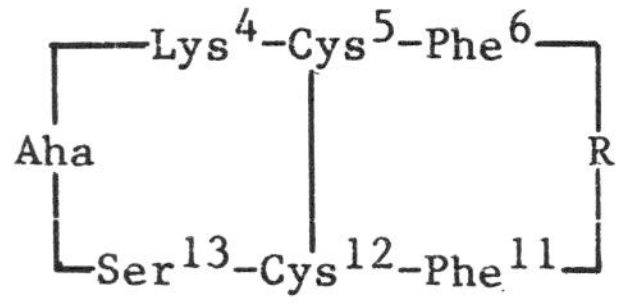

2

Aha=1-aminoheptanoic acid
R=-Phe^7-D-Trp^8-Lys^9-Thr^{10}-.

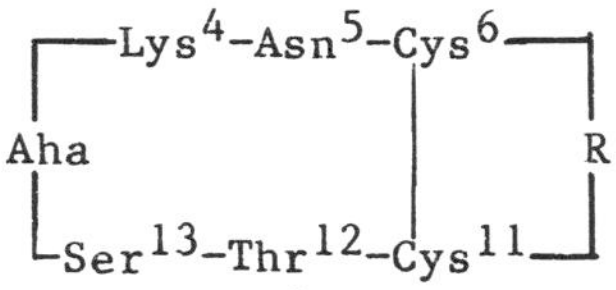

3

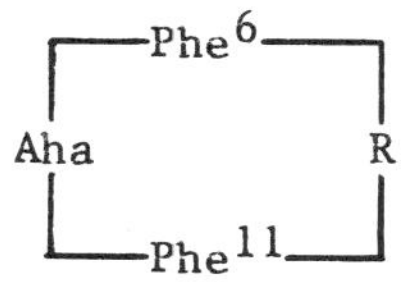

4

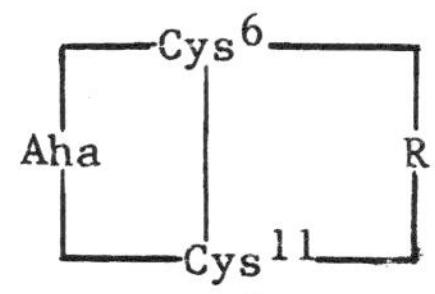

5

The synthesis of analog 1,[57] with potent and prolonged activity led to further conformational restrictions which were based upon molecular modeling.[55] Models suggested that the β-carbons of Asn^5 and Thr^{12} and the β-carbons of Phe^6 and Phe^{11} are in close proximity in space, and it was attractive to build a covalent residue between these pairs of residues via a disulfide bridge. Compounds 2 and 3 employed these modifications, and both exhibited high potency (2, GH = 0.37 and 3, GH = 0.88).[57] The high activity of compound 2 suggested that the β-carbons of residues 5 and 12 in somatostatin were in close proximity to each other upon binding to the receptor. It also showed that the 7-10 sequence of the native hormone was important for biological activity, while the amino acids in positions 5, 6, 11, and 12 probably were not. This idea was further supported by previous data, which suggested that either Lys^4 or Ser^{13} can be deleted without significant loss of activity. This led to the synthesis of 4 in which the Lys^4 and Ser^{13} are deleted in 2 and the disulfide bridge is replaced with methylene groups (GH = 0.93)[58]. Furthermore, incorporating information learned in the synthesis of 3, compound 4 was further modified by incorporating a disulfide bridge between residues 6 and 11 to form the bicyclic analog 5.[58] Compound 5 showed very high biological potency in all bioassays (GH = 1.24, glucagon inhibition = 2.66, and insulin inhibition = 3.50) with respect to somatostatin and compound 4 further supporting the hypothesis that Phe^6 and Phe^{11} serve only to stabilize the β-turn structure via stacking of the aromatic rings. This analog also showed high biological stability and prolonged activity.[58] Further analysis of the 7-10 sequence led to a series of potent analogs by locking the (7-10) tetrapeptide (D-Trp[8]) into a β-turn.[59] In most cases, low biological activity was observed, the exception being cyclo-(N-Me-Ala-Phe-D-Trp-Lys-Thr-Phe)[60] (GH = 3.5), which incorporated both Type II and VI β-turns in the Phe-D-Trp-Lys-Thr and the Thr-Phe-N-Me-Ala-Phe fragments, respectively. Conformational analysis of this compound has led to the synthesis of the retro enantiomeric analog cyclo-(D-Ala-D-Phe-D-Trp-Lys-D-Thr-N-Me-D-Phe) (GH = 0.88) which showed the full activity of the native hormone[61]. Finally, an interesting hexapeptide analog based on the potent analog cyclo-(Pro-Phe-D-Trp-Lys-Thr-Phe) (GH = 1.74), in which a Δ^{z}-Phe is substituted for Phe[6], resulted in a compound with lower activity (GH = 1.12).[62]

α-Melanotropin (α-MSH) - α-Melanotropin has been the subject of extensive classical structure-function studies.[13,63-65] α-MSH is a linear peptide (Ac-Ser-Tyr-Ser-Met-Glu-His-Phe-Arg-Trp-Gly-Lys-Pro-Val-NH_2), is highly conformationally flexible and apparently has little secondary structure in solution.[13,66] These properties and other aspects of its structure-function relationships led Schwyzer to postulate[13] that α-MSH is a "sychnologically" organized peptide hormone. Until recently most structure-activity studies were based on classical amino acid replacement studies. A major success of these studies have been the development of α-MSH (ACTH) analogs with highly potent behavioral activities (reviewed by Van Nispen and Greven[67]).

Despite the extensive structure-function studies on α-MSH, very little has been done until recently on conformation-activity relationships of α-MSH at any of its receptors. Schwyzer suggested[13] that ACTH may utilize one face of an α-helix segment spanning residues 4-10 for interacting with the corticotropin receptor and another face for interacting with the melanotropic receptors. To our knowledge, this hypothesis has not been tested by synthesis of conformationally constrained analogs. Recently, however, attempts have been made to examine the conformational requirements for α-MSH activity at its peripheral[68] and putative central nervous system (CNS) receptors.[69] These investigations have led to a number of α-MSH analogs with super potency and exceptionally prolonged *in vitro* and *in vivo* biological activities.

It was observed many years ago[70] that heat-alkali treated α-MSH displayed high potency and prolonged activity in melanocyte dispersion.[71] It was hypothesized that racemization resulted in specific conformational effects, which were responsible for the prolonged biological activity.[72] Quantitative characterization of the chemical effects of heat-alkali treatment[73] and conformational considerations[74] led to the design of [Nle[4], D-Phe[7]]-α-MSH which was found to have super potency in several assay systems[72], with extraordinarily prolonged activities both *in vitro* and *in vivo*.[75] More recently, it has been shown that the fragment analog of α-MSH, Ac-[Nle[4], D-Phe[7]]-α-MSH_{4-11}-NH_2, in which 5 of the 13 residues of α-MSH were deleted, retained most of the biological activities of the full 1-13 analog.[76,77] Other fragment analogs of varying lengths have been designed which differentiate the different receptor systems both with respect to potency and prolonged biological activity.[76-79] Analogs also have been designed[79] which have very low potency, but very prolonged activity. These results suggest that prolonged *in vitro* and *in vivo* biological activity can be a receptor-related event not dependent on enzyme stability or tissue distribution, but rather on features of structure and conformation important for reversal of bound-transduced and bound-antagonist states.[5,6,80]

These results, also led to suggestions that high potency and prolonged activites had their basis, at least in part, on specific conformational features. A conformational model for α-MSH activity was developed based on various conformational considerations,[6,7,74] and its validity was examined by preparing conformationally constrained analogs.

These considerations led to a design based on conformational restriction by cyclization and by "pseudo-isosteric" replacement of the Met-4 (side chain = $-CH_2-CH_2-S-CH_3$) and Gly-10 (side chain = H) residues by a Cys^4, Cys^{10} cyclic disulfide (side chains = $-CH_2-S-S-CH_2-$).[81] The analog obtained, [$\overline{Cys^4, Cys^{10}}$]-α-MSH, is a highly potent melanotropin in the frog skin receptor system, with high to moderate activity in the lizard skin and mouse melanoma adenylate cyclase assay systems.[81] Subsequent studies have shown that the N-terminal tripeptide Ser-Tyr-Ser has only minor effects on the potency of the α-MSH analog in all three assay systems,[82] a result compatible with the results with [Nle^4, D-Phe^7]-α-MSH fragment analogs. On the other hand, removal of the amino acids in the C-tripeptide -Lys-Pro-Val- one or more at a time had somewhat differing effects on activity than found in the [Nle^4, D-Phe^7] series.[83] In addition it has been found that both [Nle^4, D-Phe^7]-α-MSH and Ac-[$\overline{Cys^4, \text{D-}Phe^7, Cys^{10}}$]-α-$MSH_{4-12}$-$NH_2$ were very stable to degradation by serum enzymes, and both had prolonged biological activity.[77,84,85] This prolonged activity is maintained for the 4-11 analog, Ac-[Nle^4, D-Phe^7]-α-MSH_{4-11}-NH_2.[77]

Enkephalins - Many enkephalin (H-Tyr-Gly-Gly-Phe-Met(Leu)-OH) analogs have been designed with the intent of determining the active conformation.[85] Based on previous work, it appears that a bent conformation is important for binding to the opiate receptors. Therefore, rational design efforts have centered on conformational constraints which might mimic various folded conformations.

One approach has been the restriction of backbone or side chain rotations. For example, introduction of Aib in peptides has been shown to induce reverse turns. Thus, [Aib^2]-, [Aib^2, Aib^3]-, and [Aib^3, Met^5]-enkephalinamides have been synthesized (Aib = α-aminoisobutyric acid). The first two analogs, which can induce a reverse turn at the second and third residues, exhibit greater *in vivo* activity in mice than the corresponding Gly^2, Gly^3 analog. The Aib^3 analog, which centers the β-turn at the third and fourth residues, has slightly less activity.[86] Reaction of the enkephalins with acetaldehyde results in cyclization between the α-amine of Tyr^1 and the nitrogen of the Tyr^1-Gly^2 amide bond, and resulted in a considerable loss of activity in the guinea pig ileum and mouse vas deferens assays.[87] Side chain restriction of the tyrosine residue (e.g. substituting 2-amino-6-hydroxytetralin-2-carboxylic acid or 2-amino-5-hydroxyindane-2-carboxylic acid for Tyr) results in analogs that are inactive when administered s.c. or i.c.v. in the mouse hot-plate test. However, while the indane analog shows considerably less activity in the guinea pig ileum and mouse vas deferens assays than [Leu^5]enkephalin, the tetralin analog is more active in the guinea pig ileum (μ-receptor) assay and much less active in the mouse vas deferens (δ-receptor) assay than [Leu^5]enkephalin, implying an increase of receptor specificity by this conformational restriction.[88] The use of dehydro amino acids to induce rigidity and increase hydrophobicity of the enkephalins has resulted in the μ-selective [ΔAla^2, Leu^5]enkephalin and the δ-selective [D-Ala^2, ΔPhe^4, Leu^5]enkephalin.[89-92] Cyclo-[Leu^5]-enkephalin which adopts a β & γ or γ, γ turn structure in DMSO did not have sufficient aqueous solubility to allow biological testing.[93,94] Cyclization via side chain moieties, not involving the tyrosine amino group, has proven to be more fruitful. The disulfide bridged [D-Cys^{2}, D-Cys^{5}]-enkephalinamide as well as a series of cyclo [D-α,ω-diamino acid^{2}, Leu^{5}]enkephalins exhibited μ-receptor selectivity as well as considerable activity in the guinea pig ileum assay. It is interesting to note that these structures are incompatible with a 4-1 or 5-2 hydrogen bonded β-bend in the enkephalin.[95-97] In contrast to these compounds, the more conformationally constrained [D Pen^{2}, D-Cys^{5}]-, [D-Pen^{2}, Cys^{5}]-, [D-Pen^{2}, D-Pen^{5}]-, and [D-Pen^{2}, Pen^{5}]-enkephalins exhibit exceptional δ receptor selectivity.[98,99] Conformational differences associated with the Pen residues appear to be responsible for the high receptor selectivity. Finally, a series of cyclic retro-inverso enkephalins have been designed[100] to incorporate favorable conformational restriction and increased resistance to enzymatic degradation.

6

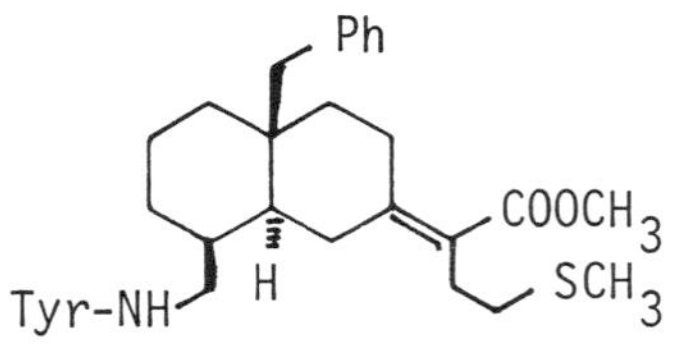

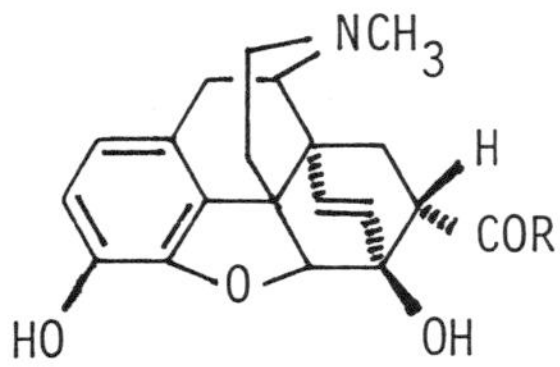

7

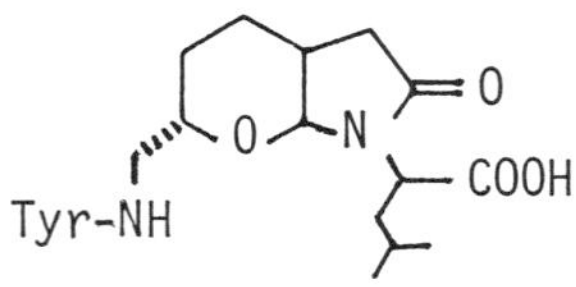

8a R= -Leu-OEt

8b R= -Phe-Leu-OEt

8c R= -Gly-Phe-Leu-OEt

Another approach to this problem is the synthesis of non-peptide enkephalin analogs, which are intended to mimic conformationally restrained enkephalins. Based on existing structure-activity relationships for the enkephalins, computer modeling and comparison with opiates, one group deduced a β-turn model for the active enkephalin conformation. Utilizing the Merck Modeling System and the Compare program[55], a perhydronaphthalene enkephalin derivative (6) was designed to mimic this proposed "active" conformation. This approach yielded analogs which exhibited only weak opiate receptor binding.[101] Another group, synthesized a semi-rigid [Leu^{5}]enkephalin analog, (7) which mimics a 5-2 β-bend conformation and obtained weak activity in the guinea pig ileum assay.[102] Finally, a series of Bentley compounds (8 a-c) were synthesized in an attempt to compare how the rigid opiates related to the enkephalins active conformation. Compound 8a was active in several of the standard opiate assays, but still doesn't resolve the question of how the rigid opiates topologically resemble the enkephalins.[103]

Leutienizing Hormone Releasing Hormone (LHRH) - Structure-activity relationships of LHRH (pGlu-His-Trp-Ser-Tyr-Gly-Leu-Arg-Pro-Gly-NH_2) have been examined by determining the effects of systematic modification of each residue on biological activity.[104-109] The possible conformational effects of these modifications have been examined,[110-111] but only a few attempts have been made at specific conformational restriction.

A γ-lactam ring has been incorporated at the Gly^{6}, Leu^{7} amide bond to provide a backbone restriction into LHRH. Computer modeling has shown that this modification will induce a Gly^{6}-Leu^{7} centered β-bend into the molecule. This LHRH analog exhibited 8.9 times the LH releasing activity of LHRH *in vitro* and 2.4 times the LH releasing activity of LHRH in adult ovariectomized female rats.[112,113] In order to test conformational predictions based on minimum energy calculations, a series of α-methyl amino acid substituted and disulfide cyclized LHRH analogs were synthesized.

While none of the analogs displayed enhanced activity over the relatively less conformationally constrained analogs which they mimicked, they did serve to show that an α-helical bend at positions 2 and 3 is compatable with LHRH antagonist activity.[114]

It has been shown that the potencies of LHRH agonists correlate with their retention times on reverse-phase high performance liquid chromatography (RP-HPLC), which relates to their overall hydrophobic character. With this in mind, Hansch analysis was applied to LHRH antagonists to optimize those factors required for receptor recognition. This approach yielded a series of potent analogs whose retention times on RP-HPLC correlate with in vivo activity better than binding affinity and in vitro activity. This implies that the effect of lipophilicity in LHRH antagonists is related to their availability to the receptor, whereas the receptor itself may have different requirements.[115] Based on the importance of lipophilicity for LHRH antagonists, a series of N^G, $N^{G'}$-dialkyl-homoArg^6 analogs were synthesized resulting in [N-Ac-D-Nal(2)^1, D-pCl-Phe^2, D-Trp^3, D-homoArg-(Et_2)^6, D-Ala^{10}] LHRH, a highly potent antagonist.[116]

Conclusion - Development of rational approaches to the design of peptide hormone analogs with useful biological and physical properties is still in its infancy. Yet, conformational and topological considerations already have provided new insights into the chemical-physical basis for the biological activity of these compounds. However, even for the most comprehensive systems studied to date, any model for a "biologically active" conformation at a receptor must be viewed as tentative. On the one hand, even the most rigid peptide hormone analogs still possess some conformational flexibility both with respect to backbone and to side chain conformations. Thus on interacting with the receptor, specific interactions between the peptide hormone and the receptor could lead to unexpected conformational perturbations of the hormone. On the other hand, we have little or no information regarding the conformational and dynamic properties of receptors, and progress in this area is urgently needed. Nonetheless, new insights into the chemical-physical basis for biological information transfer in these systems has emerged from the conformational approach. Continued development of the general structural and conformational principles of design reviewed here, and of new principles which will emerge as these approaches are expanded, should provide the basis for an increasingly rational approach to the design of peptide hormones and neurotransmitters.

Acknowledgement - Supported by grants from the U. S. Public Health Service and the National Science Foundation.

References

1. G.R. Marshall, F.A. Gorin and M.L. Moore, Annu. Rep. Med. Chem., 13, 227 (1978).
2. D. Veber in "Peptides: Structure and Biological Function," E. Gross and J. Meienhofer, Eds., Pierce Chem. Co., Rockford, 1979, p. 409.
3. G.I. Chipens, A. Krikis and L.K. Polevaja in "Biophysical and Biochemical Information Transfer in Recognition," J.G. Vassileva-Popova and E.V. Jensen, Eds., Plenum Press, N.Y., 1979, p. 23.
4. J.-P. Meraldi, D. Yamamoto, V.J. Hruby and A.I.R. Brewster in "Peptides: Chemistry, Structure and Biology," R. Walter and J. Meienhofer, Eds., Ann Arbor Science Publ., Ann Arbor, 1975, p. 803.
5. V.J. Hruby in "Perspectives in Peptide Chemistry," A. Eberle, R. Geiger and T. Wieland, Eds., S. Karger, Basel, 1981, p. 207.
6. V. J. Hruby in "Drug Design Based on Peptide and Nucleic Acid Conformational Structure," J. Vida, Ed., ACS Monograph Series, 1984, in press.
7. V.J. Hruby and M.E. Hadley in "XVIIth International Solvay Conference on Chemistry," 1984, in press.

8. P. Schiller and J. Di Maio in "Peptides: Structure and Function," V.J. Hruby and D.H. Rich, Eds., Pierce Chem. Co., Rockford, IL, 1983, p. 269.
9. E.T. Kaiser and F.J. Kézdy, Science, 223, 249 (1984).
10. R. Schwyzer in "XVIII International Solvay Conference on Chemistry," 1984, in press.
11. R.M. Epand, Molec. Cellul. Biochem., 57, 41 (1983).
12. J. Rudinger in "Drug Design," Vol. 2, E.J. Ariens, Ed., Academic Press, New York, 1971, p. 319.
13. R. Schwyzer, Ann. N. Y. Acad. Sci., 297, 3 (1977).
14. V.J. Hruby in "Topics in Molecular Pharmacology," A.S.V. Burgen and G.C.K. Roberts, Eds., Elsevier/North-Holland, Amsterdam, 1981, p. 99.
15. V.J. Hruby, Life Sciences, 31, 189 (1982).
16. R. Walter, I.L. Schwartz, J.H. Darnell and D.W. Urry, Proc. Nat. Acad. Sci. U.S.A., 68, 1355 (1971).
17. A.I.R. Brewster and V.J. Hruby, Proc. Natl. Acad. Sci. U.S.A., 70, 3806 (1973).
18. R. Walter, Fed. Proc., 36, 1872 (1977).
19. A.F. Spatola, Annu. Rep. Med. Chem., 16, 199 (1981).
20. V.J. Hruby and H.I. Mosberg in "Hormone Antagonists", M. K. Agarwal, Ed., W. de Gruyter & Co., Berlin, 1982, p. 433.
21. V.J. Hruby, T.W. Rockway, V. Viswanatha and W.Y. Chan, Int. J. Peptide Protein Res., 21, 24 (1983).
22. G.R. Marshall and H.E. Bosshard, Circ. Res. Suppl. II, 30/31, 143 (1972).
23. J. Roy, M. Johnson, D. Gazis and I.L. Schwartz, Int. J. Peptide Protein Res., 16, 55 (1980).
24. J. Roy, D. Gazis and I.L. Schwartz, ibid., 16, 106 (1980).
25. J. Roy, D. Gazis, R. Shokman and I.L. Schwartz, ibid., 20, 35 (1982).
26. U. Roy, D. Gazis, I.L. Schwartz and J. Roy, ibid., 22, 398 (1983).
27. U. Roy, D. Gazis, G.D. Pan, I.L. Schwartz and J. Roy, ibid., 22, 525 (1983).
28. V.J. Hruby, D.A. Upson, D.M. Yamamoto, C.W. Smith and R. Walter, J. Am. Chem. Soc., 101, 2717 (1979).
29. V.J. Hruby, H.I. Mosberg and V. Viswanatha, J. Am. Chem. Soc., 104, 837 (1982).
30. M. Manning, J. Lowbridge, J. Haldar and W.H. Sawyer, J. Med. Chem., 19, 376 (1976).
31. Z. Procházka, M. Lebl, T. Barth, J. Hlavácek and K. Jost in "Peptides 1982," K. Bláha and P. Malon, Eds., W. de Gruyter & Co., Berlin, 1983, p. 441.
32. Z. Grzonka, F. Kasprzykowski, B. Lammek, D. Gazis and I.L. Schwartz, in "Peptides 1982," K. Bláha and P. Malon, Eds., W. de Gruyter & Co., Berlin, 1983, p. 445.
33. V.J. Hruby, H.I. Mosberg, M.E. Hadley, W.Y. Chan and A.M. Powell, Int. J. Peptide Protein Res., 16, 372 (1980).
34. V.J. Hruby and H.I. Mosberg in "Neurohypophyseal Peptide Hormones and Other Biologically Active Peptides," D.H. Schlesinger, Ed., Elsevier/North-Holland, N.Y., 1981, p. 227.
35. H.I. Mosberg, V.J. Hruby and J.-P. Meraldi, Biochemistry, 20, 2822 (1981).
36. V.J. Hruby, H.I. Mosberg, J.W. Fox and A.T. Tu, J. Biol. Chem., 257, 4916 (1982).
37. J.-P. Meraldi, V.J. Hruby and A.I.R. Brewster, Proc. Natl. Acad. Sci. U.S.A., 74, 1373 (1977).
38. A.S.V. Burgen, G.C.K. Roberts and J. Feeney, Nature, 253, 753 (1975).
39. V.J. Hruby and H.I. Mosberg, Peptides, 3, 329 (1982).
40. P. Simek, T. Barth, F. Brtník, J. Slaninová and K. Jost in "Peptides 1982," K. Bláha and P. Malon, Eds., W. de Gruyter & Co., Berlin, 1983, p. 461.
41. M. Lebl, T. Barth, L. Servítová, J. Slaninová and K. Jost, ibid., p. 457.
42. V.J. Hruby, K.K. Deb, D.M. Yamamoto, M.E. Hadley and W.Y. Chan, J. Med. Chem., 22, 7 (1979).
43. R. Walter, C.W. Smith, P.W. Mehta, S. Boonjarern, J.A.L. Arruda and N.A. Kurtzman in "Disturbances in Body Fluid Osmolality," T.E. Andreoli, J. Gratham and F.C. Rector, Eds., Amer. Physiol. Soc., Bethesda, MD, 1977, p. 1.
44. W.H. Sawyer, P.K.T. Pang, J. Seto, M. McEnroe, M.B. Lammek and M. Manning, Science, 212, 49 (1981).
45. M. Manning, B. Lammek, A.M. Kolodziejczyk, J. Seto and W.H. Sawyer, J. Med. Chem., 24, 701 (1981).
46. P. Melin, S. Lundin, H. Vilhardt, G. Lindeberg, L.-E. Larsson and V. Pliska, Eur. J. Pharm., 67, 173 (1980).
47. M. Manning, W.A. Klis, A. Olma, J. Seto and W.H. Sawyer, J. Med. Chem., 25, 414 (1982).
48. M. Manning, A. Olma, W.A. Klis, A.M. Kolodziejczyk, J. Seto and W.H. Sawyer, J. Med. Chem., 25, 45 (1982).
49. M. Manning, A. Olma, W.A. Klis, J. Seto and W.J. Sawyer, ibid., 26, 1607 (1983).
50. W.H. Sawyer and M. Manning, Fed. Proc., 43, 87 (1984).
51. M. Knappenberg, A. Michel, A. Scarso, J. Brison, J. Zanen, K. Hallenga, P. Deschrijver and G. Van Binst, Biochim. Biophys. Acta, 700, 229 (1982).
52. B.H. Arison, R. Hirschmann, W.J. Paleveda, S.F. Brady and D.F. Veber, Biochem. Biophys. Res. Commun., 100, 1148 (1981).
53. K. Hallenga, G. Van Binst, A. Scarso, A. Michel, M. Knappenberg, C. Dremier, J. Brinson and J. Dirkx, FEBS Lett., 119, 47 (1980).

54. H. Kessler, M. Bernd, H. Kogler, J. Zarbock, O.W. Sorensen, G. Bodenhausen and R.R. Ernst, J. Am. Chem. Soc., 105, 6944 (1983).
55. P. Gund, J.D. Andose, J.B. Rhodes and G.M. Smith, Science, 208, 1425 (1980).
56. D. Veber and R. Saperstein, Ann. Rep. Med. Chem., 14, 209 (1979).
57. D.F. Veber, R.G. Strachan, S.J. Bergstrand, F.W. Holly, C.F. Homnick, R. Hirschmann, M. Torchiana and R. Saperstein, J. Amer. Chem. Soc., 98, 2367 (1976).
58. D.F. Veber, F.W. Holly, R.F. Nutt, S.J. Bergstrand, S.F. Brady, R. Hirschmann, M.S. Glitzer and R. Saperstein, Nature, 280, 512 (1979).
59. D.F. Veber in, "Peptides, Synthesis-Structure-Function," D.H. Rich and E. Gross, Eds., Pierce Chemical Co., Rockford, IL, 1981, p. 685.
60. R.M. Freidinger, D. Perlow, W.C. Randall, R. Saperstein, B.H. Arison and D.F. Veber, Int. J. Peptide Protein Res., 23, 142 (1984).
61. R.M. Freidinger, C.D. Colton, D.S. Perlow, W.L. Whitter, W.J. Paleveda, D.F. Veber, B.H. Arison and R. Saperstein in "Peptides, Structure and Function," V.J. Hruby and D.H. Rich, Eds., Pierce Chemical Co., Rockford, IL, 1983, p. 349.
62. S.F. Brady, D.W. Cochran, R.F. Nutt, F.W. Holly, C.D. Burnett, W.J. Paleveda, P.E. Curley, B.H. Arison, R. Saperstein and D.F. Veber, Int. J. Peptide Protein Res., 23, 212 (1984).
63. K. Hofmann, Handbook Physiol., Sec. 7, Endocrinol., 2, 29 (1974).
64. K. Medzihradszky, Medical Res. Rev., 2, 247 (1982).
65. V.J. Hruby, B.C. Wilkes, W.L. Cody, T.K. Sawyer and M.E. Hadley, Peptide Protein Rev., in press (1984).
66. B.J. Rawson, J. Feeney, B.J. Kimbar and H.M. Greven, J. Chem. Soc. Perkin Trans. II, 1471 (1982) and references therein.
67. J.W. van Nispen and H.M. Greven, Pharmacol. Ther., 16, 67 (1982).
68. M.E. Hadley, C.B. Heward, V.J. Hruby, T.K. Sawyer and Y.C.S. Yang in "Peptides of the Pars Intermedia," Ciba Found. Symposium 81, D. Evered and G. Lawrenson, Eds., Pitman Medical, London, 1981, p. 244.
69. T.L. O'Donohue and D.M. Dorsa, Peptides, 3, 353 (1982).
70. P.E. Smith and J.B. Graesler, Anatomical Record, 27, 187 (1924).
71. A.M. Bool, G.H. Gray II, M.E. Hadley, C.B. Heward, V.J. Hruby, T.K. Sawyer and Y.C.S. Yang, J. Endocr., 88, 57 (1981) and references therein.
72. T.K. Sawyer, P.J. Sanfilippo, V.J. Hruby, M.H. Engel, C.B. Heward, J.B. Burnett, and M.E. Hadley, Proc. Natl. Acad. Sci. U.S.A., 77, 5754 (1980).
73. M.H. Engel, T.K. Sawyer, M.E. Hadley and V.J. Hruby, Anal. Biochem., 116, 303 (1981).
74. T.K. Sawyer, V.J. Hruby, M.E. Hadley and M.H. Engel, Amer. Zool., 23, 529 (1983).
75. M.E. Hadley, B. Anderson, C.B. Heward, T.K. Sawyer and V.J. Hruby, Science, 213, 1025 (1981).
76. B.C. Wilkes, T.K. Sawyer, V.J. Hruby and M.E. Hadley, Int. J. Peptide Protein Res., 22, 313 (1983).
77. B.C. Wilkes, V.J. Hruby, H.I. Yamamura, K. Akiyama, A.M. de L. Castrucci, M.E. Hadley, J.R. Edwards and Y.-P. Wan, Life Sciences, 34, 977 (1984).
78. T.K. Sawyer, V.J. Hruby, B.C. Wilkes, M.T. Draelos, M.H. Hadley and M. Bergsneider, J. Med. Chem., 25, 1022 (1982).
79. B.C. Wilkes, T.K. Sawyer, V.J. Hruby and M.E. Hadley, Int. J. Peptide Protein Res., in press (1984).
80. V.J. Hruby, J.J. Knittel, H.I. Mosberg, T.W. Rockway, B.C. Wilkes and M.E. Hadley in "Peptides 1982," K. Bláha and P. Malon, Eds., W. de Gruyter, Berlin, 1983, p. 19.
81. T.K. Sawyer, V.J. Hruby, P.S. Darman and M.E. Hadley, Proc. Natl. Acad. Sci. U.S.A., 79, 1751 (1982).
82. J.J. Knittel, T.K. Sawyer, V.J. Hruby and M.E. Hadley, J. Med. Chem., 26, 125 (1983).
83. W.L. Cody, B.C. Wilkes, B.J. Muska, V.J. Hruby, A.M. de L. Castrucci and M.E. Hadley, J. Med. Chem., in press (1984).
84. A.M. de L. Castrucci, M.E. Hadley, E.I. Yorulmazoglu, B.C. Wilkes, T.K. Sawyer and V.J. Hruby, Pigment Cell, in press (1984).
85. P.W. Schiller in "The Peptides," Vol. 7, S. Udenfriend and J. Meienhofer, Eds., Academic Press, New York, N.Y., in press (1984).
86. T.S. Sudha and P. Balaram, Int. J. Pept. Prot. Res., 21, 381 (1983).
87. M.J. Gidley, L.D. Hall, J.K.M. Sanders and M.C. Summers, Biochemistry, 20, 3880 (1981).
88. T. Deeks, P.A. Crooks and R.D. Waigh, J. Med. Chem., 26, 762 (1983).
89. Y. Shimohigashi and C.H. Stammer, Int. J. Pept. Prot. Res., 19, 54 (1982).
90. Y. Shimohigashi, M.L. English and C.H. Stammer, Biochem. Biophys. Res. Commun., 104, 583 (1982).
91. Y. Shimohigashi and C.H. Stammer, Int. J. Pept. Prot. Res., 20, 199 (1982).
92. Y. Shimohigashi, C.H. Stammer, T. Costa and P.F. Vonvoigtlander, Int. J. Pept. Prot. Res., 22, 489 (1983).
93. H. Kessler and G. Holzemann, Ann., 2028 (1981).
94. H. Kessler and G. Holzemann, Angew. Chem. Int. Ed. Engl., 20, 124 (1981).
95. P.W. Schiller, G. Eggimann, J. DiMaio, C. Lemieux and Nguyen, J.M.-D., Biochem. Biophys. Res. Commun., 101, 337 (1981).
96. J. DiMaio, C. Lemieux and P.W. Schiller, Life Sciences, 31, 2253 (1982).

97. J. DiMaio, T.M.-D. Nguyen, C. Lemieux and P.W. Schiller, J. Med. Chem., 25, 1432 (1982).
98. H.I. Mosberg, R. Hurst, V.J. Hruby, J.J. Galligan, T.F. Burks, K. Gee and H.I. Yamamura, Biochem. Biophys. Res. Commun., 106, 506 (1982).
99. H.I. Mosberg, R. Hurst, V.J. Hruby, K. Gee, H.I. Yamamura, J.J. Galligan and T.F. Burks, Proc. Natl. Acad. Sci. U.S.A., 80, 5871 (1983).
100. J.M. Berman, M. Goodman, T.M.-D. Nguyen and P.W. Schiller, Biochem. Biophys. Res. Commun., 115, 864 (1983).
101. P.C. Belanger, C. Dufresne, J. Scheigetz, R.N. Yang, J.P. Springer and G.I. Dmitrienko, Can. J. Chem., 60, 1019 (1982).
102. J.L. Krstenansky, R.L. Baranowski and B.L. Currie, Biochem. Biophys. Res. Commun., 109, 1368 (1982).
103. H.C. Beyerman, T.S. Lie, L. Maat and M. Noordam-Wiessdorf, Recl. Trav. Chim. Pays-Bas, 101, 455 (1982).
104. I. Mezo, J. Sperodi, J. Erchegyi, I. Teplan, M. Kovacs and B. Flerko, Peptides, 4, 149 (1983).
105. A. Horvath, D.H. Coy, M.V. Nekola, E.J. Coy, A.V. Schally and I. Teplan, Peptides, 3, 969 (1982).
106. D.H. Coy, A. Horvath, M.V. Nekola, E.J. Coy, J. Erchegyi and A.V. Schally, Endocrinology, 110, 1145 (1982).
107. J.J. Nestor, Jr., T.L. Ho, R.A. Simpson, B.L. Horner, G.H. Jones, G.I. McRae and B.H. Vickery, J. Med. Chem., 25, 795 (1982).
108. S.H. Nakagawa, D.C. Yang and G. Flouret, J. Med. Chem., 24, 221 (1981).
109. K. Folkers, J. Humphries and C.Y. Bowers, Z. Naturforsch., 37b, 246 (1982).
110. F.A. Momany, J. Amer. Chem. Soc., 98, 2990, 2996 (1976).
111. P. Manavalan and F.A. Momany, Biopolymers, 19, 1943 (1980).
112. R.M. Freidinger, D.F. Veber, D.S. Perlow, J.R. Brooks and R. Saperstein, Science, 210, 656 (1980).
113. R.M. Freidinger, D.S. Perlow and D.F. Veber, J. Org. Chem., 47, 104 (1982).
114. R.W. Roeske, G.M. Anantharamaiah, F.A. Momany and C.Y. Bowers in "Peptides: Structure and Function," V.J. Hruby and D.H. Rich, Eds., Pierce Chem. Co., Rockford, IL, 1983, p. 333.
115. J. Rivier, C. Rivier, M. Perrin, J. Porter and M. Vale in "Peptides 1980," K. Brunfeldt, Ed., Scriptor, Copenhagen, 1980, p. 556.
116. J.J. Nestor, Jr., R. Tahilramani, T.L. Ho, G.I. McRae and B.H. Vickery in "Peptides: Structure and Function," V.J. Hruby and D.H. Rich, Eds., Pierce Chem. Co., Rockford, IL, 1983, p. 861.

Section VII. Worldwide Market Introductions

Editor: Richard C. Allen, Hoechst-Roussel Pharmaceuticals, Inc.
Somerville, New Jersey 08876

Chapter 31. To Market, To Market - 1983

Richard C. Allen, Hoechst-Roussel Pharmaceuticals Inc., Somerville, NJ 08876

The ultimate goal of each of us engaged in the multidisciplinary fields of drug discovery and development is the introduction of the fruit of our efforts into the marketplace. Such a milestone may represent the validation of a scientific hypothesis and perhaps an indication of a new era in therapy, or simply, the emergence of an alternative to existing therapy, with or without measurable advantages. In all cases, such a market introduction represents an accomplishment of some magnitude in the face of formidable odds.

The new chemical entities (NCEs) recently introduced into the US marketplace are by no means an accurate and timely reflection of these milestones; of the 20 NCEs appearing on the US market in 1983, only four represent first time introductions into the world marketplace—some were introduced 5-10 years ago in other markets and are certainly not an indication of what is new!

An attempt has therefore been made to compile information on the first market introductions of NCEs for human therapeutic use in the world as a whole during 1983. It is hoped that this information will be useful to the reader as an accurate reflection of past accomplishments and perhaps future directions. A short summary of the profile, advantages/uniqueness, and utility of each compound is given; several leading references are offered for the reader interested in additional information.

Acetohydroxamic Acid (hypoammonuric)[1,2]

Country of Origin: **USA**
Originator: **Research Organics; Baylor Univ.**
First Introduction: **USA**
Introduced by: **Uro-Research; Mission Pharmacal**
Trade Name: **LITHOSTAT**

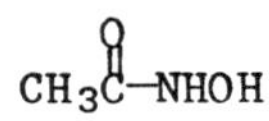

Acetohydroxamic acid is a potent, non-competitive and irreversible inhibitor of bacterial urease ($Ki \approx 10^{-7}M$). This enzyme, which is widely distributed in plants and bacteria, but not in mammalian cells, catalyzes the decomposition of urea to ammonia. Elevated urinary ammonia levels can reduce the antibacterial effectiveness of a number of agents. Thus, acetohydroxamic acid is useful as adjunctive therapy to decrease urinary ammonia and alkalinity in patients with chronic urea-splitting urinary infection. Such infections are a leading cause of recurring complications and death in paraplegics.

Afloqualone (muscle relaxant)[3,4]

Country of Origin: **Japan**
Originator: **Tanabe**
First Introduction: **Japan**
Introduced by: **Tanabe**
Trade Name: **AROFUTO**

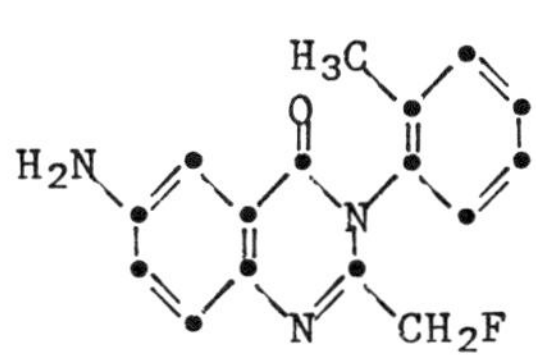

ISBN 0-12-040519-9

Afloqualone is a centrally acting muscle relaxant useful in the management of various spastic conditions, including cerebral palsy, cervical spondylosis, and multiple sclerosis. It is closely related to the hypnotic/sedative methaqualone.

Alfentanil Hydrochloride (analgesic)[5,6]

Country of Origin: **Belgium**
Originator: **Janssen**
First Introduction: **Netherlands**
Introduced by: **Janssen**
Trade Name: **RAPIFEN**

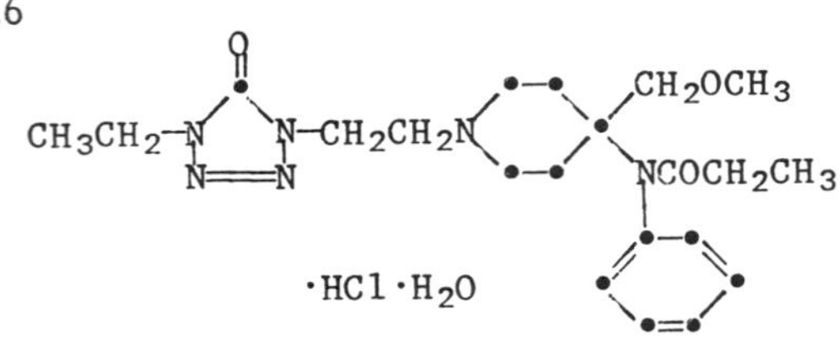

Alfentanil is a narcotic analgesic with a more rapid onset and shorter duration of action than its structural relative fentanyl. The primary utility of alfentanil is in surgical analgesia/anesthesia, especially for cardiac compromised patients and in procedures of short duration.

Alminoprofen (analgesic/antiinflammatory)[7]

Country of Origin: **France**
Originator: **Labs. Dr. E. Bouchara**
First Introduction: **France**
Introduced by: **Labs. Dr. E. Bouchara**
Trade Name: **MINALFENE**

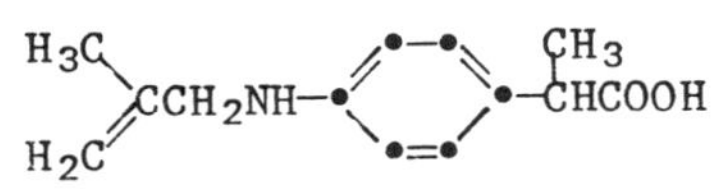

Alminoprofen is an arylpropionic acid analgesic/antiinflammatory agent indicated for the short term management of dental, traumatic and postpartum pain.

Amrinone (cardiotonic)[8,9]

Country of Origin: **USA**
Originator: **Sterling-Winthrop**
First Introduction: **Phillipines**
Introduced by: **Sterling Drug**
Trade Name: **INOCOR**

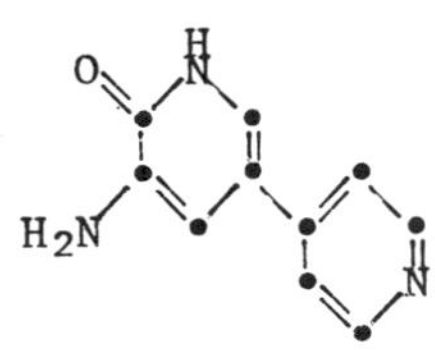

Amrinone is a positive inotropic agent useful in the management of severe congestive heart failure. It is effective even in unresponsive, fully digitalized patients.

Astemizole (antihistamine)[10,11]

Country of Origin: **Belgium**
Originator: **Janssen**
First Introduction: **United Kingdom**
Introduced by: **Janssen**
Trade Name: **HISMANAL**

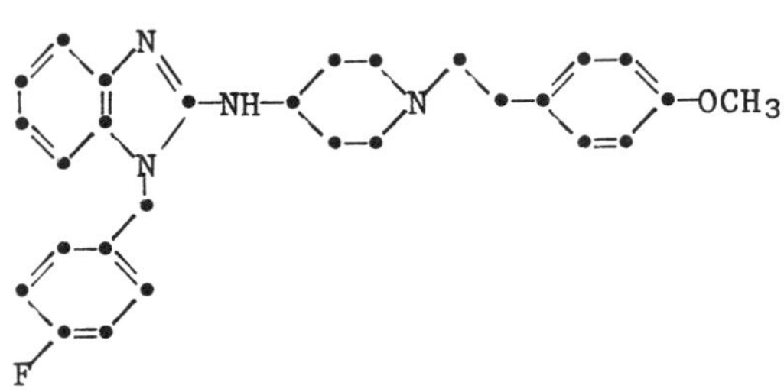

Astemizole belongs to the second-generation class of non-sedating, non-anticholinergic antihistamines. Its non-sedating properties appear to result from its poor penetration of the blood brain barrier. As a result it shows no potentiation of CNS depressants, including alcohol. Its long half-life allows once-daily dosing.

Auranofin (chrysotherapeutic)[12,13]

Country of Origin: **USA**
Originator: **Smith Klein & French**
First Introduction: **W. Germany**
Introduced by: **Smith Klein & French**
Trade Name: **RIDAURA**

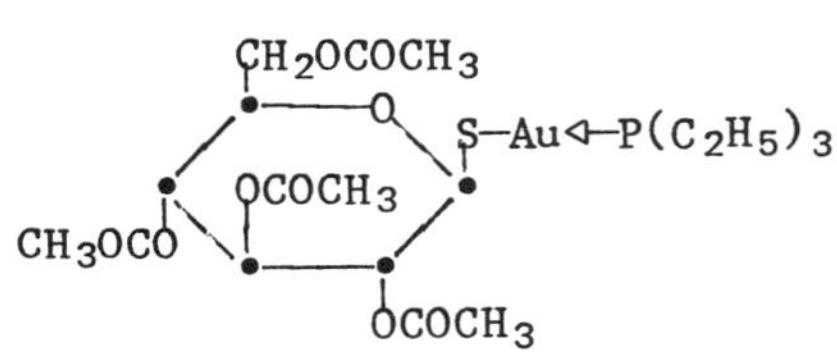

Auranofin is the first orally effective gold compound to be marketed for the treatment of severe rheumatoid arthritis. It is better tolerated and more convenient than gold sodium thiomalate, which is administered intramuscularly.

Befunolol Hydrochloride (antiglaucoma)[14,15]

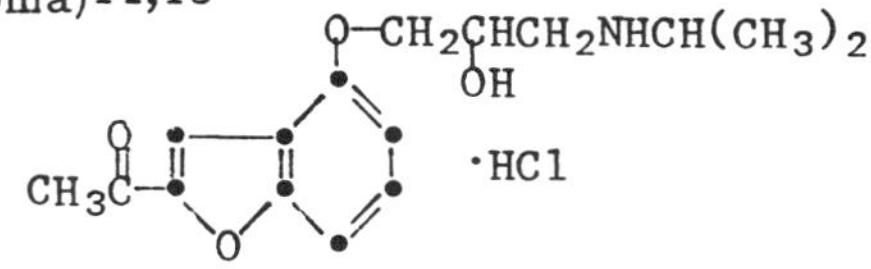

Country of Origin: **Japan**
Originator: **Kaken**
First Introduction: **Japan**
Introduced by: **Kaken**
Trade Name: **BENTOX**

Befunolol hydrochloride is the newest of the β-adrenergic blockers to be introduced for the treatment of glaucoma. It is reportedly devoid of intrinsic sympathomimetic activity, with lower membrane-stabilizing activity than propranolol.

Betaxolol Hydrochloride (β-adrenergic blocker; antihypertensive)[16,17]

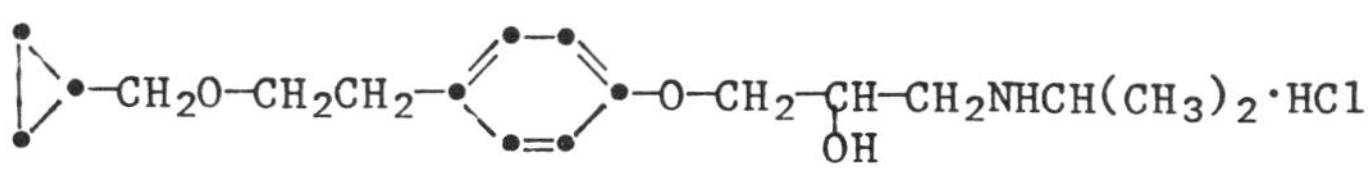

Country of Origin: **France**
First Introduction: **France**
Trade Name: **KERLONE**
Originator: **Synthelabo**
Introduced by: **Robert & Carriere**

Betaxolol hydrochloride is a cardioselective β-adrenergic blocker, reportedly devoid of intrinsic sympathomimetic and membrane stabilizing properties. Its long duration of action permits once-daily dosing in mild to moderate hypertension. It is also being evaluated in glaucoma.

Bifonazole (antifungal)[18,19]

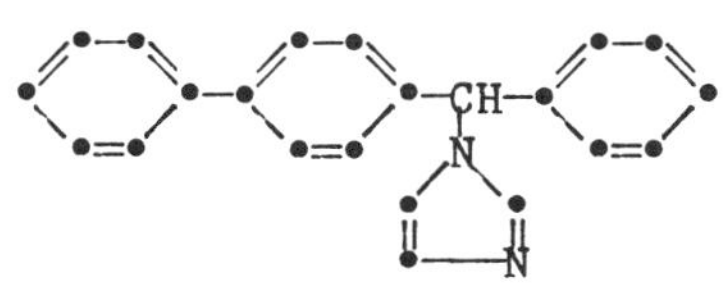

Country of Origin: **W. Germany**
Originator: **Bayer**
First Introduction: **W. Germany**
Introduced by: **Bayer**
Trade Name: **MYCOSPOR**

Bifonazole represents the first topical broad spectrum antimycotic approved for once daily administration. Its in vitro activity appears equivalent to its structural relative clotrimazole, being effective against dermatophytes, other filamentous fungi, dimorphic fungi and yeasts.

Brotizolam (hypnotic/sedative)[20,21]

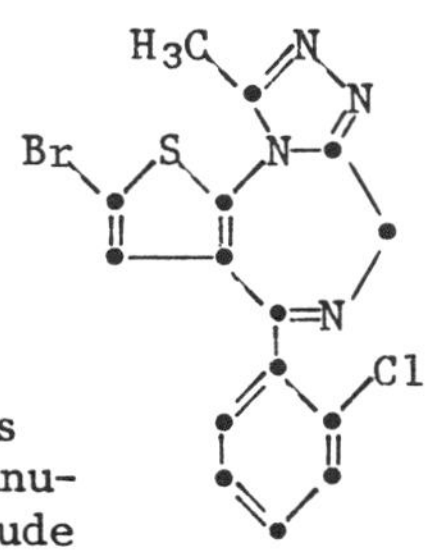

Country of Origin: **W. Germany**
Originator: **Boehringer Ingelheim**
First Introduction: **Switzerland**
Introduced by: **Boehringer Ingelheim**
Trade Name: **LENDORMIN**

Brotizolam is a hypnotic/sedative approximately 60-100 times more potent than flurazepam. As with other short-acting, annulated benzodiazepines (e.g., triazolam), cited advantages include lack of adverse effects on sleep and performance after arousal.

Budralazine (antihypertensive)[22,23]

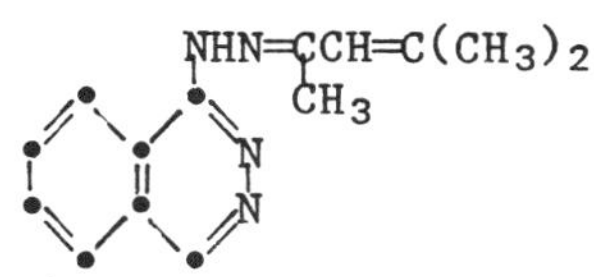

Country of Origin: **Japan**
Originator: **Daiichi**
First Introduction: **Japan**
Introduced by: **Daiichi**
Trade Name: **BUTERAZINE**

Budralazine is an antihypertensive agent which is somewhat less potent than structurally related hydralazine, and reportedly produces less tachycardia.

Butyl Flufenamate (topical antiinflammatory)[24]

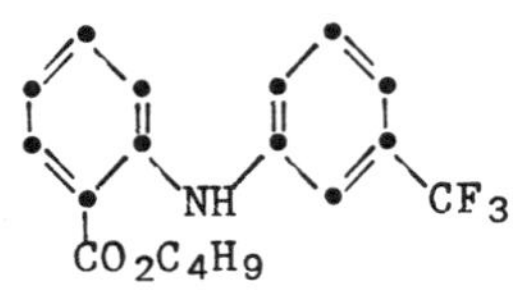

Country of Origin: **Japan**
Originator: **Hokuriku**
First Introduction: **Japan**
Introduced by: **Hokuriku; Tokyo Tanabe**
Trade Names: **FENAZOLE; COMBEC**

Butyl flufenamate is a topical, non-steroidal antiinflammatory agent indicated for the treatment of acute and chronic eczema; contact, seborrheic, and atopic dermatitis; and herpes zoster.

Cadexomer Iodine (wound healing agent)[25]

Country of Origin: **United Kingdom**
First Introduction: **United Kingdom**
Trade Name: **IODOSORB**
Originator: **Perstorp**
Introduced by: **Stuart**

Cadexomer Iodine is a hydrophilic, modified starch polymer containing 0.9% iodine within the helical structure; it is used in the treatment of decubitus and venous leg ulcers. Applied to a wound surface as a powder, it is reported to accelerate healing, stimulate granulation, clean the ulcer surface, relieve pain and reduce bacterial counts.

Cefmenoxime Hydrochloride (antibiotic)[26,27]

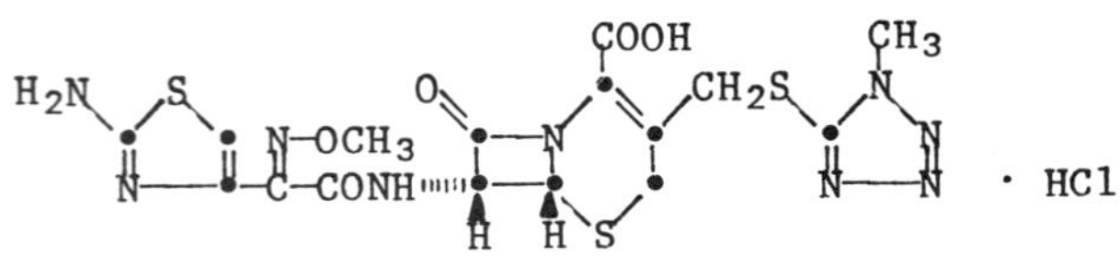

Country of Origin: **Japan**
First Introduction: **Japan**
Trade Names: **TACEF; BESTCALL**
Originator: **Takeda**
Introduced by: **Takeda; Nippon Roche**

Cefmenoxime hydrochloride is a third generation cephalosporin antibiotic. Structurally, it possesses the (1-methyl-1H-tetrazol-5-yl)thiomethyl moiety in the 3-position; like several other compounds containing this structural element (moxalactam, cefoperazone, cefamandol), bleeding linked to vitamin K interaction has been reported. Cefmenoxime has activity similar to cefotaxime, ceftizoxime and moxalactam against E. coli, C. diversus, Klebsiella, P. Mirabilis, Salmonella, Shigella, Neiseria spp., S. pyogenes, S. Pneumoniae, and H. influenzae. It is relatively ineffective against Pseudomonas and Bacteroides.

Ceftazidime (antibiotic)[28,29]

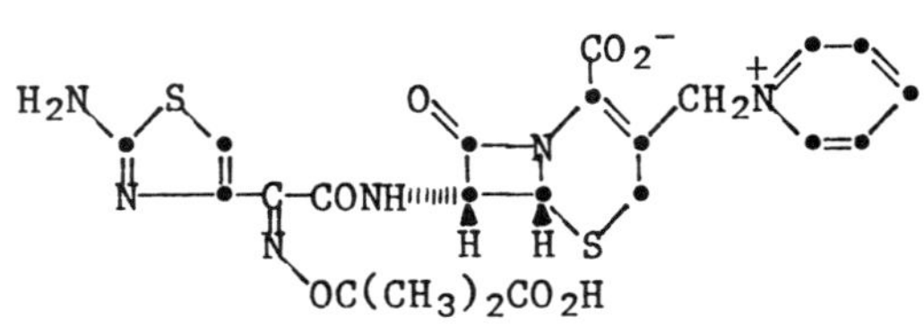

Country of Origin: **United Kingdom**
Originator: **Glaxo**
First Introduction: **United Kingdom**
Introduced by: **Glaxo**
Trade Name: **FORTAM**

Ceftazidime is the latest third generation cephalosporin to reach the market. It has one of the broadest spectrums of the cephalosporins, similar in many regards to that of cefotaxime. It is particularly active against Pseudomonas aeruginosa, being perhaps 4-5 times more potent in vitro than moxalactam and cefotaxime.

Celiprolol Hydrochloride (β-adrenergic blocker)[30,31]

Country of Origin: **Austria**
Originator: **Chemie Linz**
First Introduction: **Austria**
Introduced by: **Chemie Linz**
Trade Name: **SELECTOL**

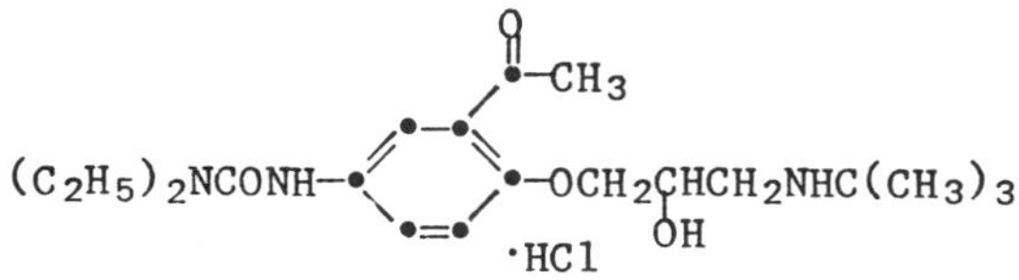

Celiprolol hydrochloride is a once-daily, cardioselective β-adrenergic blocker useful in the management of hypertension, angina pectoris and hyperkinetic heart syndrome. It is also being evaluated in glaucoma.

Chenodiol (anticholelithogenic)[32,33]

Country of Origin: **USA**
Originator: **Rowell**
First Introduction: **USA**
Introduced by: **Rowell**
Trade Name: **CHENIX**

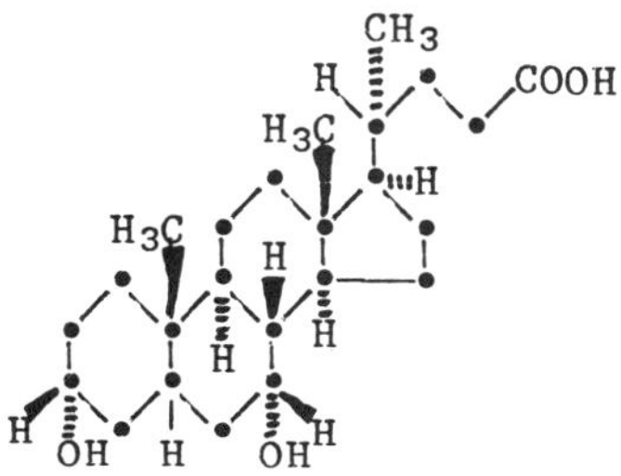

Chenodiol is the first agent to be introduced into the US market for the treatment of radiolucent gallstones. Large scale clinical trials have demonstrated the safety and efficacy of this agent. Chenodiol reduces the biliary concentration of cholesterol relative to that of bile acids and phospholipid, reducing the saturation and thus the lithogenicity of the bile. Success rates in dissolving gallstones are in the range of 50-70% within 4-24 months of treatment. Continuation of the drug after stone dissolution may be required to prevent reoccurrence. Chenodiol is the 7α-isomer of ursodeoxycholic acid which was introduced into the European market in 1978.

Cyclosporine (ciclosporin) (immunosuppressant)[34,35]

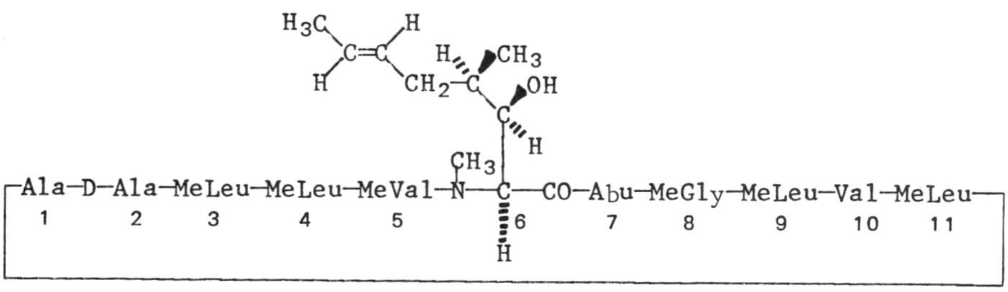

Country of Origin: **Switzerland**
First Introduction: **Switzerland**
Trade Name: **SANDIMMUN**
Originator: **Sandoz**
Introduced by: **Sandoz**

Cyclosporine is a cyclic polypeptide with potent, partially selective immunosupressive activity. Isolated from the species Cylindrocarpon lucidium and Trichoderma polysporum, cyclosporine is useful in the prevention and treatment of graft/host disease and the prevention of rejection following organ transplantation. It appears to act by preferentially suppressing T-lymphocytes. Cyclosporine lacks myelotoxicity, although impaired renal and liver function have been observed. Initial administration is via the intravenous route, followed by oral maintenance therapy.

Enocitabine (antineoplastic)[36,37]

Country of Origin: **Japan**
Originator: **Asahi**
First Introduction: **Japan**
Introduced by: **Toyo Jozo**
Trade Name: **SUNRABIN**

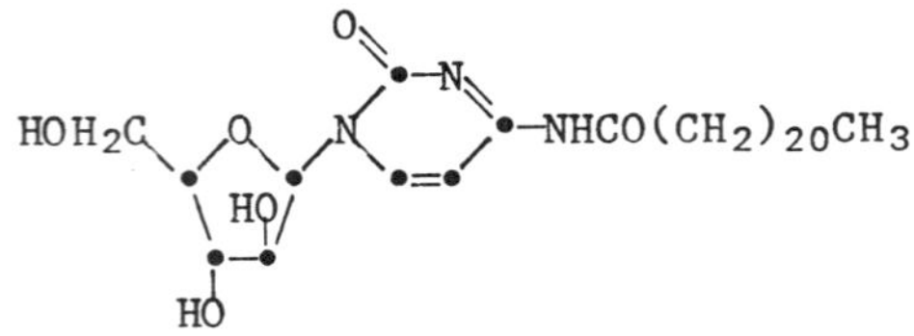

Enocitabine is an antileukemic agent closely related to cytarabine. It appears more resistant to deamination than cytarabine, thus allowing greater in vivo phosphorylation into an active cytotoxic metabolite.

Eperisone Hydrochloride (muscle relaxant)[38,39]

Country of Origin: **Japan**
Originator: **Eisai**
First Introduction: **Japan**
Introduced by: **Eisai**
Trade Name: **MYONAL**

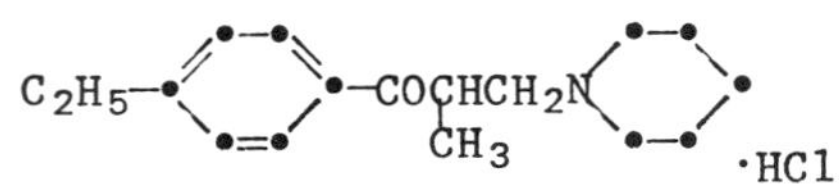

Eperisone hydrochloride is a centrally acting muscle relaxant useful in the management of various spastic conditions including cervical spondylosis and cerebral palsy. It is structurally related to tolperisone.

Epoprostenol Sodium (platelet aggregation inhibitor)[40,41]

Country of Origin: **United Kingdom**
Originator: **Burroughs Wellcome**
First Introduction: **United Kingdom**
Introduced by: **Burroughs Wellcome; Upjohn**
Trade Names: **FLOLAN; CYCLO-PROSTIN**

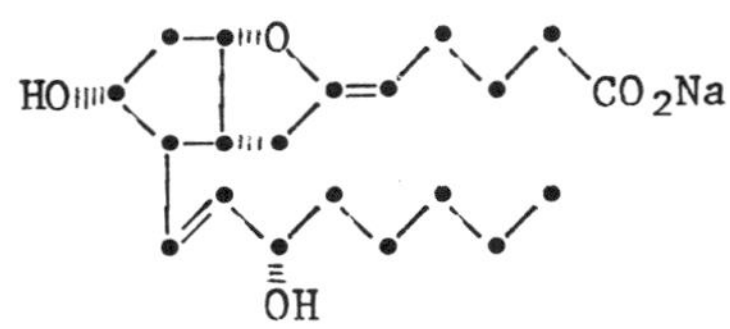

Epoprostenol sodium (prostacyclin) is a naturally occurring prostaglandin indicated for the preservation of platelet function during cardiopulmonary bypass, prevention of platelet aggregation during charcoal hemoperfusion of patients in hepatic failure, and as an alternative to heparin during renal dialysis.

Fenbuprol (choleretic)[42,43]

Country of Origin: **W. Germany**
Originator: **Klinge Pharma**
First Introduction: **W. Germany**
Introduced by: **Rhom Pharma**
Trade Name: **VALBIL**

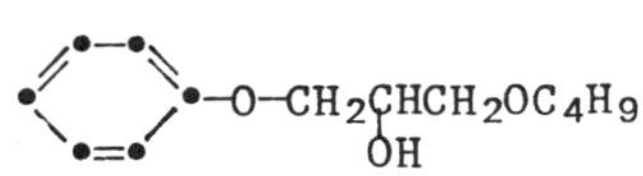

Fenbuprol produces an increase in the volume of bile secretion. It is useful in patients having symptomatology associated with biliary tract dysfunction.

Flutamide (antineoplastic)[44,45]

Country of Origin: **USA**
Originator: **Schering**
First Introduction: **Chile**
Introduced by: **Schering**
Trade Name: **DROGENIL**

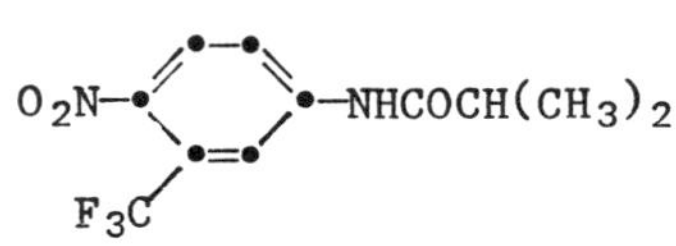

Flutamide is an orally active, non-steroidal antiandrogen indicated for the treatment of prostatic cancer in both castrates and noncastrates.

Fluvoxamine Maleate (serontonergic antidepressant)[46-48]

Country of Origin: **Netherlands**
Originator: **Duphar**
First Introduction: **Switzerland**
Introduced by: **Kali-Duphar**
Trade Name: **FLOXYFRAL**

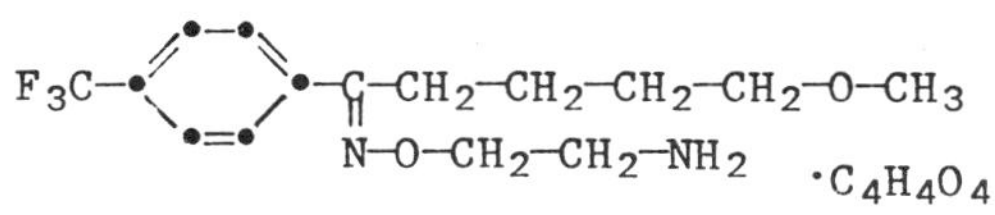

Fluvoxamine maleate is the most recent of the serotonin-specific antidepressants to reach the market. In vitro and in vivo animal experiments have shown fluvoxamine to have a marked effect on 5-HT mediated processes and little effect on norepinephrine. Clinical trials suggest similar efficacy to imipramine and clomipramine with a somewhat lower incidence of side effects, especially anticholinergic effects. Fluvoxamine, in contrast to the tricyclic antidepressants, does not appear to produce heart rate increase, postural hypotension or prolongation of the intraventricular conduction time and QT interval.

Gallopamil Hydrochloride (antianginal)[49]

Country of Origin: **W. Germany**
Originator: **Knoll**
First Introduction: **W. Germany**
Introduced by: **Chem. Werke Minden**
Trade Name: **PROCORUM**

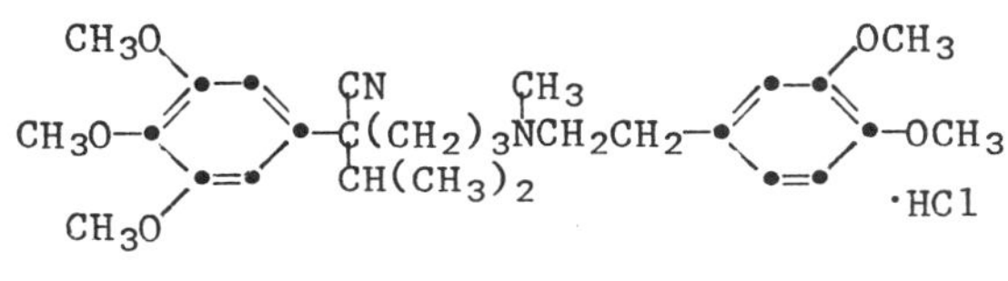

Gallopamil hydrochloride is a somewhat more potent methoxy analog of calcium channel blocker verapamil with a similar profile. It is useful in the treatment of angina and auricular arrhythmia.

Gemeprost (abortifacient)[50,51]

Country of Origin: **Japan**
Originator: **Ono**
First Introduction: **Malaysia/Singapore**
Introduced by: **May & Baker**
Trade Name: **CERVAGEM**

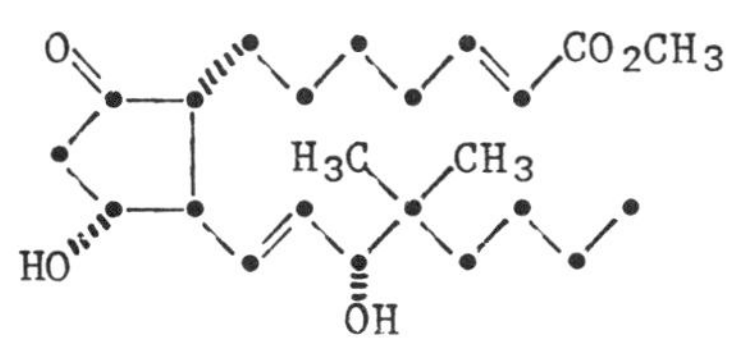

Gemeprost, a metabolically stabilized analog of PGE_1, has been demonstrated to reliably induce termination of early pregnancy when administered as a vaginal suppository. Its effect is presumably due to both uterine contraction and rapid decline of steroid hormone levels. Minimal side effects have been reported.

Guanadrel Sulfate (antihypertensive)[52,53]

Country of Origin: **USA**
Originator: **Cutter**
First Introduction: **USA**
Introduced by: **Pennwalt**
Trade Name: **HYLOREL**

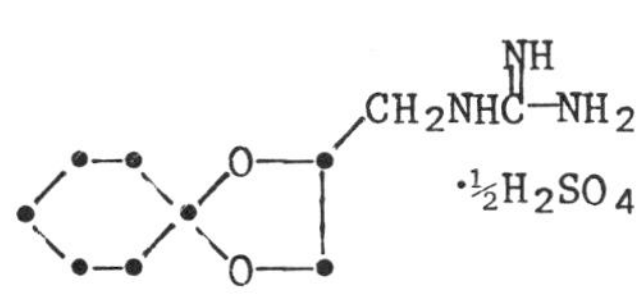

Guanadrel sulfate is an antihypertensive belonging to the class of adrenergic neuron blocking drugs. It diminishes sympathetic vasoconstriction by inhibiting norepinephrine storage and release from neuronal storage sites. It appears similar in effectiveness and side effect profile to its structural relative guanethidine.

Halometasone (topical antiinflammatory)[54,55]

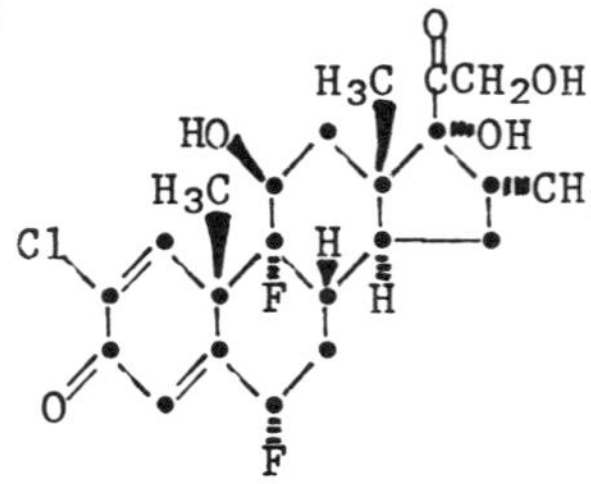

Country of Origin: **Switzerland**
Originator: **Ciba-Geigy**
First Introduction: **Switzerland**
Introduced by: **Ciba-Geigy**
Trade Name: **SICORTEN**

Halometasone is a potent, topical steroid useful in a variety of acute and chronic eczematous dermatoses and psoriasis. It is reportedly devoid of skin toxicity and systemic effects.

Hydrocortisone Butyrate Propionate (topical antiinflammatory)[56]

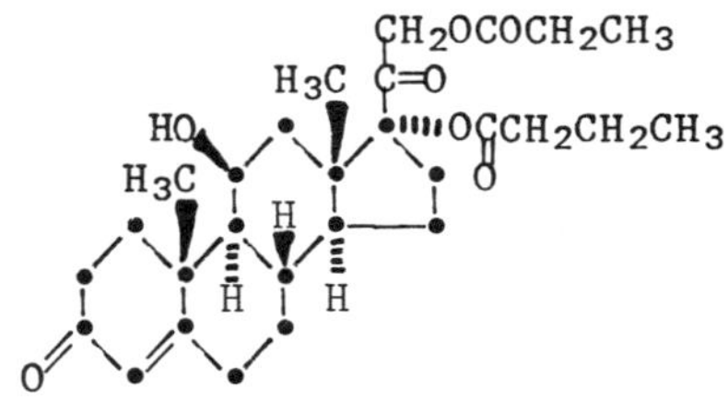

Country of Origin: **Japan**
Originator: **Taisho**
First Introduction: **Japan**
Introduced by: **Taisho**
Trade Name: **PANDEL**

Hydrocortisone butyrate propionate is a potent, topical steroid reported to possess minimal systemic side effects. It is indicated in the treatment of various acute and chronic contact, eczematous, and atopic dermatoses and psoriasis.

Indalpine (serotonergic antidepressant)[57-59]

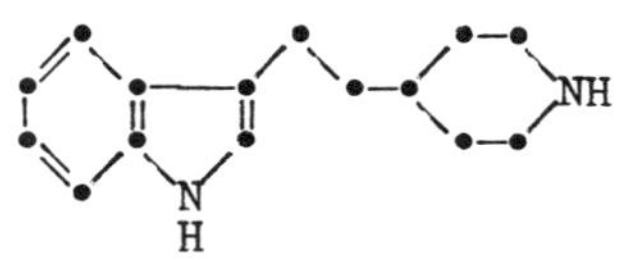

Country of Origin: **France**
Originator: **Pharmuka**
First Introduction: **France**
Introduced by: **Labs. Fournier Freres**
Trade Name: **UPSTENE**

Indalpine is a non-tricyclic antidepressant with a serotonin selective profile. It is 6-7 times more potent than fluoxetine and clomipramine in inhibiting serotonin reuptake in vitro in rat brain synaptosomes. Statistically significant clinical effects within one week of onset of treatment have been reported. An anxiolytic effect may accompany the antidepressant effect. Indalpine appears devoid of anticholinergic and cardiovascular side effects and does not promote weight gain or affect appetite.

Isoxicam (antiinflammatory)[60-62]

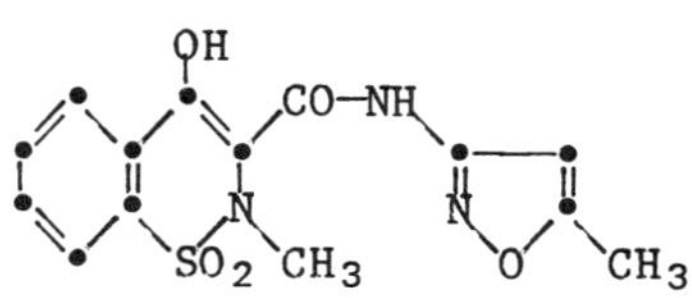

Country of Origin: **USA**
Originator: **Warner-Lambert**
First Introduction: **W. Germany**
Introduced by: **Warner-Lambert**
Trade Name: **PACYL**

Isoxicam is a non-steroidal antiinflammatory agent useful in the treatment of various forms of rheumatoid arthritis, osteoarthritis and musculoskeletal disorders. It is about one-tenth as potent as its structural relative sudoxicam; its similar long $T_{1/2}$ (>30 hrs.) allows once-daily dosing.

Loprazolam Mesylate (hypnotic/sedative)[63,64]

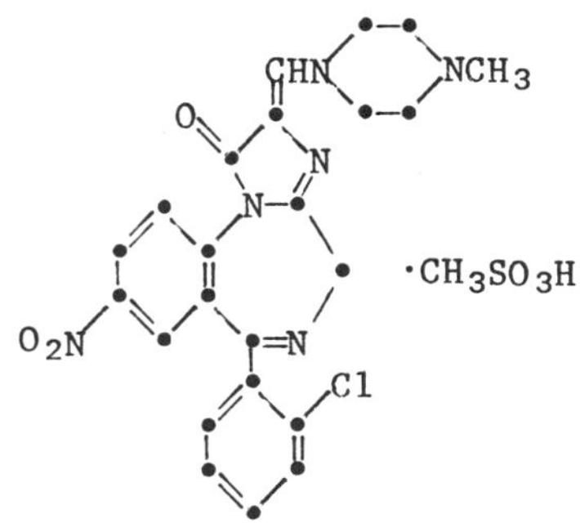

Country of Origin: **United Kingdom**
Originator: **Roussel**
First Introduction: **United Kingdom**
Introduced by: **Roussel**
Trade Name: **DORMONOCT**

Loprazolam mesylate is a potent hypnotic/sedative belonging to the second generation of annulated-1,4-benzodiazepines. The T½ of loprazolam (~6 hr.) is longer than those of triazolam and midazolam, but shorter than the "effective" T½ of flurazepam.

Meglutol (hypolipidemic)[65,66]

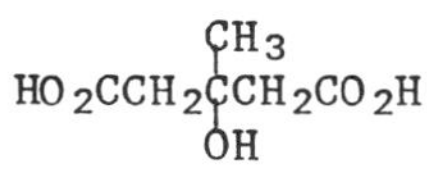

Country of Origin: **India**
Originator: **Aligarh Muslim Univ.**
First Introduction: **Italy**
Introduced by: **Ausonia**
Trade Name: **LIPOGLUTAREN**

Meglutol (β-hydroxy-β-methylglutaric acid; HMG) is a hypolipidemic agent that acts <u>via</u> the inhibition of cholesterol synthesis at the stage of HMG-CoA-reductase, blocking the conversion of HMG-CoA to mevalonate. This conversion is the rate-limiting step in cholesterol biosynthesis and is the point of physiological feedback control; as such, it appears to be the ideal locus of action for a hypolipidemic agent. Investigational compounds that act by this mechanism include the natural products compactin and mevinolin.

Meptazinol Hydrochloride (analgesic)[67-69]

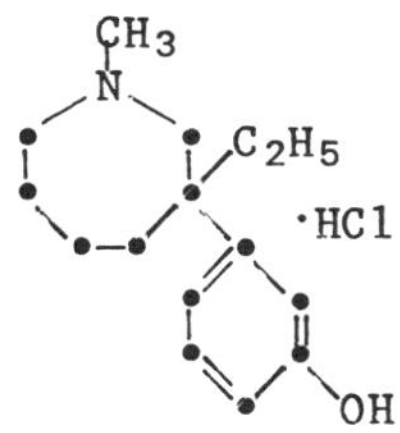

Country of Origin: **United Kingdom**
Originator: **Wyeth**
First Introduction: **United Kingdom**
Introduced by: **Wyeth**
Trade Name: **MEPTID**

Meptazinol hydrochloride is an injectable narcotic analgesic with antagonist properties; it is similar in potency to meperidine. Meptazinol appears to have a low propensity toward respiratory depression and other opiate-like side effects, possibly due to selective interaction with the mu-1 receptor. Also somewhat unique for an analgesic, it interacts with central cholinergic receptors. An oral form of meptazinol is under development.

Muzolimine (diuretic)[70-72]

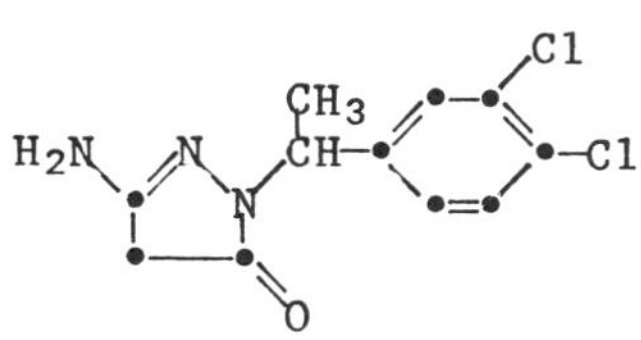

Country of Origin: **W. Germany**
Originator: **Bayer**
First Introduction: **Italy**
Introduced by: **Bayer**
Trade Name: **EDRUL**

Muzolimine is a structurally novel, pyrazolone diuretic with a high-ceiling profile. It is somewhat slower in onset than furosemide, but has a more prolonged effect, similar to the thiazides. Muzolimine has been shown to be effective in edema of cardiac, hepatic and renal origin. It also appears to be effective as an anti-hypertensive agent.

Nitrefazole (alcohol deterrent)[73,74]

Country of Origin: **W. Germany**
Originator: **E. Merck**
First Introduction: **W. Germany**
Introduced by: **E. Merck**
Trade Name: **ALTIMOL**

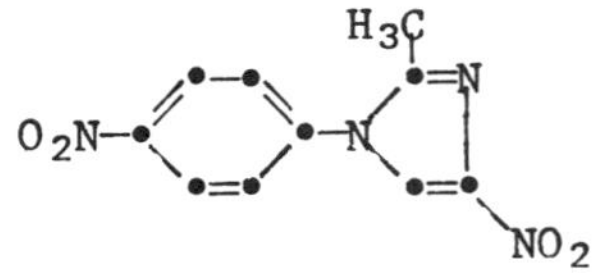

Nitrefazole is an inhibitor of aldehyde dehydrogenase useful in the management of alcoholism. Cited advantages over disulfiram include longer duration of action, higher specificity (no inhibition of dopamine β-hydroxylase) and fewer side effects. It is generally recognized that the efficacy of such agents with respect to permanent abstinence is low. However, temporary abstinence and/or controlled drinking may facilitate supportive psychotherapy and/or delay the onset of severe alcoholic diseases.

Norfloxacin (antibacterial)[75,76]

Country of Origin: **Japan**
Originator: **Kyorin**
First Introduction: **Italy**
Introduced by: **Merck**
Trade Name: **NOROXIN**

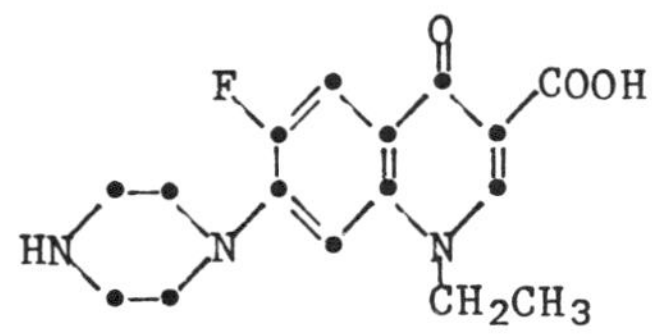

Norfloxacin is the first of the third generation nalidixic acid analogs to reach the marketplace. It exhibits potent in vitro and in vivo activity against Pseudomonas, enteric gram-negative rods and gram-positive cocci. Norfloxacin is orally effective in the treatment of urinary tract infections, including those due to organisms refractory to many other agents.

Oxaprozin (antiinflammatory)[77,78]

Country of Origin: **United Kingdom**
Originator: **Wyeth**
First Introduction: **Portugal**
Introduced by: **Wyeth**
Trade Name: **DURAPROX**

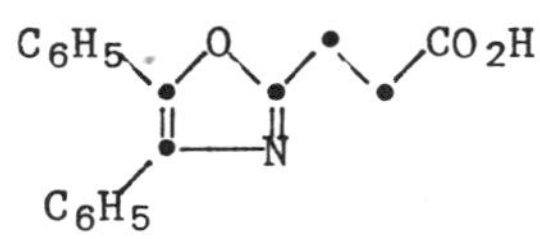

Oxaprozin is a non-steroidal antiinflammatory agent indicated for use in various forms of rheumatoid arthritis, osteoarthritis, and ankylosing spondylitis. A long $T_{1/2}$ allows once/twice daily dosing. Structurally it is somewhat unique, being a 3-substituted, rather than a 2-substituted propionic acid.

Oxiconazole Nitrate (antifungal)[79]

Country of Origin: **Switzerland**
Originator: **Siegfried**
First Introduction: **Switzerland**
Introduced by: **Sauter; Siegfried**
Trade Names: **OCERAL; MYFUNGAR**

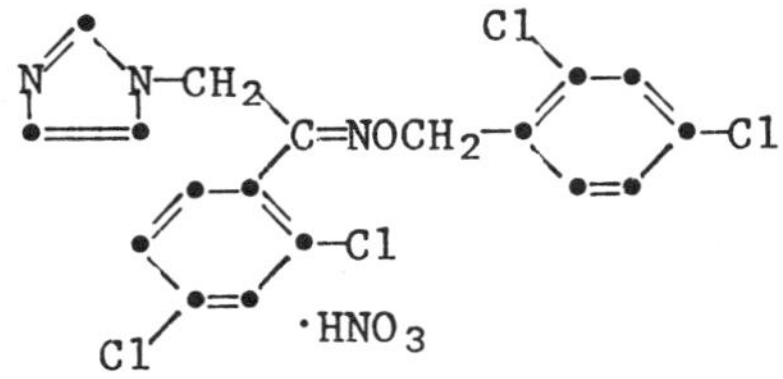

Oxiconazole nitrate is a broad-spectrum antifungal agent indicated for the treatment of skin infections due to yeasts, dermatophytes, yeast-like fungi and molds, and gram positive organisms.

Oxitropium Bromide (bronchodilator)[80-82]

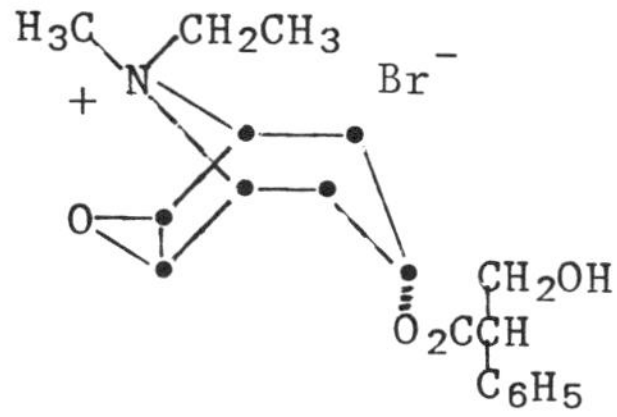

Country of Origin: **W. Germany**
Originator: **Boehringer Ingelheim**
First Introduction: **W. Germany**
Introduced by: **Dieckmann**
Trade Name: **TERSIGAT**

Oxitropium bromide is an analog of the anticholinergics methscopolamine bromide and ipratropium bromide, useful in the treatment of bronchial asthma. Cited advantages include a long duration of action and lack of cardiovascular effects.

Promegestone (progestogen)[83,84]

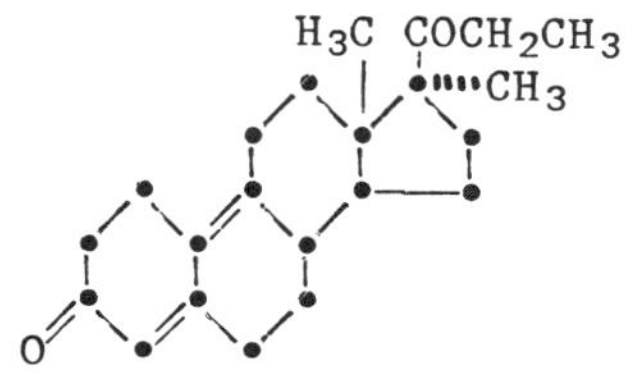

Country of Origin: **France**
Originator: **Roussel**
First Introduction: **France**
Introduced by: **Labs. Cassenne**
Trade Name: **SURGESTONE**

Promegestone is a potent progesterone-like agent devoid of androgenic properties, and thus masculinizing side effects. It is useful in the treatment of various gynecological conditions due to luteal insufficiency, such as premenopausal disorders, dysmenorrhea, and premenstrual syndrome.

Sodium Cellulose Phosphate (hypocalciuric)[85]

Country of Origin: **USA**
First Introduction: **USA**
Trade Name: **CALCIBIND**
Originator: **University of Texas**
Introduced by: **Mission Pharmacal**

Sodium cellulose phosphate (SCP) is an insoluble, non-absorbable ester of cellulose containing 34% inorganic phosphate and 11% sodium. It is capable of binding calcium in the intestinal tract, reducing absorption of this ion, as well as magnesium. SCP is indicated only for the treatment of absorptive hypercalciuria type I with recurrent calcium oxalate or calcium phosphate nephrolithiasis.

Sufentanil (analgesic)[86,87]

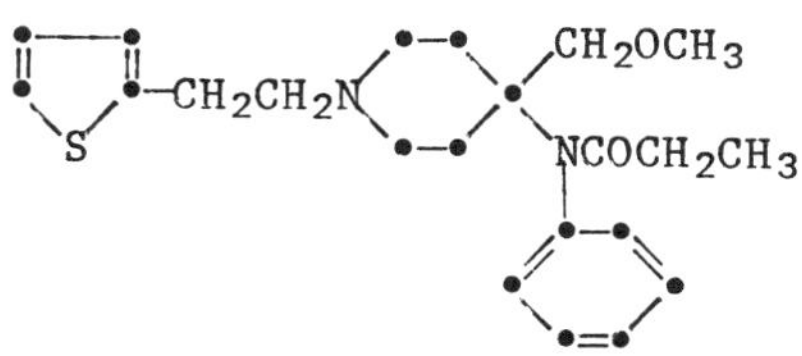

Country of Origin: **Belgium**
Originator: **Janssen**
First Introduction: **Netherlands**
Introduced by: **Janssen**
Trade Name: **SUFENTA**

Sufentanil is a narcotic analgesic with a greater potency and therapeutic ratio than its structural relative fentanyl. It appears to produce fewer cardiac effects and less respiratory depression than fentanyl, and thus is especially useful as an analgesic/anesthetic in open heart surgery.

Suprofen (analgesic/antiinflammatory)[88-90]

Country of Origin: **Belgium**
Originator: **Janssen**
First Introduction: **Switzerland**
Introduced by: **Cilag**
Trade Name: **MALDOCIL**

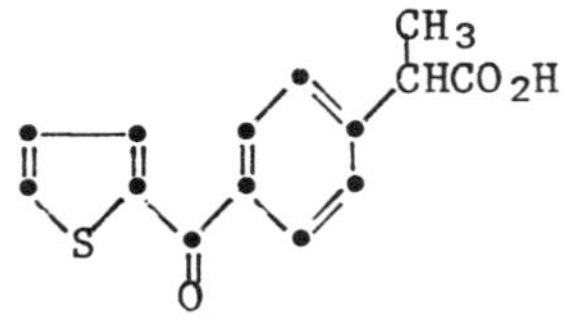

Suprofen is an arylpropionic acid analgesic/antiinflammatory agent with close structural resemblance to ketoprofen. It is more potent in many assays than indomethacin and ketoprofen and appears better tolerated.

Terconazole (antifungal)[91]

Country of Origin: **Belgium**
Originator: **Janssen**
First Introduction: **Switzerland**
introduced by: **Cilag**
Trade Name: **FUNGISTAT**

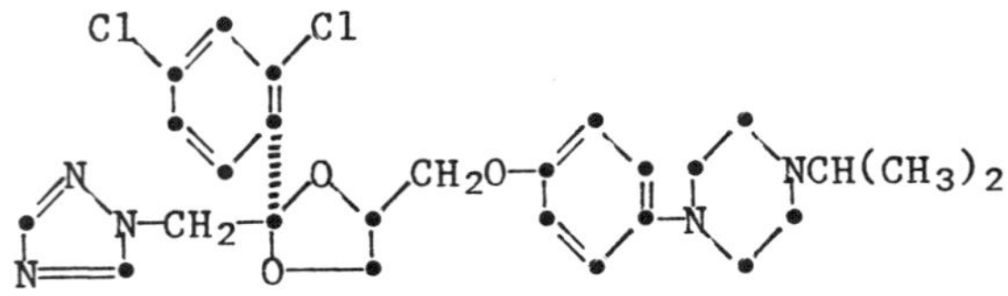

Terconazole is an antifungal agent somewhat more potent than clotrimazole and useful in the topical treatment of vaginal dermatophytosis and candidiasis.

Tianeptine Sodium (serotonergic antidepressant)[92,93]

Country of Origin: **France**
Originator: **Sci. Union et Cie**
First Introduction: **France**
Introduced by: **Servier**
Trade Name: **STABLON**

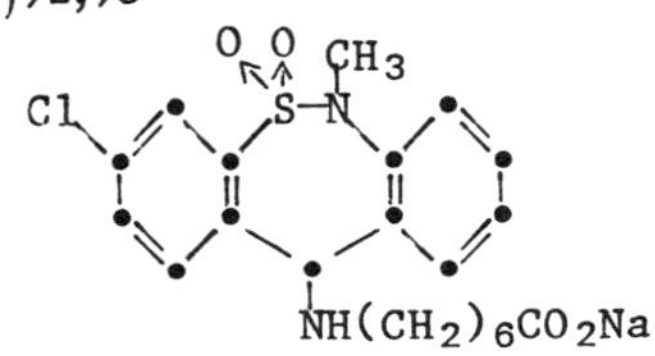

Tianeptine sodium is a structurally novel, serotonin specific antidepressant. It is useful in the treatment of neurotic and reactive depressions, as well as depressive states accompanied by anxiety.

Tioconazole (antifungal)[94]

Country of Origin: **United Kingdom**
Originator: **Pfizer**
First Introduction: **Switzerland**
Introduced by: **Pfizer**
Trade Name: **TROSYL**

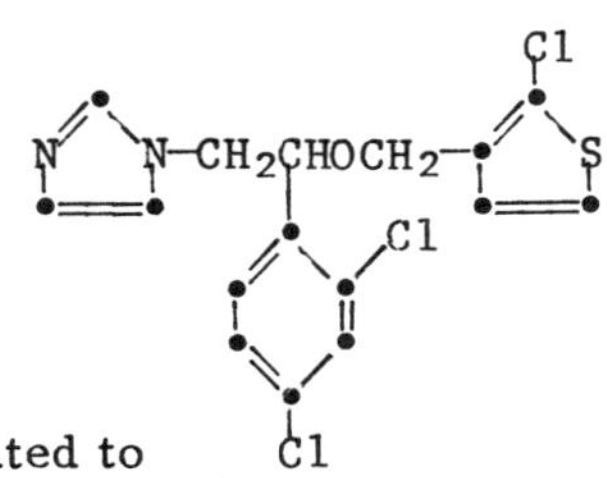

Tioconazole is an antifungal agent, closely related to miconazole. Tioconazole is effective in the topical treatment of superficial fungal infections and appears to be more potent than miconazole against <u>Candida</u> and <u>Trichophyton</u> species.

Tiropramide Hydrochloride (antispasmodic)[95-97]

Country of Origin: **Italy**
Originator: **Rotta Research**
First Introduction: **Italy**
Introduced by: **Rotta Research; Rorer**
Trade Name: **MAIORAD; ALFOSPAS**

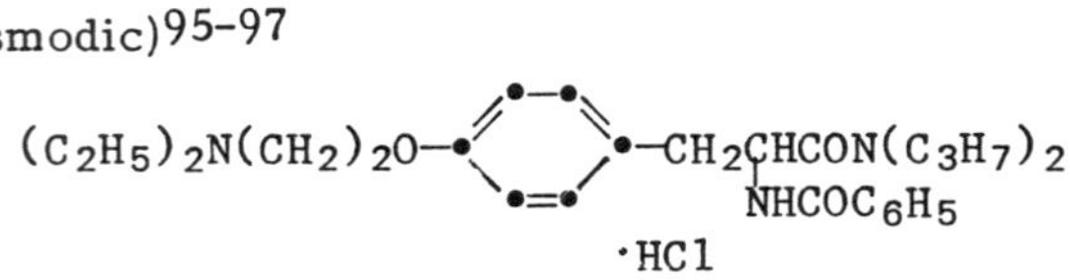

Tiropramide hydrochloride is a smooth muscle relaxant indicated for the treatment of spastic conditions of the gastrointestinal and urogenital systems. It appears to act by increasing intracellular levels of c-AMP.

References

1. G. R. Gale, Drugs Fut., 5, 599 (1980).
2. J. R. Prous, ed., Annu. Drug Data Rep., 3, 2 (1981).
3. J. R. Prous, ed., Annu. Drug Data Rep., 5, 4 (1983).
4. A. J. Sweetman, Drugs Fut., 7, 539 (1982).
5. J. R. Prous, ed., Annu. Drug Data Rep., 3, 8 (1981).
6. R. T. Owen, Drugs Today, 20, 10 (1984).
7. J. R. Prous, ed., Annu. Drug Data Rep., 4, 10 (1982).
8. K. Hillier, Drugs Fut., 4, 245 (1979).
9. J. R. Prous, ed., Annu. Drug Data Rep., 3, 14 (1981).
10. K. Hillier, Drugs Fut., 7, 10 (1982).
11. J. R. Prous, ed., Annu. Drug Data Rep., 5, 12 (1983).
12. J. A. Bogan, Drugs Fut., 1, 451 (1976).
13. J. R. Prous, ed., Annu. Drug Data Rep., 5, 13 (1983).
14. J. R. Prous, ed., Annu. Drug Data Rep., 3, 27 (1981).
15. M. Takase, Jpn. J. Ophthalmol., 26, 88 (1982).
16. K. Hillier, Drugs Fut., 4, 867 (1979).
17. J. R. Prous, ed., Annu. Drug Data Rep., 3, 31 (1981).
18. Drugs Fut., 7, 87 (1982).
19. J. R. Prous, ed., Annu. Drug Data Rep., 4, 24 (1982).
20. H. Koch, Drugs Fut., 4, 85 (1979).
21. J. R. Prous, ed., Annu. Drug Data Rep., 3, 36 (1981).
22. K. Hillier, Drugs Fut., 2, 788 (1977).
23. J. R. Prous, ed., Annu. Drug Data Rep., 5, 26 (1983).
24. J. R. Prous, ed., Annu. Drug Data Rep., 3, 120 (1981).
25. Drugs Today, 19, 426 (1983).
26. Drugs Today, 19, 429 (1983).
27. J. R. Prous, ed., Annu. Drug Data Rep., 5, 33 (1983).
28. A. J. Sweetman, Drugs Fut., 6, 612 (1981).
29. J. R. Prous, ed., Annu. Drug Data Rep., 4, 44 (1982).
30. H. Koch, Drugs Today, 18, 632 (1982).
31. J. R. Prous, ed., Annu. Drug Data Rep., 5, 37 (1983).
32. A. C. Sullivan, L. Cheng and J. C. Hamilton, Annu. Rep. Med. Chem., 12, 196 (1977).
33. M. N. Cayen, Annu. Rep. Med. Chem., 14, 205 (1979).
34. J. R. Prous, ed., Annu. Drug Data Rep., 5, 42 (1983).
35. Drugs Today, 19, 665 (1983).
36. F. Cabanillas, Drugs Fut., 5, 603 (1980).
37. J. R. Prous, ed., Annu. Drug Data Rep., 3, 28 (1981).
38. Drugs Fut., 7, 105 (1982).
39. J. R. Prous, ed., Annu. Drug Data Rep., 4, 84 (1982).
40. K. Hillier, Drugs Fut., 2, 688 (1977).
41. J. R. Prous, ed., Annu. Drug Data Rep., 5, 68 (1983).
42. P. St.Janiak, Drugs Fut., 3, 191 (1978).
43. U. Ritter and H. J. Kyrein, Arzneim. Forsch., 33, 891 (1983).
44. P. Thorpe, Drugs Fut., 1, 108 (1976).
45. J. R. Prous, ed., Annu. Drug Data Rep., 2, 105 (1979/1980).
46. P. Thorpe, Drugs Fut., 3, 288 (1978).
47 J. R. Prous, ed., Annu. Drug Data Rep., 2, 107 (1979/1980).
48 F. H. Geukes-Foppen, S. J. Papworth and A. W. Peck, eds., Br. J. Clin. Pharmacol., 15 (Suppl. 3), 347S-450S (1983).
49. J. R. Prous, ed., Annu. Drug Data Rep., 5, 85 (1983).
50. P. Leeson, Drugs Fut., 4, 38 (1979).
51. J. R. Prous, ed., Annu. Drug Data Rep., 5, 85 (1983).
52. D. M. Paton, Drugs Today, 16, 140 (1980).
53. J. R. Prous, ed., Annu. Drug Data Rep., 4, 100 (1982).
54. K. J. Schwarz, M. Konzelmann, S. J. Yawalkar and P. M. Schoenenberger, Br. J. Clin. Pract., 36, 192 (1982).
55. J. E. Murphy, ed., J. Int. Med. Res., 11 (Suppl. 1), 1-57 (1983).
56. J. R. Prous, ed., Annu. Drug Data Rep., 5, 200 (1983).
57. R. T. Owen, Drugs Fut., 4, 873 (1979).
58. J. R. Prous, ed., Annu. Drug Data Rep., 5, 95 (1983).
59. B. Shopsin, C. Lefebvre, and C. Maulet, Curr. Ther. Res., 34, 239 (1983).

60. E. Arrigoni-Martelli, Drugs Fut., 1, 123 (1976).
61. J. R. Prous, ed., Annu. Drug Data Rep., 5, 99 (1983).
62. G. DiPasquale, Drugs Today, 19, 119 (1983).
63. P. J. Thorpe, Drugs Fut., 5, 144 (1980).
64. J. R. Prous, ed., Annu. Drug Data Rep., 3, 143 (1981).
65. Z. H. Beg and M. Siddiqi, Experientia, 23, 380 (1967).
66. P. J. Lupien, S. Moorjani, D. Brun, and P. Bielmann, J. Clin. Pharmacol., 19, 120 (1979).
67. M. Thorpe, Drugs Fut., 1, 68 (1976).
68. J. R. Prous, ed., Annu. Drug Data Rep., 2, 141 (1979/1980).
69. O. W. Lever, Jr., K.-J. Chang and J. D. McDermed, Annu. Rep. Med. Chem., 18, 52 (1983).
70. K. Hillier, Drugs Fut., 2, 387 (1977).
71. J. R. Prous, ed., Annu. Drug Data Rep., 5, 129 (1983).
72. L. deAngelis, Drugs Today, 19, 267 (1983).
73. J. R. Prous, ed., Annu. Drug Data Rep., 4, 146 (1982).
74. D. E. McMillan, Drug Dev. Res., 3, 193 (1983).
75. J. R. Prous, ed., Annu. Drug Data Rep., 3, 12 (1981).
76. J. R. Prous and N. E. Mealy, Drugs Today, 19, 341 (1983).
77. E. Arrigoni-Martelli, Drugs Fut., 3, 539 (1978).
78. J. R. Prous, ed., Annu. Drug Data Rep., 5, 144 (1983).
79. J. R. Prous, ed., Annu. Drug Data Rep., 3, 184 (1981).
80. A. P. Dharma, Drugs Fut., 4, 117 (1979).
81. H. Koch, Drugs Today, 19, 444 (1983).
82. J. R. Prous, ed., Annu. Drug Data Rep., 5, 145 (1983).
83. P. J. Roberts, Drugs Fut., 3, 469 (1978).
84. J. R. Prous, ed., Annu. Drug Data Rep., 5, 156 (1983).
85. C. P. Robinson, Drugs Today, 19, 372 (1983).
86. E. Arrigoni-Martelli, Drugs Fut., 2, 334 (1977).
87. J. R. Prous, ed., Annu. Drug Data Rep., 2, 206 (1979/1980).
88. S. S. Chatterjee, Drugs Fut., 1, 148 (1976).
89. J. R. Prous, ed., Annu. Drug Data Rep., 2, 211 (1979/1980).
90. S. S. Chu, Drugs Today, 19, 675 (1983).
91. J. R. Prous, ed., Annu. Drug Data Rep., 5, 190 (1983).
92. J. R. Prous, ed., Annu. Drug Data Rep., 5, 193 (1983).
93. M. Neuman, Drugs Today, 19, 306 (1983).
94. J. R. Prous, ed., Annu. Drug Data Rep., 2, 226 (1979/1980).
95. L. deAngelis, Drugs Fut., 7, 413 (1982).
96. J. R. Prous, ed., Annu. Drug Data Rep., 5, 195 (1983).
97. L. deAngelis, Drugs Today, 19, 271 (1983).

ISBN 0-12-040519-9

ISBN 0-12-040519-9